MINNESOTA STUDIES IN THE PHILOSOPHY OF SCIENCE

Minnesota Studies in the
PHILOSOPHY OF SCIENCE

RONALD N. GIERE, GENERAL EDITOR

HERBERT FEIGL, FOUNDING EDITOR

VOLUME XIII
Scientific Explanation

EDITED BY

PHILIP KITCHER AND WESLEY C. SALMON

UNIVERSITY OF MINNESOTA PRESS, MINNEAPOLIS

Published by the University of Minnesota Press
2037 University Avenue Southeast, Minneapolis MN 55414.
Printed in the United States of America.

Library of Congress Cataloging-in-Publication Data

Scientific explanation / edited by Philip Kitcher and Wesley C.
Salmon.
p.   cm. − (Minnesota studies in the philosophy of science;
v. 13)
ISBN 0-8166-1773-2
1. Science−Philosophy. 2. Science−Methodology. I. Kitcher,
Philip, 1947-   . II. Salmon, Wesley C. III. Series.
Q175.M64 vol. 13
501 s−dc20
[501]
89-20248
CIP

For
Carl G. Hempel,
and in memory of
Herbert Feigl,
who made the whole enterprise possible

Contents

Preface

Is a new consensus emerging in the philosophy of science? This is the question to which a year-long workshop was devoted at the Minnesota Center for the Philosophy of Science during the academic year 1985–86. Throughout the fall term our discussions were directed almost exclusively to the issue of consensus regarding the nature of scientific explanation. This is the topic to which the present volume is addressed.

To ask whether a *new* consensus is emerging in philosophy of science strongly suggests that there was an old consensus. We believe, indeed, that there was one. It can be identified with what might be called *the hegemony of logical empiricism* which reached its peak in the 1950s and 1960s. With respect to scientific explanation, it seems reasonable to single out Carl G. Hempel's *Aspects of Scientific Explanation and Other Essays in the Philosophy of Science* (1965) as the pinnacle of the old consensus. The main foundation of that structure is the classic article "Studies in the Logic of Explanation" (1948), co-authored by Hempel and Paul Oppenheim. A large preponderance of subsequent philosophical work on scientific explanation flows directly or indirectly from this epoch-making essay.

The initial essay in the present volume, "Four Decades of Scientific Explanation," serves as an introduction in two senses. First, it is intended to acquaint readers who are not specialists in this area with the main issues, viewpoints, and arguments that have dominated the philosophical discussion of the nature of scientific explanation in recent decades. Hence, this volume does *not* presuppose prior knowledge of its main topics. Second, if we want to try to decide whether a new consensus is emerging, it is important to look at the developments leading up to the present situation. "Four Decades of Scientific Explanation" is also a historical introduction that describes the old consensus, its breakup, and subsequent developments. To understand the current state of things, we need to know how we got from there to here.

Although the present volume emerges from an NEH institute conducted at the Minnesota Center for the Philosophy of Science, it is in no sense a 'proceeding' of that workshop. Three of the contributors—Philip Kitcher, Merrilee Salmon, and Wesley Salmon—participated actively during the entire term. Three others—Paul Humphreys, David Papineau, and Peter Railton—paid brief visits. The remaining three—Nancy Cartwright, Matti Sintonen, and James Woodward—were invited to contribute papers because of their special interests in the problems to which the workshop was devoted.

We should like to express our deepest gratitude to the National Endowment for the Humanities for their support. We should also like to thank C. Wade Savage, co-director, with Philip Kitcher, of the NEH institute, and all of the other participants in the workshop. Finally, special appreciation is due to Candy Holmbo, without whose organizational talents we would have had far less time to think about scientific explanation.

P. K.
W. C. S.

MINNESOTA STUDIES IN THE PHILOSOPHY OF SCIENCE

Four Decades of Scientific Explanation

Introduction

The search for scientific knowledge extends far back into antiquity. At some point in that quest, at least by the time of Aristotle, philosophers recognized that a fundamental distinction should be drawn between two kinds of scientific knowledge — roughly, knowledge *that* and knowledge *why*. It is one thing to know *that* each planet periodically reverses the direction of its motion with respect to the background of fixed stars; it is quite a different matter to know *why*. Knowledge of the former type is descriptive; knowledge of the latter type is explanatory. It is explanatory knowledge that provides scientific understanding of our world.

Nevertheless, when Aristotle and many of his successors down through the centuries tried to say with some precision what constitutes scientific explanation they did not meet with great success. According to Aristotle, scientific explanations are deductive arguments; as we shall see, this idea has been extraordinarily influential. But as Aristotle clearly recognized, not all deductive arguments can qualify as explanations. Even if one accepts the idea that explanations are deductive arguments, it is no easy matter to draw a viable distinction between those arguments that do qualify and those that do not.

Forty years ago a remarkable event occurred. Carl G. Hempel and Paul Oppenheim published an essay, "Studies in the Logic of Explanation," which was truly epoch-making. It set out, with unprecedented precision and clarity, a characterization of one kind of deductive argument that, according to their account, does constitute a legitimate type of scientific explanation. It came later to be known as *the deductive-nomological model*. This 1948 article provided the foundation for the *old consensus* on the nature of scientific explanation that reached its height in the 1960s. A large preponderance of the philosophical work on scientific explanation in the succeeding four decades has occurred as a direct

I should like to express my sincere thanks to Marc Lange for expert bibliographical assistance, and my heartfelt gratitude to Paul Humphreys, Philip Kitcher, and Nicholas Rescher for extremely valuable comments on an earlier draft of this essay. My greatest debt is to Philip Kitcher for his psychological support and intellectual stimulation, without which it would never have been written.

or indirect response to this article. If we wish to assess the prospects for a *new consensus* on scientific explanation, this is where we must start. To understand the present situation we need to see how the old consensus came together and how it came apart.

0.1 A Bit of Background

I recall with amusement a personal experience that occurred in the early 1960s. J. J. C. Smart, a distinguished Australian philosopher, visited Indiana University where I was teaching at the time. Somehow we got into a conversation about the major unsolved problems in philosophy of science, and he mentioned the problem of scientific explanation. I was utterly astonished — literally, too astonished for words. At the time I considered that problem essentially solved by the deductive-nomological (D-N) account that had been promulgated by R. B. Braithwaite (1953), Carl G. Hempel (Hempel and Oppenheim 1948), Ernest Nagel (1961), and Karl Popper (1935, 1959), among many others — supplemented, perhaps, by Hempel's then recent account of statistical explanation (Hempel 1962). Although this general view had a few rather vocal critics such as N. R. Hanson (1959) and Michael Scriven (1958, 1959, 1962, 1963) it was widely accepted by scientifically minded philosophers; indeed, it qualified handily as the received view. What is now amusing about the incident is my naïveté in thinking that a major philosophical problem had actually been solved, but my attitude did reflect the then current almost complete consensus.

On one fundamental issue the consensus has remained intact. Philosophers of very diverse persuasions continue to agree that a fundamental aim of science is to provide explanations of natural phenomena. During the last forty years, few (if any) have voiced the opinion that the sole aims of science are to describe, predict, and control nature — that explanation falls into the domains of metaphysics or theology. It has not always been so. Twentieth-century scientific philosophy arose in a philosophical context dominated by post-Kantian and post-Hegelian German idealism. It was heavily infused with transcendental metaphysics and theology. The early logical positivists and logical empiricists saw it as part of their mission to overcome such influences. As philosophers of science they were eager to expunge from science any contamination by super-empirical factors arising out of these philosophies. One such item was teleology, whether in the form of an appeal to the will of a supernatural being who created and continues to direct the course of nature, or in the form of such empirically inaccessible agencies as entelechies and vital forces. In that historical context many metaphysically inclined philosophers argued that there could be no *genuine* explanation of any fact of nature that did not involve an extra-empirical appeal. They thought of explanation anthropomorphically in terms of the sort of 'human understanding' that always appeals to purposes. Many scientific philosophers (as well as philosophical scientists) reacted to this attitude by denying that science is in any way concerned with explanation. Those who did admit that science can offer explanations were eager to make it clear that explanation is nothing more than some special kind of description—it does not demand anything beyond the sphere of empirical knowledge.[1] The classic 1948 Hempel-Oppenheim paper, which will serve as our main point of departure, clearly illustrates this approach.

In recent decades there has been quite general agreement that science can tell us not only *what*, but also *why*. It is possible—in principle and often in practice—to furnish scientific explanations of such facts as the destruction of the space-shuttle *Challenger*, the extinction of the dinosaurs, the coppery color of the moon during total eclipse, and countless other facts, both particular and general. By means of these explanations, science provides us with genuine understanding of the world.

The philosophers who were most instrumental in forging the old consensus—the *logical empiricists*—looked upon the task of philosophy as the construction of *explications* of fundamental concepts. The clearest expression of that goal was given by Rudolf Carnap (1950, 1962, chap. 1; see also Coffa 1973). The concept we are attempting to explicate—in our case, *scientific explanation*—is known as the *explicandum*. This concept, which is frequently used by scientists and by others who talk about science, is vague and, possibly, ambiguous; the job of the philosopher is to provide a clear and exact concept to replace it. The resulting concept is known as the *explicatum*. The process of explication has two stages: first, the explicandum must be clarified sufficiently for us to know what concept it is that we are trying to explicate; second, an exact explicatum must be precisely articulated. Carnap specifies four criteria according to which explications are to be judged:

(1) Similarity to the explicandum. If the explicatum does not match the explicandum to a sufficient degree, it cannot fulfill the function of the concept it is designed to replace. A perfect match cannot, however, be demanded, for the explicandum is unclear and the explicatum should be far more pellucid.

(2) Exactness. Unless the explicatum is precise it does not fulfill the purpose of explication, namely, the replacement of an imprecise concept by a precise one.

(3) Fruitfulness. The new concept should enable us to say significant things and have important insights. One of the main benefits of philosophical analysis should be to deepen our understanding of the nature of science.

(4) Simplicity. The explicatum should be as simple as requirements (1)-(3) permit. Simplicity often accompanies systematic power of concepts. At any rate, simplicity aids in ease of application and avoidance of errors in application.

As Carnap emphatically notes, requirement (1) should not be applied too stringently. The aim is to provide a concept that is useful and clear. In the case of scientific explanation, it is evident that scientists use this concept in a variety of ways, some clear and some confused. Some scientists have claimed, for example, that explanation consists in showing how some unfamiliar phenomenon can be reduced to others that are already familiar; some have equated explanation with something that produces a feeling of intellectual satisfaction. We cannot hope, nor do we want, to capture all of these usages with complete fidelity. The logical empiricists do not indulge in 'ordinary language analysis'—even the ordinary language of scientists—except, perhaps, as a prolegomenon to philosophical analysis.

As already noted, requirement (4) is subservient to its predecessors. Thus, (2) and (3) take precedence: we seek philosophically useful concepts that are formulated with precision. Our discussion of the classic 1948 Hempel-Oppenheim paper in the next section will nicely exemplify the logical empiricist notion of explication. There are, however, several points of clarification that must be made before we turn to consideration of that paper.

First, we must be quite clear that it is *scientific* explanation with which we are concerned. The term "explanation" is used in many ways that have little or nothing to do with scientific explanation (see W. Salmon 1984, 9–11). Scriven once complained that one of Hempel's models of explanation could not even accommodate the case in which one explains with gestures what is wrong with one's car to a Yugoslav garage mechanic who knows no English. Hempel answered, entirely appropriately, that this is like complaining that a precise explication of the term "proof" in mathematics does not capture the meaning of that word as it occurs in such contexts as "86 proof Scotch" and "the proof of the pudding is in the eating" (Hempel 1965, 413). Suitable clarification of the explicandum should serve to forestall objections of that sort.

To seek an explanation for some fact presupposes, of course, that the phenomenon we endeavor to explain did occur—that the putative fact is, indeed, a fact. For example, Immanuel Velikovsky (1950) attempted to 'explain' various miracles reported in the *Old Testament*, such as the sun standing still (i.e., the earth ceasing to rotate) at Joshua's command. Those who are not dogmatically committed to the literal truth of some holy writ will surely require much stronger evidence that the alleged occurrence actually took place before surrendering such basic physical laws as conservation of angular momentum in an attempt to 'explain' it.[2]

To avoid serious confusion we must carefully distinguish between offering an explanation for some fact and providing grounds for believing it to be the case. Such confusion is fostered by the fact that the word "why" frequently occurs in two distinct types of locutions, namely, "Why did X occur?" and "Why should one believe that X occurred?" As an example of the first type, we might ask why Marilyn Monroe died. An answer to this *explanation-seeking* why-question is that she took an overdose of sleeping pills. A full explanation would, of course, identify the particular drug and describe its physiological effects. As an example of the second type, we might ask why we believe that she died. The answer to this *evidence-seeking* why-question, for me at least, is that it was widely reported in the press. Similarly, to take a more scientific example, it is generally believed by cosmologists that the distant galaxies are receding from us at high velocities. The main *evidence* for this hypothesis is the fact that the light from these galaxies is shifted toward the red end of the spectrum, but this red-shift does not *explain* why the galaxies are traveling away from us. The recession of the galaxies is explained on the basis of the "big bang" – the primordial explosion that sent everything flying off in different directions – *not* by the red shift.

It might be supposed that a confusion of evidential facts with explanatory facts is unlikely to arise, but this supposition would be erroneous. In recent years there has been quite a bit of discussion of the so-called *anthropic principle*. According to certain versions of this principle, earlier states of the universe can be explained by the fact that they involved necessary conditions for the later occurrence of life – particularly human life – as we know it. For example, there must have been stars capable of synthesizing nuclei as complex as carbon. It is one thing to infer, from the undisputed fact that human life exists and would be impossible without carbon, that there is some mechanism of carbon synthesis from hydrogen and helium. It is quite another to claim that the existence of human life at present *explains why* carbon was synthesized in stars in our galaxy.[3]

Another fact that sometimes tends to foster the same confusion is the structural similarity of Hempel's well-known *deductive-nomological (D-N)* model of *scientific explanation* (to be discussed in detail in the next section) and the traditional *hypothetico-deductive (H-D)* schema for *scientific confirmation*. It must be kept in mind, however, that the fundamental aims of these two schemas are quite distinct. We *use* well-confirmed scientific hypotheses, laws, or theories to explain various phenomena. The idea behind deductive-nomological explanation is that, given the truth of all of the statements involved – both those that formulate the explanatory facts and the one that asserts the occurrence of the fact-to-be-explained – the logical relation between premises and conclusion shows that the former explain why the latter obtained. The function of the explanation is not to establish (or support) the truth of its conclusion; that is already presupposed when we accept it as a correct explanation. The idea behind the hypothetico-deductive method, in contrast, is that the given logical schema can be employed to provide evidential support for a hypothesis whose truth is being questioned. The statement that is supposed to be supported by hypothetico-deductive reasoning is not the conclusion in the schema, but rather, one of its premises.[4]

Another, closely related, possible source of confusion is the recent popularity of the slogan "inference to the best explanation." As Gilbert Harman has pointed out, we sometimes use the fact that a certain statement, if true, would explain something that has happened as evidence for the truth of that statement (Harman 1965). A detective, attempting to solve a murder, may consider the possible explanations of the crime, and infer that the 'best' one is true. To describe what is going on here it will be useful to appeal to a distinction (made by Hempel and Oppenheim) between potential explanations and actual explanations. A potential explanation has all of the characteristics of a correct – i.e., actual – explanation, except possibly for the truth of the premises. Harman maintains that we canvass the available potential explanations and infer that the 'best' of these is the actual explanation. As in the case of hypothetico-deductive inference, this kind of inference supports the premises of an explanatory argument, not its conclusion, whose truth is taken for granted from the outset. Given the fact that the whole point of the present essay is to discuss a wide variety of views on the nature of scientific explanation, we are hardly in a position at this stage of our investigation to say much of anything about what constitutes 'the best explanation.' And application of this principle of inference obviously presupposes some explication of *explanation*.

0.2 The Received View

Our story begins in 1948 with the publication of the above-mentioned classic article, "Studies in the Logic of Explanation," by Hempel and Oppenheim. This landmark essay provides the initial document of the old consensus concerning the nature of scientific explanation that emerged around the middle of the twentieth century. It is the fountainhead from which the vast bulk of subsequent philosophical work on scientific explanation has flowed – directly or indirectly.

According to that account, a D-N explanation of a particular event is a valid deductive argument whose conclusion states that the event to be explained did occur. This conclusion is known as the *explanandum-statement*. Its premises – known collectively as the *explanans* – must include a statement of at least one general law that is essential to the validity of the argument – that is, if that premise were deleted and *no other change* were made in the argument, it would no longer be valid. The explanation is said to subsume the fact to be explained under these laws; hence, it is often called "the covering law model." An argument fulfilling the foregoing conditions qualifies as a *potential explanation*. If, in addition, the statements constituting the explanans are true, the argument qualifies as a *true explanation* or simply an *explanation* (of the D-N type).

From the beginning, however, Hempel and Oppenheim (1948, 250–51) recognized that not all legitimate scientific explanations are of the D-N variety; some are probabilistic or statistical. In "Deductive-Nomological vs. Statistical Explanation" (1962) Hempel offered his first account of statistical explanation; to the best of my knowledge this is the first attempt by any philosopher to give a systematic characterization of probabilistic or statistical explanation.[5] In "Aspects of Scientific Explanation" (1965) he provided an improved treatment. This account includes two types of statistical explanation. The first of these, the *inductive-statistical (I-S)*, explains particular occurrences by subsuming them under statistical laws, much as D-N explanations subsume particular events under universal laws. There is, however, a crucial difference: D-N explanations subsume the events to be explained deductively, while I-S explanations subsume them inductively. An explanation of either kind can be described as *an argument to the effect that the event to be explained was to be expected by virtue of certain explanatory facts.* In a D-N explanation, the event to be explained is deductively certain, given the explanatory facts (including the laws); in an I-S explanation the event to be explained has high inductive probability relative to the explanatory facts (including the laws).

On Hempel's theory, it is possible to explain not only particular events but also general regularities. Within the D-N model, universal generalizations are explained by deduction from more comprehensive universal generalizations. In the second type of statistical explanation, the *deductive-statistical (D-S)*, statistical regularities are explained by deduction from more comprehensive statistical laws. This type of statistical explanation is best regarded as a subclass of D-N explanation.

Table 1 shows the four categories of scientific explanations recognized by Hempel in "Aspects." However, in their explication of D-N explanation in 1948, Hempel and Oppenheim restrict their attention to explanations of particular facts, and do not attempt to provide any explication of explanations of general regularities. The reason for this restriction is given in the notorious footnote 33:

> The precise rational reconstruction of explanation as applied to general regularities presents peculiar problems for which we can offer no solution at present. The core of the difficulty can be indicated by reference to an example: Kepler's laws, K, may be conjoined with Boyle's law, B, to [form] a stronger law $K.B$; but derivation of K from the latter would not be considered an explanation of the regularities stated in Kepler's laws; rather, it would be viewed as representing, in effect, a pointless "explanation" of Kepler's laws by themselves. The derivation of Kepler's laws from Newton's laws of motion and

Explananda Laws	Particular Facts	General Regularities
Universal Laws	D-N Deductive-Nomological	D-N Deductive-Nomological
Statistical Laws	I-S Inductive-Statistical	D-S Deductive-Statistical

Table 1

gravitation, on the other hand, would be recognized as a genuine explanation in terms of more comprehensive regularities, or so-called higher-level laws. The problem therefore arises of setting up clear-cut criteria for the distinction of levels of explanation or for a comparison of generalized sentences as to their comprehensiveness. The establishment of adequate criteria for this purpose is as yet an open problem. (Hempel and Oppenheim 1948, 273; future citations, H-O 1948)

This problem is not resolved in any of Hempel's subsequent writings, including "Aspects of Scientific Explanation."

Chapter XI of Braithwaite's *Scientific Explanation* is entitled "Explanation of Scientific Laws," but it, too, fails to address the problem stated in the Hempel-Oppenheim footnote. Indeed, on the second page of that chapter Braithwaite says,

To explain a law is to exhibit an established set of hypotheses from which the law follows. It is not necessary for these higher-level hypotheses to be established independently of the law which they explain; all that is required for them to provide an explanation is that they should be regarded as established and that the law should follow logically from them. It is scarcely too much to say that this is the whole truth about the explanation of scientific laws . . . (Braithwaite 1953, 343)

It would appear that Braithwaite is prepared to say that the deduction of Kepler's laws from the conjunction of Kepler's laws and Boyle's law — or the conjunction of Kepler's laws and the law of diminishing marginal utility of money (if you accept the latter as an established law) — is a bona fide explanation of Kepler's laws. However, inasmuch as Braithwaite's book does not contain any citation of the Hempel-Oppenheim paper, it may be that he was simply unaware of the difficulty, at least in this precise form. This problem was addressed by Michael Friedman (1974); we shall discuss his seminal article in §3.5 below. It was also treated by John Watkins (1984); his approach will be discussed in §4.10. Since the same problem obviously applies to D-S explanations, it affects both sectors in the right-hand column of Table 1.

The 1948 Hempel-Oppenheim article marks the division between the prehistory and the history of modern discussions of scientific explanation.[6] Hempel's 1965 "Aspects" article is the central document in the hegemony (with respect to scientific explanation) of logical empiricism, which held sway during roughly the third quarter of the present century. Indeed, I shall use the phrase *the received view* to refer to accounts similar to that given by Hempel in "Aspects." According to the received view, I take it, every legitimate scientific explanation belongs to one of the four sectors of Table 1. As we have seen, the claim of the received view to a comprehensive theory of scientific explanation carries a large promissory note regarding explanations of laws.

The First Decade (1948–57)
Peace in the Valley
(but Some Trouble in the Foothills)

With hindsight we can appreciate the epoch-making significance of the 1948 Hempel-Oppenheim paper; as we analyze it in detail we shall see the basis of its fertility. Nevertheless, during the first decade after its appearance it had rather little influence on philosophical discussions of explanation. To the best of my knowledge only one major critical article appeared, and it came at the very end of the decade (Scheffler 1957); it was more a harbinger of the second decade than a representative of the first. Indeed, during this period not a great deal was published on the nature of scientific explanation in general (in contrast to explanation in particular disciplines).

Braithwaite's (1953) might come to mind as a possible major exception, but we should not be misled by the title. In fact this book contains hardly any explicit discussion of the topic. Braithwaite remarks at the outset that "to understand the way in which a science works, and the way in which it provides explanations of the facts which it investigates, it is necessary to understand the nature of scientific laws, and what it is to establish them" (1953, 2). He then proceeds to discuss at length the nature of scientific deductive systems, including those that involve statistical laws, as well as those that involve only universal laws. Throughout this detailed and illuminating discussion he seems to be assuming implicitly that scientific explanation consists in somehow embedding that which is to be explained in such a deductive system. In adopting this view he appears to be anticipating the Friedman-Kitcher global unification approach, which will be discussed in §3.5 below. However, he has little to say explicitly about the relationship between deductive systems and scientific explanation.[1]

The final two chapters take up some specific issues regarding scientific explanation, and in the course of these chapters Braithwaite makes a few general remarks in passing. For example, the penultimate chapter opens with the statement, "Any proper answer to a 'Why?' question may be said to be an explanation of a sort" (319). In the final chapter he remarks, similarly, "an explanation, as I understand the use of the word, is an answer to a 'Why?' question which gives some intellectual satisfaction" (348–49). These comments are, without doubt, intended for construal in terms of his foregoing discussion of formal systems, but he does not spell out the connections. In the absence of explicit analyses of the nature of why questions, of what constitutes a "proper answer," or of the notion of "intellectual satisfaction," such passing remarks, however suggestive, leave much to be desired. The fact that Braithwaite's book nowhere cites the Hempel-Oppenheim article is eloquent testimony to the neglect of that essay during the first decade.

During this decade interest focused chiefly on two sets of special issues that had been sparked by earlier work. One of these concerned the nature of historical explanation, and the question of whether historical explanations must involve, at least implicitly, appeals to general laws. Much of this discussion took as its point of departure an earlier paper by Hempel (1942). The other dealt with the question of teleological or functional explanation; it came out of the longstanding controversy over mechanism vs. teleology (see H-O 1948, §4). On this specific issue, as we shall see, Braithwaite's book does provide significant contributions. We shall return to these special topics in §1.2 and §1.3, respectively, and we shall find an important connection between them.

1.1 The Fountainhead: The Deductive-Nomological Model

The 1948 Hempel-Oppenheim paper makes no pretense of explicating anything other than D-N explanations of particular occurrences – represented by the upper left-hand sector of Table 1. It will be useful to look in some detail at their treatment of this case. We must distinguish, in the first place, between the general conditions of adequacy for any account of this type of explanation, as laid down in Part I, and the actual explication spelled out in Part III.

The general conditions of adequacy are divided into two groups, logical and empirical. Among the logical conditions we find

(1) the explanation must be a valid deductive argument,

(2) the explanans must contain essentially at least one general law,

(3) the explanans must have empirical content.

The only empirical condition is:

(4) the sentences constituting the explanans must be true.

Although these criteria may seem simple and straightforward, they have been called into serious question. We shall return to this matter a little later.

The general notion of D-N explanation can be represented in the following schema offered by Hempel and Oppenheim, where the arrow signifies deductive entailment. It should also be noted that these criteria of adequacy are meant to apply to D-N explanations of general regularities even though Hempel and Oppenheim do not attempt to provide an explicit explication of explanations of this type. Since the derivation of a narrower generalization (e.g., the behavior of double stars) from a more comprehensive theory (e.g., celestial mechanics) does not

$C_1, C_2, \ldots, C_k$ Statements of antecedent conditions

$L_1, L_2, \ldots, L_r$ General laws

Explanans

E Description of the empirical phenomenon to be explained

Explanandum

require any antecedent conditions, they deliberately refrain from requiring that the explanans contain any statements of antecedent conditions.

One of the most vexing problems arising in this context is the characterization of law-sentences. It obviously has crucial importance for the D-N model, as well as for any covering law conception of scientific explanation. Following a strategy introduced by Nelson Goodman (1947), Hempel and Oppenheim (1948, 264–70) attempt to define the broader notion of a *lawlike sentence*. Only true sentences are classified as law-sentences; lawlike sentences have all the characteristics of law-sentences, with the possible exception of truth. Thus every law-sentence is a lawlike sentence, but not all lawlike sentences are laws. Informally, lawlike sentences have four properties:

(1) they have universal form,
(2) their scope is unlimited,
(3) they do not contain designations of particular objects, and
(4) they contain only purely qualitative predicates.

Let us consider the reasons for requiring these characteristics. With regard to (1) and (2) it is intuitively plausible to expect laws of nature to be general laws whose variables range over the entire universe. Newton's laws of universal gravitation and motion apply to all bodies in the universe, and their scope is not restricted in any way. These are paradigms of lawlike statements. However, an apparently universal statement, such as "All Apache pottery is made by women," would not qualify as lawlike because its scope is restricted. Likewise, the statement, "All living things contain water," if tacitly construed to be restricted to living things on earth, would not qualify as lawlike. In contrast, however, "All pure gold is malleable"–though it may appear to have a scope limited to golden objects–is nevertheless a universal generalization of unlimited scope, for it says of each object in the universe that, if it consists of gold, it is malleable. The distinction among the foregoing examples between those that qualify as lawlike and those that do not relates to characteristic (3). The statement about Apache pottery makes explicit reference to a particular group of people, the Apache. The statement about living things, if construed as suggested, refers implicitly to our particular planet.[2]

Why does it matter, with respect to lawlikeness, whether a statement refers to a particular of some sort—a particular time, place, object, person, group, or nation? Consider a simple example. Suppose it happens to be true (because I like golden delicious apples) that all of the apples in my refrigerator are yellow. This statement involves reference to a particular person (me), a particular thing (my refrigerator), and a particular time (now). Even given my taste in apples it is not impossible for my refrigerator to contain apples of different colors. Moreover, there is no presumption that a red delicious apple would turn yellow if it were placed in my refrigerator.

The problem that arises in this context is to distinguish between laws and *accidental generalizations*. This is a crucial issue, for laws have explanatory force, while accidental generalizations, *even if they are true*, do not. It obviously is no explanation of the color of an apple that it happens to reside in my refrigerator at some particular time.

If a statement is to express a law of nature it must be true. The question is, what characteristics, in addition to truth, must it possess? *Generality* is one such characteristic: laws must apply universally and they must not contain special provisions or exceptions for particular individuals or groups. The *ability to support counterfactuals* is another: they must tell us what would happen if. . . . If this table salt were placed in water, it would dissolve. If this switch were closed, a current would flow in this circuit.[3] *Modal import* is another: laws delineate what is necessary, possible, or impossible. We are not talking about logical modalities, of course; we are concerned with what is *physically* necessary, possible, or impossible. According to relativity theory it is physically impossible to send a signal faster than light *in vacuo*; according to the first law of thermodynamics it is physically impossible to construct a perpetual motion machine (of the first type). Accidental generalizations, even if true, do not support counterfactuals or possess modal import.

Even if a given statement does not contain explicit designations of particular objects, it may involve implicit reference to one or more particulars. Such references may be hidden in the predicates we use. Terms like "lunar," "solar," "precolumbian," and "arctic," are obvious examples. Because such terms refer to particulars they do not qualify as purely qualitative. By stipulating, in property (4) above, that laws contain only purely qualitative predicates, this sort of implicit reference to particulars, is excluded. Properties (3) and (4) are designed to rule out as accidental those universal generalizations that contain either explicit or implicit reference to particulars.

As Hempel and Oppenheim are fully aware, the prohibition against reference to particulars they impose is extremely stringent. Under that restriction, neither Galileo's law of falling bodies (which refers explicitly to the earth) nor Kepler's laws of planetary motion (which refer explicitly to our solar system) would qualify as laws or lawlike statements. As we shall see, because of this consideration they distinguish between fundamental and derived laws. The foregoing re-

FOUR DECADES OF SCIENTIFIC EXPLANATION

strictions apply only to the fundamental laws. Any universal statement that can be deduced from fundamental laws qualifies as a derived law.

Yet, in spite of their careful attention to the problem of distinguishing between lawful and accidental generalizations, Hempel and Oppenheim did not succeed in explicating that distinction. Consider the following two statements:

(i) No signal travels faster than light.
(ii) No gold sphere has a mass greater than 100,000 kg.

Let us suppose, for the sake of argument, that both are true. Then we have two true (negative) universal generalizations. Both have universal form. Neither is restricted in scope; they refer, respectively, to signals and gold spheres anywhere in the universe at any time in its history—past, present, or future. Neither makes explicit reference to any particulars. Both statements satisfy characteristics (1)-(3). One might argue that the predicate "having mass greater than 100,000 kg" is not purely qualitative, since it contains a reference to a particular object—namely, the international prototype kilogram. But this difficulty can be avoided by expressing the mass in terms of atomic mass units (which refer, not to any particular object, but to carbon-12 atoms in general). Thus, with (ii) suitably reformulated, we have two statements that satisfy characteristics (1)-(4), one of which seems patently lawful, the other of which seems patently accidental. The contrast can be heightened by considering

(iii) No enriched uranium sphere has a mass greater than 100,000 kg.

Since the critical mass for enriched uranium is just a few kilograms, (iii) must be considered lawful.

Both statements (i) and (iii) have modal import, whereas (ii) does not. It is physically impossible to send a message faster than light and it is physically impossible to fabricate an enriched uranium sphere of mass greater than 100,000 kg. It is not physically impossible to fabricate a gold sphere of mass greater than 100,000 kg.[4] Likewise, statements (i) and (iii) support counterfactuals, whereas (ii) does not. If something were to travel faster than light it would not transmit information.[5] If something were a sphere with mass greater than 100,000 kg it would not be composed of enriched uranium. In contrast, we cannot legitimately conclude from the truth of (ii) that if something were a sphere with mass greater than 100,000 kg, it would not be composed of gold. We cannot conclude that if two golden hemispheres with masses greater than 50,000 kg each were brought together, they would explode, suffer gravitational collapse, undergo severe distortion of shape, or whatever, instead of forming a sphere.

Lawfulness, modal import, and support of counterfactuals seem to have a common extension; statements either possess all three or lack all three. But it is extraordinarily difficult to find criteria to separate those statements that do from those that do not. The three characteristics form a tight little circle. If we knew which statements are lawful, we could determine which statements have modal import and support counterfactuals. But the way to determine whether a statement has modal import is to determine whether it is a law. The same consideration applies to support of counterfactuals; to determine which statements support counterfactuals we need to ascertain which are laws.[6] The circle seems unbroken. To determine to which statements any one of these characteristics applies we need to be able to determine to which statements another of them applies.

There are, of course, a number of differences between statements (i) and (iii) on the one hand and statement (ii) on the other. For example, I am much less confident of the truth of (ii) than I am of (i) or (iii). But this is a psychological statement about my state of belief. However, we are assuming the truth of all three statements. Given that all three are true, is there any objective difference in their status, or is the sole difference psychological? Again, (i) and (iii) fit closely with a well-integrated body of physical theory, while (ii) does not.[7] But given that all three are true, is this more than an epistemic difference?[8] Further, there are differences in the ways I might come to know the truth of (ii), as opposed to coming to know the truth of (i) and (iii). But is this more than an epistemic or psychological difference? Still further, I would much more readily give up my belief in (ii) than I would my belief in (i) or (iii).[9] But is this more than a pragmatic difference? The unresolved question is this: is there any objective distinction between laws and true accidental generalizations?[10] Or is the distinction wholly psychological, epistemic, or pragmatic?

In his 1953 book, Braithwaite places considerable emphasis upon the nature of laws and their place in science. He writes, "In common with most of the scientists who have written on philosophy of science from Ernst Mach and Karl Pearson to Harold Jeffreys, I agree with the principal part of Hume's thesis — the part asserting that universals of law are objectively just universals of fact, and that in nature there is no extra element of necessary connexion" (1953, 294). In chapter IX he defends the view that "the difference between universals of law and universals of fact [lies] in the different roles they play in our thinking rather than in any difference in their objective content" (294–95).[11]

The most ambitious attempt by any of the logical empiricists to deal with these problems concerning the nature of laws was given by Hans Reichenbach (1954),[12] the year just after the publication of Braithwaite's book. It had been anticipated by his discussion of the same topics in his symbolic logic book (1947, chap. VIII). Reichenbach's very first requirement on law-statements makes the distinction between laws and accidental generalizations an epistemic one, for it refers explicitly to the types of evidence by which such statements are supported. It should be remarked, incidentally, that Reichenbach was not addressing these problems in the context of theories of scientific explanation.

The problem of characterizing law-statements is one that has not gone away. Skipping ahead to subsequent decades, we may note that Ernest Nagel's magnum opus on scientific explanation, published near the beginning of the second decade, has a sensitive and detailed discussion of this problem, but one that remains inconclusive.[13] Around the beginning of the third decade, Nicholas Rescher's book *Scientific Explanation* offers an extended discussion which concludes that lawfulness does not reflect objective factors in the world, but rather rests upon our imputations, and is consequently mind-dependent (1970, 97–121; see also Rescher 1969). In the fourth decade, to mention just one example among many, Brian Skyrms (1980) offers a pragmatic analysis. The fifth decade will see the publication of an extremely important work on the subject, *Laws and Symmetries*, by Bas van Fraassen. But let us return to the first decade.

To carry out their precise explication, Hempel and Oppenheim introduce a formal language in which scientific explanations are supposed to be formulated. It is a standard first order functional calculus without identity, but no open sentences are allowed. All individual variables are quantified, so generality is always expressed by means of quantifiers. Two semantical conditions are imposed on the interpretation of this language: First, the range of the individual variables consists of all physical objects in the universe or of all spatio-temporal locations; this ensures that requirement (2) on lawlike statements – that their scope be unlimited – will be fulfilled, for there is no limit on the range of the variables that are universally (or existentially) quantified. Second, the primitive predicates are all purely qualitative; this feature of the interpretation of the language is, of course, a direct reflection of the fourth requirement on lawlike statements. The explication of D-N explanation of particular occurrences is given wholly in semantical terms.

Before going into the details of the formal language, we must acknowledge a fundamental problem regarding the second of the foregoing semantical conditions, namely, the concept of a purely qualitative predicate. In his well-known book *Fact, Fiction, and Forecast* (1955), Nelson Goodman poses what he calls "the new riddle of induction" in terms of two predicates, "grue" and "bleen," that he constructs for that purpose. Select quite arbitrarily some future time t (say the beginning of the twenty-first century). "The predicate 'grue' applies to all things examined before t just in case they are green but to other things just in case they are blue" (1955, 74). "Bleen" applies to things examined before t just in case they are blue but to other things just in case they are green (1955, 79). The question Goodman poses is whether we should inductively project that twenty-first century emeralds will be green or that they will be grue. The same problem had originally been posed by Goodman in 1947.

In an answer to Goodman's query, Carnap maintained that "grue" and "bleen," in contrast to "blue" and "green," are not purely qualitative predicates, because of the reference to a particular time in their definitions. He proposes to resolve Goodman's problem by restricting the predicates of his languages for confirmation theory to purely qualitative ones (Carnap 1947).[14] Goodman demurs:

. . . the argument that the former but not the latter are purely qualitative seems to me quite unsound. True enough, if we start with "blue" and "green," then "grue" and "bleen" will be explained in terms of "blue" and "green" and a temporal term. But equally truly, if we start with "grue" and "bleen," then "blue" and "green" will be explained in terms of "grue" and "bleen" and a temporal term; "green," for example, applies to emeralds examined before time *t* just in case they are grue, and to other emeralds just in case they are bleen. Thus qualitativeness is an entirely relative matter.(1947)

It is now generally conceded that Carnap's attempt to characterize purely qualitative predicates was inadequate to deal with the problem Goodman raised. Many philosophers (including this one – (W. Salmon 1963)) have tried to make good on the distinction Carnap obviously had in mind. Whether any of these other efforts have been successful is a matter of some controversy; at any rate, no particular solution has gained general acceptance.

Our discussion, so far, has been largely preparatory with respect to the official Hempel-Oppenheim explication. We may now return to the formal language offered by Hempel and Oppenheim. Several different types of sentences must be distinguished. To begin, an *atomic sentence* is one that contains no quantifiers, no variables, and no sentential connectives. It is a sentence that attributes a particular property to a given individual (e.g., "George is tall") or asserts that a particular relation holds among two or more given individuals (e.g., "John loves Mary"). A *basic sentence* is either an atomic sentence or the negation of an atomic sentence; a basic sentence contains no quantifiers, no variables, and no binary sentential connectives. *Singular* (or *molecular*) *sentences* contain no quantifiers or variables, but they may contain binary sentential connectives (e.g., "Mary loves John or Mary loves Peter"). A *generalized sentence* contains one or more quantifiers followed by an expression containing no quantifiers (e.g., "All humans are mortal"). Since any sentence in first order logic can be transformed into *prenex normal form*, any sentence containing quantifiers can be written as a generalized sentence. *Universal sentences* are generalized sentences containing only universal quantifiers. A generalized (universal) sentence is *purely generalized (universal) if it contains no proper names of individuals. A generalized (universal)* sentence is an *essentially generalized (universal)* sentence that is not equivalent to any singular sentence. With these definitions in hand we can proceed to explicate the fundamental concepts involved in scientific explanation.

The first concept with which we must come to terms is that of a law of nature, and, as we have seen, it is one of the most problematic. Hempel and Oppenheim distinguish between *lawlike sentences* and genuine laws, and also between *fundamental* and *derivative* laws. The following series of definitions is offered:[15]

(7.3a) A *fundamental lawlike sentence* is any purely universal sentence; a *fundamental law* is purely universal and true.

(7.3b) A *derivative law* is a sentence that is essentially, but not purely, universal and is deducible from some set of fundamental laws.

(7.3c) A *law* is any sentence that is either a fundamental or a derived law.

We have already canvassed the fundamental problems encountered in this characterization of laws.

Interestingly, the concept of law does not enter into the formal explication of D-N explanation; instead, the notion of *theory* is employed.

(7.4a) A *fundamental theory* is any sentence that is purely generalized and true.

(7.4b) A *derivative theory* is any sentence that is essentially, but not purely, generalized and is derivable from fundamental theories.

(7.4c) A *theory* is any fundamental or derivative theory.

Note that the concept of a theory-like sentence is not introduced.

According to the foregoing definitions, every law is a theory and every theory is true. As the term "theory" is used in this context, there is no presumption that theories refer to unobservable entities, or that they involve any sort of special theoretical vocabulary. The difference between laws and theories is simply that theories may contain existential quantifiers, while laws contain only universal quantifiers. Clearly, many of the scientific laws or theories that are employed in explanation contain existential quantifiers. To say, for example, that every comet has a tail, that every atom has a nucleus, or that every mammal has a heart, involves a universal quantifier followed by an existential quantifier – i.e., for every x there is a y such that . . . Hempel and Oppenheim say nothing about the order in which quantifiers must occur in theories. That leaves open the interesting question of whether explanatory theories may have existential quantifiers preceding all of the universal quantifiers, or whether explanatory theories need contain any universal quantifiers at all.[16] It is perhaps worth explicit mention in this context that universality and generality are not coextensive. Existentially quantified statements are general in the sense that they involve variables having the universe as their range. To say, "there exists an x such that . . . " means that within the whole domain over which x ranges there is at least one object such that. . . . Such statements have generality without being universal. The question remains whether universality is a necessary requirement for explanatory theories, or whether generality is sufficient. As we shall see in connection with the next set of formal definitions, Hempel and Oppenheim are willing to settle for the latter alternative.

We have finally arrived at the stage at which Hempel and Oppenheim offer their formal explication of scientific explanation. The concept of a potential explanation comes first:

(7.5) $<T,C>$ is a potential explanans of E (a singular sentence) *only if*
  (1) T is essentially general and C is singular, and
  (2) E is derivable from T and C jointly, but not from C alone.

It would be natural to suppose that (7.5) would constitute a definition of "potential explanans," but Hempel and Oppenheim are careful to point out that it provides only a necessary condition. If it were taken as sufficient as well, it would leave open the possibility that "any given particular fact could be explained by means of any true lawlike sentence whatever" (H-O 1948, 276). They offer the following example. Let the explanandum-statement E be "Mount Everest is snowcapped" and let the theory T be "All metals are good conductors of heat." Take a singular sentence T_s that is an instance of T—e.g., "If the Eiffel Tower is metal it is a good conductor of heat." Now take as the singular sentence C the sentence T_s implies E—i.e., "If the fact that the Eiffel Tower is made of metal implies that it is a good conductor of heat, then Mount Everest is snowcapped." Because E is true, C must be true, for C is a material conditional statement with a true consequent. Thus,

$\vdash C = T_s \supset E$	definition
$T_s \supset E$	assumption
$\vdash T \supset T_s$	instantiation
$T \supset E$	hypothetical syllogism
$\vdash C \supset (T \supset E)$	conditional proof
$\vdash C . T \supset E$	tautology

It is evident that C does not, by itself, entail E. Therefore $<T,C>$ satisfies (7.5). But it is manifestly absurd to claim that the law about metals being good conductors of heat is the key law in the explanation of snow on Mount Everest.

The obvious difficulty with this example is that C's truth can be fully certified only on the basis of the truth of E. Evidently, some restriction must be placed on the singular sentence C that is to serve as the statement of antecedent conditions in the explanans. If knowing that the explanandum-statement is true is the only way to establish the truth of C, then in some important sense, in appealing to C, we are simply using E to explain E. Indeed, given that T is true, there must be some way to establish the truth of C without appealing to E. Hempel and Oppenheim formulate the needed restriction as follows:

  (3) T must be compatible with at least one class of basic sentences which has C but not E as a consequence.

That is to say, given that the theory T is true, there must be some way to verify that C is true without also automatically verifying E as well. Adding (3) to the necessary conditions stated in (7.5) gives

(7.8) $<T,C>$ is a potential explanans of E (a singular sentence) *iff*
  (1) T is essentially general and C is singular, and

(2) E is derivable from T and C jointly, but not from C alone.

(3) T must be compatible with at least one class of basic sentences which has C but not E as a consequence.[17]

With this definition of "potential explanans" it is a small step to the official explication of "explanans," and hence, "explanation."

(7.6) $<T,C>$ is an explanans of E (a singular sentence) *iff*

(1) $<T,C>$ is a potential explanans of E

(2) T is a theory and C is true.

Taken together, the explanans $<T,C>$ and the explanandum E constitute an explanation of E. This completes the Hempel-Oppenheim explication of *D-N explanation of a particular fact*.

Given the great care with which the foregoing explication was constructed, it would be easy to surmise that it is technically correct. Jumping ahead to the next decade for a moment, we find that such a supposition would be false. As Rolf Eberle, David Kaplan, and Richard Montague (1961) showed (roughly), on the foregoing explication any theory T can explain any fact E, where T and E have no predicates in common, and are therefore, intuitively speaking, utterly irrelevant to one another. Suppose, for example, that T is "(x)Fx" (e.g., "Everyone is imperfect.") and E is "Ha" (e.g., "C. G. Hempel is male.").[18] We can formulate another theory T′ that is a logical consequence of T:

$$T' =_{df} (x)(y)[Fx \lor (Gy \supset Hy)]$$

T′ is of purely universal form, and, on the assumption that T is true, it is true as well. As a singular sentence, take

$$C =_{df} (Fb \lor \sim Ga) \supset Ha$$

For the sake of our concrete interpretation, we can let "Gx" mean "x is a philosopher" and let "b" stand for W. V. Quine. It can now be shown that $<T',C>$ constitutes an explanans of E.

(1)	$(x)(y)[Fx \lor (Gy \supset Hy)]$	premise (T′)
(2)	$(Fb \lor \sim Ga) \supset Ha$	premise (C)
(3)	$(Fb \supset Ha) . (\sim Ga \supset Ha)$	equivalent to (2)
(4)	$Fb \supset Ha$	simplification (3)
(5)	$\sim Ga \supset Ha$	simplification (3)
(6)	$Fb \lor (Ga \supset Ha)$	instantiation (1)
(7)	$\sim Fb \supset (Ga \supset Ha)$	equivalent to (6)
(8)	$Fb \lor \sim Fb$	tautology
(9)	$Ha \lor (Ga \supset Ha)$	dilemma (4, 7, 8)
(10)	$Ha \lor \sim Ga \lor Ha$	equivalent to (9)
(11)	$\sim Ga \lor Ha$	equivalent to (10)

(12)	Ga $\supset$ Ha	equivalent to (11)
(13)	Ga v ~ Ga	tautology
(14)	Ha	dilemma (5, 12,13)

As we have seen, T' is essentially general, C is singular, and E is derivable from T' and C. Hence, conditions (1) and (2) of (7.8) are satisfied. Now, consider the set of basic sentences {~ Fb, Ga}; obviously it does not entail E (i.e., Ha). But it does entail C, as follows:

(1)	~ Fb . Ga	premise
(2)	(~ Fb . Ga) v Ha	addition (1)
(3)	~ (~ Fb . Ga) $\supset$ Ha	equivalent to (2)
(4)	(Fb v ~ Ga) $\supset$ Ha	DeMorgan (3)

Thus, condition (3) of (7.8) is also fulfilled; < T',C > is an explanans for E. We should remember that < T,C > has not been shown to be an explanans for E, so it has not been shown that any arbitrarily chosen theory explains any particular fact (as long as they share no predicates). But what has been shown is something like this: from the law of diminishing marginal utility of money (which, for the sake of argument, I take to be true), we can deduce a theory which, together with certain singular statements, provides an explanation of the explosion of the *Challenger* space-shuttle vehicle. It can hardly be doubted that Eberle, Kaplan, and Montague offered a recipe (in fact, several recipes) for constructing counterexamples that are damaging to the Hempel-Oppenheim explication.

To assess the nature of the damage, let us go back to Hempel and Oppenheim's set of necessary conditions (7.5), which at first blush looks like a suitable explication of *potential D-N explanans*. Had they offered it as such, their explication would have been technically defective, for it would have been vulnerable to the Mount Everest counterexample. However, it would not have posed a profound problem for their enterprise, since the technical defect could have been repaired by technical tinkering, namely, by adding the third condition of (7.8). Fortunately, they anticipated the problem and found a solution before their essay was published.[19]

The same attitude should, I believe, be taken to the problem discovered 13 years later by Eberle, Kaplan, and Montague. They pointed out a genuine technical defect, but again, it was one that could be repaired by technical tinkering. Indeed, in the same volume of the same journal, Kaplan (1961) provided one way of making the repair. Shortly thereafter another was offered by Jaegwon Kim (1963).

It is worth noting, I believe, that both the Mount Everest example and the Eberle-Kaplan-Montague type examples exploit a well-known feature of standard truth-functional logic. This feature is the principle of addition that allows one to deduce "p v q" from "p" — where "q" can be any arbitrary sentence whatever. This

principle is closely related to one of the so-called paradoxes of material implication, namely, the fact that if "p" is true then it is materially implied by any arbitrary sentence whatever—e.g., "~q"—so that, given "p" we can infer
"~q ⊃ p" (which is equivalent to "p v q"). In the Mount Everest example, we
chose C as T_s ⊃ E, which we were prepared to assert on the basis of the truth
of E. In the Eberle-Kaplan-Montague example we chose C as "(Fb v
~Ga) ⊃ Ha"—where "Ha" is E. In addition, from the theory "(x)Fx" we derived
"(x)(y)[Fx v (Gy ⊃ Hy)]" which is equivalent to "(x)Fx v (y)(Gy ⊃ Hy)."
Clearly the source of difficulty is that these moves are precisely the sorts that allow the introduction of irrelevancies—they are the very principles that are excluded in relevance logics. The technical problem with the Hempel-Oppenheim
explication is simply to find ways of blocking them in the context of explanation.
Kaplan and Kim have shown how to do that. Kim's revision consists in adding
a further requirement to those already included in (7.8):

(4) E must not entail any conjunct of the conjunctive normal form of C.[20]

In 1965 Hempel could be quite sanguine about the technical details of the corrected explication (Hempel 1965, 294–95).

It should be emphatically noted that the official explication—as given in the
augmented definition (7.8) is very different from the set of requirements of adequacy given in the first section of the Hempel and Oppenheim essay. This is to
be expected. The informal conditions of adequacy are part of the clarification of
the explicandum; the explication is the formal definition of the improved concept
that is to replace the original vague concept of explanation. What is perhaps surprising is the complexity of the formal explication.

Now that we have looked at the nitty-gritty details of the Hempel-Oppenheim
explication, let us return to a consideration of the fundamental philosophical issues to which it gives rise. Where does this discussion leave us? First, there is
a two-part agenda, explicitly stated in that essay, of items that are needed to complete the received view:

(1) Explications of one or more models of probabilistic or statistical explanation.

(2) An explication of D-N explanation of laws.

Second, there are, as we have seen, two serious lacunae in the Hempel-
Oppenheim explication of D-N explanation of particular facts:

(3) A satisfactory explication of the concept of a purely qualitative predicate.

(4) A satisfactory explication of the concept of a law of nature.

Third, as we shall see in greater detail when we discuss the second decade,
the Hempel-Oppenheim essay advances several important philosophical theses
that have been the subject of much controversy:

(5) The inferential conception of scientific explanation—i.e., the thesis that all legitimate scientific explanations are arguments of one sort or another. This thesis is involved in Hempel and Oppenheim's first condition of adequacy for scientific explanations. A number of subsequent authors including Michael Scriven, Richard Jeffrey, Bas van Fraassen, Peter Railton, and I—have rejected this inferential conception.

(6) The covering law conception of scientific explanation[21]—i.e., the view that every fully articulated adequate scientific explanation contains one or more laws in an essential fashion. This thesis is set forth in Hempel and Oppenheim's second condition of adequacy. As we shall see, this view is also rejected by a number of authors—including Scriven and van Fraassen. Thomas S. Kuhn strenuously denies it.[22] So do many others who are primarily concerned with the nature of explanation in history.

(7) The explanation/prediction symmetry thesis. According to the strong form of this thesis, as set forth in the Hempel-Oppenheim article, any correct D-N explanation could serve, in appropriate circumstances, as a scientific prediction; conversely, any deductive scientific prediction could, in appropriate circumstances, serve as a D-N explanation. This thesis, which Hempel later extended to I-S explanation as well, has been widely disputed. Hanson, Israel Scheffler, Scriven, van Fraassen, and I—among many others—have been unwilling to accept it.

(8) The role of causality in scientific explanation. Hempel and Oppenheim casually identify causal explanation with D-N explanation (H-O 1948, 250), but their official explication makes no reference to any causal requirements. In "Aspects," Hempel explicitly rejects the idea that causality plays any essential explanatory role (1965, 352). The question is whether to follow Hempel in "Aspects" in expunging causality from explanation altogether, or to find a place for causality, as Scriven, Railton, James H. Fetzer, Paul Humphreys, and I—among many others—have urged. As we shall see, this issue has been the subject of considerable subsequent discussion.

(9) Literal truth of the explanans. This is Hempel and Oppenheim's fourth condition of adequacy. Many authors have maintained that this requirement is too strong—that it should be replaced by a requirement of high degree of confirmation, or by a requirement of approximate truth. For various reasons this requirement has been rejected by Nancy Cartwright, Kuhn, and Larry Laudan.

(10) The possibility of quasi-formal models of explanation. The early logical empiricists seemed confident that many fundamental methodological concepts—such as confirmation and explanation—are amenable to formal semantical explication, more or less as attempted in the Hempel-Oppenheim paper. It now seems to many—including Peter Achinstein and me—that explications of that sort are bound to turn out to be inadequate. Even Hempel in his later writings on scientific

explanation eschews the degree of formality employed in the Hempel-Oppenheim explication.

Our review of the essay by Hempel and Oppenheim leaves no doubt whatever about its richness as a source of material for philosophical discussions of the nature of scientific explanation in the second half of the twentieth century. Can anyone seriously dispute the propriety of its designation as the fountainhead?

1.2 Explanation in History and Prehistory

In the decade following the publication of the Hempel-Oppenheim paper, as I pointed out above, not a great deal appeared in print on general problems concerning the nature of scientific explanation.[23] During this time, however, quite a good deal was written on the nature of explanation in history, stimulated, to a large degree, by Hempel's essay, "The Function of General Laws in History" (1942). The focus of much of the discussion was the question of whether historical explanations can or should embody general laws. This literature is directly pertinent to our discussion only if history is considered a science. Since I am not firmly committed to any particular view on this matter, and know very little about history, I shall not pursue the issue here. I refer the reader to Hempel's 1942 article, to §7 of "Aspects," and to the excellent bibliography of that essay.

Whether history is classified as a science or not, there can be no doubt that some sciences have essential historical aspects. Cosmology, geology, and evolutionary biology come immediately to mind. Archaeology (often called "prehistory") is somewhat ambiguous, for it has been pursued, not infrequently, as a humanistic, rather than scientific, discipline.

During the 1950s, an important development occurred in archaeology, namely, the origin of an influential movement called the *New Archaeology*. The main thrust of this movement is the attempt to make archaeology a bona fide science. When Lewis Binford, one of its major founders, was a student at the University of Michigan, Leslie White—a distinguished cultural anthropologist and a teacher of Binford—told him that, to find out what it means to be scientific, he should read some philosophy of science. He did (Binford 1972, 7–8). Although Binford's published work does not contain extensive discussions of philosophical issues, the effect upon him was profound, and he profoundly influenced other archaeologists. The most explicit expression of the effect of philosophy of science upon archaeology was given in Watson, LeBlanc, and Redman's widely used text, *Explanation in Archaeology* (1971, 1984). This book relies heavily on Hempel's work on explanation, especially his *Philosophy of Natural Science* (1966) as well as "Aspects of Scientific Explanation." It is hardly an exaggeration to say that, for these authors, the hallmark of genuine science is the use of hypothetico-deductive confirmation and D-N explanation. They urge their students and colleagues to adopt this kind of methodology. The New Archaeology, which had its heyday in the late 1960s and throughout the 1970s, provides an example of an outstanding influence of philosophy of science upon an actual science. It is still a strong movement. Whether the influence was good or bad is a matter of some controversy. A detailed account of the relationships between the New Archaeology and philosophy of science is provided by Merrilee H. Salmon in *Philosophy and Archaeology* (1982).

1.3 Teleology and Functional Explanation

As we have seen, logical empiricism arose in a philosophical context in which teleology, final causes, purposes, and ends played vital roles. Those scientific philosophers who—like Braithwaite, Hempel, and Nagel—had not abandoned the notion that science can provide legitimate explanations of various kinds of phenomena were deeply concerned to provide accounts that would require admissible scientific explanations to have empirical content. They strove to prohibit pseudo-explanations that appeal to entelechies or final causes.

Nevertheless, there appeared to be cases in the empirical sciences and in common sense of bona fide explanations that are framed in terms of ends or goals. One often explains one's own behavior in this way. Why did I go to the Groceria Italiana? To get fresh pasta. In cases of this sort, however, it is not the future state of procuring pasta that explains my act; it is my preceding desire for the pasta along with the concurrent belief that that particular grocery store was a good place to get it. The explanation would be correct even if I failed to get the pasta because they were all sold out.

Midway through the first decade, Braithwaite addressed the problem of teleological explanation (1953, chap. X).[24] He saw no problem in teleological explanations of actions that are the result of conscious intention, but he also pointed out that there are cases of goal-directed behavior in which *conscious* intent seems to be absent. Among the examples Braithwaite mentions are the behavior of rats in mazes and the operation of homing torpedoes. The basic philosophical problem in these cases, given the absence of conscious intent (or any kind of intent on the part of the torpedo), involves the question of whether a future state can legitimately be said to explain a present fact.

Braithwaite notices two striking features of goal-directed behavior, *plasticity* and *variancy*. Plasticity refers to the fact that, in many cases, the agent has more than one way to achieve the goal; if one means is frustrated another will be tried. An animal trying to arrive at a place where food is located will try different routes if the usual one is blocked. Variancy refers to the fact that the goal can successfully be reached under a wide variety of initial and background conditions. A homing torpedo can—within limits—find its target regardless of the evasive action taken and—again, within limits—regardless of weather or conditions of the sea.

It has been pointed out that discussions of teleological or functional explanation often employ either or both of two models: (1) *goal-directed* behavior, as illustrated by the foregoing examples of the homing torpedo and the rat, and (2) *self-regulating* behavior, as illustrated by a thermostatically controlled heat-pump or maintenance of body temperature by a human or other animal (Canfield 1966, Scheffler 1958). A thermostatically controlled heat-pump, for example, will keep the temperature of a house within a certain temperature range whether the initial temperature is below or above that range by a little or a lot. It will do so under a wide range of outdoor temperatures, wind velocities, conditions of precipitation, etc. The human body also is capable of maintaining an internal body temperature within a narrow normal range under a wide variety of external circumstances.

In some cases, such as the homing torpedo and the heat-pump, we can give a full causal account of the apparently purposeful behavior. In an influential article, Rosenblueth, Wiener, and Bigelow (1943) gave a cybernetic explanation of the behavior of the homing torpedo in terms of negative feedback, and they suggested that the apparently goal-directed behavior of animals can be understood in similar terms. These are Braithwaite's kind of example as well. Ernest Nagel, who dealt mainly with teleological or functional explanation in physiology, focused much more intensively on the *homeostatic* or self-regulating type of example (1956; 1961, chap. 12). Just as we can provide a complete causal account of the way in which the heat-pump controls the house temperature, starting with any set of background conditions within its range of operation, so also, he maintained, could we give a complete causal account of the way in which the human body regulates its internal temperature.

According to Braithwaite, if we give a teleological explanation of the operation of a homing torpedo in terms of the goal of reaching its target, or the heat-pump/thermostat system in terms of the goal of maintaining a given house temperature, that is merely a shorthand for a much more complicated causal explanation, and philosophically it is not very interesting. In other cases, such as the rat in the maze, we may not be able to give a complete causal explanation in physico-chemical terms, without reference to goal-seeking. In that case a teleological explanation is appropriate, legitimate, and interesting. But we have no reason to believe that the causal explanation in purely physico-chemical terms is impossible in principle. Indeed, Nagel argued the stronger point that, in many physiological cases, such causal explanations are already in hand — and that, in such cases, any teleological explanation can, in principle, be replaced by an equivalent causal explanation. Thus, we have no basis for supposing that in any instance of goal-directed behavior is a present fact ultimately explainable only in terms of a future result.

If we stretch the temporal bound just a little beyond the end of the first decade, we can include two other significant contributions to the discussion of teleological or functional explanation—namely, the paper by Israel Scheffler (1958)[25] and a characteristically clear treatment by Hempel (1959). Hempel's discussion brings the D-N model to bear directly on the problem of functional explanation.

Hempel's philosophical concern is rather different from Braithwaite's; he does not focus particularly on the issue of final causation—on the problem of explaining present facts in terms of future goals. Given his later suggestion that subsequent events might sometimes explain earlier events (Hempel 1965, 353–54), that issue might prove embarrassing. He has, instead, a logical point in mind. The kinds of cases in which he is interested are often referred to as *functional explanations*; let us consider some simple examples. In physiology, for instance, the presence of a particular component in a given species of organisms is explained in terms of its ability to perform a function that is indispensable to the continued life or health of organisms of that sort. In humans, the blood contains hemoglobin to convey oxygen from the lungs to the other parts of the body. Without this transport of oxygen a human being could not survive.

In evolutionary biology, for another example, a feature of a species of animal is explained in terms of its enhancement of the chances of survival and reproduction. The large ears of the jackrabbit, which inhabits very hot regions, enable the animal to control its body temperature. When its body temperature rises too high, the animal seeks shade and dilates the many blood vessels in the ears. Blood coming from other parts of the body brings heat which is radiated into the environment, thereby cooling the animal. I have heard that the ears of elephants function in much the same way. In biological evolution adaptation to the environment plays a crucial role, and this appears to involve the attainment of goals.

Anthropology and sociology provide many additional cases of functional explanation. In the study of primitive societies, for instance, such prominent anthropologists as A. R. Radcliffe-Brown (1952) and B. Malinowski (1954) maintain that many—if not all—institutions, customs, and rituals are to be explained on the basis of the social functions they fulfill. The performance of a rain dance, for example, may provide an occasion for social interaction that contributes to the cohesiveness of the society. The influential sociologist R. K. Merton (1950, 1957) also advocates functional analysis in the study of human institutions.[26] He distinguishes carefully between *latent function* and *manifest function*. The rain dance has the manifest function of bringing rainfall, and it may be altogether unsuccessful in fulfilling that function; it has the latent function, however, of promoting social cohesiveness in times of distress, and it may fulfill that function quite effectively. In such cases, the latent function *explains* the survival of a practice that fails miserably to fulfill its manifest function.

Freudian psychoanalysis (which may or may not deserve to be classified as a science) is another rich source of functional explanations. For example, Freud claims that dreams serve as wish-fulfillments to prevent interruption of sleep, which is essential to the health of any human being. He also offers functional explanations of slips of the tongue or pen, and for the presence of neurotic symptoms.

The statements involved in the foregoing accounts — with the possible exception of psychoanalysis — are empirically testable and scientifically legitimate. Even if one denies that psychoanalysis is a genuine science, the basis for that judgment should not be the presence of functional explanations; the crucial issue lies in the empirical testability of its claims. It appears, then, that we have a number of different sciences that make legitimate use of functional explanations. The problem, as far as Hempel is concerned, is that functional explanations do not fit any of the logical patterns for scientific explanation, including those for inductive or statistical explanation, recognized by the received view.[27] It is *very* significant that Hempel entitled his article "The Logic of Functional Analysis" rather than "The Logic of Functional Explanation."

The fundamental problem Hempel faces can easily be seen by schematizing a simple case. He chooses the example

(3.1) The heartbeat in vertebrates has the function of circulating blood through the organism. (1959, 305)

In attempting to understand the meaning of such a statement we might be tempted simply to substitute the word "effect" for the word "function," since the circulation of blood is an effect of the beating of the heart. But the beating of the heart also has other effects that we would be unwilling to consider functions — for example, it has the effect of producing heart sounds, but that is not one of its functions. The general idea is that a function must be important to the health or survival of the organism. To do more justice to the significance of (3.1) he reformulates it as

(3.3) The heartbeat has the effect of circulating the blood, and this ensures the satisfaction of certain conditions (supply of nutriment and removal of waste) which are necessary for the proper working of the organism. (1959, 305)

To formulate the situation more abstractly, Hempel offers the following schema:

(3.4) *Basic pattern of a functional analysis*: The object of the analysis is some "item" i, which is a relatively persistent trait or disposition (e.g., the beating of the heart) occurring in a system s, (e.g., the body of a living vertebrate); and the analysis aims to show that s, is in a state, or internal condition, c_i and in an environment representing certain external conditions c_e such that under conditions c_i and c_e (jointly referred to as c) the trait i has effects which satisfy some "need" or "functional requirement" of s, i.e., a condition n which is necessary for the system's remaining in adequate, or effective, or proper, working order. (1959, 306)

It is clear from this characterization that Hempel, like Nagel, is thinking primarily in terms of the furnace analogy rather than that of the homing torpedo.

Given the fact that Hempel regards explanations as arguments, he seeks an argument that, under the conditions outlined in (3.4), would constitute an explanation of item *i* (the heartbeat). The crux of the problem that arises is this. We can assert (1) that if *i* is present and the system is in normal conditions *c* then *n* (it will be operating normally). We see (2) that *n* obtains (it is operating normally in these conditions). If, however, we attempt to deduce the presence of *i* from premises (1) and (2), we will be guilty of committing the fallacy of affirming the consequent. Given *i* in the circumstances we can deduce the normal operation *n*, but given the normal operation *n* we cannot validly deduce the presence of *i*. Moreover, Hempel recognizes, there is no inductive argument with (1) and (2) as premises which establishes the presence of *i* with high inductive probability; thus it cannot qualify as an acceptable I-S explanation either.

The problem to which Hempel points arises quite generally with functional explanations; it is the problem of *functional equivalents*. When we identify some item as fulfilling a function, we recognize that it is sufficient to produce some result in a certain situation. But usually we cannot claim that it is the only possible device that would fulfill that function. It is not necessary for the realization of the goal. As an obvious example, consider the large ears of the jackrabbit. This is one effective mechanism for cooling the animal, but other animals use other mechanisms. Humans perspire and dogs pant. These alternative devices are sufficient to bring about the result in question. Given only that the jackrabbit is an animal that inhabits hot regions, and thus must have some mechanism for reducing body heat, it does not follow deductively, or with high inductive probability, that the jackrabbit has extra-large ears.

Similar remarks apply to the other examples we have considered. Given that a particular ceremony enhances social cohesiveness in a given primitive society, it does not follow that no other ceremony could achieve the same end. Given that a particular dream fulfills a particular unconscious wish, it does not follow that no other dream could have done the job. Given that hemoglobin transports oxygen from one part of the body to another, it does not follow that no other chemical substance could fulfill the same function. Given that the heartbeat causes the blood to circulate, it does not follow that there is no other way for that result to come about. The problem about functional explanation can be stated simply in the following terms. In a correct D-N explanation the explanans is logically sufficient for the explanandum. In the typical functional explanation the explanandum is, given the conditions, sufficient for the explanans. From Hempel's standpoint that is just the wrong way around.

In his discussion of physiological examples, Nagel attempts to avoid the problem of functional equivalents by arguing that, if we specify with sufficient precision the kind of organism we are dealing with, only one kind of mechanism will do the trick. To cite Hempel's example, perhaps there are, in principle, mechanisms other than a beating heart that could circulate blood. Nevertheless, given the stage of evolution of homo sapiens, that is the only mechanism available. Hence if we have a healthy human — not someone undergoing surgery with a heart-lung machine attached — we know that blood is circulating, and if the blood is circulating, we know that the heart is beating. Nagel's prime example is the presence of chlorophyll in green plants (1961, 403–6). To take my example of the jackrabbit's ears, it could be claimed that in the circumstances in which that trait evolved, perspiring and panting were not biologically available. In a hot *dry* climate — the habitat of the jackrabbit — the conservation of water is absolutely critical; both perspiration and panting deplete the animal's supply of water. Whether other heat regulatory mechanisms might be available I do not know.

In view of this treatment of the problem of functional equivalents, Nagel has offered a solution to Hempel's problem about the failure of functional explanations to fit the received models. He can say quite straightforwardly that they do fit the deductive model (Nagel 1961, 405). In 1956 Nagel seemed to claim that Merton's brand of functional analysis would fit the same pattern as functional explanation in physiology, but in 1961 he had doubts that the problem of functional equivalents could be that easily dismissed in the realm of social sciences (Nagel 1961, 533–35). It seems to me, therefore, that Nagel has not succeeded in eliminating the problem of functional equivalents for teleological or functional explanation in general.

Hempel's analysis of functions is, I think, logically impeccable. If an admissible explanation of any fact must be an argument to the effect that the fact-to-be-explained *was to be expected* by virtue of the explanatory facts, then functional 'explanations' are not admissible explanations. But I have often noticed that, in philosophy as well as other human endeavors, one person's counterexample is another's modus ponens. Hempel concludes from his discussion that functional analysis cannot qualify as an admissible type of explanation; at best, it has heuristic value. Others, myself included, would take the moral to be that, since functional explanations play a legitimate scientific role, explanations cannot always be arguments of the sorts endorsed by the received view.[28] I considered it a virtue of the statistical-relevance model that it did not encounter similar problems in connection with functional explanation (W. Salmon 1982).

It turns out that there is a deep connection between the problem of functional or teleological explanation and the problem of explanation in history. Explanations in human history, as well as in the (other?) sciences of human behavior, make frequent appeals to conscious purposes and goals. As Braithwaite pointed out, such cases pose no serious problem for the philosophical theory of scientific explanation.[29] Explanations in these disciplines may, in other cases, make use of unconscious purposes, as in the example of the rain dance. In evolutionary biology functional considerations play a crucial role, and — since the time of Darwin

—it has been appropriate to deny that such appeals to functions involve the conscious purposes of a creator, or any other sort of final causation. The basic idea is that we can understand the evolutionary process in terms of efficient causes in conjunction, perhaps, with chance occurrences. Evolutionary biology thus requires a causal explication of function.

Furthermore, many authors who deal with explanation in human history insist that there is a *narrative* or *genetic* type of explanation that consists in telling the story leading up to the event to be explained.[30] Since the mere recital of just any set of preceding occurrences may have no explanatory value whatever, the narrative must involve events that are causally relevant to the explanandum if it is to serve as an explanation. This, again, demands some form of causal explanation. In human history, as in evolutionary biology, we need an account of what has come to be known as *etiological explanation* that will encompass both genetic and functional explanation.[31] In my opinion, the real breakthrough on functional explanation was provided by Larry Wright (1976) when he advanced an explicitly causal account of functional or teleological explanation.[32] Nagel took the subject up again in 1977. We shall return to this topic in §3.8.

The Second Decade (1958–67)
Manifest Destiny—Expansion and Conflict

As we have already remarked, the first decade after the publication of the Hempel-Oppenheim paper saw little published criticism — or acknowledgment — of it. Quite possibly this portion of the received view — the box in the upper left corner of Table 1 — was accepted with considerable satisfaction for the most part by the philosophy of science community. The situation changed rather dramatically around 1958. This was the year in which the second volume of *Minnesota Studies in the Philosophy of Science* (Feigl et al. 1958) was published, containing Scriven's first article attacking the D-N model. Hanson's *Patterns of Discovery* (1958) appeared during the same year.

In the next few years following 1958 a great many papers on scientific explanation appeared, devoted mainly to issues we have already mentioned. These included, for example, debates on the covering law conception and on the explanation/prediction symmetry thesis. The critiques of the Hempel-Oppenheim account fall into three main categories. First, as we have seen, the Eberle-Kaplan-Montague (1961) critique and replies to it come under the heading of sympathetic efforts to find and eliminate any technical flaws in that explication. Unlike the following two types, these were critiques of the formal explication given in Part III of the Hempel-Oppenheim paper, rather than objections to the preliminary conditions of adequacy advanced in Part I. Second, as we shall see in §2.1, the attacks by Hanson, Scriven, and others were motivated by deep philosophical disagreements with anything resembling the logical empiricist point of view. Third, there were constructive efforts by philosophers such as Bromberger (1962, 1963, 1966) and Scheffler (1957, 1963) who generally accepted something like the received view, and sought ways to improve and perfect it. As we shall see in §2.3, many of the problems raised under the latter two headings were formulated with the aid of putative counterexamples that have since become standard in the literature.

During this time there were also attempts to further elaborate or defend the received view. One major effort in that direction can be found in May Brodbeck's (1962) defense of deductive explanation in her contribution to the third volume in the *Minnesota Studies* series. More significant still is Ernest Nagel's magnum opus, *The Structure of Science* (1961) which contains a great wealth of factual material on explanation in many different branches of science – physical, biological, and social – as well as in history. Nagel begins with a series of examples that include explanations in mathematics, explanations of laws (universal and statistical), deductive explanations of particular facts, probabilistic explanations, functional explanations, and genetic explanations. He also provides searching discussions of such issues as the nature of laws and the status of scientific theories. Although he does call attention to the importance of probabilistic explanation (1961, 22–23), and does discuss examples of probabilistic explanation in history (1961, 550–63), he does not provide an analysis of it. That task was not undertaken in a serious way until Hempel did so in 1962.

This period was a time of intense activity in philosophy of science in America. The Minnesota Center for the Philosophy of Science, under the direction of Herbert Feigl, had been founded in 1953, but had devoted most of its attention during the first few years to problems in the foundations of psychology (see Feigl and Scriven 1956). In the years just before 1958 the topics of discussion became more general (see Feigl et al. 1958). In 1959 Feigl and Grover Maxwell organized an ambitious program for Section L (History and Philosophy of Science) of the American Association for the Advancement of Science, the proceedings of which were published in (Feigl and Maxwell 1961). The Center for the Philosophy of Science at the University of Pittsburgh, was founded in 1960, under the direction of Adolf Grünbaum, and in the same year the Department of History & Philosophy of Science at Indiana University was created with Hanson as Chair. In addition, Robert S. Cohen and Marx Wartofsky of Boston University initiated the Boston Colloquium for the Philosophy of Science in 1960. These entities are still functioning actively. During the two academic years 1961–63, the University of Delaware conducted its Philosophy of Science Seminar (see Baumrin 1963). A great deal of important work in philosophy of science around this time was sponsored by these institutions and was published in preceedings of one kind or another. Scientific explanation was by no means the only topic – nor even the main topic, for the most part – but it was the focus of considerable attention.

Perhaps the most momentous development of the second decade was the clear articulation of theories of statistical explanation. Although a number of authors had already called attention to the need for a model of explanation in which the explanandum followed from the explanans with something less than complete deductive certainty, no real theory of that sort of explanation existed before 1962.[1] Hempel made the first attempt in that year (Hempel 1962), but followed it up with a greatly improved account in (Hempel 1965a, §3, 376–411). Much of the work of the following two decades leans heavily upon this achievement.

2.1 A Major Source of Conflict

Although Hanson was an American by birth and Scriven an Australian, both received their advanced training in England. Ludwig Wittgenstein's *Philosophical Investigations* had been published in 1953, and his influence at Cambridge and Oxford was formidable. During that era a strong opposition developed between the Wittgensteinians, who practiced ordinary language analysis, and the logical empiricists, who might be characterized as artificial language analysts. The Hempel-Oppenheim article is an outstanding example of the use of an artificial formal language for purposes of explicating a fundamental scientific concept. After Hanson and Scriven moved to America, Stephen Toulmin, who was English by birth, was the best known philosopher of science of the ordinary language school remaining in England.[2] Hanson, Scriven, and Toulmin were the most conspicuous Wittgensteinian opponents of the received view of scientific explanation during the second decade.

The basic reason for the opposition between these two schools can be seen quite readily. There is a widely accepted tripartite classification of domains in the study of languages, natural or artificial. *Syntactics* is the area in which we study merely the relationships among symbols, without regard for their relationships either to the users of the symbols or to the objects to which the symbols refer. In ordinary English it is a syntactic rule, I believe, that every sentence must begin with a capital letter and have a dot at the end (recalling that both the exclamation mark and the question mark contain dots). In a formal language, the rules for forming well-formed-formulas (wffs) are part of syntactics. *Semantics* is concerned with the relationships between symbols and the objects to which they refer, but without taking language users into account. Among its important concepts are designation, meaning, and truth. Deductive validity can be construed either syntactically or semantically; for purposes of our discussion it does not matter much which interpretation is chosen. *Pragmatics* takes account of the users of symbols, as well as the interrelations among the symbols and the relationships between the symbols and the entities they stand for. It emphasizes, among many other things, the context in which a statement is made, the purpose of the person who makes it, and the presuppositions that are shared in that context.

The explications offered by the logical empiricists were usually constructed entirely in syntactical and/or semantical terms. The point is well illustrated by the Hempel-Oppenheim treatment of scientific explanation. They start with a standard first order logic, which can be characterized in purely syntactic terms, and proceed to offer semantic rules for its interpretation. All of the key notions – e.g., lawlike sentence, law, theory, explanans – are defined semantically. Pragmatics plays hardly any role at all. Writing in the third decade, Nicholas Rescher (1970, 6–8) points to the many pragmatic features of scientific explanation, but maintains that, for purposes of logical analysis, it is best to abstract from them.

For anyone who focuses primarily on the ordinary uses of language, the pragmatic aspects will be most conspicuous. Some explanations, for example, may be requested by formulating why-questions. The answers are to be judged in terms of the interests and background knowledge of the questioner. An explanation of a particular phenomenon that is entirely satisfactory for one person may be totally inappropriate for another.

It is easy to see why deep philosophical conflict arose in that situation. The ordinary language philosopher finds the logical empiricist insensitive to human needs and interests. The logical empiricist finds the ordinary language philosopher unappreciative of the objective features that determine whether a proffered explanation is a bona fide scientific explanation. The logical empiricists employed formal techniques; the ordinary language philosophers tended to deprecate and avoid them. In those days formal pragmatics did not exist in any highly developed form. As we shall see in the fourth decade, formal pragmatics now plays a crucial role in various approaches to explanation, especially that of Bas van Fraassen.

Hempel was aware from the beginning that when scientists and others offer explanations they often omit parts that are obvious. Often the omitted part is the law. Such explanations are unobjectionable to Hempel if the law is obvious to both the questioner and the respondent; it would be needlessly pedantic to insist on mention of the obvious. Partial explanations and explanation sketches are frequently acceptable. To this degree, at least, Hempel acknowledges the pragmatic aspects of explanation (see 1965a, §4–5, regarding these and other pragmatic considerations). Nevertheless, Hempel insists, when you spell out the correct explanation in complete detail, it will always contain a law.

Scriven takes an opposite attitude. He notices that in many—if not most—cases, an explanation of one particular fact consists in citing another particular fact. Why did the automobile radiator rupture? Because the car was left outside overnight without antifreeze, and the temperature fell to 10° F during the night. If this explanation is questioned, the law concerning the expansion of water upon freezing may be cited, but, according to Scriven, not as a suppressed part of an incomplete explanation. The explanation, as given, was satisfactory. If the law is invoked it serves to justify the explanation, not to complete it.

This view is in strict accord with a thesis about laws that had been shared by some early logical positivists and the ordinary language philosophers, namely, that laws are not statements of fact, but rather, rules of inference. On that view, the sentence "Silver is an electrical conductor" is not a true generalization; it is an 'inference ticket' that entitles one to conclude from "This object is made of silver" that "This object is capable of conducting electricity." Hempel and Oppenheim insisted that a correct D-N explanation be a valid deductive argument, but they did not suggest that the argument include a statement of the rules of deduction to which it conforms. If anyone questions the validity of a given explanation, we can trot out the rules to demonstrate its validity. Philosophers who, like Toulmin, regarded laws as 'inference tickets,' likewise objected to the idea that these rules of inference be included as parts of the explanation itself. They stand on the side, so to speak, to be called up if the correctness of a given explanation is challenged. In making this point about explanation, Scriven refers to the *role-justifying function* of laws (1962, 200, 207, 211).

2.2 Deeper Linguistic Challenges

Not long after Hanson and Scriven had fired their opening salvos against the Hempel-Oppenheim approach, a rather different sort of linguistic approach was initiated by Sylvain Bromberger. Informed by empirical linguistics, which at the time had had virtually no impact on ordinary language philosophers, Bromberger advanced a much more detailed and precise account of explanation than was offered by any of them.

At the beginning of his first paper, "An Approach to Explanation," Bromberger (1962, 72–73) invites consideration of three statements that involve explanation:

(1) The kinetic theory of gases explains why the state equation of vapours near condensation differs markedly from that of an ideal gas.
(2) Someone explained to somebody how World War II might have been avoided.
(3) Newton explained a long time ago the variations of the tides.

He selects statements like (2) for primary attention as a matter of convenience, he says, and he readily extends his considerations to statements like (3). It should be noted, however, that when he gets to the end of this essay he has little to say about statements like (1), and he admits to having no theory that adequately handles them. At the close of the essay he remarks, "The account of the nature of explanation just given falls short of what is eventually wanted: it fails to provide the sort of insight that can be translated into explicit standards and into a patttern of analysis applicable to all explanations and capable of deciding their correctness; it fails to make explicit the criteria that make correct explanations *correct* explanations" (1962, 104–5). So I think the choice of statements like (2) – at least as opposed to those like (1) – is based on more than convenience. What Bromberger admittedly failed to achieve is just the sort of thing Hempel and Oppenheim were trying to accomplish.

From a linguistic standpoint, a statement like (2) is especially tractable because it implies the occurrence of a particular sort of linguistic performance involving two particular individuals or groups of people at some particular time. In addition to the people and the time, such statements incorporate some form of the verb "to explain," and 'something' that is explained, "where the 'something' can be specified by means of an indirect question" (1962, 73). It is close to what Bas van

Fraassen, in the fourth decade, refers to as *the topic* of a question. More precisely, Bromberger is concerned with statements of the form "A E to B W," where "A" and "B" indicate places that can be occupied by terms designating individuals or groups of people, "E" indicates a place where a tensed form of "to explain" occurs, and "W" indicates a place where an indirect question occurs. The heavily linguistic character of this approach is shown by the fact that this characterization could not even purport to apply to explanations in any language other than English. The Hempel-Oppenheim model, by contrast, could apply to explanations offered in myriad different languages.

Bromberger's linguistic analysis proceeds by classifying the verb "to explain," according to a taxonomy provided by Zeno Vendler, as an *accomplishment* term. Other verbs fall into the categories of *activity* terms, *state* terms, or *achievement* terms. Activity terms and accomplishment terms apply to doing something that occupies a span of time, but an accomplishment term refers to an activity that can result in a completion. Explaining and reading are activities that take place in that way, but when an activity term is used in the simple past, it does not imply any completion. If I say I read last night, that does not mean I finished a story, an article, or a book. It is an activity in which I could be engaged and which I could have stopped doing without finishing anything. It means merely that I spent some time engaged in that activity. If I say that I explained in class yesterday, that claim would not be correct unless I had completed an explanation. If no explanation was completed, I could correctly say that I was trying to explain, or that I got part way through an explanation, but not that I explained anything.

> The difference between activity terms and accomplishment terms is readily seen when we compare their simple past tense. Both types have a simple past tense which implies that the continuous present was applicable at some moments in the past. The simple past tense of an activity term is applicable as soon as such moments have passed, and implies only the existence of such moments in the past. Aristotle walked. This implies that during some moments in the past Aristotle was walking. It does not tell whether or not Aristotle is through walking. The simple past tense of accomplishment terms implies more. It implies that relevant activities took place in the past, but furthermore that they have come to an end. And not to a mere stop, but to a conclusion. In other words, the simple past tense of accomplishment verbs entails that something has been finished, completed, that might, in principle, have been left unfinished, incomplete, that might have stopped before coming to its logical end. (1962, 75)

State terms, such as "to know" and "to love" (in their non-episodic senses), truly apply in the simple present tense to individuals at different times even though the individual is in that state just once—e.g., he loves her, said at any time throughout their 35 years of marriage. Achievement terms can be truly applied in the simple present at more than one time only if the individual achieves more than once. She wins at tennis at time t_1 and she wins at tennis at t_2 (where t_1 is different from t_2) only if she wins two different games (or sets, or matches).

Since "to explain" is an accomplishment term, "A good analysis should therefore make explicit the nature of the completion implied by statements in 'A E to B W' form in which 'to explain' occurs in the *simple past tense*; it should bring out what must be the case for a statement in 'A explain*ed* to B W' form to be true' " (1962, 76).

Another item of important concern about statements of the form 'A E to B W' is the nature of the indirect question at the place indicated by "W." Clearly, Bromberger emphasizes, certain sorts of questions are appropriate, while other kinds are inappropriate.

Many kinds of indirect question can occupy the position indicated by 'W', and they may open on a variety of interrogatives[3] – 'Why', 'How', 'Whence', 'Whither', 'What' – but not *every* indirect question is at home there; some would be out of place, awkward, reminiscent of Eisenhower prose, *e.g.* 'what the distance in miles between London and Paris is' or 'whether it will rain tomorrow' or 'what time it is' or 'which sister Sam married'. A good analysis should show why some indirect questions do not sit well in these contexts. (1962, 74)

Bromberger invites us to consider two questions:

(A) What is the height of Mt. Kilimanjaro?
(B) Why do tea kettles emit a humming noise just before the water begins to boil? (1962, 80)

Each of these is a sound question – each has a correct answer and neither has any false presuppositions.

Although he does not know a correct answer for either, he knows quite a bit about what would constitute a correct answer. A correct answer to (A) might be a positive integer followed by an expression, such as "feet," designating a unit of length. If the answer is given in feet, he knows that the number will be greater than 100 and less than 30,000; if some other unit is used, it will be possible to convert the answer into feet, and the same numerical limits will apply. He knows enough about the answer to be able to exclude many expressions – e.g., "12 feet," "Morton White," "19,321 pounds." Although he does not know a correct answer, there is a straightforward sense in which he has thought of a correct answer, for he could write down and understand any integral numeral between "100" and "30,000."

A correct answer for (B) would be statable as a sentence (which may be a conjunction of sentences) following the word "because." This sentence would include mention of something that happens whenever water is just about to boil, and it would include mention of something that creates vibrations of air of suitable frequency and amplitude. Other requirements on an answer might be offered, and they would serve to exclude many expressions as possible answers, but altogether they do not add up to a correct answer. In the case of (B), however, unlike (A), he has not thought of any answer that could possibly be a correct answer.

This is a key feature of explanation. Someone could tell Bromberger the height of Kilimanjaro, but it would be incorrect usage to say anyone had explained it to him. Perhaps there is a geophysical explanation of the height of Kilimanjaro, but question (A) was not a request for it. Question (B) was a request for an explanation. He recapitulates:

(i)   I take both (A) and (B) to be sound questions, to admit of a right answer.

(ii) I know, or believe I know, enough about each answer to be able to eliminate a number of possible utterances, *i.e.* of expressions, as not being formulations of it.

(iii) In the case of (A) I can think of some possible utterances that I cannot eliminate in this way.

(iv) In the case of (B) I can think[4] of no expression that I cannot eliminate in this way. (1962, 82)

On the basis of these considerations, Bromberger defines the notion of a *p-predicament* as follows:

S is in a p-predicament with regard to question Q if and only if, on S's views, Q admits of a right answer, but S can think of no answer to which, on S's views, there are no decisive objections. (1962, 82)

He points out that one could not be in a p-predicament with regard to an indirect question beginning with "whether" that requires a "yes" or "no" answer—e.g., whether it is raining here just now. Even if one does not know the correct answer, it surely has been thought of. Similarly, for reasons that have already been mentioned, one would not be in a p-predicament regarding indirect questions like (A), beginning with "what is," that call for quantitative answers. Likewise, one cannot be in a p-predicament with respect to indirect questions, beginning with "which," that call for a selection from a well-defined set of alternatives. In my present state of knowledge I may not know which planet in the solar system has the highest surface temperature, but I believe I have thought of the possible candidates, and I suppose it is either Mercury or Venus.

To arrive at a characterization of explanation, Bromberger also defines the notion of a b-predicament:

S is in a b-predicament with regard to Q if and only if the question mentioned in it admits of a right answer, but that answer is beyond what the person men-

tioned can conceive, can think of, can imagine, i.e. is something that that person cannot remember, cannot excogitate, cannot compose. (1962, 90, italics not in original)

The main difference between a p-predicament and a b-predicament is that the latter refers to a right answer, whereas the former refers to possible answers that cannot be eliminated.

Bromberger offers a series of four hypotheses concerning the nature of explanations, and he rejects the first three. In stating these hypotheses he refers to the person or persons doing the explaining as the *tutor* and to the recipient as the *tutee*. We shall look only at the fourth hypothesis:

The essential characteristics of explaining episodes are the following:

(a) the question is sound, i.e. admits of a right answer;

(b) the tutor is rational and knows the right answer to the question at the time of the episode;

(c) during the episode the tutor knows, or believes, or at least assumes that at the beginning of the episode the tutee was in a p-predicament with regard to the question,

or that, at the beginning of the episode the tutee was in a b-predicament with regard to the question,

or that, at the beginning of the episode the tutee was in either a p-predicament or a b-predicament with regard to the question;

(d) in the course of the episode the tutor presents the facts that, in his opinion, the tutee must learn to know the right answer to the question;

(e) in the course of the episode the tutor also provides the tutee with such instruction as he (the tutor) thinks necessary to remove the basis of whichever of the states mentioned in (c) he deems the tutee to be in;

(f) at the end of the episode all the facts mentioned in (d) and (e) have been presented to the tutee by the tutor. (1962, 94–95)

Bromberger acknowledges that this lengthy characterization contains redundancies: (a) is entailed by (b); (c) and (f) are entailed by (d) and (e). Removing the redundancies, we have

The essential characteristics of explaining episodes are the following:

(b) the tutor is rational and knows the right answer to the question at the time of the episode;

(d) in the course of the episode the tutor presents the facts that, in his opinion, the tutee must learn to know the right answer to the question;

(e) in the course of the episode the tutor also provides the tutee with such instruction as he (the tutor) thinks necessary to remove the basis of whichever of the predicaments he deems the tutee to be in.

All the kinds of explanation we have discussed so far fit the schema "A E to B W," where "W" indicates a place where an indirect question occurs. After presenting this account, Bromberger goes on to discuss explanations that fit the schema "A E to B X," where "A," "B," and "E" have the same significance, but "X" is the location of something other than a question, for example, a noun phrase. He rather easily shows how these can be handled in terms of the preceding type, and remarks on the desirability of keeping clear on "the essential connection that links explaining to questions." (1962, 100)

Statement (3) of Bromberger's initial set, it will be recalled, is "Newton explained a long time ago the variations of the tides." This statement does not imply an explaining episode of the sort we have been discussing. Here "explained" is not an accomplishment term; it is, rather, an *achievement* term. The statement means that Newton solved the problem. His solution of the problem is such that, for anyone who understands it, that person can provide an explanation of the sort we previously discussed. "The connection, then, between the truth-conditions of 'A E W' and 'A E to B W' is that to have explained something in the sense now under consideration is to have become able to explain something in the [previous] sense . . . as a result of one's own endeavours and ingenuity." (1962, 101)[5]

Bromberger's first example of a statement involving "to explain" – the one about the ability of the kinetic theory to explain why certain vapors do not behave as ideal gases – has the form "T explains W," where "T" does not indicate a place for a term referring to a person or group of people. In this sentence, "to explain" functions as a state term; it refers to the continuing ability of a theory to function in explanations of the sort we have already discussed. As I remarked at the outset, Bromberger does not claim to have an adequate treatment of this kind of explanation.

I have discussed Bromberger's approach in considerable detail and I have presented lengthy direct quotations to give a clear flavor of what he is attempting, and to contrast it with the sort of thing Hempel and Oppenheim were trying to do. Clearly, Bromberger has great sensitivity to the nuances of language. He goes to considerable length to understand the usages of "to explain" – for instance, to discover the differences between explanation-seeking questions and other kinds of questions. In his principal examples the explanation-seeking question was a why-question, but he denies that all and only why-questions are explanation-seeking. His identification of p-predicaments and b-predicaments involves deep insight into the nature of questions. It also exhibits clear appreciation of many of the pragmatic aspects of explanation.

In the end, however, as Bromberger is clearly aware, his characterization of explanation employs the unanalyzed notion of a right answer to a why-question or any other kind of explanation-seeking question. As I said above, what Hempel and Oppenheim were trying to do is to analyze the notion of a correct answer to an explanation-seeking question. Although they undoubtedly made significant progress on their enterprise, they offered no criteria at all to enable us to distinguish explanation-seeking from non-explanation-seeking questions. Bromberger's paper, which he aptly titled "An Approach to Explanation" (rather than, say, "A Theory of Explanation") provides a valuable prolegomenon to the Hempel-Oppenheim project, but, as he recognized, not a substitute for it.

Toward the end of the second decade, Bromberger (1966) returned to this issue and attempted to give a partial answer. Although he still insists that not every request for an explanation can be given in the form of a why-question, he does try to characterize correct answers to explanation-seeking why-questions. He offers several counterexamples – similar in import to the flagpole (CE-2, to be discussed in the next section) – to show that not every argument that satisfies the criteria for D-N explanation, as set out by Hempel and Oppenheim, qualifies as an acceptable answer to a particular why-question. However, he does not dispute the thesis that these conditions are necessary for satisfactory answers – for satisfactory *deductive* answers, we should say, for he does not deny that inductive or statistical explanations are possible. The theory of why-questions can thus be seen as a friendly amendment to the Hempel-Oppenheim account.[6] He does not, however, discuss any of the issues surrounding the theories of statistical explanation that were published by Hempel shortly before the appearance of this article by Bromberger.

As his point of departure, Bromberger rejects the covering law conception that identifies explanation, or correct answers to why-questions, with subsumption under true lawful generalizations. Consider the question, "Why does this live oak keep its leaves during the winter?" The appropriate response, he suggests, is "All live oaks do!" "Because all live oaks do" is, he claims, not correct. Since he maintains that answers to why-questions are sentences that follow "because," the correct response is not an answer, but a rejection of the question (1966, 102). The question suggests that there is something atypical about this particular tree; the response denies that supposition.

Bromberger's basic idea is that why-questions arise "when one believes that the presupposition is true, views it as a departure from a general rule, and thinks that the conditions under which departures from the general rule occur can be generalized" (1966, 100). To develop this thesis we need several definitions.

(1) A *general rule* is simply a *lawlike* generalization of the form

$$(x)[(F_1x \cdot F_2x \ . \ . \ . \ F_jx) \supset (S_1x \cdot S_2x \ . \ . \ . \ S_kx)] \qquad (j, k \geq 1)$$

Examples are: All gold is malleable, no rubber bands are brittle, and the velocity of an object never changes. Obviously, a general rule need not be true or even plausible (1966, 97).

(2) Bromberger introduces the concept of a *general abnormic law*, but the definition is quite complex, and we do not need all of the details for our discussion. He remarks that every general abnormic law is equivalent to a conjunction of spe-

cial abnormic laws. It should, however, be emphatically noted that general abnormic laws are true (1966, 97–99).

(3) A *special abnormic law* is a true lawlike generalization that contains an unless-clause. Examples are: The velocity of an object does not change unless the net force on it is not equal to zero; no rubber band is brittle unless it is very old or very cold. Special abnormic laws have the following form:

$$(x)\{(F_1x \cdot F_2x \ldots F_jx) \supset [-Ex \equiv (A_1x \, A_2x \ldots A_nx)]\} \quad (j, n \geq 1)$$

which can obviously be reformulated as

$$(x)\{(F_1x \cdot F_2x \ldots F_jx) \supset [Ex \equiv -(A_1x \, A_2x \ldots A_nx)]\} \quad (j, \ n \geq 1)$$

They must also fulfill further restrictions that need not be spelled out for our purposes.[7]

(4) Bromberger defines the *antonymic predicates of an abnormic law,* but we need only the concept of *antonymic predicates of a special abnormic law.* These are the predicates that occur in the place of "E" and its negation in the foregoing forms. For example, "brittle" and "nonbrittle" would be the antonymic predicates in "No rubber bands are brittle unless they are old or cold" (1966, 99).

(5) Finally, we need the concept of the *completion of a general rule by an abnormic law.* An abnormic law is the completion of a general rule if and only if the general rule is false and is obtainable from the abnormic law by dropping the "unless"-clause (1966, 99).

With this terminology in place we can now characterize the concept of a correct answer to a why-question:

Here then is the relation: *b* is the *correct answer* to the question whose presupposition is *a* if and only if (1) there is an abnormic law *L* (general or special) and *a* is an instantiation of one of *L* 's antonymic predicates; (2) *b* is a member of a set of premises that together with *L* constitute a deductive nomological explanation whose conclusion is *a* ; (3) the remaining premises together with the general rule completed by *L* constitute a deduction in every respect like a deductive nomological explanation except for a false lawlike premise and false conclusion, whose conclusion is a contrary [or contradictory] of *a*. (1966, 100)[8]

Consider one of Bromberger's examples. Suppose we have a simple pendulum that is oscillating with a particular period p. It consists of a bob suspended by a string of a given length. From the law of the simple pendulum, the acceleration of gravity at the earth's surface g, and the length l of the string, one can deduce the period p of our pendulum. This argument would qualify as a deductive nomological explanation according to the Hempel-Oppenheim account, and it would be regarded as a correct explanation of the period. However, from the same law, the same value for the acceleration of gravity, and the period of the pendulum we can deduce the length of the string. This would not seem to constitute a good answer to the question of why the string has its particular length l.

Consider the general rule, "No simple pendulum oscillates with period p." It is, of course, false. It can, however, be completed by the abnormic law, "No simple pendulum has the period p unless[9] it has a suspension of length l." From the abnormic law and the fact that the suspension has length l, we can deduce that it has period p. From the general rule and the fact that we have a simple pendulum we can deduce that it is not oscillating with period p. Thus, it is correct, according to Bromberger's account, to answer the why-question about the period of oscillation with facts about the length of the suspension and the local acceleration of gravity, and the law of the simple pendulum. However, to answer the question, "Why does this pendulum have a suspension of length l?" it would be manifestly incorrect to answer, "Because it is oscillating with period p." For consider the putative abnormic law, "No pendulum has a suspension of length l unless it is oscillating with period p." This is patently false, for a pendulum that is not oscillating may have just such a suspension. Bromberger thus attempts to solve a nagging problem about asymmetries of explanation.

Bromberger closes this essay with the following remark:

> It may seem odd that abnormic laws should be associated with a special interrogative. But they are, after all, the form in which many common-sense generalizations that have been qualified through the ages are put. They are also a form of law appropriate to stages of exploratory theoretical developments when general rules are tried, then amended, until finally completely replaced. We are always at such a stage. (1966, 107)

What Bromberger has noticed, I believe, is a deep fact about scientific explanations. To provide an adequate explanation of any given fact, we need to provide information that is relevant to the occurrence of that fact—information that *makes a difference* to its occurrence. It is not sufficient simply to subsume an occurrence under a general law; it is necessary to show that it has some special characteristics that account for the features we seek to explain. As we shall see, this notion of explanatory relevance plays a key role in the development of theories of statistical explanation.

As I have already remarked, and as we shall see in greater detail, the formulation of theories of inductive or statistical explanation was one of the most significant—perhaps the most significant—development in the second decade. Being confined to deductive types of explanation, Bromberger's treatment did not encompass this broader range of issues. Moreover, as an incisive critique by Paul Teller (1974) showed, it does not even achieve its aim within its intended range of application. Nevertheless, it provided a fertile basis for subsequent work on the pragmatics of explanation, especially that of Achinstein and van Fraassen during the fourth decade. In addition, it has served as a major stimulus to Philip Kitcher in developing the novel unification approach to explanation embodied in "Explanatory Unification and the Causal Structure of the World" in this volume (Kitcher & Salmon 1989) especially (Bromberger 1963).

2.3 Famous Counterexamples to the Deductive-Nomological Model

At the conclusion of §1.1 we raised a number of general issues arising out of the Hempel-Oppenheim treatment of scientific explanation of particular occurrences. During the second decade many of them were vividly posed in terms of counterexamples that have since become quite standard. The existence of such a standard set is, in itself, a tribute to the solidity of the received view. Let us take a look at some of the best known.

One rather obvious problem has to do with the temporal relations between the explanatory facts (as expressed by the singular sentences in the explanans) and the fact-to-be-explained (as expressed by the explanandum-sentence). In the schema reproduced above (p. 13; H-O 1948, 249) the Cs are labeled as "antecedent conditions," but in the formal explication no temporal constraints are given. Indeed, no such temporal constraints are mentioned even in the informal conditions of adequacy. This issue has been posed in terms of the explanation of an eclipse.

(CE-1) The eclipse. Going along with the D-N model, we might, for example, regard a total lunar eclipse as satisfactorily explained by deducing its occurrence from the relative positions of the earth, sun, and moon at some prior time in conjunction with the laws of celestial mechanics that govern their motions. It is equally possible, however, to deduce the occurrence of the eclipse from the relative positions of the earth, sun, and moon at some time after the eclipse in conjunction with the very same laws. Yet, hardly anyone would admit that the latter deduction qualifies as an explanation.[10] One might suppose that the failure to impose temporal restrictions was merely an oversight that could be corrected later, but Hempel (1965a, 353) raises this question explicitly and declines to add any temporal constraint.[11]

Another issue, closely related to the matter of temporal priority, has to do with the role of causality in scientific explanation. Our commonsense notion of explanation seems to take it for granted that to explain some particular event is to identify its cause and, possibly, point out the causal connection. Hempel and Oppenheim seem to share this intuition, for they remark, "The type of explanation which has been considered here so far is often referred to as causal explanation" (H-O 1948, 250). In "Aspects of Scientific Explanation," while admitting that some D-N explanations are causal, Hempel explicitly denies that all are (1965a, 352–54).

The problems that arise in this connection can readily be seen by considering several additional well-known examples.

(CE-2). Bromberger's flagpole example. A vertical flagpole of a certain height stands on a flat level piece ground.[12] The sun is at a certain elevation and is shining brightly. The flagpole casts a shadow of a certain length. Given the foregoing facts about the height of the flagpole and the position of the sun, along with the law of rectilinear propagation of light, we can deduce the length of the shadow. This deduction may be accepted as a legitimate D-N explanation of the length of the shadow. Similarly, given the foregoing facts about the position of the sun and the length of the shadow, we can invoke the same law to deduce the height of the flagpole. Nevertheless, few people would be willing to concede that the height of the flagpole is explained by the length of its shadow.[13] The reason for this asymmetry seems to lie in the fact that a flagpole of a certain height causes a shadow of a given length, and thereby explains the length of the shadow, whereas the shadow does not cause the flagpole, and consequently cannot explain its height.

(CE-3) The barometer. If a sharp drop in the reading on a properly functioning barometer occurs, we can infer that there will be a storm — for the sake of argument, let us assume that there is a law that whenever the barometric pressure drops sharply a storm will occur. Nevertheless, we do not want to say that the barometric reading explains the storm, since both the drop in barometric reading and the occurrence of the storm are caused by atmospheric conditions in that region. When two different occurrences are effects of a common cause, we do not allow that either one of the effects explains the other. However, the explanation of the storm on the basis of the barometric reading fits the D-N model.

(CE-4) The moon and the tides. Long before the time of Newton, mariners were fully aware of the correlation between the position and phase of the moon and the rising and falling of the tides. They had no knowledge of the causal connection between the moon and the tides, so they had no explanation for the rising and falling of the tides, and they made no claim to any scientific explanation. To whatever extent they thought they had an explanation, it was probably that God in his goodness put the moon in the sky as a sign for the benefit of mariners. Nevertheless, given the strict law correlating the position and phase of the moon with the ebb and flow of the tides,[14] it was obviously within their power to construct D-N explanations of the behavior of the tides. It was not until Newton furnished the causal connection, however, that the tides could actually be explained.

One of the most controversial theses propounded by Hempel and Oppenheim is *the symmetry of explanation and prediction*. According to this view, the very same logical schema fits scientific explanation and scientific prediction; the sole difference between them is pragmatic. If the event described by E has already occurred, we may ask why. A D-N explanation consisting of a derivation of E from laws and antecedent conditions provides a suitable response. If, however, we are in possession of the same laws and antecedent conditions before the occurrence of E, then that same argument provides a prediction of E. Any D-N explanation is an argument that, were we in possession of it early enough, would enable us to anticipate, on a sound scientific basis, the occurrence of E. Since every D-N explanation involves laws, a hallmark of explanations of this type is that they provide *nomic expectability*. [15]

In discussing the symmetry of explanation and prediction in the preceding paragraph, I was tacitly assuming that the so-called antecedent conditions in the explanans are, in fact, earlier than the explanandum event. However, in view of Hempel's rejection of any requirement of temporal priority, the symmetry thesis must be construed a bit more broadly. Suppose, for example, that the explanandum-event E occurs before the conditions C in the explanans. Then, as I construe the symmetry thesis, we would be committed to the view that the D-N explanation is an argument that could be used subsequent to the occurrence of the explanatory conditions C to retrodict E. It is quite possible, of course, that E has occurred, but that we are ignorant of that fact. With knowledge of the appropriate laws, our subsequent knowledge of conditions C would enable us to learn that E did, in fact, obtain. Parallel remarks could be made about the case in which C and E are simultaneous. Thus, in its full generality, the symmetry thesis should be interpreted in such a way that "prediction" is construed as "inference from the known to the unknown."[16]

As Hempel later pointed out in "Aspects of Scientific Explanation," the symmetry thesis can be separated into two parts: (i) Every D-N explanation is a prediction—in the sense explained in the preceding paragraph—and (ii) every (nonstatistical) scientific prediction is a D-N explanation. It is worthwhile, I think, to distinguish a *narrower symmetry thesis*, which applies only to D-N explanations of particular facts, and a *broader symmetry thesis*, which applies to both D-N and I-S explanations of particular facts. According to the narrower thesis, every *nonstatistical* prediction is a D-N explanation; according to the broader thesis, every prediction is an explanation of either the D-N or I-S variety. Given the fact that statistical explanation is not explicated in the Hempel-Oppenheim article, only the narrower symmetry thesis is asserted there. The broader thesis, as we shall see, was advocated (with certain limitations) in "Aspects."

Nevertheless, various critics of the Hempel-Oppenheim article failed to take sufficient notice of the explicit assertion that not all legitimate scientific explanations are D-N—that some are statistical. Scriven (1959) strongly attacked subthesis (i)—that all explanations could serve as predictions under suitable pragmatic circumstances—by citing evolutionary biology and asserting that it furnishes explanations (of what has evolved) but not predictions (of what will evolve). If, as I believe, evolutionary biology is a statistical theory, then Scriven's argument applies at best to the broader, not the narrower symmetry thesis. Although this argument was published in 1959,[17] it does, I believe, pose a serious problem for the theory of statistical explanation Hempel published three years later. In the same article Scriven set forth a widely cited counterexample:

(CE-5) Syphilis and paresis. Paresis is one form of tertiary syphilis, and it can occur only in individuals who go through the primary, secondary, and latent stages of the disease without treatment with penicillin. If a subject falls victim to paresis, the explanation is that it was due to latent untreated syphilis. However, only a relatively small percentage – about 25% – of victims of latent untreated syphilis develop paresis. Hence, if a person has latent untreated syphilis, the correct prediction is that he or she will not develop paresis. This counterexample, like the argument from evolutionary biology, applies only to the broader symmetry thesis.

When the narrower symmetry thesis is spelled out carefully, it seems impossible to provide a counterexample for subthesis (i) – that every explanation is a prediction (given the right pragmatic situation). That subthesis amounts only to the assertion that the conclusion of a D-N argument follows from its premises. Against subthesis (i) of the broader symmetry thesis the syphilis/paresis counterexample is, I think, quite telling.

When we turn to subthesis (ii) of the narrower symmetry thesis – i.e., that every (nonstatistical) prediction is an explanation – the situation is quite different. Here (CE-3) and (CE-4) provide important counterexamples. From the barometric reading, the storm can be predicted, but the barometric reading does not explain the storm. From the position and phase of the moon, pre-Newtonians could predict the behavior of the tides, but they had no explanation of them. Various kinds of correlations exist that provide excellent bases for prediction, but because no suitable causal relations exist (or are known), these correlations do not furnish explanations.

There is another basis for doubting that every scientific prediction can serve, in appropriate pragmatic circumstances, as an explanation. Hempel and Oppenheim insist strongly upon the covering law character of explanations. However, it seems plausible to suppose that some respectable scientific predictions can be made without benefit of laws – i.e., some predictions are inferences from particular instances to particular instances. Suppose, for instance, that I have tried quite a number of figs from a particular tree, and have found each of them tasteless. A friend picks a fig from this tree and is about to eat it. I warn the friend, "Don't eat it; it will be tasteless." This is, to be sure, low-level science, but I do not consider it an unscientific prediction. Moreover, I do not think any genuine laws are involved in the prediction. In (1965, 376) Hempel considers the acceptability of subthesis (ii) of the symmetry thesis an open question.

There is another fundamental difficulty with Hempel and Oppenheim's explication of D-N explanation; this one has to do with explanatory relevance. It can be illustrated by a few well-known counterexamples.

(CE-6) The hexed salt. A sample of table salt has been placed in water and it has dissolved. Why? Because a person wearing a funny hat mumbled some non-sense syllables and waved a wand over it—i.e., cast a dissolving spell upon it. The explanation offered for the fact that it dissolved is that it was hexed, and all hexed samples of table salt dissolve when placed in water. In this example it is *not* being supposed that any actual magic occurs. All hexed table salt is water-soluble because all table salt is water-soluble. This example fulfills the requirements for D-N explanation, but it manifestly fails to be a bona fide explanation.[18]

(CE-7) Birth-control pills. John Jones (a male) has not become pregnant during the past year because he has faithfully consumed his wife's birth-control pills, and any male who regularly takes oral contraceptives will avoid becoming pregnant. Like (CE-6), this example conforms to the requirements for D-N explanation.

The problem of relevance illustrated by (CE-6) and (CE-7) is actually more acute in the realm of statistical explanation than it is in connection with D-N explanation. Insofar as D-N explanation is concerned, it is possible to block examples of the sort just considered by any of several technical devices.[19] We will return to this issue when we discuss statistical explanation.

2.4 Statistical Explanation

In an article entitled "The Stochastic Revolution and the Nature of Scientific Explanation," Nicholas Rescher (1962) made an eloquent plea for an extension of the concept of scientific explanation beyond the limits of deductive explanation. The "stochastic revolution" yields "forcible considerations . . . that militate towards a view of explanation prepared to recognize as an 'explanation' of some fact an argument which provides a rationalization of this fact from premises which render it *not necessary but merely probable*" (ibid., 200). He adds, "To refuse to accord to such explanatory reasonings the title of 'explanation' is to set up so narrow a concept of explanation that many of the reasonings ordinarily so-called in modern scientific discussions are put outside the pale of *explanations proper* by what is in the final analysis, a fiat of definition buttressed solely by fond memories of what explanation used to be in nineteenth-century physics" (ibid. 204).

The most important development to occur in the second decade (1958–67) of our chronicle—the explicit treatment of statistical explanation—had its public inception in 1962. Although Rescher clearly recognized the inductive character of such explanations, neither he nor any of several other authors who recognized the legitimacy of statistical explanation offered an explicit model. They thought,

I suspect, that nothing was needed beyond a trivial relaxation of the requirements imposed on D-N explanations. Hempel appears to have been the first to notice the profound difficulties involved in statistical explanation, in particular, the problem of *ambiguity*, over which he labored long and hard. The first explicit model was offered in his "Deductive-Nomological vs. Statistical Explanation" (1962) in volume III of *Minnesota Studies*. It constitutes the first serious attempt (of which I am aware) by any philosopher or scientist to offer a detailed and systematic account of any pattern of statistical explanation.[20] In this paper, Hempel deals with statistical explanations of particular facts — the lower left-hand box of Table 1 — and characterizes what he later calls the *inductive-statistical (I-S)* model. In the same year (1962a) he offered a brief popular account of both deductive and inductive explanations of particular facts in history as well as in science. In section 3 of "Aspects," he offers a much improved account of the I-S model, and also introduces the *deductive-statistical (D-S)* model.

2.4.1 The Deductive-Statistical Model

In a D-S explanation, a statistical law[21] is explained by deriving it from other laws, at least one of which is statistical. There is no prohibition against the explanans containing universal laws as well. A time-honored example will illustrate how this sort of explanation goes.

In the 17th century the Chevalier de Méré posed the following question: Why is it that, when a player tosses a pair of standard dice in the standard manner 24 times, he has less than a 50–50 chance of getting double 6 at least once? The answer is as follows. On each toss of a fair die, the chance of getting 6 is 1/6. This probability is independent of the outcome of the toss of the other die. Consequently, the probability of double 6 on any throw is 1/36, and the probability of not getting double 6 is 35/36. The outcome of any toss of the pair of dice is independent of the outcomes of previous tosses; hence, the probability of not getting double 6 on n throws is $(35/36)^n$. When we calculate the value of $(35/36)^{24}$ we find it is greater than 1/2; therefore, the probability of getting double 6 in 24 tosses is less than 1/2. Assuming that there is a sharp distinction between mathematics and empirical science, we can say that the empirical statistical laws in this derivation are the generalizations about the behavior of standard dice: the probability of 6 is 1/6 and the tosses are independent of one another. The rest is simply arithmetic. The "laws of probability" — such as the multiplication rule — are not empirical laws at all; they are laws of mathematics which have no factual content.

Another example of a D-S explanation would be the derivation of the half-life of uranium-238 from the basic laws of quantum mechanics — which are statistical — and from the height of the potential barrier surrounding the nucleus and the kinetic energies of alpha particles within the nucleus. The answer is approximately 4.5×10^9 years. This example is highly theoretical, and consequently distinctly *not* empirical in the narrow sense of "empirical."

As we have seen, the Hempel-Oppenheim article attempted to characterize laws of nature. Because of the very simple formal language employed in that context, there is no possibility of formulating statistical laws within it. The reason, very simply, is that the language does not contain any numerical expressions. Whatever the philosophical merits of using such a language may be, clearly a much richer language is needed to express any important scientific content.[22] Adopting a language that would begin to be adequate for any real science would involve using one in which statistical laws could be formulated. So any language that could contain real universal scientific laws could contain statistical laws as well.

The statistical laws of empirical science are general in the same sense that universal laws are general. The simplest universal law would have the form, "All F are G." It would be formulated using a variable that ranges over all individuals in the universe. The force of the generalization is that, in the entire universe, nothing is an F and not a G or, equivalently, everything is either not an F or else it is a G. The corresponding negative generalization would mean that, in the entire universe, nothing is both F and G. The existential generalization, "Some F are G," would mean that, in the entire universe, at least one thing is an F and also a G. The simplest statistical law could be construed in either of two ways. First, it might be taken to mean, "Every F has a certain propensity to be G," in which case it is strictly analogous to a universal generalization. Second, it might be taken to mean, "A certain proportion of F are G." In that case we can construe it to mean that, among all of the individuals in the universe that are F, a certain proportion are also G. Either way it has the same sort of generality as universal or existential generalizations.[23]

In addition, the same problem about purely qualitative predicates obviously arises in connection with both universal and statistical laws. The same point about not restricting generality by making reference to particular objects, places, or times also applies to statistical as well as to universal laws. And the problem of distinguishing between lawful and accidental statistical generalizations is the same as for universal generalizations. Recalling that the official Hempel-Oppenheim explication of D-N explanation appealed to theories rather than laws, we can say that generalizations involving universal quantifiers, existential quantifiers, mixtures of the two kinds, or probabilities are on a par.

We should also remember that Hempel and Oppenheim confessed their inability to provide an explication for D-N explanation of laws because of the difficulty stated in "the notorious footnote 33." The problem was to make an appropriate distinction between really explaining a law by deducing it from a genuinely more general law and giving a pseudo-explanation by some such device as deducing it from a 'law' that consists of a conjunction of which it is one conjunct. This prob-

lem is clearly just as acute for statistical laws as for nonstatistical laws and theories. For reasons of these sorts, I suggested at the outset that we not regard D-S
explanations as belonging to a type different from D-N, but treat them rather as
a subtype of D-N. The right-hand column of Table 1 can be regarded as representing one kind of explanation, namely, D-N explanation of generalizations. Nowhere in "Aspects" does Hempel offer an answer to the problem stated in the
notorious footnote.

2.4.2 The Inductive-Statistical Model

Hempel's main concern in section 3 of "Aspects" is clearly with I-S explanation. As he says, "Ultimately, however, statistical laws are meant to be applied
to particular occurrences and to establish explanatory and predictive connections
among them" (1965, 381).[24] So let us turn to his explication of that model.

There is a natural way to try to extend the treatment of D-N explanation of
particular facts to cover statistical explanation of particular facts. Given the ex-
planation/prediction symmetry thesis, we can say that a D-N explanation of a
given fact is a deductive argument showing that the event in question was predictable had the explanatory facts been available early enough. We can then suppose
that an I-S explanation is also an argument that would render the explanandum
predictable had the explanatory facts been available early enough. In the case of
I-S explanation, the explanans must include, essentially, at least one statistical
law; as a result, it is impossible to *deduce* the explanandum statement from the
explanans. Hempel therefore requires the I-S explanation to be an *inductive* argument that would render the explanandum predictable, not with deductive certainty
but with high inductive probability, given the explanans. The simplest kind of example would fit the following schema:

$$\text{I-S} \quad \begin{array}{c} P(G\,|\,F) = r \\ Fb \\ \hline\hline \\ Gb \end{array} \quad [r]$$

The first premise of this argument is a statistical law that asserts that the relative
frequency of Gs among Fs is r,[25] where r is fairly close to 1. The double line
separating the premises from the conclusion signifies that the argument is inductive rather than deductive. The expression "[r]" next to the double line represents
the degree of inductive probability conferred on the conclusion by the premises.
Note that I-S explanations are covering law explanations in exactly the same sense
as D-N explanations are—namely, each such explanation must contain at least one
law in its explanans.

Perhaps a few examples would be helpful. Why, for instance, did Yamamoto
suffer severe physical injury in August of 1945? Because he was only a kilometer

from the epicenter of the atomic blast in Hiroshima, and any person that distance from the epicenter of an atomic explosion of that magnitude will very probably suffer severe physical injury.

Why, for another example, is the ratio of carbon-14 to other isotopes of carbon in this particular piece of charcoal about half of the ratio of C^{14} to other isotopes in the atmosphere? Because the piece of wood from which this piece of charcoal came was cut about 5730 years ago, and the half-life of C^{14} is 5730 years. The proportion of C^{14} in the atmosphere remains fairly constant, because it is replenished by cosmic radiation at about the same rate as it decays radioactively. While a tree is living it continues absorbing C^{14} from the atmosphere, but when the tree dies or is cut it no longer does so, and the C^{14} content decreases as a result of spontaneous radioactive decay. This example qualifies as I-S rather than D-N for two reasons. First, the explanandum is a particular fact, namely the C^{14} content of one particular piece of charcoal. The fact that many atoms are involved does not make the explanandum a statistical law. Second, the law governing radioactive decay is a statistical law. From this law we can conclude that it is highly probable, but not absolutely certain, that about 1/2 of any reasonably numerous collection of atoms of a given unstable isotope will decay in the period of time constituting the half-life of that particular isotope.

Hempel's main example of I-S explanation is the case of John Jones who recovered quickly from a streptococcus infection. When we ask why we are told that penicillin was administered, and that most (but not all) strep infections clear up quickly when treated with penicillin. If we supply a definite number r for the probability of quick recovery from a strep infection, given that penicillin is administered, this example is easily seen to fit the I-S schema set out above, as follows:

(1) $P(G|F.H) = r$
Fb.Hb
$\overline{\qquad\qquad}$ [r]
Gb

where F stands for having a strep infection, H for administration of penicillin, G for quick recovery, b is John Jones, and r is a number close to 1.

This example can be used to illustrate a basic difficulty with I-S explanation. It is known that certain strains of streptococcus bacilli are resistant to penicillin. If it turns out that John Jones is infected with a penicillin-resistant strain, then the probability of his quick recovery after treatment of penicillin is low. In that case, we could set up the following inductive argument:

(2) $P(G|F.H.J) = r_1$
Fb.Hb.Jb
$\overline{\qquad\qquad}$ [r_1]
Gb

where J stands for the penicillin-resistant character of the strep infection and r_1 is a number close to zero.

This case exemplifies what Hempel calls *the ambiguity of I-S explanation*. We have two inductive arguments; the premises of these arguments are all logically compatible — all of them could be true. The conclusions of these two arguments are identical. Nevertheless, in one argument the conclusion is strongly supported by the premises, whereas in the other the premises strongly undermine the same conclusion — indeed, argument (2) can readily be transformed into an argument that strongly supports the negation of the conclusion of argument (1):

(3) $P(\sim G|F.H.J) = 1 - r_1$
       Fb.Hb.Jb
       $$\overline{\rule{3cm}{0pt}} \quad [1 - r_1]$$
       $\sim Gb$

Given that r_1 is close to zero, $1-r_1$ must be close to 1. Hence, we have two strong inductive arguments with compatible premises whose conclusions contradict one another. This situation is inconceivable in deductive logic. If two valid deductive arguments have incompatible conclusions, the premises of one argument must contradict the premises of the other.

At this point we are confronting one of the most fundamental differences between deductive and inductive logic. Deductive logic has a *weakening principle*, according to which p $\supset$ q entails p.t $\supset$ q, for any arbitrary t whatever. From this it follows that, given a valid deductive argument, it will remain valid if additional premises are inserted, provided none of the original premises is deleted. Probability theory does not have any such weakening principle. Let $P(G|F)$ be as close to 1 as you like, $P(G|F.H)$ may be arbitrarily close to 0.[26] From this it follows that an inductive argument that strongly supports its conclusion may be transformed, by the addition of a new premise consistent with the original premises, into an argument that strongly undermines that conclusion. Before the discovery of Australia, for example, Europeans had strong inductive evidence, based on many observations of swans, for the conclusion "All swans are white." Nevertheless, the addition of one premise reporting the observation of a black swan in Australia not only undermines the inductive conclusion, it deductively refutes it.

Inductive logicians have long recognized this feature of inductive arguments, and have come to terms with it by means of *the requirement of total evidence*. According to this principle, an inductive argument strongly supports its conclusion only if (i) it has true premises, (ii) it has correct inductive form, and (iii) no additional evidence that would change the degree of support is available at the time. Argument (1) above clearly fails to satisfy the requirement of total evidence, for its set of premises can be supplemented as follows:

(4) $P(G|F.H) = r$
$P(G|F.H.J) = r_1$
Fb.Hb
Jb
$$======= \quad [r_1]$$
Gb

where r_1 is very different from r.

The first temptation, with regard to the ambiguity of I-S explanation, might be to impose the requirement of total evidence, and to say that any inductive argument that is to qualify as an explanation must satisfy that requirement.[27] Such a move would be disastrous, for normally when we try to explain some fact we already know that it is a fact. Hence, our body of knowledge includes the conclusion of the argument. If the conclusion is not included among the premises, the requirement of total evidence is violated. If the conclusion is included among the premises, the argument is not inductive; rather, it is a trivially valid deduction — one, incidentally, that cannot even qualify as a D-N explanation, because, if the explanandum is included in the explanans, no law statement can occur *essentially* in the explanans. Consequently, in order to deal with the ambiguity of I-S explanation, Hempel sought a weaker counterpart for the requirement of total evidence that would not rule out altogether the possibility of I-S explanations.

The requirement Hempel saw fit to impose he called *the requirement of maximal specificity (RMS)*. Suppose, referring back to the schema (I-S) given above, that s is the conjunction of all of the premises and k is the body of knowledge at the time in question. "Then," Hempel says, "to be rationally acceptable in [that] knowledge situation . . . the proposed explanation . . . must meet the following condition:

If s.k implies that b belongs to a class F_1, and that F_1 is a subclass of F, then s.k must also imply a statement specifying the statistical probability of G in F_1, say

$P(G|F_1) = r_1$

Here, r_1 must equal r unless the probability statement just cited is simply a theorem of mathematical probability theory. (1965, 400)[28]

The unless-clause in the final sentence is intended to guard against RMS being so strong that it would rule out the possibility of I-S explanation altogether. If we want to explain Gb, then presumably we know that b belongs to the class G, and if we want to use $P(G|F) = r$ as a statistical law, we know that b belongs to F; consequently, k includes the statement that b is a member of $F_1 = F.G$, which is a subclass of F. But, trivially, $P(G|F.G) = 1$; indeed, trivially, all F.Gs are Gs. But this is not an appropriate basis for condemning the original explanation.

Since, Hempel claims, all bona fide I-S explanations must satisfy RMS, and since RMS makes specific reference to a particular knowledge situation, "*the concept of statistical explanation for particular events is essentially relative to a given knowledge situation as represented by a class K of accepted statements* " (1965, 402). Hempel refers to this feature as the *epistemic relativity of statistical explanation*. This relativity has no counterpart in D-N explanation. The reason is that the requirement of maximal specificity is automatically fulfilled in the case of D-N explanation, for, given that all F are G, it follows immediately that all F_1 are G if F_1 is a subclass of F. This is just an application of the weakening principle that was cited above.

We must guard against one easy misunderstanding. Someone might claim that D-N explanations are relativized to knowledge situations because what we take to be a law depends upon what we know at any given time. It is true, of course, that what is considered a law at one time may be rejected at another, and that we can never know for certain whether a given general statement is true. At best, we can hope to have general statements that are highly confirmed and that we are justified in accepting. These considerations apply equally to universal laws, general theories, and statistical laws, and consequently they apply equally to D-N and I-S explanations. The epistemic relativity of I-S explanation refers, however, to something entirely different.

Suppose that we have two putative explanations of two different particular facts, one a D-N explanation, the other I-S. Suppose that each of them has correct logical form—deductive and inductive, respectively. Suppose further that we are prepared, on the basis of our knowledge at the time, to accept the premises of each argument as true. Then, according to Hempel, we are entitled to accept the D-N explanation as correct (recognizing at the same time that we may be mistaken). We are not, however, entitled to accept the I-S explanation as correct, on the grounds just mentioned; in addition, we have to determine whether the statistical law to which we appeal is maximally specific. Whether it is or not depends upon the content of our body of knowledge. A statistical law can be true without being maximally specific. That is why, according to Hempel, we need the requirement of maximal specificity.

With RMS in place, Hempel has provided us with two models of scientific explanation of particular facts, one deductive and one inductive. A comprehensive characterization can be given. Any explanation of a particular occurrence is an argument to the effect that the event-to-be-explained was to be expected by virtue of certain explanatory facts. The explanatory facts must include at least one general law. The essence of scientific explanation can thus be described as *nomic expectability* —that is, expectability on the basis of lawful connections (1962a).

The general conditions of adequacy for scientific explanations set out in the first section of the Hempel-Oppenheim paper can be revised to encompass statistical explanation in the following way:

Logical conditions

1. An explanation is an argument having correct logical form (either deductive or inductive).
2. The explanans must contain, essentially, at least one general law (either universal or statistical).[29]
3. The general law must have empirical content.

Hempel and Oppenheim admit that requirement 3 is vacuous, for it will be automatically satisfied by any putative explanation that satisfies conditions 1 and 2. Any putative explanation that satisfies these conditions qualifies as a potential explanation. In order to qualify as an actual explanation (or, simply, an explanation), a potential explanation must fulfill two more conditions.

Empirical condition:

4. The statements in the explanans must be true.

Relevance condition:

5. The requirement of maximal specificity.

As we have seen, this relevance requirement is automatically satisfied by D-N explanations.

At this juncture, the received view embraced two models of explanation of particular facts, and a large promissory note about D-N explanations of laws (including, of course, statistical laws). The first serious effort to pay off the debt was made by Michael Friedman in 1974 — well into the third decade[30] — and he did not attempt to deal with statistical laws. We shall examine his work in due course, noting for the moment that, even if he succeeds in his effort, the promise of the lower right corner of Table 1 remains unfulfilled.[31]

2.5 Early Objections to the Inductive-Statistical Model

My own particular break with the received view occurred shortly after the incident with Smart that I related at the beginning of this essay. In a paper (W. Salmon 1965) presented at the 1963 meeting of the American Association for the Advancement of Science, Section L, organized by Adolf Grünbaum, I argued that Hempel's I-S model (as formulated in his (1962)), with its high probability requirement and its demand for expectability, is fundamentally mistaken. Hempel's example of John Jones's rapid recovery from his strep infection immediately called to mind such issues as the alleged efficacy of vitamin C in preventing, shortening, or mitigating the severity of common colds,[32] and the alleged efficacy of various types of psychotherapy. I offered the following examples:

(CE-8) John Jones was almost certain to recover from his cold within a week because he took vitamin C, and almost all colds clear up within a week after administration of vitamin C.

(CE-9) John Jones experienced significant remission of his neurotic symptoms because he underwent psychotherapy, and a sizable percentage of people who undergo psychotherapy experience significant remission of neurotic symptoms.

Because almost all colds clear up within a week whether or not the patient takes vitamin C, I suggested, the first example is not a bona fide explanation. Because many sorts of psychological problems have fairly large spontaneous remission rates, I called into question the legitimacy of the explanation proffered in the second example. What is crucial for statistical explanation, I claimed, is not how probable the explanans renders the explanandum, but rather, whether the facts cited in the explanans *make a difference* to the probability of the explanandum.

To test the efficacy of any sort of therapy, physical or psychological, controlled experiments are required. By comparing the outcomes in an experimental group (the members of which receive the treatment in question) with those of a control group (the members of which do not receive that treatment), we procure evidence concerning the effectiveness of the treatment. This determines whether we are justified in claiming explanatory import for the treatment vis-à-vis the remission of the disease. If, for example, the rate of remission of a certain type of neurotic symptom during or shortly after psychotherapy is high, but no higher than the spontaneous remission rate, it would be illegitimate to cite the treatment as the explanation (or even part of the explanation) of the disappearance of that symptom. Moreover, if the rate of remission of a symptom in the presence of psychotherapy is not very high, but is nevertheless significantly higher than the spontaneous remission rate, the therapy can legitimately be offered as at least part of the explanation of the patient's recovery. It follows from these considerations that high probability is neither necessary nor sufficient for bona fide statistical explanation. Statistical relevance, not high probability, I argued, is the key desideratum in statistical explanation.

Henry E. Kyburg, Jr., who commented on my AAAS paper, noticed that a similar point could be made with regard to D-N explanations of particular facts. He illustrated his claim by offering CE-6, the 'explanation' of the dissolving of a sample of table salt on the basis of a 'dissolving spell' (Kyburg 1965). Once the point has been recognized, it is easy to come up with an unlimited supply of similar examples, including my favorite, CE-7, John Jones's 'explanation' of his failure to become pregnant on the basis of his consumption of oral contraceptives.[33]

Scriven's paresis example, CE-5, brought up much earlier in connection with the explanation/prediction symmetry thesis and the notion of expectability, shows that high probability is not required for probabilistic explanation. Scriven's discussion of explanation in evolutionary biology (1959) draws the same conclusion. The appeal to latent untreated syphilis to explain paresis is obviously an appeal to a statistically relevant factor, for the probability that someone with latent untreated syphilis will develop paresis, while not high, is considerably higher than the probability for a randomly selected member of the human population at large. Moreover, it can plausibly be argued that the evolutionary biologist, in explaining occurrences in that domain, also invokes statistically relevant facts. The issue of high probability vs. statistical relevance has thus been joined. It will prove to be a question of considerable importance in the further development of our story.

The Third Decade (1968–77)
Deepening Differences

The third decade is bracketed by Hempel's last two publications on scientific explanation. It opened with his emendation of the requirement of maximal specificity (RMS*), which was designed to fix a couple of technical flaws in his account of I-S explanation (1968).[1] It ended with the publication (in German) of a substantial postscript to section 3 of "Aspects" which is devoted to statistical explanation (1977). Among other things, he retracted the high-probability requirement on I-S explanations.

Insofar as published material was concerned, the issue of high probability vs. statistical relevance remained quite dormant for about five years after I had raised it rather obscurely in (1965).[2] Richard C. Jeffrey (1969) argued elegantly that statistical explanations—with the exception of certain limiting cases—are not arguments, and that the degree of probability conferred upon an explanandum by an explanans is not a measure of the goodness of the explanation. My next publication on the topic was in 1970; it contained a theory of statistical explanation, based upon statistical relevance relations, that was spelled out in considerable detail.[3] An ingenious information-theoretic account of scientific explanation, in which statistical relevance relations play a key role, was published by James G. Greeno in 1970. An account of the statistical-relevance (S-R) model of scientific explanation—based on the three papers by Greeno, Jeffrey, and me—was published as a small book the following year (W. Salmon et al. 1971).

The introduction of the inductive-statistical model of scientific explanation constituted, I believe, a crucial turning point for the received view. Before that, I suspect, many philosophers felt (as I did) quite comfortable with the basic idea of D-N explanation, and confident that an equally satisfactory statistical conception would be forthcoming. As things turned out, the I-S model gave rise to a number of fundamental problems. Three avenues presented themselves as ways of coping with the difficulties. The first was to maintain the received view—to pursue the course already laid out by Hempel—seeking to defend the I-S model against objections, and to repair the faults as they were detected. The second was to attempt to construct an alternative account of statistical explanation of particular facts, such as the S-R model, in the hope of avoiding difficulties encountered by the received view. The third was to reject altogether the possibility of providing probabilistic or statistical explanations of particular facts, thereby maintaining strict deductivism with regard to scientific explanation. The deductivist claims that the box in the lower left-hand corner of table 1 is empty; all legitimate explanations are of the D-N variety (where we persist in regarding D-S explanations as a subset of D-N). These three avenues correspond roughly to three basic conceptions that seem to me, in retrospect, to have dominated the discussion of scientific explanation from the time of Aristotle to the present.[4] Their distinctness stands out clearly, I think, only when we try to give an explicit account of statistical explanation. Statistical explanation is like a Stern-Gerlach magnet that separates the incoming notion of scientific explanation into three divergent beams—the epistemic, modal, and ontic conceptions.

3.1 The Statistical-Relevance Model

If one takes seriously the notion that statistical relevance rather than high probability is the key concept in statistical explanation, one naturally asks what role inductive arguments have in the statistical explanation of particular facts. The simple and rather obvious answer is "None."[5] Approaching Hempel's I-S model from a different angle, in a philosophical gem (1969), Richard Jeffrey argued incisively to that very conclusion. He maintained that when a stochastic mechanism—e.g., tossing of coins or genetic determination of inherited characteristics—produces a variety of outcomes, some more probable and others less probable, we understand those with small probabilities exactly as well as we do those that are highly probable. Our understanding results from an understanding of that mechanism and a recognition of the fact that it is stochastic. Showing that the outcome is highly probable, and that it was to be expected, has nothing to do with the explanation. That the outcome was probable or improbable might be added—in the margin, so to speak—as an interesting gloss on the main text. This paper also provided inspiration for some philosophers who adopted an explicitly mechanistic conception of scientific explanation in the fourth decade.[6]

At about the same time, James G. Greeno (1970) developed an information-theoretic account of statistical explanation in terms of the amount of information transmitted by a statistical theory. Transmitted information is a statistical relevance concept, and among its other virtues, it is defined quantitatively. It can be used to evaluate the explanatory power of statistical hypotheses, but it does not yield quantitative evaluations of individual explanations. Within this approach there is no suggestion that statistical explanations should be identified with inductive arguments.

At around the same time I was trying to work out an alternative to the I-S model that would be based on statistical relevance rather than high probability. The first crucial step is to notice that, whereas high probability involves just one probability value, statistical relevance involves a comparison between two probability values. To construct an explanation based upon statistical relevance we need to compare a posterior probability with a prior probability.

Consider John Jones and his quick recovery from a strep infection. Initially, Jones is simply a person with a strep infection – this is the reference class within which he is included for purposes of assessing the *prior* probability of quick recovery. We may assume, I believe, that this probability is fairly small. This reference class is not, however, homogeneous. We can partition it into two subclasses, namely, those who are treated with penicillin and those who are not. Among those who are treated with penicillin the probability of quick recovery is quite high; among those not so treated the probability is much smaller. Since Jones belongs to the subclass treated with penicillin, the *posterior* probability of his quick recovery is much greater than the prior probability. The original reference class has been relevantly partitioned.

We begin by asking why this member of the class of people with strep infections experienced quick recovery. We answer that he belonged to the subclass who received penicillin, noting that the probability in the subclass is different from the probability in the original reference class. As Hempel points out, however, we are not quite done. The subclass of people who are treated with penicillin can be further relevantly partitioned into those whose infection is of the penicillin-resistant strain, and those whose infection is not. The probability of quick recovery for those who have the penicillin-resistant infection is much smaller than is that for those who have a non-penicillin-resistant infection. We may therefore answer the why-question by stating that Jones recovered quickly from his strep infection because his infection was of the non-penicillin-resistant strain and it was treated with penicillin.

Let us define some of the terms we are using. By a *partition* of a class F we mean a set of subclasses that are mutually exclusive and exhaustive within F – that is, every member of F belongs to one and only one member of the partition. Each of these subclasses is a *cell* of the partition of F. A partition is *relevant* with respect to some attribute G if the probability of G in each cell is different from its probability in each of the other cells. The possibility is left open that the probability of G in one of the cells is equal to the probability of G in the original reference class F.[7] A class F is *homogeneous* with respect to the attribute G if no relevant partition can be made in F. It is *epistemically homogeneous* if we do not know how to make a relevant partition – if, that is, our total body of knowledge does not contain information that would yield a relevant partition. The class is *objectively homogeneous* if it is impossible in principle, regardless of the state of our knowledge, to make a relevant partition. We may then define a *homogeneous relevant partition* as a relevant partition in which each cell is homogeneous. We may then distinguish in an obvious way between *epistemically homogeneous relevant partitions* and *objectively homogeneous relevant partitions*.

Two points should be carefully noted: (1) when an objectively homogeneous relevant partition of a reference class has been given, all relevant factors have been taken into account – i.e., all relevant partitions have been effected – and (2) a relevant partition admits *only* relevant factors, since no two cells in the partition have the same probability for the attribute G.

These concepts deserve illustration. Greeno (1970) offers the example of an American teenager, Albert, who is convicted of stealing a car (D = delinquent offense).[8] Albert is a boy and he lives in an urban environment (San Francisco). Take as the original reference class for purposes of explanation the class of American teenagers (T). If we subdivide it into American teenage boys (M) and American teenage girls (F), we will make a relevant partition, for the probability that a boy will steal a car is greater than the probability that a girl will do so. Moreover, if we partition the class of American teenagers into urban dwellers (U) and rural dwellers (R), we will make another relevant partition, for the delinquency rate in urban areas is higher than it is in rural areas. Taking both of these partitions into account, we have four subclasses of the original reference class – namely, urban male, urban female, rural male, and rural female – and we find that the probability of delinquency is different in these four cells of the combined partition. Symbolically, we can write

$$P(D\,|\,T.C_i) = p_i; \quad p_i \neq p_j \text{ if } i \neq j$$

where

$$U.M = C_1; \quad U.F = C_2; \quad R.M = C_3; \quad R.F = C_4.$$

This is a relevant partition, but it is obviously not homogeneous, for there are many other factors, such as religion and socio-economic status, that are relevant to delinquency.

Hempel's example of quick recovery (Q) from a strep infection provides a different illustration. In this example, the original reference class is the class of people who have streptococcus infections (S). Clearly, the partition of this class into those who receive treatment with penicillin (T) and those who do not ($\sim T$) is relevant. Also, a partition can be made into those who have penicillin-resistant infections (R) and those whose infections are not penicillin-resistant ($\sim R$). If, however, we combine these two partitions to form the cells

$$T.R = C_1; \quad T.\sim R = C_2; \quad \sim T.R = C_3; \quad \sim T.\sim R = C_4$$

the resulting partition is not a relevant partition, for the probabilities

$$P(Q\,|\,C_i) = p_i$$

are not all different. If the infection is penicillin resistant, it makes no difference (I assume) whether penicillin is administered or not, and (I further assume) the probability of quick recovery is the same for a person with a non-penicillin-resistant infection who does not receive penicillin as it is for anyone with a penicillin-resistant strain. Thus it appears that

$$p_1 = p_3 = p_4 \neq p_2,$$

so our relevant partition of S is

$$S.C_1 = S.\sim R.T; \quad S.C_2 = S.(R.T \lor R.\sim T \lor \sim R.\sim T).$$

This partition is not likely to be homogeneous, for—as Philip Kitcher pointed out—a further partition in terms of allergy to penicillin would be relevant.

The basic structure of an S-R explanation can now be given. Although there are many different ways to ask for a scientific explanation, we can reformulate the request into an explanation-seeking why-question that has the canonical form, "Why does this (member of the class) A have the attribute B?" When an explanation is requested, the request may come in the canonical form—for example, "Why did Jones, who is a member of the class of people who have strep infections, recover quickly?" Or, it may need translation into standard form—for example, "Why did Albert steal a car?" becomes "Why did this American teenager commit a delinquent act?" A is the reference class for the prior probability, and B is the attribute. When a translation of this sort is required, pragmatic considerations determine the choice of an appropriate reference class.

An S-R explanation consists of the prior probability, a homogeneous relevant partition with respect to the attribute in question, the posterior probabilities of the attribute in cells of the partition, and a statement of the location of the individual in question in a particular cell of the partition:

$$P(B|A) = p$$
$$P(B|A.C_1) = p_1; \quad P(B|A.C_2) = p_2; \; . \; . \; .$$

$A.C_1, A.C_2, \; . \; . \; .$ constitute a homogeneous relevant partition of A
b is a member of $A.C_k$

Given our definition of a homogeneous relevant partition, we know that our explanation appeals to all of the factors relevant to the explanandum, and only to factors that are relevant.

These relevance considerations enable us to deal not only with such counterexamples to the I-S model as CE-8 (vitamin C and colds) and CE-9 (psychotherapy vs. spontaneous remission), but also with such counterexamples as CE-3 (barometer) that we encountered much earlier in connection with the D-N model. To do so, we need a relation known as *screening off*.

Consider the fact that a sharp drop in the reading on a barometer (B) is highly relevant to the occurrence of a storm in the vicinity (S). The probability of a storm, given the sharply falling barometric reading, is much higher than its probability without reference to the barometric reading, i.e.,

$P(S|B) > P(S)$.

This is the reason that the barometer is a useful instrument for predicting storms. However, as we remarked above, the reading on the barometer does not explain the storm. The fact is that a certain set of atmospheric conditions, including a sharp decrease in atmospheric pressure, are responsible both for the storm and for the reading on the barometer. Indeed, the actual drop in atmospheric pressure renders the reading on the barometer irrelevant. The probability of a storm (S), given both the drop in atmospheric pressure (A) and the drop in the reading on the barometer (B), is equal to the probability of a storm, given only the drop in atmospheric pressure—i.e.,

$P(S|A.B) = P(S|A)$.

Thus, although B is relevant to S if A is not taken into account, B is irrelevant to S in the presence of A. This is what it means to say that A screens B off from S. It should be noted that B does not screen A off from S. If the barometer gives a faulty reading because of some malfunction, or if it is placed in a vacuum chamber and thus caused to register a sharp drop, that will have no effect on the probability of a storm. Thus,

$P(S|A.B) \neq P(S|B)$.

Indeed, destroying all of the barometers inside a thousand mile radius will not affect the probability of storms in the least.

According to the S-R model, no irrelevant factors should be included in an explanation. "To screen off" *means* "to render irrelevant." Accordingly, no factor that is screened off can be invoked legitimately in an S-R explanation. The barometer example is an instance of a large and important class of cases in which a common cause gives rise to two or more effects that are correlated with one another. The atmospheric conditions cause both the barometric reading and the storm. In such cases, we want to block any attempt to explain one of these common effects in terms of others. This can often be accomplished by appealing to the fact that the common cause screens any one of the correlated effects off from any of the others.

Hempel provides another example of just this type. He points out that Koplik spots—small white blisters on the inside surface of the cheek—are a very reliable sign of a measles infection, and they can be used to predict that other symptoms, such as a skin rash or a fever, will soon appear. All of these symptoms, including the Koplik spots, are among the effects of the measles infection. According to the S-R model, then, the Koplik spots have no part in explaining the other symptoms of the disease. Hempel (1965, 374–75) suggests, however, that there might have been occasions, at a time when the nature of the underlying infection was totally unknown, on which it could have been appropriate to invoke them in order to explain the other symptoms.

In my earliest criticism of the I-S model, I suggested that positive relevance rather than high probability is the key explanatory relation. At that time I had not worked out the details of any alternative model. In developing the S-R model— leaning heavily on the work of Jeffrey and Greeno—I came to the conclusion that positive relevance is not required, but that negative relevance could also have explanatory import. The reasons can be given quite straightforwardly and simply. If one makes a relevant partition of a reference class, in some cells the posterior probability of the attribute will be higher than the prior probability, and in some the posterior probability will be lower. This point is obvious in the case of a partition with just two cells. Suppose that we have a large class of coin tosses, and that the probability of heads in that class is ½. Suppose, further, that this reference class can be partitioned into two subclasses, one consisting of tosses of a coin heavily biased toward heads (0.9), the other consisting of tosses of another coin equally biased toward tails. Among the tosses of the second coin some yield heads; the probability for this result in this cell of the partition is 0.1. However, according to an argument of Jeffrey mentioned briefly above, we understand the low-probability outcomes of any given stochastic process just as well as we understand the high-probability outcomes; we understand the heads resulting from tosses of the coin biased for tails just as well as we understand the tails, and by parity of reasoning, we understand the tails resulting from tosses of the coin biased toward heads just as well as we understand the heads. Thus, it appears, we understand all of the results in the original reference class equally well. Some have high probabilities, some have low. Some are found in cells of the relevant partition which reduce their probability, some in cells that increase their probability. Negatively relevant factors as well as positively relevant factors contribute to our understanding of the phenomena we are studying.

We are now in a position to offer a succinct contrast between the I-S and S-R models of statistical explanation.

I-S model: An explanation of a particular fact is an *inductive argument* that confers upon the fact-to-be-explained a *high inductive probability.*

S-R model: An explanation of a particular fact is an assemblage of facts *statistically relevant* to the fact-to-be-explained *regardless of the degree of probability* that results. (W. Salmon et al. 1971, 11)

Both models conform to the covering law conception of scientific explanation, for they both require laws in the explanantia. In particular, in an S-R explanation, the statements of values of prior and posterior probabilities qualify as statistical laws.

3.2 Problems with Maximal Specificity

Hempel's recognition of what he termed "the ambiguity of I-S explanation," which led to the requirement of maximal specificity and the doctrine of essential epistemic relativization of I-S explanation, gave rise to deep difficulties. J. Alberto Coffa was, to the best of my knowledge, the first to recognize their profundity. In a classic article, "Hempel's Ambiguity" (1974) he spelled them out with great care.[9]

Coffa begins by identifying the following schema

$$\text{(I)} \quad \begin{array}{l} (x)[Fx \supset Gx] \\ Fb \\ \hline Gb \end{array}$$

as the 'basic schema' for the simplest form of D-N explanation (1974, 142). Since the following schema

$$\text{(II)} \quad \begin{array}{l} P(G|F) = r \\ Fb \\ \hline Gb \end{array} \quad [r]$$

is the obvious inductive analog of schema I, we might be tempted to offer *the naïve model of inductive explanation* in which an I-S explanation is an argument of form II, where r is close to 1, the first premise is a statistical law, and the premises are true. But, as we have seen, given an instance of (II), we can often find another argument of the form

$$\text{(III)} \quad \begin{array}{l} P(\sim G|H) = s \\ Hb \\ \hline \sim Gb \end{array} \quad [s]$$

where s is close to 1, the first premise is a statistical law, and the premises are true as well. Such pairs of arguments exemplify what Hempel called "the ambiguity of I-S explanation," and its existence utterly undermines the naïve model (Coffa 1974, 143–44).

The existence of such pairs of arguments is an example of what Hempel (1960) called "inductive inconsistency"—namely, two inductive arguments with mutually consistent (indeed, true!) premises that lend strong inductive support to two mutually inconsistent conclusions. Finding this situation intolerable in the domain of scientific explanation, Hempel seeks a ground for rejecting at least one of the two 'explanations.' Having briefly suggested in his first article on statistical explanation (1962) that the requirement of total evidence be invoked, he subsequently formulated (and reformulated (1968)) the *requirement of maximal specificity* to do the job. But, Coffa points out, if the problem were really just the inductive inconsistency, a *much* simpler device would suffice. Since, as Hempel seems to agree, we explain only actual occurrences, not things that do not happen, it is easy to choose the correct explanation from the foregoing pair. It is the one with the true conclusion. Since Hempel did not adopt this obvious and easy solution, Coffa suggests, he must *not* have been exercised about *this* problem. When we look at the way he actually tried to solve the problem, we see that the *real* problem was not the problem of inconsistency, but rather the venerable *reference class problem*.

The problem of choosing the appropriate reference class arose traditionally when adherents of the frequency interpretation of probability, such as John Venn and Hans Reichenbach, attempted to apply probabilities to single cases. What is the probability, for example, that William Smith will still be alive 15 years from now? This is a question of considerable importance to his wife and children, to his employer, and to his insurance company. The problem is that he belongs to many different reference classes, and the frequencies of 15-year survival may differ greatly from one of these classes to another. Within the class of American males it is fairly high, among 40-year-old American males it is even higher, among grossly obese Americans it is not nearly as high, and among heavy cigarette-smoking individuals it is still different. If we are going to try to predict whether he will be alive in 15 years, or estimate his chances of surviving that long, we will want to take into account all available relevant evidence.

Starting with a very broad reference class, such as the class of all Americans, we should partition it in terms of such factors as sex, age, occupation, state of health, etc., until we have taken into account all known factors that are statistically relevant. By the same token, we do not want to partition the class in terms of factors that are known to be irrelevant to his survival, or in terms of factors whose relevance or lack thereof is unknown. We may say that the rule for assigning a probability to a single case is to refer it to the *broadest homogeneous reference class* available, where it is understood that the class is *epistemically* broad and homogeneous—i.e., we have not used any partitions that are not known to be relevant and we do not know how to make any further relevant partitions (see W. Salmon 1967, 90–96, and Salmon et al. 1971, 40–47). Since we are concerned here with prediction, the maxim is to use all available evidence; epistemic relativization is entirely appropriate.

The problem of choosing the correct reference class for explanatory purposes is quite different. When we want to explain some fact, such as John Jones's rapid recovery from his strep infection, we already know that his recovery has occurred—i.e., the explanandum is already part of our total evidence—so we must not appeal to that portion of our total evidence. The reference class problem becomes the problem of determining precisely to what part of our total evidence we are allowed to appeal in constructing an I-S explanation. Hempel's RMS and its successor RMS* were designed to accomplish this aim. Roughly speaking (referring back to the schema II), both versions of the requirement of maximal specificity stipulated that, *in a knowledge situation K*, every known relevant partition of F, except one in terms of G or in terms of properties *logically* related to G, be made. The idea of this requirement is to tell us what factors must be taken into account and what factors must not be taken into account in constructing I-S explanations. Hempel attempted to prove, for each version of the requirement, that it would eliminate the unwholesome ambiguity. RMS* was devised in response to a counterexample to RMS constructed by Richard Grandy (Hempel 1968).

The profound problem Coffa recognized arises, not from the observation that *many* explanations that fit schema II are subject to ambiguity, but from Hempel's assertion that *all* of them are. It is a striking fact that Hempel does not offer arguments to support this transition from many to all (Coffa 1974, 151). Yet, this is the crucial issue, for it is only from the assertion of *universal* ambiguity that the doctrine of *essential* epistemic relativity arises. So Coffa does two things. First, he attempts to reconstruct conjecturally the kind of considerations that may have led Hempel to his conclusion of essential epistemic relativity (151–55), and second, he maintains that it leads to disaster for the theory of I-S explanation (155–60).

Coffa argues that Hempel was looking for some condition to impose on the law-premise of schema II—"$P(G|F) = r$"—that would guarantee that the problem of ambiguity would not arise. RMS (later, RMS*) were the conditions he proposed. Each contains a reference to a knowledge situation, thus implying that the explanation in question would be epistemically relativized. It appears, then, that Hempel felt that no adequate substitute for RMS could exist that did not also contain reference to a knowledge situation. Consequently, according to Hempel, I-S explanation is *essentially epistemically relativized*.

Both Coffa and I agreed that epistemic relativization of probabilistic explanation could be avoided—though we took different tacks in trying to do so—but before looking at them we should consider the consequences of accepting epistemic relativity. In (1971) I thought that nothing of much importance hinged on the issue of epistemic relativity; Coffa's 1974 article changed my mind radically on this point (W. Salmon 1974).

Coffa begins his discussion of epistemic relativization of I-S explanation by analyzing the distinction between epistemic and non-epistemic concepts:

> It is an obvious fact that the meaning of some expressions or concepts can be given without referring to knowledge, whereas that of others cannot. Let me call the latter epistemic and the former non-epistemic. Examples of non-epistemic expressions are easy to find. 'Table,' 'chair,' 'electron,' according to

many, 'truth,' would be typical instances. Examples of epistemic notions are also readily available. The best known instance may be that of the concept of confirmation. Although the syntactic form of expressions like 'hypothesis *h* is well-confirmed' may mislead us into believing that confirmation is a property of sentences, closer inspection reveals the fact that it is a relation between sentences and knowledge situations and that the concept of confirmation cannot be properly defined (that is, its meaning cannot be given) without a reference to sentences intended to describe a knowledge situation. . . .

Having introduced the distinction between epistemic and non-epistemic concepts, we go on to notice that there is a further interesting distinction to be drawn within the class of epistemic notions based upon the kind of role knowledge plays in them. On the one hand there are those epistemic notions in which knowledge enters essentially as an argument in a confirmation function, or, equivalently, as an ingredient in a statement of rational belief. And then there is the obscure and largely unintelligible remainder.

In the first group we find a significant example provided by Hempel's theory of deductive explanation. After introducing his non-epistemic notion of D-N explanation Hempel went on to say that he could define now the concept of a well-confirmed D-N explanation, a well-confirmed D-N explanation in a tacitly assumed knowledge situation *K* being, in effect, an argument which in knowledge situation *K* it is rational to believe is a D-N explanation, i.e., a true D-N explanation. In precisely the same fashion we could correctly and uninterestingly define the concepts of well-confirmed table, well-confirmed chair and well-confirmed electron, given that we started by having the concepts of table, chair and electron. *Since we can only have reason to believe meaningful sentences, a confirmational epistemic predicate is an articulation of independently meaningful components.*

Of course we can understand what a well-confirmed chair is because we began by understanding what a chair is. If '*x* is a chair' had not had a meaning, we would not even have been able to make sense of the statement of rational belief made about it. Similarly, we can understand, if not appreciate, the notion of well-confirmed D-N explanation, because we were told first what kind of thing a D-N explanation is.*

Now we are in a position to state Hempel's thesis of the epistemic relativity of inductive explanation. As a consequence of his analysis of the phenomenon of ambiguity, Hempel concludes that the concept of inductive explanation, unlike its deductive counterpart, is epistemic; and he goes on to add that it is not epistemic in the sense in which well-confirmed deductive explanations are.

---

*Prof. Gerald Massey has drawn my attention towards the apparent opacity of epistemically relativized predicates. He has pointed out that this raises serious doubts concerning the possibility of viewing them as expressing properties. (Coffa's note).

The concept of inductive explanation is a non-confirmational epistemic concept. . . .

As Hempel is careful to point out, this means that there is no concept that stands to his epistemically relativized notion of inductive explanation as the concept of true D-N explanation stands to that of well-confirmed D-N explanation. *According to the thesis of epistemic relativity there is no meaningful notion of true inductive explanation.* Hence, we couldn't possibly have reasons to believe that anything is a true inductive explanation. Thus, it would be sheer confusion to see inductive explanations relative to K in Hempel's sense as those inductive arguments which in knowledge situation K it is rational to believe are inductive explanations.

It is clear that, according to Hempel, *there is a remarkable and surprising disanalogy between deductive and inductive explanations.* When somebody asks us to give an account of deductive explanations, we can do so without referring to anybody's knowledge. If asked, for instance, what sort of thing would it be to explain deductively the present position of a planet, we would refer to descriptions of certain nomic and non-nomic facts but never to anybody else's knowledge. This is a desirable feature in a non-psychologistic account of explanation. Yet, according to Hempel, when we ask what an inductive explanation of the same event would look like, there is no way in which an appropriate answer can be given without talking about knowledge; presumably the knowledge available at the time of the explanation. Even more surprisingly, this reference to knowledge does not play the standard role that such references usually play, to wit, that of providing the epistemic platform for a judgment of rational belief. What role such reference plays is a question which deserves serious attention, since here we find the Achilles' heel of Hempel's whole construction. (1974, 148–50, emphasis added)

Having established the distinction between epistemic and non-epistemic concepts, and having pointed out, within the former category, the distinction between confirmational and non-confirmational concepts, Coffa now announces the result he intends to draw.

I have argued that from a certain problem Hempel felt forced to draw the conclusion that the notion of inductive explanation is epistemically relativized. . . . Now, I would like to explain why I find Hempel's conclusion worth avoiding. I will try to convince you that *to accept Hempel's thesis of epistemic relativity amounts to accepting the claim that there are no inductive explanations*, the concept of I-S explanation relative to K functioning as a placebo which can only calm the intellectual anxieties of the uncautious user. If I am right, anyone willing to hold that there are inductive explanations will have to begin by spotting a flaw in Hempel's argument. . . .

Maybe the best way in which I can briefly convey my feelings about the oddity implicit in Hempel's theory of inductive explanation is by noting, that in my view, Hempel's decision to develop a theory of I-S explanation relative to *K* after having argued that the notion of true inductive explanation makes no sense, seems comparable to that of a man who establishes conclusively that Hegel's philosophy is strict nonsense, and then proceeds to devote the rest of his life to produce the definitive edition of Hegel's writings. (1974, 155, emphasis added)

The strategy Coffa adopts to carry out his assault on Hempel's I-S model of explanation and, indeed, any essentially epistemically relativized model of explanation, is to pose a fundamental challenge for Hempel.

Now the question I would like to put to Hempel is the following. Take any I-S explanation relative to *K,* for some given *K.* . . . What is it about this inductive argument that makes it an explanation of its last formula? What reason could anyone have to say that it is an explanation of its conclusion?

It is not difficult to answer this question when we pose it, not for the inductive, but for the deductive case. If one asks, for example, what reason we have to believe that a causal deductive explanation explains its explanandum, the answer is that its premises identify certain features of the world that are nomically responsible for the occurrence of the explanandum event.

Could we say, as in the deductive case, that I-S explanations relative to *K* explain because their premises somehow identify features of the world that are nomically responsible for the explanandum event? Certainly not. This is what we vaguely conceived to be possible while tacitly espousing the naïve model, until Hempel shattered our illusions to pieces by focusing the reference class problem on the theory of explanation. Indeed, *if there is no characterization of true inductive explanation, then it must be because there are no such things that go on in the non-epistemic world of facts that can inductively explain the event.* For if there were such non-epistemic goings on, their characterization would be a characterization of true inductive explanation. Thus, the possibility of a notion of true explanation, inductive or otherwise, is not just a desirable but ultimately dispensable feature of a model of explanation: it is the *sine qua non* of its realistic, non-psychologistic inspiration. It is because certain features of the world can be deterministically responsible for others that we can describe a concept of true deductive explanation by simply describing the form of such features. If there are features of the world which can be non-deterministically responsible for others, then we should be able to define a model of true inductive explanation. And, conversely, if we could define a model of true inductive explanation, there could be features of the world non-deterministically responsible for others. The thesis of epistemic relativity implies that, for Hempel, there are no such features. What, then, is the interest of I-S explanations relative to *K* ? Surely not, as we have seen above, that in knowledge situation K we have reason to believe that they are inductive explanations. Then what? We detect in Hempel's writings not even a hint as to what an answer to this question might be. (1974, 157–58, emphasis added)

I have quoted from Coffa at length to convey the subtlety and cogency of his argument, and to capture the style and clarity of his exposition. I think his conclusion is entirely correct.

Now that we have exhibited the consequences of the doctrine of essential epistemic relativity of I-S explanation, let us examine more closely the considerations that seem to have motivated that doctrine. Coffa has shown beyond doubt that it arises somehow from the reference class problem. Looking at the problem in the context of the frequency interpretation of probability, in which it originally arose, the problem seems to be that any reference class we choose will be inhomogeneous.

Coffa considers the example of Jones who is both a Texan and a philosopher. Given that most Texans are millionaires and that most philosophers are not millionaires, we have a naïve inductive explanation of Jones being a millionaire and of Jones not being a millionaire (1974, 144). Obviously, neither the class of Texans nor the class of philosophers is homogeneous with respect to being a millionaire, and it appears that there will always be ways of partitioning any given subclass that will yield different relative frequencies. Perhaps Jones was born at the moment a Chinese mandarin sneezed, and perhaps the class of Texas philosophers born at any such moment had a proportion of millionaires different from that in the class of Texan philosophers (1974, 154). Coffa offers a suggestion – he does not spell it out in detail – that the main difficulty can be circumvented by thinking of probabilities in terms of propensities rather than frequencies. I shall return to this idea below. In the meantime, let me follow the frequentist line a little further.

If we consider some less fanciful cases, such as Greeno's delinquency example or Hempel's example of the quick recovery from a streptococcus infection, a plausible line of argument emerges. In explaining why Albert stole a car, we saw that there were many relevant factors that could be summoned – sex, age, place of residence, socio-economic status of the family, religious background, etc. The list seems almost endless. Whenever we arrive at a narrower reference class as a result of another relevant partition, if that class still contains youths who do not commit delinquent acts, we are tempted to suppose that there is some further factor that relevantly distinguishes Albert from the non-delinquents. The process of relevant partitioning will end only when we have found a reference class in which all members are juvenile delinquents. At that point, it may be assumed, we will have a bona fide explanation of Albert's delinquency. It should be noted emphatically that *the resulting explanation is not inductive, it is deductive-nomological.* Obviously, whenever we reach a universal generalization of the form "(x)[Fx ⊃

Gx]" we have a reference class F that is trivially homogeneous with respect to the attribute G, for every F is a G, and so is every member of every subclass of F. For this reason, D-N explanations automatically satisfy RMS. A similar analysis could plausibly be offered in the case of Jones's quick recovery from his strep infection.

One reason for claiming that I-S explanations are essentially epistemically relativized is the supposition that the statistical generalization among the premises does not contain a homogeneous reference class—i.e., that the reference class can, in all cases, in principle, be relevantly partitioned, even if we do not know how to do it at the time. One obvious motivation for such a claim would be a commitment to determinism. According to the determinist, every event that happens—including Albert's theft of an auto and Jones's quick recovery from his strep infection—are completely determined by antecedent causes. Those causes define a reference class in which every member has the attribute in question. If we construct an I-S explanation, the statistical law that occurs as a premise must embody an inhomogeneous reference class—i.e., one that could be relevantly partitioned. The only reason for using such a reference class would be ignorance of the additional factors needed to effect the further partition.

If this were the situation, we could easily understand the epistemic relativity of I-S explanations. If determinism is true, all bona fide explanations are deductive. I-S explanations are not well-confirmed inductive explanations; they are simply incomplete D-N explanations. An I-S explanation is analogous to an enthymeme—a deductive argument with missing premises. By supplying missing premises we can make the argument more complete, but when all of the missing premises are furnished the argument is no longer an enthymeme. At that point it becomes a valid deductive argument. There are no valid enthymemes, for, by definition, they lack premises needed for validity.

Similarly, an I-S explanation is simply an incomplete D-N explanation as long as it embodies an inhomogeneous reference class. As more and more relevant partitions are made, it approaches more closely to a D-N explanation, but it ceases to be an I-S explanation when homogeneity is achieved because, according to the deterministic assumption, homogeneity obtains only when the statistical law has been transformed into a universal law. From the determinist's standpoint, then, there are no genuine inductive explanations; the ones we took to be inductive are simply incomplete deductive explanations. It is easy to see what epistemic relativity amounts to in the context of determinism. When I posed the situation in this way to Hempel—suggesting that he was implicitly committed to determinism (1974)—he informed me emphatically that he is not a determinist (private correspondence).

An indeterminist, it seems to me, is committed to the view that there are reference classes in bona fide statistical laws that are homogeneous—not merely epistemically homogeneous, but objectively homogeneous. One might suppose that the class of carbon-14 nuclei is homogeneous with respect to spontaneous radioactive decay within the next 5730 years (the half-life of C^{14})–that there is, in principle, no way to make a relevant partition. Unlike William Smith (with whose survival we were concerned), C^{14} atoms do not age, suffer ill health, indulge in hazardous occupations, etc. Their chances of suffering spontaneous decay are remarkably unaffected by their environments. If quantum theory is correct, there is no way to select a subclass of the class of C^{14} nuclei in which the probability of spontaneous decay within the next period of 5730 years is other than one-half. In other words, modern physics strongly suggests that there are nontrivial cases of objectively homogeneous reference classes.

If one accepts the notion that there may be objectively homogeneous reference classes, then one might construct an unrelativized model of I-S explanation along the following lines:

An argument of the form

$$(II) \quad P(G|F) = r$$
$$Fb$$
$$======= \quad [r]$$
$$Gb$$

is an I-S explanation of Gb, provided that (1) the first premise is a statistical law, (2) both premises are true, (3) r is close to 1, and (4) F is an objectively homogeneous reference class with respect to G.

Since this characterization makes no reference to a knowledge situation, it can be taken as an explication of a *true I-S explanation*. An argument of form (II) is a *well-confirmed I-S explanation* with respect to a knowledge situation K if the information in K provides good grounds for believing that it satisfies the conditions following schema (II) above. An argument of form (II) that satisfies conditions (1)–(3), but not condition (4), is–at best– an *incomplete I-S explanation*. An argument of form (II) that qualifies as an incomplete I-S explanation could be considered an *optimal I-S explanation with respect to knowledge situation K* if the reference class F is epistemically homogeneous–that is, in knowledge situation K we do not know how to effect a relevant partition (although a relevant partition is possible in principle). In the foregoing explication, the role of RMS is taken over by condition (4).

As I understood his communication, Hempel implicitly rejected the foregoing strategy, not because of a commitment to determinism, but, rather, because of deep doubts about the intelligibility of the notion of objective homogeneity. On the one hand, these doubts were surely justified as long as no one had produced a reasonably clear explication of the concept. As I learned in the process of attempting to provide such an explication, it is no easy task. I shall return to this problem below. On the other hand, there were two strong reasons for thinking that objective homogeneity is an intelligible concept. First, in the trivial case of universal generalizations (affirmative or negative), the concept is clearly applicable. Second, in a vast range of nontrivial cases, we can easily see that the negation of the concept is applicable. We can be quite certain that the class of American teen-agers is not homogeneous with respect to delinquency, and that the class of humans is not homogeneous with respect to survival for an additional 15 years (W. Salmon 1984, 53–55). It would be surprising to find that the negation of a meaningful concept were unintelligible. So one wonders whether a deterministic prejudice is not operating after all. As Peter Railton (1980) most aptly remarked, some people who do not hold the doctrine of determinism are nevertheless held by it.

It seems to me that proponents of either the I-S model or the S-R model must face the reference class problem. Since the S-R model appealed to objective homogeneity, it was clear that an explication of that concept was sorely needed, and I attempted to offer one (1977).[10] The I-S model (in Hempel's epistemically relativized version) also required a homogeneity condition, but, of course, an epistemic one. RMS was introduced for this purpose. It seems to me, however, that neither RMS nor RMS* was adequate even to Hempel's needs, and that examining its inadequacy provides a good start on the analysis of objective homogeneity. For this purpose, we will need to look at RMS in detail.

When Hempel first elaborated his I-S model of explanation, he realized that a crucial problem concerning relevance arises. He sought to deal with it in terms of RMS. The basic idea behind that requirement is that, subject to certain important restrictions, all available relevant knowledge should be brought to bear when I-S explanations are constructed. The problem concerns these restrictions.

The most obvious consideration, as we have seen, is that when we attempt to explain any given fact we already know that it is a fact—for example, when we try to explain John Jones's rapid recovery from his strep infection (Gb), we know that the rapid recovery has occurred. Thus, if we were to bring to bear all of our relevant knowledge, we would have to place him in the class of people who have strep infections (F), who are treated with penicillin (H), and who experience quick recoveries (G). Relative to that class, the probability of his rapid recovery is trivially equal to 1; indeed,

$$P(G|F.H.G) = 1$$

is a theorem of the mathematical calculus of probability. The fact that he belongs to that class clearly has no explanatory value. Let us recall the formulation of RMS. With respect to arguments that satisfy schema (II), let s be the conjunction of the premises, and let k be a statement that is equivalent to the content of our body of knowledge K. RMS says:

If s.k implies that b belongs to a class F_1, and that F_1 is a subclass of F, then s.k must also imply a statement specifying the statistical probability of G in F_1, say

$P(G|F_1) = r_1$

Here, r_1 must equal r unless the probability statement just cited is simply a theorem of mathematical probability theory. (Hempel 1965, 400)

If the restriction stated in the unless-clause were not incorporated in RMS, there could be no I-S explanations.

The unless-clause is not, however, strong enough. Suppose, for instance, that John Jones is a prominent public figure whose illnesses are newsworthy. On the evening of his recovery the local television news program reports that fact. Thus, he belongs to the class of individuals who have strep infections (F) treated by penicillin (H) whose quick recoveries are reported by reliable news media (K). Let us assume, at least for the sake of the example, that the probability that a quick recovery took place (G), given that it is so reported (K), is virtually 1; hence, it is greater than the probability of the quick recovery, given only the administration of penicillin. Since it is not a theorem of the probability calculus that $P(G|F.H.K) = 1$, Hempel's unless-clause does not block the use of that relation in connection with RMS. Thus, RMS would disqualify Hempel's original example, even though the report of Jones's recovery is irrelevant to the explanation of the recovery. Although the news report may be the source of our knowledge that Jones has recovered quickly, it plays no part in telling us why the quick recovery occurred. Though it may answer an *evidence-seeking* why-question, it contributes nothing toward an answer to the *explanation-seeking* why-question. Thus, the unless-clause needs to be strengthened.

In a probability expression of the form "$P(Y|X)$" X is known as the *reference class* and Y as the *attribute class*. The problem RMS is intended to address is that of choosing a statistical law with a suitable—i.e., *maximally specific*—reference class for an I-S explanation of any particular fact. A reference class F is maximally specific with respect to any attribute class G if our available knowledge does not provide any way of making a relevant partition of it—i.e., to pick out a subset of F in which the probability of G is different than it is in F. But we cannot allow complete freedom in the choice of properties in terms of which to make such partitions. Certainly the restriction formulated in Hempel's unless-clause is required, but as our example has just shown, additional restrictions are also needed. Even if a relevant partition of a reference class can be made—but only in terms of information not available until after the occurrence of the fact-to-be-explained—the reference class should not on that account be ruled out as not maximally specific.

A further restriction is needed, as can easily be seen in terms of another example. Suppose we want to know why, in a genetic experiment, a particular plant has a red blossom. In answer, we are told that, in this species, the color of the blossom is determined by a single gene and that red is a recessive characteristic. In addition, both parent plants had red blossoms. Therefore, barring a fairly improbable mutation, the offspring would have red blossoms. This is a legitimate I-S explanation, for given the statistical laws and antecedent conditions, the explanandum is highly probable, but not certain.

This explanation could, however, be challenged on the basis of RMS. Given certain well-known facts about human vision, we know that red is at the opposite end of the visible spectrum from violet. This is *not* a matter of definition.[11] Therefore, it could be said, we know that the blossom is on a plant both of whose parents had red blossoms, and that the blossom on this plant has a color that lies at the opposite end of the visible spectrum from violet. Given this information (which serves to rule out the occurrence of a mutation affecting the color), the probability that the color of the blossom on this plant is red is unity. Therefore, the reference class in the statistical law in the original explanation was not maximally specific and the putative explanation fails to satisfy RMS.

The patent fault in this sort of use of RMS lies in the fact that knowledge that the color of the blossom is at the opposite end of the visible spectrum from violet is tantamount to knowledge that the color is red. In this example—as I intend it to be construed—the only basis of our knowledge that the color is at the opposite end of the visible spectrum from violet is our observation that the blossom is red. We have no way of knowing independently that the color is at the opposite end from violet. Whenever the blossom has a red color, it has a color at the opposite end of the spectrum from violet. Thus, to cite the fact that the color is at the opposite end from violet as an explanatory fact is to explain the red color on the basis of itself.

The best way to block this sort of difficulty is, it seems to me, to insist that the antecedent conditions in an I-S explanation must be temporally antecedent to the explanandum. Hempel's RMS might, therefore, be amended to read as follows:

> If s.k implies that b belongs to a class F_1, and that F_1 is a subclass of F, then s.k must also imply a statement specifying the statistical probability of G in F_1, say
>
> $$P(G|F_1) = r_1$$
>
> Here, r_1 must equal r unless the probability statement just cited is simply a theorem of mathematical probability theory, *or unless b's membership in F_1 cannot be known before its membership in G has been ascertained.* [12]

This version of RMS is obviously relativized to the state of knowledge of the explainer, but Hempel's versions all had that characteristic as well. I am inclined to think that this formulation adequately expresses the intent of Hempel's require-ment.[13] It should be recalled, however, that Hempel explicitly rejected a similar temporal constraint on the 'antecedent' conditions in D-N explanations of particu-lar facts.

To make good on the homogeneity requirements for the S-R model of scientific explanation, as remarked above, I attempted to explicate a fully objective concept of homogeneity (1977).[14] The general idea is that, if determinism is true, no ex-planations of particular facts will be irreducibly statistical; every statistical expla-nation can, in principle, be supplemented with additional information so as to be transformed into a D-N explanation. If indeterminism is true, some explanations will be irreducibly statistical — that is, they will be full-blooded explanations whose statistical character results not merely from limitations of our knowledge, but rather from the fact that there are no additional factors that would make it pos-sible in principle to beef them up, thereby transforming them into D-N explana-tions. Whether determinism or indeterminism is true is a question of fact; I be-lieve that the evidence points toward indeterminism, but that is not crucial for this discussion. Whichever is true, we should try to understand clearly the notion of an irreducibly statistical explanation. Speaking roughly and intuitively, given an irreducibly statistical explanation, it must be impossible in principle to identify anything that happens before the event-to-be-explained that would physically necessitate its occurrence.

These general ideas can be stated more precisely in terms of reference classes. Given an explanation that fits form (II), it will be irreducibly statistical if the term F that occurs in the statistical law premise "$P(G \mid F) = r$" designates a class that cannot be relevantly partitioned, even in principle, with respect to the attribute G. Suppose, that is, that we want to explain why b has the attribute G (at some particular time), and that we have appealed to the fact that b belongs to the class F to do so. If the explanation is to be irreducibly statistical, then prior to the fact that b has the attribute G, b cannot have any attribute in addition to F that has any bearing upon the fact that b has attribute G.

To try to make these ideas more precise, I introduced the concept of a *selection by an associated sequence*. It is, perhaps, best explained in terms of concrete ex-amples. Consider, for instance, all of the basketball games played by the State College team. This class of events can be taken to constitute a reference class F; we are concerned, say, with the attribute G (games won by this team). The mem-bers of F (the games) can be taken in chronological order and designated x_1, $x_2, \ldots$ [15] Suppose just before each game there is a solo rendition of "The Fight Song" over the public-address system; call this class of events A, and let its mem-bers be taken in chronological order and designated $y_1, y_2, \ldots$, where y_i is the rendition of the song immediately preceding game x_i. A is an *associated se-quence* for the reference class F, because there is a one-one correspondence (sig-nified by the subscripts) between the members of F and those of A, and each y_i

in A precedes the corresponding x_i in F. Suppose that sometimes the song is sung by a male vocalist and sometimes by a female. We might wonder whether the sex of the vocalist has any bearing upon the chances of winning. So, designating the renditions by male vocalists B, we make a selection S of members of F on the basis of whether the corresponding member of A belongs to B. S consists of exactly those games that are preceded by a male's rendition of "The Fight Song." S is a *selection by an associated sequence*. Formally, x_i belongs to S iff y_i belongs to B. Someone seriously interested in the team's record might wonder whether the sex of the singer has any bearing on the team's success. If $P(G|F.S)$ differs from $P(G|F)$, the sex of the vocalist is relevant to winning. If so, the probability of winning is not invariant with respect to this particular selection by an associated sequence. Given this situation, we must acknowledge that the reference class F is not homogeneous with respect to winning, for we have just discovered a way to partition it relevantly.

Consider Greeno's example of Albert's theft of a car. The reference class F is the class of American teen-agers; let its members be arranged in some sequential order, $x_1, x_2, \ldots$. The attribute G is commission of a car theft. In this case, we should note, the reference class consists not of events but of enduring objects. Nevertheless, the attribute is an event involving a member of the reference class. (The negation of the attribute would be to reach the age of 20 without having stolen a car) We might wonder whether coming from a 'broken home' is relevant to car theft by a teenager. To construct an associated sequence we can again take the class of American teen-agers, but, so to say, at earlier times in their lives. To make the case particularly clear, we might specify that the breakup of the home had to occur before the youth in question turned 13. What is crucial is that the selective attribute (coming from a broken home) occur earlier than the attribute of interest (theft of a car). Examples of this sort are most easily handled by reformulating them to make both the original reference class and the associated sequence consist of events (the breakup of a home, the theft of a car, becoming 20 without having stolen a car, etc.).

As another example, let F be a sequence of tosses of a given coin and let G be the coin landing heads up. We wonder whether the outcome of the preceding toss has a bearing on the outcome of this toss. We form the associated sequence A consisting of tosses of the same coin, where each toss x_i is associated with its immediate predecessor $y_i = x_{i-1}$. B is the attribute of landing heads up. The selection S thus picks out the tosses immediately following a toss resulting in heads. If this is a standard coin being tossed in the standard way, then this selection S will not furnish a relevant partition in F, for $P(G|F) = P(G|F.S)$. In this example, as in the former ones, it is crucial that the event used to make the selection occur before the event consisting of the occurrence or non-occurrence of the attribute of interest.

The general idea is that a reference class F is objectively homogeneous with respect to a given attribute G iff the probability of G in F is invariant with respect to all selections by associated sequences.[16] To rule out the counterexamples offered above in connection with Hempel's RMS, we have to ensure that the selection be made on the basis of events prior to those we are trying to explain. Thus, the case of the red blossoms, we do not even have an associated sequence, for the class F under consideration (flowers) constitutes the sequence each member of which has or lacks 'both' the color red and the color at the opposite end of the visible spectrum from violet, and each member has or lacks 'both' colors at the same time. In the case of Jones's quick recovery from his strep infection, we set up another sequence — TV news reports — but it does not qualify as an associated sequence because its members come after the corresponding members of the original reference class of strep infections. To qualify as an associated sequence, the members must precede the corresponding members of the original reference class. Consequently, although the report of Jones's recovery is statistically relevant to his quick recovery, it does not signify any lack of homogeneity in the original reference class.

The most difficult aspect of the explication of objective homogeneity is to impose suitable restrictions on the properties B that may be used to effect the selection S. Consider the probability that a person who contracts pneumonia will die as a result. There are several types of pneumonia — bacterial, viral, and fungal — and I presume that the probability of death of a victim varies from one type to another. Moreover, an individual may have more than one type at the same time; according to a recent report the combination of a bacterial and a viral infection is particularly lethal. Now, if we start with the general reference class of cases of pneumonia, we should partition in terms of the various types of infection, and, no doubt, in terms of various other factors such as the age and general physical condition of the patient, and the sort of treatment (e.g., penicillin) given. But we must not partition in terms of fatal vs. nonfatal infections. The reason is that very definition of the term "fatal" depends not just on the occurrence of an infection, but also on the occurrence or non-occurrence of a subsequent event, namely, death. Once you see how to do it in one or two examples, it is easy to cook up predicates whose applicability to a given event depends upon something that occurs (or fails to occur) at some later time.

The suggestion I offered for the restriction to be imposed is that B must determine an *objectively codefined class*. Very roughly, such a class is one consisting of events whose membership in the class can, in principle, be determined at the time the event occurs. More precisely, given that y_i is the member of an associated sequence corresponding to x_i, it must be possible in principle to set up a physical detecting apparatus (connected to a computer if necessary) that can deliver the verdict of whether y_i has the property B before it receives any information about the occurrence of x_i or any other event that happens later than x_i. My hope is that the resulting characterization of objective homogeneity is unrelativized either to particular knowledge situations or to particular languages (see W. Salmon 1984, chap. 3, for details). Whether the explication as given is satisfactory or not, it does not appear impossible in principle to define an adequate concept of objective homogeneity.[17]

The utility of the notion of objective homogeneity is not confined, of course, to the S-R model of explanation. If this concept is used in conjunction with Hempel's I-S model, the problem of essential epistemic relativization, as analyzed by Coffa, evaporates, for we can then say what constitutes a true I-S explanation. The fact that almost all of our actual I-S explanations are incomplete would not pose any fundamental conceptual problem. At the end of the third decade, Hempel (1977) acknowledged that there might be some nontrivial cases of objectively homogeneous reference classes. By implication he thereby admitted that the concept of objective homogeneity is not unintelligible.

3.3 Coffa's Dispositional Theory of Inductive Explanation

After articulating his devastating critique of the doctrine of essential epistemic relativization, Coffa offered a different solution to the problem of ambiguity of inductive explanation. According to Coffa, a great deal of the trouble Hempel encountered was a result of his implicit identification of the problem of ambiguity with the reference class problem. As we have seen, the reference class problem arises out of the frequency interpretation of probability when we try to apply probabilities to single cases. The situation could have been significantly improved, Coffa suggests, if Hempel had stuck with the propensity interpretation, which can be construed as a single case physical probability concept. In his doctoral dissertation Coffa (1973, chap. IV) argues that an appeal to the propensity interpretation of probability enables us to develop a theory of inductive explanation that is a straightforward generalization of deductive-nomological explanation, and that avoids both epistemic relativization and the reference class problem. This ingenious approach has, unfortunately, received no attention, for it was never extracted from the dissertation for publication elsewhere.

Coffa begins with a critical examination of the D-N model. Consider a deductive-nomological explanation of the simplest form:

(1) $(x)[Fx \supset Gx]$
     Fa
     $\overline{}$
     Ga

for example – one cited by Carnap and Hempel – in answer to the question, "Why did this iron rod expand?" the explanation that it was heated and that "whenever a body is heated, it expands." This latter statement is, of course, the law required in the explanans — in this case, the law of thermal expansion.[18] One might suppose that this law should read

$$\Delta L = k \cdot L_o \cdot \Delta T$$

where L_o is the original length of the rod, k the coefficient of thermal expansion of the substance in question, and ΔT is the amount by which its temperature was raised. The foregoing explanation could presumably be construed as

(2) $\Delta L = k \cdot L_o \cdot \Delta T$
        Rod r increased its temperature by ΔT.
        _____

        Rod r increased its length by $k \cdot L_o \cdot \Delta T$.

This explanation is not acceptable as it stands, Coffa claims, for the statement of the law of thermal expansion is defective; a correct formulation requires what he calls an "extremal clause." It is not true, in general, that bodies expand when heated. The iron rod will not expand when heated if a sufficient compressing force is present. The correct statement is "*In the absence of other changes relevant to length*, an increase in temperature ΔT is physically sufficient for an increase in length $k \cdot L_o \cdot \Delta T$" (1973, 210). The italicized phrase is the extremal clause (EC). Only when the extremal clause is included do we have a bona fide law, and that is what is needed as the first premise of our explanation. Explanation (2) must be amended as follows:

(3)   If EC then $\Delta L = k \cdot L_o \cdot \Delta T$
        EC
        Rod r increased its temperature by ΔT.
        _____

        Rod r increased its length by $k \cdot L_o \cdot \Delta T$.

Because the first premise is a law of nature, change of temperature is *nomically relevant* to change of length. EC says that there is no other factor in the situation that is nomically relevant to the change of length. Assuming that EC is true, and that the initial condition stated in the third premise is true, Coffa claims, we now have a bona fide D-N explanation of the increase in length of the rod. Coffa's major thesis about D-N explanation is that *every law involves an extremal clause*, and consequently, *every D-N explanation must contain, as one of its premises, an assertion that the extremal clause holds on the occasion in question.*

An obvious objection can be raised at this point. If the extremal clause has to be incorporated into the law-statement, does that not make the law vacuous? Hempel and Oppenheim, we should recall, explicitly demanded that the law-statement in a D-N explanation have empirical content. The question is, does the extremal clause in the first premise rob the law of thermal expansion of its empirical content?

Coffa meets this objection squarely, citing an argument that had been given by Hempel to the effect that any law containing an extremal (or ceteris paribus) clause can always be protected from refutation—in all cases in which the law seems to fail—by maintaining that, because the extremal clause is not fulfilled, the law is actually correct. Coffa points out that any law in conditional form, even without the extremal clause, can be protected against refutation—in all cases in which it *seems* to fail—by maintaining that its antecedent was unfulfilled. Thus, any law having a conditional structure can be rendered vacuous by the same technique. He then observes that the pertinent question is not whether laws—with or without extremal clauses—*can* be treated in this way, but whether they *must* be so treated.

The answer, he argues, is negative. Consider, for example, Newton's second law of motion. It states that any body of mass m, upon which the net force is F, experiences an acceleration F/m. When this law makes reference to the *net force* acting on a given body, it is implicitly making a statement about all forces, known or unknown, detected or undetected. Nevertheless, we use Newton's laws of motion nonvacuously for predictive and explanatory purposes. In other words, we sometimes explicitly identify a set of forces acting upon a body and assert (corrigibly, of course) that there are no other forces. This latter assertion is an extremal clause.

Suppose Coffa is right in his claim that all D-N explanations involve extremal clauses. How does this fact apply to nondeductive explanations? It applies quite directly. We can think of universal laws as describing physical dispositions. Massive bodies have a disposition to accelerate when forces are applied; metal wires have a disposition to carry electrical current when a potential difference exists between the two ends. In addition to universal dispositions, the world also seems to contain probabilistic dispositions of varying strengths. A fair coin flipped by an unbiased device has a disposition of strength ½ to land heads up; a tritium atom has a disposition of strength ¾ to decay within a period of 25 years.

If we grant that there are both universal and probabilistic dispositions, then it is natural to think of both deductive and inductive explanation in terms of the following schema (1973, 273):

$$(4) \quad P(Fa \,|\, Ga) = r$$
$$Ga$$
$$EC(a,F,G)$$
$$\overline{}$$
$$Fa$$

where "EC(a,F,G)" is an extremal clause stating that nothing nomically relevant to a's having property F, other than the fact that a has the property G, is present in this explanatory situation. In the case of inductive explanation, the extremal clause does the job of Hempel's requirement of maximal specificity and of my requirement of objective homogeneity. It makes inductive explanation just as objective and epistemically unrelativized as is deductive explanation. Before accepting Coffa's model I would, of course, insist upon a requirement banning inclusion of irrelevant factors, but that requirement is easily imposed.

Coffa's theory of scientific explanation adheres strongly to the ontic conception. Proponents of this conception can speak in either of two ways about the relationship between explanations and the world. First, one can say that explanations exist in the world. The explanation of some fact is whatever produced it or brought it about. The explanans consists of certain particular facts and lawful relationships. The explanandum is also some fact. This manner of speaking will sound strange to philosophers who have been strongly influenced by the thesis that explanations are arguments, or by the deeply linguistic approaches that regard explanations as speech acts. In nonphilosophical contexts, however, it seems entirely appropriate to say such things as that the gravitational attraction of the moon explains the tides, or the drop in temperature explains the bursting of the pipes. The gravitational attraction and the drop in temperature are out there in the physical world; they are neither linguistic entities (sentences) nor abstract entities (propositions). Second, the advocate of the ontic conception can say that an explanation is something—consisting of sentences or propositions—that reports such facts. It seems to me that either way of putting the ontic conception is acceptable; one can properly say either that the explanandum-fact is explained by the explanans-facts or that the explanans-statements explain the explanandum-statement. Coffa frequently adopts the first of these two manners of speaking; nevertheless, he does identify explanations as arguments, and that would seem to commit him, on those occasions, to the second.

The fact that Coffa often identifies the explanans with what produced or brought about the explanandum strongly suggests that his model is a causal model. He also adopts a covering law conception. The premise "$P(Fa \mid Ga) = r$" is a law-statement, so there must be a nomic connection between G and F, whether the law is universal ($r = 1$) or statistical ($r < 1$). The law-premise contains a symbol standing for probability, and, as I have mentioned, Coffa construes the probabilities as propensities. Coffa maintains that explanations should appeal only to nomically relevant factors—excluding those that are merely statistically relevant. He suggests, for example, that there may be no nomic relationship whatsoever between the sneeze of a Chinese mandarin at the time of one's birth and one's being a millionaire.

It would be natural, I should think, to restrict the laws admissible in such explanations to causal laws—where we admit probabilistic as well as deterministic causes. Propensities, I would suggest, are best understood as some sort of probabilistic causes. However, Coffa does not impose that restriction. According to Hempel's covering law approach, we can explain why a gas has a certain temperature by specifying the volume, pressure, and number of moles. Such an explanation may tell us nothing at all about how the temperature was brought about; consequently, for Coffa, I believe, it should not count as an explanation.

Several comments must be made regarding Coffa's proposed theory of inductive explanation.[19] In the first place, I have great sympathy with the idea that only nomic relevance should have a place in statistical explanation. In articulating the S-R model, I always maintained that it is a covering law model, all of the probability relations appearing in any such explanation being statistical laws. It is, of course, a difficult matter to distinguish statistical laws from true accidental statistical generalizations — at least as difficult as making the corresponding distinction in the case of universal generalizations, which is, as we have seen, an unsolved problem. But switching from statistical generalizations to propensities does not help in making this crucial distinction.

Second, the propensity interpretation does not escape the problem of the single case; indeed, it faces a problem that is the precise counterpart of the reference class problem. Basic to the single case propensity interpretation is the concept of a *chance set-up*. A chance set-up is a device that produces a set of alternative outcomes, each with a determinate probability. When a fair die is rolled from a dice cup in the standard manner, for example, there are six possible outcomes, and this chance set-up has a propensity of ⅙ to produce each of them on a given roll. A chance set-up can be operated repeatedly to produce a sequence of results. However, when we describe the chance set-up, we have to include relevant features and omit various irrelevant ones to determine what constitutes an additional trial on the same chance set-up. In the example of rolling the die, it is important that the die be physically symmetrical and that a standard cup be used, but it is irrelevant whether the device is operated during the day or during the night, or whether the roll is by a left-handed or right-handed player.

Consider a famous example — due to Laplace. Suppose that a coin is about to be flipped. It is known to be asymmetrical; it has a bias either for heads or for tails, but we do not know which. Laplace claimed, on the basis of his notorious *principle of indifference*, that, because of our ignorance, the probability of heads on this toss is ½. A modern propensity theorist would maintain that this chance set-up has a propensity other than ½; it is either greater than ½ or less than ½, but we do not know what it is. I would insist that either of the foregoing claims might be correct, depending upon how we specify the chance set-up. Suppose this coin has been produced by a machine that operates in the following way. Asymmetrical disks are fed into the machine in a random fashion, and the machine stamps the insignia on these disks. Half of the coins get heads on the favored side and half get tails on the favored side. A coin has been chosen randomly from the pile of coins produced by this machine, and it is about to be flipped. If we describe the operation of the chance set-up as picking a coin and flipping it in the standard manner, we do not know whether the propensity is equal to ½ or unequal to ½. If the continued operation of the chance set-up is to flip the same coin again in the same manner, then the propensity is different from ½. If the continued operation of the chance set-up is to pick another coin from the output of the machine and flip it just once, then the propensity of heads is ½. Even for the propensity interpretation, it is necessary to deal with a problem not essentially different from the reference class problem. I suggest, then, that switching to the propensity interpretation of probability does nothing to ease Hempel's problem of ambiguity.[20]

Third, as Paul Humphreys pointed out to me in conversation, the "propensity interpretation of probability" is not an admissible interpretation of the probability calculus at all (see Humphreys 1985). In fact, Coffa had already hinted at this point in his dissertation (1973, 272–73, note 32).[21] Bayes's theorem is one of the theorems of the mathematical calculus, and it provides a way of calculating 'inverse probabilities' from 'direct probabilities.' Imagine a factory that produces corkscrews. It has two machines, one old and one new, each of which makes a certain number per day. The output of each machine contains a certain percentage of defective corkscrews. Without undue strain, we can speak of the relative propensities of the two machines to produce corkscrews (each produces a certain proportion of the entire output of the factory), and of their propensities to produce defective corkscrews. If numerical values are given, we can calculate the propensity of this factory to produce defective corkscrews. So far, so good. Now, suppose an inspector picks one corkscrew from the day's output and finds it defective. Using Bayes's theorem we can calculate the *probability* that the defective corkscrew was produced by the new machine, but it would hardly be reasonable to speak of the *propensity* of that corkscrew to have been produced by the new machine. Propensities make sense as direct probabilities (because we think of them as probabilistic causes, I suspect), but not as inverse probabilities (because the causal direction is wrong).

Fourth, the concept of probabilistic causality is fraught with difficulties. When Coffa completed his dissertation in 1973 there were three serious theories of probabilistic causality in the literature – one by Hans Reichenbach (1956, chap. IV), another by I. J. Good (1960–61), and the third by Patrick Suppes (1970). These attempts at explication are far from simple. It is not just a matter of picking out a suitable subset of probabilities and identifying them as probabilistic causes. I think that each of the theories suffers from fundamental defects. In (1980) I published a critical study of all three and offered some suggestions of my own. We shall return to this problem in the fourth decade.

Finally, I am not completely convinced that laws of nature always involve extremal clauses, but I shall not attempt to argue that issue here. It is interesting to note that in 1982 a workshop on The Limitations of Deductivism was held at the University of Pittsburgh. Both Hempel and Coffa were participants. Hempel presented a paper in which he claimed that the use of scientific laws and theories, in conjunction with initial and boundary conditions, to make predictions cannot be construed as a strictly deductive affair. The basic reason is the need for provisos – qualifications that strongly resemble Coffa's extremal clauses. The proceedings of this workshop are published in (Grünbaum and Salmon 1988). Whether or not extremal clauses are always involved when universal laws are used, they certainly can be invoked in connection with explanations that employ statistical laws. It is evident that, in such cases, establishing the extremal clause is exactly the same as establishing the objective homogeneity of a reference class. Coffa's appeal to propensities and extremal clauses does not escape the immense difficulties associated with objective homogeneity.

At the midpoint of the third decade Coffa offered profound critiques of Hempel's I-S model and my S-R model. As an alternative he produced an extremely suggestive model of causal explanation that included both deterministic and indeterministic causes. The basic unity of his treatment of deductive and inductive explanation is an appealing aspect of his approach. It certainly contains strong anticipations of developments that occurred in the fourth decade. Nevertheless, enormous problems remained. Chief among them, I believe, is the need for a detailed treatment of the causal concepts to which we must appeal if we are to have any satisfactory causal model of scientific explanation.

Another of the many partisans of propensities in the third decade is James H. Fetzer. Along with Coffa, he deserves mention because of the central place he accords that concept in the theory of scientific explanation. Beginning in 1971, he published a series of papers dealing with the so-called propensity interpretation of probability and its bearing on problems of scientific explanation (Fetzer 1971, 1974, 1974a, 1975, 1976, 1977). However, because the mature version of his work on these issues is contained in his 1971 book, *Scientific Knowledge*, we shall deal with his views in the fourth decade. The fourth decade also includes his collaborative work with Donald Nute on probabilistic causality (Fetzer and Nute 1979).

3.4 Explanation and Evidence

One of the first philosophers to attack the Hempel-Oppenheim thesis of the symmetry of explanation and prediction was Nicholas Rescher (1958). At the very beginning of the second decade, he argued that, because of certain fundamental physical asymmetries in nature, many retrodictive inferences can be made with great certainty, whereas predictions are often much more hazardous. Because of a strong commitment to the inferential conception of scientific explanation he maintained that explanations are arguments. Because of a commitment to a rather robust high-probability requirement he claimed that there is a basic difference between explanatory arguments and predictive arguments – namely, that in explanatory arguments the premises must support the conclusion with certainty or near certainty, whereas in predictive arguments we often have to settle for much lower degrees of probability. For purposes of prediction we need only know that a given occurrence is more probable than not—or, in many cases, merely more probable than any other member of a reasonable set of alternatives. Rescher does not exclude the possibility that there are predictions—such as an astronomical prediction of a solar eclipse—that can be made with high degrees of certainty. His claim is simply that many predictive inferences do not have this character. He stresses his view that this kind of asymmetry is a fact of nature rather than a truth of logic.

Given Hempel and Oppenheim's failure to impose temporal constraints on scientific explanations, they leave open the possibility that the so-called *antecedent conditions* appearing in the explanans may actually occur after the explanandum. Recall our discussion of standard counterexample CE-1 (the eclipse) in §2.3 above. Nevertheless, it seems clear that Hempel-Oppenheim, in enunciating their symmetry thesis, did *not* mean to assert that prediction and retrodiction are equally reliable. They meant rather to say that an argument from *temporally prior* antecedent conditions and laws to a subsequent fact may serve in some contexts as a prediction and in others as an explanation. This is surely the way other philosophers, such as Scheffler, construed the symmetry thesis, and Hempel continued to construe it in that way in his "Aspects" essay. Similarly, I take it, as long as he leaves open the possibility of explanation in terms of subsequent conditions, he must say that there could be cases in which an argument from subsequent 'antecedent' conditions and laws would serve as a retrodiction in some circumstances and as an explanation in others. If explanations may sometimes be given in terms of subsequent conditions, then the symmetry condition would need to be formulated in terms not of prediction in the narrow temporal sense, but, rather, in terms of inference from the observed to the unobserved, whether that inference be predictive or retrodictive.

In an article published in the middle of the second decade, Rescher (1963) further elaborates his views on the relationships among prediction, retrodiction, and explanation. To expound his position, he introduces the notion of a *discrete state system*—i.e., a physical system that may assume different states out of a finite or infinite collection of discrete states. He shows how to construct systems in which both prediction and retrodiction are possible, those in which prediction is possible but retrodiction is impossible, those in which prediction is impossible but retrodiction is possible, and those in which both are impossible. The fact that systems of all of these types can be physically realized—by digital computers, for example—undermines the symmetry thesis as he construes it.[22]

Rescher's 1963 article constitutes one main kernel of his book *Scientific Explanation* (1970), which was published near the beginning of the third decade.[23] Another main kernel comes from an article on the nature of evidence (1958). Because Rescher is strongly committed to the main tenet of the received view—that explanations are arguments—he construes the explanatory relation as an evidentiary relation. Given this view, he continues to maintain that explanations in terms of subsequent conditions are just as legitimate as explanations in terms of temporally antecedent conditions, provided a strong argument can be supplied.

In his 1953 book, we recall, Braithwaite raised certain issues about teleological explanation; he asked, in particular, whether explanations in terms of fulfillment of functions or realization of goals can be considered legitimate. He decided not to preclude them even though such explanations appeal to subsequent conditions. Rescher also raises "the much agitated issue of 'teleology' vs 'mechanism' in the theory of explanation," but he acknowledges that "[f]or purposes of the present discussion, the concept of 'teleology' (and its opposite) will be construed in a somewhat artificial chronologized sense" (Rescher 1970, 66). A teleological explanation in this sense is simply one in which the "antecedent conditions" (in the Hempel-Oppenheim meaning of the term) occur later than the event to be explained. Such explanations need not have anything to do with purposes, ends, or functions. Rescher's entire concern is to argue that putative explanations should not be rejected *solely* on the basis of the temporal relation between the explanatory facts and the fact-to-be-explained. In this regard, he supports Braithwaite's claim, but he does not try to give a theory of explanations that appeal to purposes, ends, or functions. He does not address the difficulties—raised by Hempel— regarding the logical structure of functional explanations.

One of Rescher's central concerns revolves around the nature of evidence. His treatment of this topic is considerably more sophisticated than Hempel's. According to the received view, as usually construed, the premises of a D-N explanation constitute conclusive evidence for its explanandum, and the premises of an I-S explanation provide strong evidence for its explanandum. The latter claim amounts to Hempel's high-probability requirement, and Rescher rejects it. He recognizes the need for some sort of relevance requirement in connection with the concept of evidence, and he embodies it in his theory of evidence. In developing this theory he introduces the notions of evidential presumption, supporting evidence, and confirming evidence (1970, 76–77). Briefly, presumptive evidence renders a given hypothesis more probable than its negation; this is a very weak high-probability concept (1970, 78–79). Supporting evidence is a relevance concept; supporting evidence renders the hypothesis more probable in the presence of that evidence than it was in its absence (1970, 80–83). The concept of confirming evidence incorporates these two notions:

> The statement p is *confirming evidence* for q if (1) p is a presumptive factor for q, and (2) p is supporting evidence for q. Thus confirming evidence must at once render its hypothesis more likely than before *and* more likely than not. This two-pronged concept of confirming evidence perhaps most closely approximates to the idea represented in the common usage of "evidence." (1970, 84)

It is remarkable that—to the best of my knowledge—no one (including Rescher and me) noticed any connection between what he was doing in his evidential approach to probabilistic explanation and what those of us who were concerned with statistical relevance were up to. It may have been due, in part at least, to the fact that his investigation was embedded in a theory of evidence, which is naturally taken to be part of confirmation theory, while the S-R model was developed within the theory of scientific explanation. Such compartmentalized thinking seems, in retrospect, quite shocking.

Rescher reflects another aspect of the received view by maintaining the covering law conception of scientific explanation. As he says, "Scientific explanations are invariably subsumption arguments that cite facts to establish an explanandum as a special case within the scope of lawful generalizations" (1970, 9). This leads him to undertake an extended analysis of the concept of lawfulness (1970, 97-121). He takes it as a point of departure—a point acknowledged by most authors who deal with this issue—that lawful generalizations have counterfactual and modal import. Whereas accidental generalizations state only what happens under actual circumstances, laws state what would happen under circumstances that are merely possible, i.e., under counterfactual circumstances. He then argues that empirical evidence can support only the general claim about actual cases, and, consequently, the assertion of lawfulness regarding a generalization goes beyond the empirical evidence. From this consideration he does not conclude that we have super-empirical *knowledge* of lawfulness; instead, he argues, we *impute* lawfulness on the basis of epistemic considerations. Lawfulness is not something that we discover about generalizations; it is something we supply (1970, 107). Lawfulness is mind-dependent (1970, 113-14; see also Rescher 1969). It reflects the use to which we want to put a generalization in organizing and systematizing our knowledge of the world (1970, 107). Skyrms (1980) and van Fraassen (1980), exactly a decade later, offer similarly pragmatic accounts of lawfulness.

In his discussion of the role of laws in scientific explanation, and the modal import of laws, Rescher asserts,

> A recourse to laws is indispensable for scientific explanation, because this feature of nomic necessity makes it possible for scientific explanations to achieve their task of showing not just *what* is the case, but *why* it is the case. This is achieved by deploying laws to narrow the range of possible alternatives so as to show that the fact to be explained "had to" be as it is, in an appropriate sense of this term. (1970, 13-14)

Inasmuch as this passage comes at the conclusion of a discussion of the role of universal *and probabilistic* laws in explanation, it appears that Rescher wants to attribute some kind of modal import to probabilistic laws. He does not say what it is. As we shall see, D. H. Mellor, writing shortly before the close of the third decade, also advocates a modal interpretation of probabilistic explanation (1976).

I shall explain in greater detail below why I find the notion of *degrees of necessity* quite mystifying.

In spite of his strong adherence to the thesis that explanations are subsumptive arguments, Rescher does suggest, from time to time, that scientific explanation involves an exhibition of the mechanisms by which nature works. He makes this point, for example, in connection with the explanatory power of Darwin's theory (1970, 14–15). This discussion suggests that subsumption under nomic regularities is not really all there is to scientific explanation after all; we need to appeal to the underlying mechanisms as well.

In his discussion of discrete state systems, it will be recalled, Rescher provides a variety of types in which prediction and/or retrodiction, and hence, on his view, explanation is impossible. He makes the profound observation that, in the most fundamental sense, we may nevertheless understand completely what goes on in these systems—we know how they work. The issue arises in the context of statistical explanation:

> The consideration of stochastic systems forces us to the realization that scientific understanding can be present despite an impotence to explain (predict, etc.) *even in principle* certain particular occurrences. As regards explanation, this leads us to recognize the fundamentality of description, being able to deploy laws so as to describe the modes of functioning of natural systems. . . .
>
> The root task of science ought thus to be thought of as a fundamentally *descriptive* one: the search for the laws that delineate the functioning of natural processes. . . . It is thus our grasp of natural laws, be they universal or probabilistic—not our capacity to explain, predict, and so on—that appears to be the basic thing in scientific understanding. For it is undeniable that a knowledge of the pertinent laws goes a long way with endowing us with all that we can possibly ask for in the way of an understanding of "the way in which things work"—even in those cases where the specific desiderata of explanation, prediction, etc., are, in the very nature of things, beyond our reach. (1970, 133–34)

Although it is not Rescher's intention to deny the possibility of probabilistic explanation, the foregoing passage does suggest one line of argument that has been adopted by such current deductivists as John Watkins (1984) and Kitcher (Kitcher & Salmon 1989). They maintain that we can provide deductive explanations of lawful statistical regularities by appealing to fundamental statistical laws, but that we cannot give deductive explanations of particular occurrences. Nevertheless, they can maintain, that does not matter, for our knowledge of the lawful statistical regularities provides understanding of how the world works. We have no need, they argue, for any nondeductive type of explanation.

As I shall explain in discussing the opening of the fourth decade, there are at least three main conceptions of scientific explanation: the *epistemic*, the *modal*, and the *ontic*. Although they have been present since the time of Aristotle, they had not been clearly distinguished. After the advent of models of probabilistic or statistical explanation it became crucial to distinguish them. In the foregoing discussion of Rescher's account we have noticed all three.

3.5 Explanations of Laws

Early in our story we called attention to Hempel and Oppenheim's "notorious footnote 33," in which they pointed to a basic difficulty in connection with explanations of laws. The problem is to distinguish the cases in which we genuinely explain a lawful regularity by deductive subsumption under broader regularities from such nonexplanatory moves as deducing a law from a conjunction of itself with another totally unrelated law. Because they saw no solution to this problem, they refrained from even attempting to provide a D-N account of explanations of laws. To the best of my knowledge this problem was not seriously attacked until more than a quarter of a century later—when Michael Friedman (1974) finally took it on. The delay is especially surprising given the fact that, as Friedman notes, explanations of regularities are more usual than explanations of particular facts in the physical sciences.[24]

The basic theme of Friedman's paper is that science explains various phenomena (where phenomena are general regularities) through unification. Citing the kinetic theory of gases as an example, he says,

> This theory explains phenomena involving the behavior of gases, such as the fact that gases approximately obey the Boyle-Charles law, by reference to the behavior of molecules of which gases are composed. For example, we can deduce that any collection of molecules of the sort that gases are, which obeys the laws of mechanics will also approximately obey the Boyle-Charles law. How does this make us understand the behavior of gases? I submit that if this were all the kinetic theory did we would have added nothing to our understanding. We would have simply replaced one brute fact with another. But this is not all the kinetic theory does—it also permits us to derive other phenomena involving the behavior of gases, such as the fact that they obey Graham's law of diffusion and (within certain limits) that they have the specific heat capacities that they do have, from the laws of mechanics. The kinetic theory effects a significant *unification* in what we have to accept. Where we once had three independent brute facts—that gases approximately obey the Boyle-Charles law, that they obey Graham's law, and that they have the specific heat capacities they do have—we now have only one—that molecules obey the laws of mechanics. Furthermore, the kinetic theory also allows us to integrate the behavior of gases with other phenomena, such as the motions of the planets and of falling bodies near the earth. This is because the laws of mechanics also per-

mit us to derive both the fact that planets obey Kepler's laws and the fact that falling bodies obey Galileo's laws. From the fact that *all* bodies obey the laws of mechanics it follows that the planets behave as they do, falling bodies behave as they do, and gases behave as they do. Once again, we have reduced a multiplicity of unexplained, independent phenomena to one. I claim that this is the crucial property of scientific theories we are looking for; this is the essence of scientific explanation—science increases our understanding of the world by reducing the total number of independent phenomena that we have to accept as ultimate or given. A world with fewer independent phenomena is, other things equal, more comprehensible than one with more. (1974, 14–15)

It is evident that, if Friedman is successfully to carry out his program, it will be crucial to provide a method for counting independently acceptable phenomena. By "phenomena" in this context we shall understand general uniformities; they can be represented by lawlike sentences. Since *acceptance* is a central notion, we presume that "at any given time there is a set K of *accepted* lawlike sentences, a set of laws accepted by the scientific community. Furthermore, . . . the set K is deductively closed . . . K contains all lawlike consequences of members of K " (1974, 15). Recalling the Hempel-Oppenheim footnote, we immediately ask whether we can reduce the number of phenomena by, for example, deducing the Boyle-Charles law and Graham's law from the conjunction of these two laws, thereby reducing the number from two to one. Clearly, we do not want this kind of move to count as a reduction. Friedman suggests that an appeal to the notion of *independently acceptable lawlike sentences* will enable us to deal with that problem.

Unfortunately, Friedman does not furnish any precise characterization of the of what it means for one sentence to be acceptable independently of another.

Presumably, it means something like: there are sufficient grounds for accepting one which are not also sufficient grounds for accepting the other. If this is correct, the notion of independent acceptability satisfies the following conditions:

(1) If $S \vdash Q$, then S is not acceptable independently of Q.
(2) If S is acceptable independently of P and $Q \vdash P$, then S is acceptable independently of Q.

(assuming that sufficient grounds for accepting S are also sufficient for accepting any consequence of S). (1974, 16–17)

The rationale for condition (1) is evident. By virtue of the entailment, grounds for accepting S simply *are* grounds for accepting Q. However, since S is, in many cases, a stronger statement than Q, grounds for accepting Q may well be insufficient grounds for accepting S.

The rationale for condition (2) is not quite as easy to see. Perhaps it can be made clearer by looking at a trivial reformulation:

(2') Given $Q \vdash P$, if S is not acceptable independently of Q then S is not acceptable independently of P.

Suppose that S were not acceptable independently of Q. Then grounds for accepting S would be grounds for accepting Q. Because Q entails P, however, grounds for accepting Q would ipso facto be grounds for accepting P. Hence, grounds for accepting S would be grounds for P, and S would not be acceptable independently of P. To look at the condition a bit more formally, let "$E \models X$" mean "E constitutes sufficient grounds for accepting X." Then, given $Q \vdash P$, we have

$$E \models Q \rightarrow E \models P$$
$$E \models S \rightarrow E \models Q$$

from which it follows that

$$E \models S \rightarrow E \models P.$$

This obviously establishes Friedman's second condition in the form (2').

An indispensable requirement for the implementation of this program is to develop a method for counting phenomena (as represented as lawlike sentences). To get on with the job, Friedman introduces the concept of a *partition* of a sentence S. A partition of S is a set Z of sentences such that Z is logically equivalent to S and each S′ in Z is acceptable independently of S. Note that the members of the partition do not have to be mutually exclusive. Using this definition, he defines a *K-atomic sentence S* as one that has no partition (in K). This means that there is no set of sentences $\{S_1, S_2\}$ in K such that S_1 and S_2 are acceptable independently of S and S_1 & S_2 is logically equivalent to S (1974, 17). If, for example, we took the conjunction of the Boyle-Charles law (S_1) and Graham's law (S_2), it would not be K-atomic, for $\{S_1, S_2\}$ is a partition of it. The question is whether there can be any K-atomic sentences.

Consider any sentence in K of the form

$$(x)[Fx \supset Gx] \tag{1}$$

which is equivalent to

$$(x)[Fx \cdot (Hx \vee \sim Hx) \supset Gx] \tag{2}$$

and to

$$\{(x)[Fx \cdot Hx \supset Gx], (x)[Fx \cdot \sim Hx \supset Gx]\} \tag{3}$$

If (3) is a partition of (1), then (1) is not K-atomic. But (3) is a partition of (1) unless at least one member of (3) is not a member of K or at least one member of (3) is not acceptable independently of (1). Because (1) entails each member of (3), each member of (3) must belong to K unless it fails to be lawlike.

In private correspondence I posed the foregoing problem to Philip Kitcher (who has written incisively about Friedman's proposal (1976)) as well as to Friedman. Kitcher's response went something like this: If "H" is a reasonable predicate (signifying something like a natural kind), then " ~ H" is apt not to be a reasonable predicate and the second member of (3) may fail to be lawlike.

I would be inclined to reply to Kitcher as follows: One could argue (along the lines of Hempel and Oppenheim as discussed above) that if the first member of (3) is lawlike, so will be the second. If the first member has unlimited scope so has the second; if the first makes no reference to any particular entity, neither does the second; if the first contains only purely qualitative predicates so does the second; etc.[25] Leaving all that aside, however, it is plausible to suppose that a similar partition could be based upon an exclusive and exhaustive disjunction H_1, H_2, . . . , such that, for each i,

$$(x)[Fx \cdot H_ix \supset Gx] \tag{4}$$

is a lawlike sentence. Since each sentence of the form (4) is a logical consequence of (1), each is a member of K. In that case, (4) provides a partition of (1).[26]

Here is Friedman's reply to my original problem:

> The problem you raise . . . is just the *kind* of problem I had in mind in requiring that a partition of a sentence S consist of sentences that are acceptable independently of S. In your case, it would seem in general that our ground for accepting the two conjuncts is just that they follow from 'All F are G.' If, on the other hand, we have grounds for accepting the two conjuncts independently — by testing for the two conjuncts directly, say — then it would seem that 'All F are G' is in no way an explanation of the two, but just a summary of what we already know.

I am not satisfied that Friedman's answer is adequate. Consider the generalization (in K), "All humans are mortal." Is it K-atomic? K contains "All men (i.e., human males) are mortal" and "All women (i.e., human females) are mortal." Suppose, as is plausible, that ancient Greek men were male chauvinist pigs. They noticed that each man seemed to die, so by simple induction they concluded that all men are mortal. At this point they could accept "All men are mortal" independently of "All humans are mortal." Some male then made the further generalization that all humans are mortal and it was accepted by the scientific community. Now, I take it, "All men are mortal" is still acceptable independently of "All humans are mortal," but "All women are mortal" is not, for the only basis for asserting this latter statement is that it follows from "All humans are mortal." Thus, {"All men are mortal"; "All women are mortal"} does not constitute a partition of "All humans are mortal," since the second member is not acceptable independently of it. Perhaps this general sentence, "All humans are mortal," is K-atomic.[27]

Suppose that, without changing the membership of the class K in any way, one or more of these MCP Greeks starts noticing that women also die sooner or later, thus accumulating inductive evidence sufficient to accept "All women are mortal" on that basis. Now it is possible to accept that sentence independently of "All humans are mortal," with the result that this latter sentence is not K-atomic, since it now has a partition.

Judging from remarks earlier in his essay, Friedman (1974, 14)—as well as Hempel and Oppenheim—is eager to develop an objective account of scientific explanation. In view of this consideration Friedman would, I should think, regard the order in which the different sorts of evidence for a sentence are actually acquired as irrelevant to the K-atomicity of that sentence. The foregoing example seems to violate this kind of objectivity.

Consider a more serious (quasi-historical) example. Newton's law of universal gravitation states that between any pair of material objects there exists a certain sort of attractive force. Is this a K-atomic sentence? By the time his *Principia* was published, there was considerable evidence to support this law in the domain of pairs of material bodies in which each member of the pair is a large astronomical body (earth, moon, planet, sun, etc.) and in the domain in which one member of the pair is a large astronomical body (earth) and the other a much smaller object (apple, stone, etc.). However, the only evidence for the claim that it also holds in the domain of pairs of which both members are small is the fact that it follows from the general law. Hence, the foregoing tripartite division of the set of all pairs of material bodies {both large, one large and one small, both small} could not have been used at the end of the seventeenth century to provide a K-partition of Newton's law of universal gravitation. However, after the Cavendish torsion balance experiment was performed in the following century there was independent evidence for Newton's law in the third domain of pairs. We could apparently then use as a K-partition: {"Given any two extremely massive material bodies, there exists an attractive force between them . . . "; "Given any two bodies one of which is extremely massive and the other is not, there exists a force . . . "; "Given any two material bodies, neither of which is extremely massive, there exists a force . . . "}. The result is that Newton's law is demonstrably not K-atomic.[28]

If a sentence that was K-atomic (with respect to a given class K of lawlike sentences) becomes non-K-atomic (with respect to the same class), that fact has a profound bearing upon its explanatory import. As we shall see, Friedman offered two definitions of scientific explanation. As we shall also see, Kitcher proved that, under Friedman's first definition of explanation, only K-atomic sentences explain.[29] Consequently, we find a strong presumption that the performance of the Cavendish experiment (along with others, perhaps) robs Newton's law of gravitation of its explanatory import. My personal intuition is that the explanatory power of Newton's law was enhanced as a result of such experiments.

Be that as it may, I must return to my original question, namely, on Friedman's characterization, can there be any K-atomic sentences. I have suggested that among simple generalizations of the form "All As are Bs" there cannot be any. I have not tried to extend the argument to lawlike generalizations of all sorts, but I should think that, if there are any K-atomic sentences, some of them would have this simple form. *If there are no K-atomic sentences, Friedman's program of characterizing scientific explanation in terms of the reduction of the number of independently accepted laws cannot get off the ground.*

It strikes me that Friedman's resolution of the problem is excessively pragmatic for his purposes.[30] If *there exists* evidence sufficient for the acceptance of a generalization having form (1) above, then *there exists* evidence for the independent acceptance of both of the generalizations in the partition (3) above. Which pieces of evidence are found first and which lawlike generalizations are formulated first should be irrelevant if our aim is to formulate an *objective* theory of scientific explanation.

While the difficulties in characterizing K-atomic statements seem serious, I do not mean to suggest that they are insuperable. Let us therefore try to see what Friedman has accomplished if the concept of K-atomic sentence turns out to be viable. The answer, I believe, is that he has solved the problem originally stated in the famous Hempel-Oppenheim footnote. Without going into all of the technical details of his explication, I shall attempt to sketch the development.

Given the notion of a K-atomic sentence, the *K-partition* of a set of sentences Y can be defined as a set Z of K-atomic sentences such that Y $\leftrightarrow$ Z. In general there may be more than one K-partition of a given set Y. The *K-cardinality* of a class Y can be taken as the number of members of the smallest K-partition of Y (i.e., the greatest lower bound of the cardinality of K-partitions of Y). A sentence S *reduces* a class Y if the K-cardinality of the union of {S} with Y is smaller than the K-cardinality of Y. Friedman then proceeds to define explanation:

> How can we define *explanation* in terms of these ideas? If S is a candidate for explaining some S' in K, we want to know whether S permits a reduction in the number of independent sentences. I think that the relevant set we want S to reduce is the set of *independently acceptable* consequences of S ($con_K(S)$). For instance, Newton's laws are a good candidate for explaining Boyle's law, say, because Newton's laws reduce the set of their independently acceptable consequences – the set containing Boyle's law, Graham's law, etc. On the other hand, the conjunction of Boyle's law and Graham's law is not a good candidate, since it does not reduce the set of its independently acceptable consequences. This suggests the following definition of explanation between laws:

> (D1) S_1 explains S_2 iff $S_2 \in con_K(S_1)$ and S_1 reduces $con_K(S_1)$.

(1974, 17)

Friedman is not satisfied with this definition, for he finds it too restrictive.

Actually this definition seems to me to be too strong; for if S_1 explains S_2 and S_3 is some independently acceptable law, then S_1 & S_3 will not explain S_2 — since S_1 & S_3 will not reduce $con_K(S_1$ & $S_3)$. This seems undesirable — why should the conjunction of a completely irrelevant law to a good explanation destroy its explanatory power? So I will weaken (D1) to

(D1') S_1 explains S_2 iff there exists a partition Z of S_1 and an $S_i \in Z$ such that $S_2 \in con_K(S_i)$ and S_i reduces $con_K(S_i)$.

(1974, 17)

As Kitcher points out (1976, 211), (D1') is not just a liberalization of (D1). Having demonstrated that, according to (D1), *only* K-atomic sentences can explain, he points to the obvious fact that, according to (D1'), *no* K-atomic sentences can explain. (D1) and (D1') are simply different sorts of definition. Kitcher suggests that Friedman's intent might best be captured by a definition whose definiendum consisted of the disjunction of the right-hand sides of (D1) and (D1'). My response to this situation is different. Having already canvassed the many difficulties that arise if we allow explanations to contain irrelevancies, I believe (D1') should simply be rejected.

If we continue to assume that the problem of K-atomic sentences can be solved or circumvented, then Friedman's theory provides a way for the received view to fill the upper right-hand sector of Table 1 — the D-N explanation of universal laws. It is not clear whether Friedman's approach could be extended to handle the lower right-hand sector — what Hempel called D-S explanation. Since I have maintained above that D-S explanation should be regarded as a species of D-N explanation, the hope would be that Friedman's approach could handle this sector as well. As the end of the third decade approached, the received view had theories of D-N explanation of particular facts, I-S explanation of particular facts, D-N explanation of universal generalizations, and hopes for a theory of D-N explanation of statistical generalizations. Although each of these models suffered from severe difficulties, the received view was at least approaching a complete articulation.

Assuming that Friedman's account of D-N explanation of laws could handle the problem of K-atomic sentences, it appears to be just what Hempel and Oppenheim were looking for in answer to the problem stated in their "notorious footnote 33." However, as Kitcher showed in his critique of Friedman's paper, that theory of explanation of laws is not what we should have wanted, for it is subject to some severe shortcomings.

Kitcher claims that Friedman's definition (D1) is vulnerable to two sorts of counterexamples.

Counterexamples of the first type occur when we have independently acceptable laws which (intuitively) belong to the same theory and which can be put together in genuine explanations. The explanantia that result are not K-atomic and hence fail to meet the necessary condition derived from Friedman's theory.

Consider, for example, the usual derivation of the law of adiabatic expansion of an ideal gas, given in books on classical thermodynamics. The explanans here is the conjunction of the Boyle-Charles law and the first law of thermodynamics. These laws are acceptable on the basis of quite independent tests, so their conjunction is not K-atomic. However, the derivation of the law of adiabatic expansion from the conjunction is, intuitively, a genuine explanation. (1976, 209–10)

Kitcher's second type of counterexample concerns the explanation of complex phenomena where, typically, many different laws or theories are brought to bear. He cites, as one instance, the explanation of why lightning flashes are followed by thunderclaps. "The explanation utilizes laws of electricity, thermodynamics, and acoustics, which are independently acceptable."[31]

Kitcher goes on to show—convincingly, in my opinion—that an appeal to Friedman's other definition (D1') does not help in dealing with the foregoing sorts of counterexamples. He concludes, therefore, that Friedman's account of explanation by unification is unsatisfactory. Nevertheless, Kitcher is firmly committed to the general thesis (with which I am strongly inclined to agree) that unification is a fundamental goal of explanation—that unification yields genuine scientific understanding. In the fifth decade Kitcher articulates his version of the unification theory in Kitcher and Salmon (1989).

3.6 Are Explanations Arguments?

Early in the third decade Jeffrey (1969) argued persuasively that, in many cases at least, statistical explanations are not arguments. His article as well as his thesis was incorporated into the publication in which the S-R model was first set forth in detail (W. Salmon et al. 1971). By the end of the third decade (1977c) I had labeled the claim that scientific explanations are arguments a *third dogma of empiricism* and urged its wholesale rejection.[32] Since the doctrine that scientific explanations are arguments—deductive or inductive—is absolutely central to the received view, my intent was to undermine the orthodox position as deeply as possible. In my zeal to rebut the claim that all explanations are arguments, I argued that no explanations are arguments. This view now seems too extreme; as it seems to me now, some are and some are not.[33] The challenge took the form of the following three questions I hoped would prove embarrassing to devotees of the received view. What I failed to notice until recently is that, while these questions *are* acutely embarrassing with regard to explanations of particular facts, they are innocuous for explanations of laws.

QUESTION 1. Why are irrelevancies harmless to arguments but fatal to explanations?

In deductive logic, irrelevant premises are pointless, but they have no effect whatever on the validity of the argument. Even in the Anderson-Belnap relevance logic, p & q ⊢ p is a valid schema. Consider the following variation on a time-honored argument:

All men are mortal.
Socrates is a man.
Xantippe is a woman.
_____

Socrates is mortal.

It is strange, somewhat inelegant, and possibly, mildly amusing, but it is obviously valid. In contrast, as we have seen, the appearance of an irrelevancy in a D-N explanation can be disastrous.

> The rooster who explains the rising of the sun on the basis of his regular crowing is guilty of more than a minor logical inelegancy. So also is the person who explains the dissolving of a piece of sugar by citing the fact that the liquid in which it dissolved is *holy* water. So also is the man who explains his failure to become pregnant by noting that he has faithfully consumed birth control pills. (1977c, 150)

The situation is no different when we consider inductive relations. There is a *requirement of total evidence* that requires inductive inferences to contain all relevant premises, but it does not ensure against the inclusion of irrelevant premises. When a detective attempting to establish the identity of a murderer comes across evidence that may or may not be relevant—i.e., about whose relevance he or she cannot as yet be sure—it behooves him or her to include it. If it is irrelevant it can do no harm; an irrelevant premise simply does not change the probability of the conclusion. If it turns out to be relevant it may be very helpful.

Where inductive explanation is concerned, again the situation is entirely different. If psychotherapy is irrelevant to the remission of neurotic symptoms, it should not be invoked to explain a patient's psychological improvement. If massive doses of vitamin C are irrelevant to the rapidity of recovery from a common cold, such medication has no legitimate place in the explanation of a quick recovery.

Inference, whether inductive or deductive, demands a requirement of total evidence—a requirement that *all* relevant evidence be mentioned in the premises. This requirement, which has substantive importance for inductive inferences, is automatically satisfied for deductive inferences. Explanation, in contrast, seems to demand a further requirement—namely, that *only* considerations relevant to the explanandum be contained in the explanans. This, it seems to me, constitutes a deep difference between explanations and arguments. (1977c, 151)

The second query comes in two forms that are so closely related as to constitute one question:

QUESTION 2:   Can events whose probabilities are low be explained?

QUESTION 2′:   Is genuine scientific explanation possible if indeterminism is true?

The close relationship between these two questions rests upon a principle of symmetry that we have discussed above. Suppose, in a genuinely indeterministic situation, there are two possible outcomes, one highly probable, the other quite improbable. Jeffrey argued, and I fully agree, that in such circumstances, when we understand that both are results of the same stochastic process, we understand the improbable outcome (when it occurs) just as well as we understand the probable outcome (when it occurs). But if probabilistic explanations are inductive arguments, an undesirable asymmetry arises. The explanans confers upon the probable explanandum a high inductive probability, so the explanation is a strong inductive argument. In the case of the improbable outcome, the explanans provides a strong inductive argument for the *non-occurrence* of the event to be explained. The fact that some events have low probabilities seems to cast doubt upon all statistical explanations of individual events if explanations are viewed as being, in essence, arguments.

QUESTION 3:   Why should requirements of temporal asymmetry be imposed upon explanations (while arguments are not subject to the same constraints)?

Among the earliest counterexamples brought against the D-N model of Hempel and Oppenheim was (CE-1), the case of the eclipse. We noted that the occurrence of a given eclipse could be deductively inferred, on the basis of the same laws, either from conditions prior to the eclipse or from conditions subsequent to the eclipse. Only one of these inferences—that from prior conditions—might possibly count as an explanation. A similar point can be made in terms of (CE-2), the flagpole example. Given the angle of elevation of the sun in the sky, the length of the shadow can be inferred from the height of the flagpole, or the height of the flagpole can be inferred from the length of the shadow. The height of the flagpole explains the length of the shadow because the interaction between the sunlight and the flagpole occurs before the interaction between the sunlight and the ground.[34] The length of the shadow does not explain the height of the flagpole because the temporal relation is wrong.

As Rescher had already pointed out, it would be quite mistaken to suppose that inference is in general temporally symmetrical. In a wide variety of cases it is possible to make quite reliable inferences from subsequent facts to earlier occurrences. It is often difficult to predict whether precipitation will occur at a particular place and time; it is easy to look at the newspaper to determine whether rain or snow occurred there yesterday. Dendrochronologists can infer with considerable precision and reliability from the examination of tree rings the relative annual rainfall in certain areas for thousands of years in the past. No one can predict the relative annual rainfall with any reliability for even a decade into the future. The situation is obvious. We can have records – human or natural – of things that have happened in the past. We do not have records of the future, or any comparable resource for making predictions.

We see, then, that explanation involves a temporal asymmetry, and that inference in many cases also involves a temporal asymmetry. What is striking is that the two asymmetries are counterdirected. This would be very strange indeed if explanations are, in essence, arguments. The reason for the contrast is easy to fathom. We explain effects in terms of their causes, but in many cases we infer causes from their effects. Recognizing the crucial role of causal considerations in relation to the third question, I declared that the time had come to put "cause" back into "because" (1977c, 160).

Proponents of the received view had various possible answers to these questions. In response to the first question they could simply impose some sort of requirement that would block irrelevancies. No one (to the best of my knowledge) ever claimed that *all* logically correct arguments with true premises are explanations. As we saw, Hempel and Oppenheim worked hard to characterize the kinds of valid deductive arguments that can qualify as explanations. Advocates of the received view could simply admit that a further requirement, excluding irrelevancies, should be added to the D-N model.[35] Similarly, there are various ways of dealing with irrelevancies in the theory of I-S explanations. A simple way, as I remarked above, would be to amend the requirement of maximal specificity to make it into the requirement of *the maximal class of maximal specificity.*[36]

Several sorts of response to the second question are available to champions of the received view. In the first place, one could simply deny the symmetry principle offered by Jeffrey et al., insisting that we can explain highly probable occurrences, but not those with smaller probabilities. D. H. Mellor (1976) has defended this position. I must reiterate my strong intuition that this response perpetuates a highly arbitrary feature of Hempel's I-S model.

In the second place, one could deny that there are such things as probabilistic explanations of particular facts, insisting that all legitimate explanations are of the D-N variety (perhaps including the D-S type). For reasons that appear to be somewhat different from one another, Wolfgang Stegmüller (1973) and G. H. von Wright (1971) take this tack. I worry about whether this position reflects an anachronistic hankering after determinism, but I shall discuss this thesis in detail in connection with developments occurring in the fourth decade, especially §4.9.

It may turn out, in the end, to be the most promising way to avoid a host of problems associated with statistical explanation.

In the third place, Hempel (private correspondence) suggested the possibility of getting around the problem posed in the second question by reconstruing the concept of an inductive argument. During the late 1940s Carnap developed a detailed and rigorous system of inductive logic that was published in his monumental work *Logical Foundations of Probability* (1950, 2nd ed., 1962). Hempel was strongly influenced by Carnap's theory. One of the peculiarities of Carnap's system is that it has no place for *rules of acceptance* in inductive logic. In deductive logic we think of an argument as a group of statements consisting of a set of premises and a conclusion. If the argument is valid, then, if we accept the premises as true (or well-founded) we should be willing to accept the conclusion as well. It is natural to think of inductive arguments in an analogous way. Under this conception, an inductive argument would be a set of statements consisting of premises and a conclusion. If the argument has a correct inductive form, then, if we accept the premises as true (or well-founded) *and if they comprise all available evidence relevant to the conclusion* we should be willing to accept the conclusion as well. The schemas Hempel has offered in his various discussions of D-N and I-S explanation strongly suggest just such an analogy. In Carnap's inductive logic no such analogy holds. His denial of rules of acceptance for inductive logic means that *there are no inductive arguments* in this straightforward sense. Inductive logic furnishes only degree of confirmation statements of the form "$c(h,e) = r$" where the hypothesis h may be any statement, evidence e any consistent statement, and r any real number between 0 and 1 inclusive. Inductive logic tells us how to compute r. If r has been calculated correctly, it is the degree of confirmation of hypothesis h on evidence e. If e comprises all available evidence relevant to h, then r may be interpreted as a fair betting quotient; that is, $r:1 - r$ constitute fair odds for a bet on the truth of h. But no matter how close r is to 1, we are not permitted to extract the hypothesis h from the degree of confirmation statement and assert it separately as a statement we accept. There are no arguments of the form

$c(h,e) = r$

e

e contains all relevant evidence

$$\overline{\qquad\qquad\qquad\qquad\qquad} \quad [r]$$

h

in Carnap's inductive logic.[37]

If one adopts this sort of inductive logic, it is not altogether clear how there could be any such thing as I-S explanation. For one thing, we could not literally construe I-S explanations as inductive arguments. For another, we would have to abandon the covering law conception, for we could never have any *accepted* law statements to include in the explanans of *any* explanation, D-N or I-S. But one benefit would accrue. There would no longer be any need to impose the high-probability requirement. In Carnap's inductive logic high probabilities do not enjoy special virtues. If inductive logic yields only betting quotients, determining fair odds for wagering, what counts is having a *correct* value, not a *high* value. It is not necessarily unreasonable to bet on improbable outcomes, provided the odds are right. Thus, by adopting this special Carnapian construal of "inductive argument" Hempel can drop the high-probability requirement and avoid the asymmetry between explanations of high-probability and low-probability occurrences. But the price, with respect to the received view of scientific explanation, is rather large (see W. Salmon 1977a).

Question 3, which ties in directly with the role of causality in scientific explanation, raises the most profound issues, I think. The notion that scientific explanation and causality are intimately connected is not new; it goes back at least to Aristotle. As we have seen, some of the earliest objections to Hempel's approach, voiced by Scriven, criticized the received view for its virtual neglect of causality. A little later, when the S-R model was being developed, it seemed clear to me that any adequate theory of scientific explanation must accord to causality a central role. The fact that we were dealing with statistical explanation, in contexts that might in some cases be indeterministic, did not preclude causality. After reading the nearly complete manuscript of Reichenbach's *The Direction of Time* (1956) in the summer of 1952 I was convinced that probabilistic causality is a viable notion. I hoped that such probabilistic concepts as his screening off and conjunctive forks would be helpful in explicating probabilistic causality, but none of it was really worked out at that time (W. Salmon et al. 1971, 76, 81).

Two major factors forced me to focus more carefully upon causality. The first was a brief article by Hugh Lehman (1972) that nicely pointed out the limitations, with respect to scientific explanation, of statistical relevance relations alone. Showing how distinct causal factors could give rise to identical statistical relevance relations, his argument strongly suggested the necessity of appealing to causal mechanisms.

The second factor involved theoretical explanation. Everyone recognizes the obvious fact that among our most impressive scientific explanations are many that appeal to theories that apparently make reference to such unobservable entities as molecules, atoms, and subatomic particles. There is no obvious way in which statistical relevance relations by themselves can account for this feature of scientific explanation. Greeno made a stab at doing something along that line, but his attempt just did not work (see Greeno 1971 and Salmon 1971). I decided to have a try at analyzing theoretical explanation in a paper (whose ambitious title was chosen before the paper was written) for the Conference on Explanation at the University of Bristol in 1973 (W. Salmon 1975). In attempting to cope with that task I found myself deeply involved with causal connections and causal mechanisms. The paper fell far short of its intended goal. But it reinforced in my mind the need to come to grips in a serious way with causal concepts. Theoretical explanation had to wait for the fourth decade (W. Salmon 1978).

3.7 The Challenge of Causality

With causality demanding attention, it was impossible to ignore the issues Hume had raised regarding the nature of causal relations. If we are to succeed in putting "cause" back into "because," we need to understand causal connections. In a great many cases, in everyday life and in science, we take one event to be a cause of another even though they are not spatio-temporally contiguous. In such cases we expect to be able to find some physical connection between them. We flick a switch on the wall by the door and a light on the ceiling at the center of the room lights up. They are connected by an electrical circuit, and a current flows when the switch is closed. Ammonia is spilled in the laundry room and its odor is detected in the hallway. Ammonia molecules diffuse through the air from one place to another. As long as we steer clear of quantum mechanical phenomena, the notion of continuous physical connections between causes and effects seems assured. Hume, in his characterization of cause-effect relations, mentioned priority, contiguity, and constant conjunction. He failed to find any additional feature of the situation that constitutes a causal connection. If the cause and effect are not contiguous, the natural move is to interpolate intermediate causes so as to form a causal chain with contiguous links. But every time a new link is found and inserted, the Humean question arises all over again: what is the *connection* between these intermediate causes and their effects? It seems that the Humean question never gets answered.

Oddly enough, the special theory of relativity offers a useful clue. According to that theory, light travels in a vacuum at a constant speed c, and no signal can be transmitted at any greater speed. Material particles can be accelerated to speeds approaching c, but they can never attain that speed or any greater speed.[38] It is often said, roughly, that nothing can travel faster than light, but that statement must be treated with great care. What does the "thing" in "nothing" refer to? Certainly it covers such entities as trains, baseballs, electrons, spaceships, photons, radio waves, and sound waves. But what about shadows? Are they *things* in the relevant sense of the term? The answer is negative; shadows can travel at arbitrarily high speeds. Their speed is not limited by the speed of light. Are there other examples? Yes. Consider the spot of light that is cast by a moving flashlight on the wall of a dark room. Like the shadow, this spot can travel at an arbitrarily high speed. There are many other examples. They all share the characteristic of being incapable of transmitting messages. If signals could be sent faster than light Einstein's famous *principle of relativity of simultaneity* would be undermined.

The best way to explain the difference illustrated by the examples in the preceding paragraph is, I think, to make a distinction between genuine *causal processes* and *pseudo-processes*. I shall use the unqualified term "process" in a general way to cover both the causal and pseudo varieties, but I shall not attempt to define it; examples will have to suffice. The general idea is quite simple. An event is something that happens in a fairly restricted region of spacetime. The context determines how large or small that region may be. A process is something that, in the context, has greater temporal duration than an event. From the standpoint of cosmology a supernova explosion—such as occurred in our cosmic neighborhood in the last year of the fourth decade—could be considered an event; the travel of a photon or neutrino from the explosion to earth (requiring thousands of years) would be a process. In ordinary affairs a chance meeting with a friend in a supermarket would normally be considered an event; the entire shopping trip might qualify as a process. In microphysics a collision of a photon with an electron would constitute an event; an electron orbiting an atomic nucleus would qualify as a process. A process, whether causal or pseudo, will exhibit some sort of uniformity or continuity. Something that, in one context, would be considered a single process (such as running a mile) would often be considered a complex combination of many processes from another standpoint (e.g., that of a physiologist).[39]

We must now try to distinguish between genuine causal processes and pseudo-processes. Material particles in motion (or even at rest[40]), radio waves, and photons are examples of genuine causal processes. All of them can be used to send signals; they can transmit information. Consider an ordinary piece of paper. You can write a message on it and send it through the mail to another person at another location. Or consider radio waves. They can be used to send information from a radio station to a receiver in someone's home, or from a command center to a space vehicle traveling to distant planets. Shadows and moving spots of light cast on walls are *not* causal processes; they are incapable of transmitting information. By means of another example I shall try to show why.

A lighthouse on a promontory of land sends a white beam of light that can be seen by ships at a great distance. This beam can be modified or marked. By fitting the light in the tower with a red lens, the beam becomes red. By inserting the red glass in the beam *at one place*, we can change the beam of light from white to red *from that point on*. It does not matter where in the beam the red filter is placed; wherever it is inserted it changes the beam from white to red from that point on. A genuine *causal process* is one that can transmit a mark; if the process is modified at one stage the modification persists beyond that point *without any additional intervention*.[41]

The lighthouse beacon rotates, so that it can be seen by ships in all directions. On a cloudy night, the rotating beacon casts a moving white spot on the clouds. This spot moves in a regular way; it is some sort of process, but it is *not* a causal process. If somehow the spot on the cloud is changed to red at one place—by someone in a balloon holding a piece of red cellophane right at the surface of the cloud, or by encountering a red balloon at the cloud's surface – it will suffer a modification or mark at that place, but the modification or mark will not persist beyond that place without further interventions. A 'process' that can be marked at one place, but without having any such modification persist beyond the point at which the mark is made, cannot *transmit* marks.[42] Such 'processes' are *pseudo-processes*.[43]

The moving spot of light is like the shadow in that it can travel at an arbitrarily high speed. This is easy to see. If the beacon rotates quite rapidly, its spot of light will traverse the clouds at high speed. Imagine now that the beacon continues to rotate at the same rate, but that the clouds are moved farther away. The farther away the clouds, the greater the linear velocity of the spot will be – since it has to traverse a circle of greater circumference in the same amount of time. A dramatic example is furnished by the pulsar in the Crab nebula, which rotates 30 times per second, sending out a continuous beam of electromagnetic radiation. It is about 6500 light years away. Its 'spot' of radiation sweeps by us at approximately 4×10^{13} times the speed of light.

There are many familiar examples of pseudo-processes. The scanning pattern on a cathode ray tube – a television receiver, a computer terminal, or an oscilloscope – furnishes one. The screen is made of a substance that scintillates when electrons strike it. An electron gun shoots electrons at it. As a result of the impinging electrons a spot of light moves rapidly back and forth across the screen. Another is the action as viewed at the cinema. Genuine causal processes – such as a horse running across a plain – are depicted, but what is seen on the screen is a pseudo-process created by light from a projector passing through a film and falling upon the screen. The light traveling from the projector to the screen is a whole sequence of causal processes – we might say that the light passing through each separate frame is a causal process. The motion of the horse on the screen is a pseudo-process. By momentarily shining a red light on the image of the horse on the screen it is possible to put a red spot on the horse, but if the red light does not continue to impinge on the image of the horse, the image of the horse will not continue to exhibit the red spot.[44]

The distinction between causal processes and pseudo-processes is important to our discussion of causality because processes that are *capable of transmitting marks* are processes that can also transmit information. Such processes transmit energy; they also transmit *causal influence*. They provide the *causal connections* among events that happen at different times and places in the universe.

Transmission is the key concept here. We can say, intuitively, that a process has some sort of structure – for instance, a light beam contains light of certain frequencies. A filter in its path changes that structure by removing light of some of these frequencies. Other modifications can be made by means of polarizers or mirrors. If the mark persists beyond the point in the process at which it is imposed, the structure has been modified in a lasting way *without further interventions*. If the change in structure is transmitted, the structure itself is being transmitted. Processes that can transmit marks *actually do transmit their own structure*.

Still, we should remind ourselves, the concept of transmission is a causal concept, and it must confront the Humean question. The terminology of mark transmission strongly suggests that the earlier parts of the marked process have the power to produce or reproduce the marked structure. In what does this power consist? How does the mark *get from* an earlier place in the process to a later one? Putting the question in this way suggests a strong analogy. More than 2500 years ago Zeno of Elea posed the famous paradox of the flying arrow. How does the arrow *get from* point A to point B in its trajectory? Zeno seems to have suggested that, since the arrow simply occupies the intervening positions, it must always be at rest. If it is always at rest it can never move; motion implies a contradiction.

Early in the twentieth century Bertrand Russell offered what is, I believe, a completely satisfactory resolution of the arrow paradox in terms of the so-called *at-at theory of motion*. Motion, Russell observed, is nothing more than a functional relationship between points of space and moments of time. To move *is* simply to occupy different positions in space at different moments of time. Considering any single point of space and single instant of time, an object simply is at that point of space at that instant of time. There is no distinction between being in motion and being at rest as long as we consider only that one point and that one instant.[45] To get from point A to point B consists merely of being *at* the intervening points of space *at* the corresponding moments of time. There is no further question of how the moving arrow gets from one point to another.

There was, it seemed to me, a similar answer to the question about mark transmission. A mark that is imposed at point A in a process is *transmitted* to point B in that same process if, *without additional interventions*, the mark is present at each intervening stage in the process. The difference between a process transmitting a mark and not transmitting that mark is that in the latter case the mark is present at the later stages in the process only if additional interventions occur reimposing that mark (W. Salmon 1977).

For a process to qualify as causal, it is not necessary that it actually be transmitting a mark; it is sufficient that it be capable of transmitting marks. Moreover, a causal process may not transmit every mark that is imposed upon it; a light wave that encounters another light wave will be modified at the point of intersection, but the modification will not persist beyond the point of intersection. That does not matter. The fact that it will transmit some kinds of marks is sufficient to qualify a light wave as a causal process.

With the *at-at theory of mark transmission* and the distinction between causal processes and pseudo-processes we have, it seemed to me, a satisfactory answer to Hume's basic question about the nature of causal connections. Although quite a bit more is required to develop a full-blown explication of causal explanation, this appeared to me to be the essential key (for further details see W. Salmon 1984, chap. 5).

3.8 Teleological and Functional Explanation

As we saw at the outset, a deep concern about teleology was a major impetus to the philosophical study of scientific explanation that has developed in the twentieth century. In roughly the first decade of our chronicle several of the principal early contributors to that literature—Braithwaite, Hempel, Nagel, and Scheffler—explicitly addressed the problem of teleological and/or functional explanation. This concern was instigated in large part by vitalism in biology.

Until the beginning of the third decade philosophy of biology was not pursued very actively by philosophers of science, but, as this decade opened, the situation began to change. Morton Beckner's classic, *The Biological Way of Thought*, appeared in 1968, and during the early 1970s the field began to expand dramatically with the work of such people as David Hull, Michael Ruse, and William Wimsatt. It has continued to flourish and grow right down to the present.[46] The problem of teleological/functional explanation is, of course, central to biology. Toward the end of the third decade, Larry Wright (1976) provided an account that strikes me as fundamentally correct, and it applies to many other areas in addition to biology. Eleven years later, at the very end of the fourth decade, John Bigelow and Robert Pargetter (1987) offer a theory of functional explanations that may be an improvement over Wright's, but, as we shall see, it is not *fundamentally* such a very different account.

Wright takes as his point of departure the view that teleological explanations occur frequently in science, and he maintains—contra Hempel and Beckner—that many of them are sound. He regards teleological explanations as causal in a straightforward sense—in a sense that does *not* require causes that come after their effects. At the same time, nevertheless, he considers them to be future-oriented; teleology by definition involves goal-seeking behavior. To capture both of these insights, he offers an account in terms of what he calls a *consequence-etiology*.

The basic idea of a consequence-etiology is as ingenious as it is simple. A particular bit of behavior B occurs because B has been *causally efficacious in the past* in achieving a goal G. A cat, hunting for prey, is clearly engaged in goal-directed behavior. It stalks in a typically catlike way because such stalking has resulted in the procurement of food. It is not caused by the future catching of this particular mouse, for (among other problems) in this instance he may not succeed in catching his prey. But such behavior has worked often enough to have conferred an evolutionary advantage on the members of the species. Similarly, a human being searching for water in an arid region may look for cottonwood trees, since past experience shows that they grow only when there is water in the vicinity. Such intentional behavior relies on a conviction that this action B is a suitable means of achieving G. The behavior of a homing torpedo also occurs because in the past such actions B have resulted in hitting the target (goal G). Roughly speaking, the causal efficacy of B in bringing about G in the past is, itself, an indispensable part of the cause of the occurrence of B on this occasion. It is a consequence-etiology because the consequences of doing B are a crucial part of the etiology of the doing of B.

In the case of deliberate action on the part of a human agent, it may be that direct experience of the consequence G of action B is what makes the agent do B on this occasion. Perhaps he or she has spent a great deal of time in arid regions and has found that locating cottonwood trees has been a successful strategy in finding water. Perhaps the agent has not had any such direct experience but has heard or read about this strategy. Perhaps this person has neither had direct experience, nor learned of it through reports, but, rather, inferred that it would work. Whatever the actual situation, the agent does B because he or she has reason to believe, before performing that act on this particular occasion, that B is an appropriate method for getting G.

A homing torpedo is an artifact created by a human designer to behave in a certain way. If it is being used in warfare there is a reasonable presumption (we hope) that this type of device was tested and found to perform in the desired way. The fact that similar behavior by similar objects in the past has resulted in the reaching of the goal etiologically causes the present behavior of this particular torpedo.

Examples from evolutionary biology are, I think, the clearest illustrations of Wright's model. Certain kinds of behavior, such as the stalking by a cat, become typical behavior for a species of animals because it confers an advantage with respect to the goal of survival or of reproduction. As Wright explicitly notes, the behavior may have arisen in either of two ways—deliberate creation by a supernatural agency or as a result of natural selection. The analysis of the concept of *teleology* does not tell us which of the two possible etiologies is the actual one. It only ensures that there is an etiology; empirical science must furnish the answer to which it is.

According to Wright's analysis, human conscious intentional behavior, the behavior of human artifacts that have been designed to do some particular job, and behavior that has resulted from natural selection all qualify as teleological. He offers the following schematic statement (T):

S does B for the sake of G iff:

(i) B tends to bring about G.

(ii) B occurs because (i.e., is brought about by the fact that) it tends to bring about G. (1976, 39)

The term "tends" signifies the fact that B need not *always* succeed in bringing about G. Indeed, it may be that B never brings about G, as long as it has a disposition to do so under suitable conditions. Wright offers this gloss:

> teleological behavior is behavior with a consequence-etiology: and behavior with a consequence-etiology is behavior that occurs because it brings about, is the type of thing that brings about, tends to bring about, is required to bring about, or is in some other way appropriate for bringing about some specific goal. (1976, 38–39)

Although they are obviously closely related, Wright does not identify teleological explanations with functional ascriptions or explanations. The basic reason is that only behavior is teleological – goal-directed. But in many cases something fulfills a function just by "being there." A piece of newspaper, stuffed under a door, fulfills the function of preventing a draft just because of its location. A vinyl cover lying on a playing field fulfills the function of keeping the ground dry. The function of an entity is distinguished from all sorts of other things that might result from its being where it is by the fact that its location has a consequence-etiology. The vinyl cover is where it is because it keeps the field dry. It may also catch pools of water in which children play, but that is an accidental result. Wright offers the following schematic formulation (F):

The function of X is Z iff:

 (i) Z is a consequence (result) of X's being there,
 (ii) X is there because it does (results in) Z. (1976, 81)

He comments,

> The ascription of a function simply *is* the answer to a 'Why?' question, and one with etiological force. . . . So, not only do functional explanations provide consequence-etiologies, just like explanations in terms of goals, but the simple attribution of a function ipso facto *provides* that explanation (ascription-explanation), just as does the simple attribution of a goal to behavior. This displays the enormous parallel that obtains between goals and functions, and possibly accounts for the tendency in the philosophical literature to run them together. (1976, 81)

Functional ascriptions are, of course, frequently made in physiology and in evolutionary biology. The function of the long neck of the giraffe is to enable it to reach food other animals cannot. The function of the stripes on the tiger is to provide camouflage. The function of the heart is to pump blood. The function of chlorophyll in green plants is to enable them to produce starch by photosynthesis. In the course of evolution certain attributes arise in the first instance as a result of a mutation. In such a first occurrence the attribute does not have a function, for it has no consequence-etiology at that stage. The function is established only when the recurrence or persistence of the attribute occurs because of a consequence of its previous presence.

Wright asserts explicitly that his account of teleological explanation and functional ascription is not inconsistent with a completely mechanistic explanation of goal-directed behavior (1976, 57–72). The fact that we can give a completely mechanistic account of the motion of a homing torpedo does not render it non-teleological. The fact that we can give a completely mechanical account of the operation of a governor on a steam engine does not deprive it of the function of regulating the engine's speed. He maintains, moreover—contra Charles Taylor (1964) and others—that, even if it should turn out to be possible to provide a completely physico-chemical account of the behavior of plants and animals (including the conscious behavior of human animals), that would not eliminate its teleological character. Teleological and functional explanations are on as firm ground as any explanations in any science.

The fourth-decade proposal of Bigelow and Pargetter (1987) is a "propensity theory" that they claim to be more forward-directed, temporally, than Wright's consequence-etiology. It relies heavily upon the tendency or disposition of a given item to do whatever is its function. As we have seen (schema F above), Wright refers explicitly to what the item *does*, not to what it tends to do, though on the same page, immediately after the schema, he allows for cases in which an item has a function even if it never successfully performs it. What Bigelow and Pargetter fail to note is that Wright's analysis of function supplements his analysis of teleological behavior (schema T above); like the majority of philosophers to whom Wright referred, they seem to run teleology and function together. Whether they are correct in so doing is an issue I shall not argue. The point is that Wright's analysis of teleological behavior is as much a propensity theory, and is just as forward-looking, as is the Bigelow-Pargetter theory of functions. And Wright has emphasized, as we have seen, the extremely close connection between teleology and function. I am inclined to think that, on Wright's account, every case of a function presupposes some bit of teleological behavior, but he might well disagree.[47] In any case, as Bigelow and Pargetter seem in some places to concede, their theory is not very different from Wright's. The main difference may be a disagreement over whether an item has a function the first time it occurs, or whether, as Wright maintains, it can properly be said to have a function only subsequently. If this is the *main difference*, it is a matter of fine-tuning, or as Larry used to say regarding engines of racing cars, "demon tweeking."

While I agree wholeheartedly with Wright's complete causal grounding of teleological and functional explanations, I do have one major philosophical disagreement with him. Following in the footsteps of his teacher, Michael Scriven, he declines to offer any characterization of causality itself; instead, he maintains that we have perfectly objective ways of recognizing causal relations, and that is enough. In my view, this is simply evasion of a fundamental philosophical problem. As we shall see in discussing the fourth decade, I have devoted considerable effort to the explication of causality, and I do not believe the unanalyzed notion is nearly as unproblematic as Scriven and Wright have claimed.

In 1977 – the final year of the third decade – in his John Dewey Lectures, Nagel again addressed the issues of teleological and functional explanation (1977). In the course of his discussion he offers criticisms of Hempel (Nagel 1977, 305–9) and Wright (ibid., 296–301)[48] as well as others.[49] His main response to Hempel consists in a challenge to the notion of functional equivalents. According to Nagel, if one specifies with sufficient precision the nature of the organism or system within which an item has a function, it is in many cases – including Hempel's example of circulation of blood as the function of the heart in a normal human being – necessary for the fulfillment of the function. Nagel is not moved by other possible devices, such as Jarvik artificial hearts, inasmuch as they do not circulate blood in a *normal* humans. He is equally unmoved by fictitious possibilities that are not realized in nature. He therefore concludes that the normal functioning *n* in Hempel's schema entails the presence of item *i*. Nagel does acknowledge the fact that in many actual cases more than one item fulfill the same function; for example, a normal human has two ears for hearing, either one of which will do the job fairly adequately. In cases of this sort, the fulfilling of the function entails the existence of a nonempty set which may contain more than one member.

The conclusion Nagel draws from this argument is that functional explanations fit the deductive pattern. From the fact that the organism is in a normal state we can deduce the existence of the item *i* (or set of such items). If one accepts Nagel's argument about functional equivalents he overcomes the main problem that led Hempel to conclude that so-called 'functional explanations' are not real explanations at all.[50] Nagel considers that result desirable, for he maintains that many sciences, including biology in particular, do provide legitimate teleological or functional explanations. In accepting this conclusion, however, he embraces the major difficulty voiced by Braithwaite, Scheffler, and others. He allows that various kinds of facts can be explained by appeal to subsequent conditions:

> What then is accomplished by such explanations? They make explicit *one* effect of an item *i* in system *S*, as well as that the item must be present in *S* on the assumption that the item does have that effect. In short, explanations of function ascriptions make evident one role some item plays in a given system. But if this is what such explanations accomplish, would it not be intellectually more profitable, so it might be asked, to discontinue investigations of the *effects* of various items, and replace them by inquiries into the *causal* (or antecedent) conditions for the occurrence of those items? The appropriate answer, it seems to me, is that inquiries into effects or consequences are just as legitimate as inquiries into causes or antecedent conditions; that biologists as

well as other students of nature have long been concerned with ascertaining the effects produced by various systems and subsystems; and that a reasonably adequate account of the scientific enterprise must include the examination of both kinds of inquiries. (1977, 315)

No one would deny, I trust, that inquiries into effects are as legitimate and important as inquiries into causes. It does not follow from this that effects have as much explanatory import with respect to their causes as causes have with respect to their effects.

3.9 The End of a Decade/The End of an Era?

So, how did things stand at the end of the third decade? It had opened, we recall, with Hempel's (1968) attempt to tidy up the theory of statistical explanation he had offered in (1965), and it ended with his 1976 "Nachwort" (published in 1977), in which he retracted both the doctrine of *essential epistemic relativization of I-S explanation* and the *high-probability requirement*. He published no works on scientific explanation between these two. At the time Hempel admitted that there might be objectively homogeneous reference classes, I offered a serious — though flawed — attempt to explicate that concept (1977b). In relinquishing the high-probability requirement, I think, he rendered the I-S model indefensible. One could say, at the end of the third decade, that, if there is to be an admissible model of statistical explanation, its fundamental explanatory relation would be some sort of relevance relation rather than a relation of high probability of the explanans relative to the explanandum. The key doctrine of the received view — that explanations are arguments — stood on shaky ground.

At the same time there was, I think, a growing realization, for a number of reasons, that causality must play a central role in scientific explanation. To my mind, the *at-at theory* of causal influence provided a fundamental building block required for fuller elaboration of a causal theory of scientific explanation. Further development of that project occupied considerable attention in the fourth decade. As we shall see, it eventuated in the abandonment of the S-R model as an autonomous form of scientific explanation. That model became completely subservient to causal theories.

The Fourth Decade (1978–87)
A Time of Maturation

The history of the last decade is the hardest to recount, especially for one who has been deeply involved in the discussions and controversies. It is difficult to achieve perspective at such close range. Nevertheless, I think certain features are discernible. It is a period during which several different lines of thought achieved relatively high degrees of maturity. For example: (1) The role of causality in scientific explanation has been pursued in far greater detail than previously. (2) Views on the nature of statistical explanation and its relationship to causal explanation have become much more sophisticated. (3) Our understanding of the appeal to unobservables for purposes of explanation has been considerably advanced. (4) The pragmatics of explanation – which posed fundamental points of controversy from the beginning of the second decade – has been investigated more deeply and with more precision than ever before. (5) The question of the relationship between descriptive knowledge and explanatory knowledge has been examined more closely. (6) Perhaps most important, the explicandum has received significant and much-needed clarification during the decade just passed. Let us begin with this last item.

4.1 New Foundations

During the third decade of our chronicle – the one entitled "Deepening Differences" – there was an increasing awareness of the difficulties in the received view. As the hegemony crumbled philosophers looked more closely at the foundations. I recall vividly my own feeling that, in discussions of scientific explanation, the explicandum was in serious need of clarification. Too often, I felt, those who wrote on the subject would begin with a couple of examples and, assuming that these particular cases made the concept clear enough, would proceed to the task of explicating it. Referring to the exemplary job Carnap had done on clarification of the explicandum in *Logical Foundations of Probability* (chaps. I, II, and IV), I declared that the same sort of thing needed to be done for scientific explanation.

I was by no means the only philosopher to feel an urgent need for clarification of the explicandum in the mid-1970s. Michael Friedman's "Explanation and Scientific Understanding" (1974) was a seminal contribution. So also were two articles published near the close of the third decade – one by D. H. Mellor (1976), the other by Alberto Coffa (1977). Largely as a result of these latter two articles, I became acutely aware of the need to distinguish three fundamentally distinct conceptions of scientific explanation – modal, epistemic, and ontic (first presented in W. Salmon 1982).

Mellor adopts a *modal conception* of scientific explanation. Suppose some event E occurs; for all we know at the moment, it might or might not have happened. We explain it by showing that, given other circumstances, it *had to happen*. The modal conception has been advocated by a number of philosophers, but almost always in a deterministic context. If determinism is true, according to most who embrace this view, then all events are in principle explainable. If indeterminism is true, some events will not be amenable to explanation, even in principle; only those events that are necessitated by preceding conditions can be explained. In either case, however, it is not sufficient to show merely *that* the event-to-be-explained had to happen. By citing antecedent circumstances and universal laws one shows *why it had to happen* – by virtue of what it was necessitated.

But Mellor puts a novel twist on the modal conception. He claims that there can be explanations of events that are irreducibly probabilistic. By showing that an event has a high probability relative to preceding circumstances, Mellor claims, we close the gap to some extent, so to speak. There are degrees of necessitation. If an event is completely necessitated, its occurrence is fully entailed by laws and explanatory facts. If it is not fully necessitated, its occurrence is partially entailed by explanatory facts. The greater the degree of partial entailment, the better the explanation.

The concept of *partial entailment* is intuitively appealing to many philosophers who want to follow the Carnapian strategy of constructing an inductive logic on a logical interpretation of probability. The idea is to build inductive logic on the relation of partial entailment in much the same way as deductive logic can be built on the relation of full entailment. Although Carnap never used this concept in his precise explications, he does make informal reference to it (1950, 297). This concept strikes me as one of dubious value. In order to have a measure of degree of partial entailment, one is required to select an a priori measure that is tantamount to assigning prior probabilities to all of the statements that can be formulated in the language to which the inductive logic is to be applied. Given the wide range of choices for such a measure – a nondenumerable infinity (Carnap 1952) – the a priori character of this choice makes it egregiously arbitrary.[1] Thus, it seems to me, partial entailment cannot be construed as degree of necessitation in any way that is useful to the modal conception of scientific explanation. The modal conception therefore appears to require the domain of legitimate scientific explanations to be restricted to deductive explanations. Whether this restriction is tolerable in an era in which physics gives strong indication that we live in an indeterministic universe is an extremely serious question. We shall have to consider this issue in greater detail in §4.9 on deductivism. In my view it is an untenable position.

In the year following the appearance of Mellor's paper, Coffa offered a perceptive contrast between the I-S (inductive-statistical) and the S-R (statistical-relevance) models of explanation (1977).[2] On Hempel's account, we recall, an explanation – deductive or inductive – is an argument. In an I-S explanation, the relation of explanans to explanandum is an inductive or epistemic probability. In a good I-S explanation that probability is high, the higher the better. Explanations – both deductive and inductive – show that the event to be explained *was to be expected*. In view of these considerations, Coffa described Hempel's conception as *epistemic*. The S-R model is conceived in terms of objective probabilities; my formulation was given in terms of relative frequencies. In his closely related dispositional theory of inductive explanation Coffa appealed to propensities.[3] On either account, high probability per se has no particular virtue; what matters is to get the objective probabilities right. Consequently, Coffa characterized the conception underlying these models as *ontic*.

The first year of the fourth decade saw my first serious effort in print at *clarification of the explicandum*. It was embodied in "Why Ask, 'Why?'? – An Inquiry Concerning Scientific Explanation" (1978), my Presidential Address to the Pacific Division of the American Philosophical Association. Attention was focused on two basic concepts of explanation, the inferential (à la Hempel) and the causal (à la Scriven). I tried to exhibit the powerful intuitions that underlie each of them. The most appealing examples from the standpoint of the former are those in which one or more regularities are explained by derivation from more comprehensive regularities – the Newtonian explanation of Kepler's laws, or the Maxwellian explanation of the laws of optics on the basis of electromagnetic theory. Yet the vast majority of examples given by Hempel and other supporters of this conception are explanations of particular facts. Moreover – a point I did not make but should have emphasized – neither the Hempel-Oppenheim article nor Hempel's "Aspects" essay even attempts to offer an account of explanations of laws.

The most persuasive examples from the standpoint of the causal conception are explanations of particular occurrences, often in the context of practical applications. Explanations of airplane crashes and other disasters provide a wealth of instances. The main shortcoming of the causal conception, I argued, was its lack of any adequate analysis of causality. I tried to provide the foundations of such an account in terms of *causal processes, conjunctive forks*, and *interactive forks*. It was a theme to which I have returned a number of times during the last ten years.

Coffa's distinction between the epistemic and ontic conceptions of explanation arose in the context of statistical explanation. In "Why Ask, 'Why?'?" I sought to apply it more generally to the distinction between the inferential conception and the causal conception. As it seemed to me, the epistemic conception is oriented toward the notion of scientific expectability, while the ontic conception focuses upon the fitting of events into natural regularities. Those regularities are sometimes, if not always, causal. Thinking in terms of this distinction between the epistemic and ontic conceptions, I was surprised to find that, in "Aspects of Scientific Explanation," Hempel offers brief characterizations of explanation in general near the beginning and at the end. The two do not agree with each other.

In his initial discussion of D-N (deductive-nomological) explanation, Hempel says that an explanation of this type "may be regarded as an argument to the effect that the phenomenon to be explained . . . was to be expected in virtue of certain explanatory facts" (1965, 336). A bit later this conception is applied to statistical explanation as well. This characterization obviously reflects the epistemic conception. At the conclusion of this essay he sums up his theory of scientific explanation in these terms: "The central theme of this essay has been, briefly, that all scientific explanation involves, explicitly or by implication, a subsumption of its subject matter under general regularities; it seeks to provide a systematic understanding of empirical phenomena by showing that they fit into a nomic nexus" (1965, 488). For the sake of causal explanation, I would be inclined to rephrase the general characterization slightly: it seeks to provide a systematic understanding of empirical phenomena by showing *how* they fit into a *causal* nexus. Nevertheless, either way, the ontic conception is being expressed. Hempel did not, I suspect, notice the differences between his initial and final formulations.

In his doctoral dissertation — a work that is, in my opinion, quite possibly the best thing written on scientific explanation since Hempel's "Aspects" essay — Peter Railton (1980) makes an observation regarding Hempel's theory that is closely related to Coffa's distinction between epistemic and ontic conceptions. Approaching the received view in an extraordinarily sensitive way, Railton shows that Hempel's thesis to the effect that explanations explain by conferring *nomic expectability* on the explanandum cannot be maintained. The problem is that nomic expectability involves two components, *nomicity* and *expectability*, that can conflict with each other. A particular event, such as a spontaneous radioactive decay, may be rather improbable, yet we know the ineluctably statistical laws that govern its occurrence. The nomic side is fulfilled, but the expectability side is not. Hempel chose to reject, as nonexplanatory, any account that renders the event-to-be-explained improbable.[4] Railton argues that it would be better to accept such accounts as explanatory, provided they fulfill certain general conditions, and to take nomicity rather than expectability as the key to scientific explanation. If one takes that tack — as both Coffa and Railton clearly realized — it amounts to relinquishing the epistemic conception in favor of the ontic.

According to the ontic conception, the events we attempt to explain occur in a world full of regularities that are causal or lawful or both. These regularities may be deterministic or irreducibly statistical. In any case, the explanation of events consists in fitting them into the patterns that exist in the objective world. When we seek an explanation, it is because we have not discerned some feature of the regular patterns; we explain by providing information about these patterns that reveals how the explanandum-events fit in. Along with Coffa, both Railton and I endorse the ontic conception. We all maintain that explanations reveal the mechanisms, causal or other, that produce the facts we are trying to explain. The greatest difference between Railton and me concerns the degree to which explanations must be causal. His view is more lenient than mine with regard to noncausal explanation.

During the latter part of 1978 I had the great privilege of visiting Australia and offering a seminar on scientific explanation at the University of Melbourne in the Department of History and Philosophy of Science (the oldest such department in the world, I believe). This visit afforded the opportunity to think through more fully these foundational questions and to discuss them with a number of colleagues and students at Melbourne and at several other Australian universities as well. During this visit to Australia I composed the first drafts of several chapters of *Scientific Explanation and the Causal Structure of the World* (1984).

Taking the cues provided by Coffa and Mellor, it seemed to me that we could distinguish three basic conceptions of scientific explanation—modal, epistemic, and ontic—that could be discerned in Aristotle, and that have persisted down through the ages. In the context of Laplacian determinism they seem to merge harmoniously; in the context of statistical explanation they diverge dramatically. The modal conception (I claim, *pace* Mellor) precludes statistical explanation— except, perhaps, as some sort of incomplete explanation. The epistemic conception, as it occurs within the received view, requires high inductive probabilities. The ontic conception demands objective probabilities, whether high, middling, or low.[5] I offered a brief general discussion of these differing conceptions in "Comets, Pollen, and Dreams: Some Reflections on Scientific Explanation," a rather popularized account that was published in a collection of essays that grew out of my trip to Australia (Salmon 1982).

The foregoing tripartite division was a fairly serviceable crude classification scheme for theories of scientific explanation, but it required refinement. In the far more detailed treatment of the three basic conceptions (1984, chap. 4), I pointed out that the epistemic conception has three different versions—inferential, information-theoretic, and erotetic. Because of his insistence that explanations are arguments, it is appropriate to dub Hempel's conception the *inferential version* of the *epistemic conception*. This version represents the received view of scientific explanation.

Early in the third decade, James G. Greeno (1970) and Joseph Hanna (1969) began offering information-theoretic accounts of scientific explanation. Greeno took the notion of *transmitted information* as a basis for evaluating the explanatory adequacy of statistical theories. Given the fact that information is the key concept, he clearly is adopting an epistemic approach; an appropriate designation would be the *information-theoretic version* of the *epistemic conception*. A crucial difference between his model and Hempel's I-S model is that information transmitted reflects a relevance relation. Another major difference is that Greeno evaluates the explanatory power of laws or theories, but he does not provide a method for evaluating particular explanations of particular facts. This global feature constitutes an affinity between Greeno's information-theoretic account and Friedman's unification account. It seems to me that, if one wants to maintain an epistemic conception, even in the face of serious criticisms of Hempel's models, Greeno's approach is the most promising. Kenneth Sayre (1977) and Joseph Hanna (1978, 1981, 1983) have made subsequent important contributions to it.

Still another version—the *erotetic version*—of the epistemic conception must be distinguished. The term "erotetic" was chosen because the logic of questions has been known traditionally as erotetic logic. This version was suggested by Braithwaite when he remarked that "an explanation, as I understand the use of the word, is an answer to a 'Why?' question which gives some intellectual satisfaction" (1953, 348–49), but he does not develop this approach. Bromberger's work on why-questions (1966), in contrast, involves a sustained effort to elaborate the nature of why-questions and their relations to scientific explanation. But the best-known articulation of this version of the epistemic conception can be found in Bas van Fraassen's provocative work *The Scientific Image* (1980, chap. 5). This approach has obvious connections with Bromberger's earlier work, but a few major differences should be noted. First, although Bromberger presented a theory of explanations as answers to why-questions, he did not consider it a comprehensive treatment, for he flatly denied that all requests for explanations can be phrased as why-questions. Along with a number of other philosophers, Bromberger claims that some explanations are answers to (among others) how-possibly-questions, and that these are different from explanations that are answers to why-questions. Since erotetic logic deals with all kinds of questions, that fact does not disqualify Bromberger as a representative of the erotetic version of the epistemic approach. Van Fraassen, in contrast, affirms the view that all explanations can be considered answers to why-questions. Second, Bromberger made no attempt to deal with statistical explanations; van Fraassen's theory is intended to include explanations of this type. We shall discuss van Fraassen's theory in detail in §4.4.

4.2 Theoretical Explanation

Another major topic of "Why Ask, 'Why?'?" involved the appeal to unobservable entities for purposes of scientific explanation. The issue was by no means new. Within the hegemony of logical empiricism there was what might well be termed a "received view of theories." The main idea was that at the most basic level we have the particular empirical facts revealed by observation, at the next level are empirical generalizations concerning observables, and at the next higher level are theories that seem to make reference to unobservable entities. According to the received view of scientific explanation, the empirical laws explain the observed phenomena, and the theories explain the empirical laws. For example, various observed facts about pressures, temperatures, and volumes of gases are explained by the empirical ideal gas *law*, and that law is explained by the molecular-kinetic *theory*. There may, of course, be still higher level theories that explain the lower level theories. Wilfrid Sellars disparagingly dubbed this account of observable facts, empirical laws, theories, and the explanatory relationships among them the "layer cake" account. It was spelled out rather explicitly in the first four chapters of Braithwaite's *Scientific Explanation*.

During the nineteenth century there had been a good deal of resistance on the part of many scientists and philosophers to the notion that such microentities as atoms or molecules actually exist, or, at any rate, to the notion that we could possibly know anything about them if they do exist. Even those scientists who recognized the utility of the molecular-kinetic theory sometimes regarded such entities merely as useful fictions. This viewpoint is known as *instrumentalism*. The theory is a useful instrument for making scientific predictions (see Gardner 1979).

In the early part of the twentieth century these concerns about our ability to have knowledge of unobservable entities were transformed into questions about the meaningfulness of theories. Operationists and logical positivists denied that utterances putatively about unobservable objects could be scientifically meaningful. Such logical empiricists as Carnap and Hempel made serious efforts to show how — without abandoning empiricist principles — scientific theories could be meaningfully construed. An excellent account of these developments can be found in Hempel's classic essay, "The Theoretician's Dilemma" (1958). At the end of this paper he argues that theories are required for "inductive systematization," but I am not sure that he can establish more than the heuristic value of theories. Even the instrumentalist can cheerfully admit that theories are extremely useful in science. The basic question is whether the unobservable entities to which they seem to refer actually exist.

It is imperative, I believe, to separate the question of the existence of entities not directly observable by means of the unaided human senses from the issue of the meaningfulness of a theoretical vocabulary. Logical empiricists like Carnap and Hempel suggested that the terms of our scientific language can be subdivided into two parts — an observational vocabulary, containing such terms as "table," "dog," "red," "larger than," etc., and a theoretical vocabulary containing such terms as "electron," "atom," "molecule," "gene," "excited state (of an atom)," etc. The viability of a sharp observational-theoretical distinction was frequently called into question, but that particular problem need not detain us now. The instrumentalism issue can be formulated without reference to any such distinction in the scientific vocabulary.

Consider, for example, the epoch-making work of Jean Perrin on Brownian movement in the first decade of the twentieth century. To conduct his experiments he created tiny spheres of gamboge (a bright yellow resinous substance) less than a micrometer in diameter, and he accounted for their motions in terms of collisions with even smaller particles. These experiments were, as we shall see, crucial to Perrin's argument regarding the reality of molecules. Notice that I have formulated the key statements about unobservable entities without going beyond the observational vocabulary.[6]

By the time the statistical-relevance model of scientific explanation had been fairly completely articulated (circa 1970) I was aware of the fact that it was not obviously capable of accommodating theoretical explanation. At that time, James Greeno, who had developed an information-theoretic approach to statistical explanation, presented a paper (1971) in a Philosophy of Science Association symposium in which he tried to show how appeal to theories could yield an increase in information. In my comments in that symposium (1971) I showed that his approach, attractive as it was, would not work. I then set about trying to provide one that would.

In 1973 a conference on explanation (not just scientific explanation) was held at the University of Bristol; the proceedings were published in (Körner 1975). Upon receiving an invitation, but (regrettably) before writing the paper, I proposed the title "Theoretical Explanation" for my contribution. In the end I found I had written a paper on causal explanation, in which I discussed at some length appeals to continuous causal processes and common cause arguments, but which failed to yield any solid result about unobservables. In "Why Ask, 'Why?'?" I thought I had the fundamentals of an approach that would work.

Unconvinced by the various arguments about theoretical realism that had been offered by philosophers from the 1930s to the 1970s, I undertook to find out what considerations convinced natural scientists of the existence of such unobservables as atoms and molecules. Without having been aware at that time of the historical importance of Avogadro's number N, I did recognize that it provided a crucial link between the macrocosm and the microcosm. With the help of N, one could calculate micro-quantities from macro-quantities and conversely. From the mass of a mole of any given substance, for example, N gives us immediately the mass of a molecule of that substance. Thus, it seemed to me, the ascertainment of N was a good place to start.

A first and most obvious way to get at the value of N is through the phenomenon of Brownian movement. According to the molecular-kinetic theory of gases, as Einstein and Smoluchowski showed in 1905–06, the motion of a Brownian particle suspended in a gas is the result of random bombardment by the molecules of the gas. Assuming that the gas and the Brownian particles are in thermal equilibrium, it follows that the average kinetic energy of the Brownian particles is equal to the average kinetic energy of the molecules. As Perrin remarked at about the same time, the dance of the Brownian particles represents qualitatively the random motion of the molecules. By ascertaining the mass of the Brownian particle, the average velocity of the Brownian particle, and the average velocity of the molecules, one can compute the mass of the molecule directly. In practice the situation is a bit more complicated. Although the mass of the Brownian particle and the average velocities of the molecules are quite directly measurable, the average velocity of the Brownian particle is not, for it changes its direction of motion too rapidly. But by indirect means—basically, observation of rates of diffusion of Brownian particles—what amounts to the same method can be applied. As a result we know the mass of a molecule of a gas and Avogadro's number N, the number of molecules in a mole of that gas. This type of experiment was done in the early years of the twentieth century by Perrin.[7]

The instrumentalist can easily reply to the preceding consideration by pointing out that since the advent of the molecular-kinetic theory we have known that it is useful to think of a gas as composed of little particles that move at high speeds and collide with one another and with the walls of the container. Now, it can be added, the Brownian particle behaves as if it is being bombarded by these little particles, and, indeed, we can say that the gas behaves as if it is composed of a certain number of these tiny particles. But all of this does not prove the reality of molecules; it shows that the molecular kinetic theory is an excellent instrument for predicting the behavior of gases (and of the Brownian particles suspended in them). Thus, the manifest success of the molecular-kinetic theory did not constitute compelling evidence for the existence of molecules. As a matter of historical fact many serious and knowledgeable physical scientists at the turn of the century did not believe that atoms and molecules are real.

The way out of this difficulty lies in the fact that N can be ascertained in a variety of ways. In "Why Ask, 'Why?'?" I mentioned a determination by means of electrolysis as an example of a totally different experimental approach. If an electric current passes through a solution containing a silver salt, an amount of metallic silver proportional to the amount of electric charge passing through the solution is deposited on the cathode. The amount required to deposit one mole of a monovalent metal (such as silver) is known as a faraday. A faraday is found by experiment to be 96,487 coulombs. If that charge is divided by the charge on the electron, empirically determined by J. J. Thomson and Robert Millikan to be -1.602×10^{-19} coulombs, the result is N. A faraday is simply Avogadro's number of electron charges. Superficially, the phenomena involved in these two experiments are entirely different, yet they agree in the numerical value they yield for N. I suggested that this agreement, within experimental error, of the value derived via the study of Brownian movement with the value derived via the electrolysis experiment suggests that the particulate character of matter—the reality of such things as molecules, atoms, ions, and electrons—is a common cause of this agreement. I realized at the time that there are many other ways of ascertaining N, but mentioned just two in order to illustrate the common cause character of the argument.

At some time during 1978 I became aware of Mary Jo Nye's superb historical account of the work of Jean Perrin (1972) and, through that, of Perrin's own semipopular treatment (1913, English translation 1923). In the period between 1905 and 1913 Perrin had done a spectacular set of experiments on Brownian movement and the determination of N. Combining his findings with those of other workers, Perrin presents a table, near the close of his book, listing the results of *thirteen distinct methods* for ascertaining N, and notes the striking agreement among them. Immediately after the table he remarks:

> Our wonder is aroused at the very remarkable agreement found between values derived from the consideration of such widely different phenomena. Seeing that not only is the same magnitude obtained by each method when the conditions under which it is applied are varied as much as possible, but that the numbers thus established also agree among themselves, without discrepancy, for all the methods employed, the real existence of the molecule is given a probability bordering on certainty. (1923, 215–16)

This was, indeed, the argument that convinced virtually every physical scientist of the existence of unobservable entities of this sort.

In *Scientific Explanation and the Causal Structure of the World* (213–27) I tried to spell out this common cause argument in some detail. However, instead of dealing with thirteen different ways of ascertaining N, I confined my attention to five: Brownian movement, electrolysis, alpha radiation and helium production, X ray diffraction by crystals, and blackbody radiation. They constitute a highly diverse group of experiments. I thought it fitting, given St. Thomas Aquinas's five ways of demonstrating the existence of God, that I should cite the same number to establish the existence of atoms and molecules. I consider the argument that convinced the scientists in the early part of the twentieth century philosophically compelling.

4.3 Descriptive vs. Explanatory Knowledge

What good are explanations? This question has been asked—explicitly or implicitly—on innumerable occasions over the years, and it has received a variety of answers. During the first half of the twentieth century, scientific philosophers were concerned to refute such answers as, "Explanations inform us of the ultimate purposes for which the world was created," "Explanations reveal the underlying essences of the things we find in the world," or "Explanations exhibit the vital forces or entelechies within living beings." As I remarked near the beginning, the heavy theological and metaphysical involvements of philosophy during that period led some scientists and philosophers to reject altogether the notion that science has anything to do with explanation. Those who did not want to relinquish the claim that science can furnish legitimate explanations were at pains to make it clear that explanatory knowledge is part of our empirically based descriptive knowledge of the natural world. The classic paper by Hempel and Oppenheim, as well as Hempel's "Aspects" paper, were efforts to delineate exactly what sort of descriptive knowledge constitutes explanatory knowledge. These essays were designed to show that legitimate scientific explanations could be had without appealing to superempirical facts or agencies. This is what the received view was all about.

At the same time, these treatments of explanation rejected such psychologistic answers as "Explanations increase our understanding by reducing the unfamiliar to the familiar," or "Explanations ease our intellectual discomforts and make us feel more at home in the world" (see H-O 1948, §4; W. Salmon 1984, 12–15). Great caution must be exercised when we say that scientific explanations have value in that they enable us to understand our world, for *understanding* is an extremely vague concept. Moreover—because of the strong connotations of human empathy the word "understanding" carries—this line can easily lead to anthropomorphism. The received view was concerned also to show that scientific explanations can be given without indulging in that intellectual vice either.

These considerations lead, however, to a deeply perplexing puzzle. If explanatory knowledge does not exceed the bounds of descriptive knowledge, in what *does* it consist? What do we have when we have explanations that we did not already have by virtue of our descriptive knowledge? What is the nature of the understanding scientific explanations are supposed to convey? The full force of this problem did not strike me until about 1978, and I addressed it explicitly in "Why Ask, 'Why?'?" (1978). At about the same time, van Fraassen was asking the same question. We shall consider his answer, which is quite different from mine, below.

The way I posed the problem is this. Suppose you were Laplace's demon, possessing a complete description of the world at one particular moment, knowing all of the laws, and having the ability to solve the mathematical problems required to predict or postdict everything that ever has happened or ever will happen. Such knowledge would appear to be descriptively complete; what else would be involved in having explanations?

I now think this was a particularly inept way to put the question, for one obvious answer is that the demon would have no occasion to ask "Why?" Van Fraassen would, I suspect, agree with this response. As our discussion of Bromberger's work has revealed, one asks a why-question only as a result of some sort of perplexity—recall his p- and b-predicaments. The demon would not be in any such predicament. So we should drop the fantasy of the Laplacian demon. Why-questions are raised only in contexts in which our knowledge is incomplete. Nevertheless, the fundamental question remains, and remains important. What sort of information is explanatory information? How, if at all, does explanatory knowledge differ from other types of descriptive knowledge? The three basic conceptions discussed in §4.1 offer distinct answers to this question.

For the modal conception a straightforward answer is available. Explanatory knowledge adds a modal dimension to our descriptive and predictive knowledge. Explanatory knowledge is knowledge of what is necessary and what is impossible.[8]

In "Why Ask, 'Why?'?" I attempted to respond on behalf of the ontic conception. Even though the question was badly put, the answer points, I think, in the right direction. I suggested that, in addition to purely descriptive knowledge, one would need causal knowledge: recognition of the difference between causal and noncausal laws, the difference between causal processes and pseudo-processes, and the difference between causal interactions and mere spatio-temporal coincidences. At present I would be inclined to phrase the answer somewhat differently in terms of laying bare the underlying mechanisms, but the basic idea seems to me sound. According to the ontic conception, explanatory knowledge is knowledge of the causal mechanisms, and mechanisms of other types perhaps, that produce the phenomena with which we are concerned.

Since it comes in three distinct versions, the epistemic conception requires three answers. According to the received view, explanations involve the subsumption of the fact to be explained under some kind of lawful regularity, universal or statistical. In that way they provide *nomic expectability*. But, one might ask, do we not already have nomic expectability as part of our descriptive and predictive knowledge?[9] The received view seeks assiduously to avoid any incursion into metaphysics or theology for purposes of scientific explanation, but it is hard to see just what constitutes the explanatory import of an explanation on that conception. Indeed, given the original strong version of Hempel's thesis of the symmetry between explanation and prediction, this point obviously holds. The *only* difference between explanation and prediction is pragmatic. If we know that the fact described in the conclusion of an argument conforming to one of the models of explanation actually obtains, the argument constitutes an explanation of that fact. If we do not already know that this fact obtains, the very same argument constitutes a prediction.

As Israel Scheffler pointed out around the beginning of the second decade, it is highly implausible to maintain that every legitimate scientific prediction can serve, under suitable pragmatic conditions, as an explanation. One major reason, as we saw, is that inductive arguments from particulars to particulars—not involving any law—can provide reliable predictions. The symmetry thesis must therefore be amended, at the very least, to the claim that predictions based upon laws can function as explanations in certain epistemic contexts. Thus, the appeal to *nomic regularities* is the crucial feature of explanation. Since it seems reasonable *prima facie* to claim that law statements constitute a crucial part of our descriptive knowledge, we still must ask in what way explanations, as characterized by the received view, have *explanatory import*.

It is useful to recall, in this connection, the fundamental point made by Railton concerning nomic expectability—namely, sometimes nomicity and expectability conflict. If one goes with expectability, thus fully preserving the epistemic character of the conception, descriptive and explanatory knowledge seem indistinguishable. If one opts for nomicity instead, one ends up—as Hempel did at the conclusion of "Aspects"—in the ontic conception.

Before going on to consider the other two versions of the epistemic conception, let us pause to compare the three responses already given. When we appeal to modality—to necessity or impossibility—it may involve either of two distinct approaches. First, one might look upon the modalities as metaphysical categories, residing in a realm separate from the domain that can be empirically investigated. This construal would exile explanation from science. Second, one might say that nothing beyond physical necessity and impossibility is involved, and that these come directly from the laws of nature. If law-statements constitute part of our descriptive knowledge, then the modal conception makes no appeal to anything beyond our descriptive knowledge. Notice that, on this interpretation, the modal conception and the received view are in complete agreement on the status of D-N explanation. The received view differs from the modal conception only insofar as it admits some form of statistical explanation.

When we look at the inferential version of the epistemic conception—involving *nomic* expectability—we see immediately that the status of laws becomes crucial. Consequently, we must ask, what is the difference between true lawlike generalizations (of either the universal or statistical variety) and true accidental generalizations? Recall our previous example comparing statements about massive spheres of gold with statements about massive spheres of uranium. Is there any objective difference between lawful and accidental generalizations, or is it merely a matter of our greater confidence in one as opposed to the other? Alternatively, do we want to characterize laws in terms of relations among universals? If so, this takes the distinction between laws and nonlaws, and, consequently, the characterization of explanation, out of the realm of empirical science. It appears, therefore, that the choice, for the adherent of the received view, is between a heavily pragmatic theory (such as we found in Rescher) and an extrascientific metaphysical one. Neither choice, I believe, captures the intent of the received view.

According to the ontic conception, there is a further gap between explanation and prediction. As we noted much earlier in connection with the famous barometer example, the sharply falling barometric reading is a satisfactory basis for predicting a storm, but contributes in no way to the explanation of the storm. The reason is, of course, the lack of a direct causal connection. For the ontic conception, therefore, mere subsumption under a law is not sufficient for explanation. There must be, in addition, a suitable causal relation between the explanans and the explanandum—at least as long as we steer clear of quantum mechanical phenomena.

The basic question, for this conception, is the status of causality. If we construe causal relations in extrascientific metaphysical terms we will banish explanation from science. If we follow a purely Humean tack, construing causality strictly in terms of constant conjunction (see Mackie 1974),[10] we will make the ontic approach identical with the received view.[11] The approach I adopted in "Why Ask, 'Why?'?" involved an appeal to causal processes and causal interactions (see §3.6 above). Causal processes are distinguished from pseudo-processes in terms of the ability to transmit marks. I attempted to give an entirely empirical construal of the causal concept of transmission in terms of the *at-at* theory. I distinguished causal interactions from mere spatio-temporal intersections of processes in terms of mutual modification of the processes involved. To my great regret, I found no way of carrying out these explications without the use of counterfactual conditionals (W. Salmon 1984, 148–50, 171–74). I have no analysis of counterfactuals, though I do offer a method for testing the truth of the kinds of counterfactuals that are invoked in this context. I am not terribly dissatisfied with this state of affairs, but doubtless other philosophers will not be as easily satisfied on this score.

We have known for a long time that three sets of issues are tightly intertwined: modalities, laws, and counterfactuals (see W. Salmon, 1976). It seemed that any two of the three could be satisfactorily analyzed in terms of the third, but it is difficult to produce a satisfying analysis of any one of them that did not invoke at least one of the other two. Circularity is, consequently, always a serious threat. The most popular current approach to counterfactuals seems to be one that appeals to possible worlds (see Lewis 1973).[12] It has two major shortcomings in my opinion. First, the postulation of the existence of myriad possible worlds, distinct from our actual world, takes us deep into the superempirical.[13] Second, evaluation of the similarity of possible worlds—which is essential to the analysis of counterfactuals—requires an appeal to laws. So, it has not broken us out of the circle. It is quite interesting to note, then, that (1) the modal conception attaches itself to modality, (2) the epistemic conception, inferential version, depends upon laws, and (3) the ontic conception—as far as I can see, at least—appeals to counterfactuals. Each conception has its cross to bear. Yet in that cross may lie the 'something extra' beyond sheer descriptive knowledge that is supposed to provide scientific understanding.

Let us now return to the remaining two versions of the epistemic conception. The information-theoretic version faces almost the same difficulty as did the inferential version in explicating the distinction between purely descriptive knowledge and explanatory knowledge. According to the information-theoretic approach the explanatory value of a law or theory is measured in terms of its efficacy in improving predictive power. That is just what information transmitted amounts to. There is, nevertheless, a possible approach that can be shared by the inferential version and the information-theoretic version. In discussing explanation, prediction, the status of theories, and other related concepts, Hempel has often referred generally to *systematization* — both *deductive systematization* and *inductive systematization*. This suggests the possibility of construing explanatory force, not in terms of some extra type of knowledge, but rather, in terms of the organization of the descriptive knowledge we have or can procure.

This way of looking at explanatory force bears a striking resemblance to Friedman's unification theory of explanation. According to that view, explanations improve our understanding through the unification of our knowledge. Our understanding is increased when we can reduce the number of independent assumptions we have to make about the world. Another way to look at unification is in terms of information theory. When a great deal of information about the world is contained in a short message, we have increased understanding. Either way, what is crucial to explanation is not some particular kind of explanatory knowledge, but, rather, the way in which our descriptive knowledge is organized.

According to the erotetic version of the epistemic conception, as developed by van Fraassen, there is no fundamental difference between descriptive knowledge and explanatory knowledge. On van Fraassen's theory an explanation is simply an *answer* to a why-question; it is nothing other than descriptive information that, *in a given context*, answers a particular type of question. Whether a piece of information constitutes explanatory knowledge depends solely upon the context in which it is furnished. Thus, whatever distinction there is between descriptive and explanatory knowledge is entirely pragmatic.

When we use the term "descriptive knowledge," there is serious danger of equivocation. In one meaning of the term, describing the world or any part of it consists only in reporting what is apparent to the senses. Thus, one could describe a given volume of air as warm or cold, calm or windy, clear or hazy, etc. One might also include such readily detectable features as moisture content and pressure. Such a description would *not* include the fact that air consists of molecules of various gases such as nitrogen, oxygen, and carbon dioxide, or that these molecules are made up of atoms having certain characteristics. A complete description of this sort would be a description solely of appearances. In another sense, a complete description of an entity would include all kinds of facts about it, directly observable or not. Such a description of the same body of air, if complete, would specify how many molecules there are of each type, the atomic constitution of molecules of each type, and facts about the collisions of molecules. This is the sort of description Laplace's demon would have possessed. When we raise the question about the relationship between descriptive knowledge and explanatory knowledge, we must be careful to indicate in which of the foregoing two senses we are construing the word "descriptive." This has a crucial bearing upon the theory of scientific explanation.

In the first chapter of his doctoral dissertation, Coffa discussed in detail the nature of explication, and he devoted serious attention to the clarification of the explicandum he was trying to explicate. He called attention to

> a centuries-old philosophical tradition, sometimes referred to by the name of 'instrumentalism', that has denied the claim that science has explanatory power. For instrumentalists there are no scientific explanations. Science is acknowledged to have a number of virtues, but none of them is associated with the production of a better understanding of what goes on in the world. Science could be a source of predictive power, the foundation of technology, an unlimited fountainhead of aesthetic pleasure, the maximal organizational principle of experience; but it would not be a source of understanding. For the instrumentalist science is a prediction machine with pleasant, but only psychological side-effects.
>
> The notion of explanation we want to explicate is one whose correct applicability to a scientific argument is denied by instrumentalists. An alleged theory of explanation (i.e., an explication of explanation) that elucidates the concept of explanation in such a way that the instrumentalist may consistently agree that science contains explanations in the explicated sense, will not be an explication of any of the explicanda described here. (1973, 61–62)

Coffa's main object, in raising the instrumentalism-realism issue was to apply it to the received view. According to Hempel's account, a deductive-nomological explanation is an argument that contains essentially a lawful generalization. This law may be a generalization concerning observables only—for example, our old friend the ideal gas law. The subsumption of a fact under that generalization constitutes a satisfactory explanation on Hempel's account. If we go on to explain the ideal gas law in terms of the kinetic-molecular theory, that is a distinct explanation, and its existence does nothing to impair the status of the foregoing explanation. As the received view has it, there are legitimate scientific explanations that do not appeal to theories about unobservables; indeed, it is possible in principle (in a world very different from ours), though contrary to fact, for all legitimate explanations to be given wholly in terms of laws governing observable phenomena.[14] Thus, it is possible in principle for the instrumentalist to embrace the received view without qualms. To Coffa's mind, this constitutes a severe shortcoming of any account of scientific explanation. I agree.

To see how these considerations apply to the ontic conception, let us focus upon causal explanation. The Humean tradition suggests that the world is full of constant conjunctions among observable phenomena, but that there is serious doubt as to whether there is anything in the external world that distinguishes accidental conjunctions from genuine causal relations. Hume seems to say that the distinction is in the mind — in the imagination. I do not believe that causality resides in the mind; moreover, I think it is a distinctly nontrivial matter to distinguish genuine cause-effect relations from accidental correlations or from cases in which the correlation is produced by a common cause. Furthermore, in many cases in which we seek a cause we do not even have a constant conjunction *among observables* to appeal to. For example, the nearly simultaneous illness of a number of people who had attended the American Legion convention in Philadelphia in 1976 could not be satisfactorily explained until the *Legionella* bacillus was identified, and its source in Philadelphia located.

As I said above, in "Why Ask, 'Why?'?" I attempted to spell out in some detail the nature of the causal mechanisms that seem to exist in our world. They include causal processes, causal interactions (interactive forks), and conjunctive forks. Our casual observation of phenomena seldom reveals the causal mechanisms; indeed, careful scrutiny of the observable phenomena does not generally reveal their presence and nature. So our efforts at finding causal relations and causal explanations often — if not always — take us beyond the realm of observable phenomena. Such knowledge is empirical knowledge,[15] and it involves descriptive knowledge of the hidden mechanisms of the world, but it does go beyond descriptive knowledge of the observable phenomena. There is no *logical* necessity in the fact that causal mechanisms involve unobservables; that is just the way our world happens to work.

As we have already noted, Coffa is a staunch defender of the ontic conception of scientific explanation, and his theory of explanation reflects this attitude. An explanation of any occurrence is a set of objective facts and relationships. For Coffa, what explains an event is whatever produced it or brought it about. Explanations are nomic dispositions of universal ($p = 1$) or less than universal ($p < 1$) strength. The linguistic entities that are often called 'explanations' are statements reporting on the actual explanation. Explanations, in his view, are fully objective and, where explanations of nonhuman facts are concerned, they exist whether or not anyone ever discovers or describes them. Explanations are not epistemically relativized, nor (outside of the realm of human psychology) do they have psychological components, nor do they have pragmatic dimensions. Since the mechanisms that operate in our world are frequently hidden, the true explanation (on Coffa's view) is often something whose existence the instrumentalist denies.

Traditionally, instrumentalism has been opposed to one sort of realism or another. The instrumentalist in physics recognizes that the molecular-kinetic theory of gases provides a useful tool for establishing regular relationships among such observables as temperature, pressure, volume, and mass, but denies that such things as molecules actually exist. The instrumentalist in psychology sees the function of psychological theories as providing relationships between observable stimuli and observable responses, without appealing to such unobservable entities as feelings of hunger, anxiety, or pain.[16] For the instrumentalist, our descriptive knowledge of the world is confined to knowledge of whatever is more or less directly observable. The instrumentalist cannot appeal to unobservables for purposes of explaining observed fact, for he or she denies that any such things exist.

The realist, in contrast, makes at least a rough demarcation between descriptive knowledge of observables and descriptive knowledge of unobservables. According to the realist, we can, in principle, have knowledge of observables by direct observation—though, in fact, much of our knowledge even of observables is rather indirect. The color, shape, size, and surface texture of a satellite of Pluto, for instance, can at present be known only by means of complex theoretical inferences. In addition, the realist claims, we can have descriptive knowledge of unobservables on the basis of theoretical inferences. Although some realists might maintain that our alleged knowledge of unobservables transcends the empirical realm, I claim that we have fully empirical knowledge of them. My thesis is that realism and empiricism are entirely compatible, for we can confirm or disconfirm theories about unobservables on the basis of observational evidence.[17] The realist constructs explanatory theories that are intended and believed to make reference to unobservable entities. The realist asserts that such things as atoms, molecules, ions, subatomic particles, and microorganisms actually exist, and that we can explain a vast range of physical, chemical, and biological phenomena in terms of their behavior. As we shall see, this claim is sharply denied by van Fraassen, who is *not* an instrumentalist, but who shares a number of important views with philosophers of that persuasion.

For the proponent of the ontic conception of scientific explanation, realism provides a straightforward answer to the question of the distinction between descriptive and explanatory knowledge. Taking "description" in the narrower sense which includes only description of appearances, the realist can say that explanatory knowledge is knowledge of the underlying mechanisms—causal or otherwise—that produce the phenomena we want to explain. To explain is to expose the internal workings, to lay bare the hidden mechanisms, to open the black boxes nature presents to us.

The foregoing discussion of the relationship between descriptive and explanatory knowledge has focused almost exclusively on the intellectual value of scientific explanations. We should not forget that explanations have practical value as well. Finding scientific explanations of various types of occurrences often points to useful ways of controlling important features of our world. It may help to eliminate such undesirable events as epidemics and airplane crashes. It may help to bring about such desirable results as a greater healthy life span for humans. It may also help to alleviate superstitious fears.[18] But, to return to our main theme, our discussion has left us with three apparently viable answers to the question of the intellectual value of scientific explanations. Such explanations enhance our understanding of the world. Our understanding is increased (1) when we obtain knowledge of the hidden mechanisms, causal or other, that produce the phenomena we seek to explain, (2) when our knowledge of the world is so organized that we can comprehend what we know under a smaller number of assumptions than previously, and (3) when we supply missing bits of descriptive knowledge that answer why-questions and remove us from particular sorts of intellectual predicaments. Which of these is *the* function of scientific explanation? None *uniquely* qualifies, I should say; all three are admissible. But not everyone would agree. And even among those who do agree, some will say that one of the three is fundamental, and the others have a distinctly derivative status.

4.4 The Pragmatics of Explanation

The most articulate and prominent anti-realist of the fourth decade is Bas van Fraassen. The full statement of his position appears in *The Scientific Image* (1980), but it had been anticipated to some extent in articles that appeared near the close of the third decade (1976, 1977). It should be carefully noted that van Fraassen is *not* an instrumentalist; his position is *constructive empiricism*. Like a sophisticated instrumentalist, van Fraassen recognizes that theories, which at least appear to make reference to unobservable entities, have played an indispensable role in the development of modern science, and he recognizes that in all likelihood they will continue to do so. The most fundamental difference between the instrumentalist and the constructive empiricist is that the former *denies* the existence of unobservables while the latter remains *agnostic* with respect to their existence.

According to van Fraassen, accepted scientific theories are *accepted as empirically adequate*, but they need not be *believed to be true*. Thus, he claims, when we accept a theory that seems to make reference to unobservables for various purposes—including use in giving explanations—we are committed to claiming that it yields true statements about observables, but we are not committed to claiming that what it says about unobservables is true. Likewise, however, we are not committed to claiming that what it says about unobservables is false. Van Fraassen's thesis is that one cannot be convicted of irrationality for disbelief in such things as molecules and one cannot be convicted of irrationality for belief in them. Regarding the name he has chosen for his position, he says, "I use the adjective 'constructive' to indicate my view that scientific activity is one of construction rather than discovery: construction of models that must be adequate to the phenomena, and not discovery of truth concerning the unobservable" (1980, 5).

In spite of his anti-realism, van Fraassen offers a theory of explanation that fits with his overall conception of the nature of science. Contrasting his view with the more traditional approaches (including, of course, the received view), he says,

The discussion of explanation went wrong at the very beginning when explanation was conceived of as a relationship like description: a relation between theory and fact. Really it is a three-term relation, between theory, fact, and context. No wonder that no single relation between theory and fact ever managed to fit more than a few examples! (1980, 156)

This statement contains a complete rejection of the conception upon which Coffa focused his attention, an ontic conception that located explanations in the external world and which is totally unavailable to instrumentalists.

The theory of explanation van Fraassen offers is *not* intended to be a theory only of scientific explanation; it should encompass other kinds of explanations as well. On his view, an explanation is *an answer to a why-question*. A scientific explanation is one that relies essentially on scientific knowledge. We might ask, for example, why Hitler invaded Russia during World War II. It is a historical question, and it calls for a historical answer. Whether or not history is a science, an answer to this question will be an explanation, and it will be within the purview of van Fraassen's theory.[19] Similarly, one might ask why the cat is sitting in front of the door. The common-sense answer is that he wants to go outside. Again, van Fraassen's theory is meant to handle such questions and answers. History and common sense are, after all, closely related to science whether or not they qualify as parts of science. It is not too unreasonable to expect explanations in those domains to resemble scientific explanations.

There are, however, why-questions that do not seem to be calls for explanations. One might ask, in a time of grief, why a loved one had died. Such a question is not intended to evoke an explanation; it is a cry for sympathy. Exclamations and tears may constitute a far better response than would any factual proposition. Indeed, the grieving individual may be fully aware of the scientific explanation of the demise. Other why-questions seem best interpreted as requests for moral justification. The question has been raised in courts of law as to why a member of a minority group was admitted to medical school to the exclusion of some non-minority candidate whose qualifications were somewhat better. The point at issue is the ethical basis for that decision. It appears that many why-questions are not requests for explanations; consequently, to sustain the claim that explanations are answers to why-questions we would have to distinguish *explanation-seeking* why-questions from other kinds of why-questions. Van Fraassen does not undertake that task, but one might reasonably claim that contextual cues, which play a central role in his theory, should enable us to sort them out.

Given that not all why-questions seek to elicit explanations, the next question is whether all explanations—at least, all scientific explanations—are sought by means of why-questions. Clearly the answer to this question is negative. It has been suggested, however, that any request for a scientific explanation—no matter how it is actually formulated—can be appropriately rephrased in the form of a why-question. Inasmuch as van Fraassen claims that explanations are answers to why-questions, he is patently committed to this view. A number of philosophers have denied it; for instance, Bromberger, William Dray, and Frederick Suppe have argued that some explanations are answers to questions of *how-possibly*. Dray had raised this issue (1957) and it is discussed by Hempel (1965, 428–30).

Consider a concrete example. There is an old saying that when a cat falls (from a sufficient height) it always lands on its feet. We know, however, that angular momentum is conserved. How is it then possible for a cat, dropped from an adequate height with zero angular momentum and with its legs pointing upward, to land on its feet? Your first reaction might be to suppose that the old saying is simply not true — that it is an old wives' tale. But that is not correct. Experiment has shown that the cat can twist its body in various ways while the net angular momentum remains zero to achieve the desired position upon landing.[20] A diver who does a twist, as distinguished from a somersault, achieves a similar feat. Hempel's response to such examples is to admit that there is a pragmatic difference between why- and how-possibly-questions, in that the person who poses a how-possibly question is under the mistaken impression that the occurrence is either physically impossible or highly improbable. The appropriate response is to expose the misapprehension and produce either a D-N or an I-S explanation of the phenomenon in question. The original question could thus have been rephrased, "Why did this event (which I initially regarded as impossible or highly improbable) occur?" (Hempel 1965, 428–30).

In (1984, 10) I also expressed the claim that all requests for scientific explanations can be formulated as why-questions, but I now suspect that it is mistaken. Hempel's response seems inadequate for two reasons. First, a how-possibly question does not require an actual explanation; any potential explanation not ruled out by known facts is a suitable answer. For example, a DC-9 jet airplane recently crashed upon takeoff at Denver's Stapleton Airport during a snowstorm. One peculiar feature of this accident is that the plane flipped over onto its back. There are many explanations of a crash under the circumstances, but I wondered how it could have flipped over. Two how-possibly explanations were mentioned in the news reports. One is that it encountered wing-tip turbulence from another airplane just after it became airborne. Another was suggested by the report of a survivor, who claimed that the plane was de-iced three times during its wait for departure, but that on the latter two of these occasions one wing, but not the other, was treated. If one wing had an accumulation of ice on its leading edge while the other did not, the difference in lift provided by the two wings might have been sufficient cause for the plane to flip over. As I write this paragraph I have not yet heard the final determination regarding the cause of this crash. Both potential explanations I have mentioned are satisfactory answers to the how-possibly question, but we do not know the correct answer to the why-question.

Second, improbable events do occur. Not long before the Denver crash, another DC-9 crashed on take-off at Detroit's Metropolitan Airport. Investigators have concluded, I believe, that the pilot failed to extend the wing flaps for take-off. It is extremely unlikely that an experienced pilot would make such an error, that the co-pilot would fail to notice, and that the warning signal would fail to be sounded or would be ignored. But apparently that is what happened. Hempel suggests that we are obliged to find additional factors that would make the errors highly probable. It seems to me that, even if one insists on high probabilities for explanations that answer why-questions, no such thing is required for answers to how-possibly questions. It is sufficient to show that the probability is different from zero.

Still other explanations may be answers to *how-actually* questions. How did there come to be mammals (other than bats) in New Zealand? They were humans, who came in boats, and who later imported other mammals. This is, I think, a genuine scientific explanation. It is not an explanation of why they came; rather, it is an explanation of how they got there.

Having posted some caveats about simply identifying explanations as answers to why-questions, I shall now turn to a discussion of some of the details of van Fraassen's theory of explanation. Before doing so, it is worth noting that, although this issue is crucial for an advocate of the erotetic conception, it has little – if any – genuine significance for the proponent of the ontic conception. According to this latter conception the search for explanations is a search for underlying mechanisms, and the form of the question requesting them is not very important.

The development of van Fraassen's theory of why-questions and their answers in the fourth decade was greatly facilitated by progress in formal pragmatics and the publication, near the end of the third decade, of Belnap and Steel's *The Logic of Questions and Answers* (1976). Although this is a landmark work in erotetic logic, it contains hardly any treatment of why-questions. To begin, it is essential to realize that, as van Fraassen sets forth his theory, questions and answers are abstract entities. An answer is a proposition, and a given proposition can be expressed by means of many different declarative sentences. Moreover, a particular sentence, uttered on different occasions, can express different propositions. For any given sentence, the context determines which proposition it expresses. "I am here," for example, always expresses a true proposition, but the proposition it expresses depends on who utters the sentence, and where and when. Similarly, a given question may be expressed by many different interrogative sentences, and a particular interrogative may pose different questions on different occasions. When an interrogative sentence is uttered the context determines which question is being asked.

According to van Fraassen, we can think of the standard form of a why-question as

Why (is it the case that) P_k?   (Q)

where P_k states the fact to be explained (the explanandum phenomenon). Such a question can be identified with an ordered triple

$Q = <P_k, X, R>$

where P_k is the *topic* of the question, $X = \{P_1, P_2, \ldots, P_k, \ldots\}$ is the *contrast class*, and R is the *relevance relation*. To take a familiar example, consider the question,

Why did the Bunsen flame turn yellow?

The topic is

The Bunsen flame turned yellow.

The contrast class is

The Bunsen flame remained blue (P_1)
The Bunsen flame turned green (P_2)
The Bunsen flame turned orange (P_3)

.

.

.

The Bunsen flame turned yellow (P_k)

.

.

.

The relevance relation R is the relation of cause to effect. Hempel's answer, we recall, is that a piece of rock salt was placed in the flame, rock salt is a sodium compound, and all sodium compounds turn Bunsen flames yellow.

A crucial feature of van Fraassen's account is its emphasis upon the fact that the same interrogative sentence – the same group of words – can express different questions. This can easily happen if different contrast classes are involved. He invites consideration of the interrogative

Why did Adam eat the apple?

By the inflection or emphasis of the speaker, or by other contextual clues, we might find that any of three different questions is being expressed. It might mean,

Why did *Adam* eat the apple?

where the contrast class = {Eve ate the apple, the serpent ate the apple, the goat ate the apple, etc.}. At the same time, it might mean

Why did Adam *eat* the apple?

where the contrast class = {Adam ate the apple, Adam threw the apple away, Adam gave the apple back to Eve, Adam fed the apple to the goat, etc.}. Also, it might mean

> Why did Adam eat the *apple*?

where the contrast class = {Adam ate the apple, Adam ate the pear, Adam ate the pomegranate, etc.}. The context determines which is the appropriate contrast class.

Another feature of questions is that they generally come with *presuppositions*, and why-questions are no exception. The presupposition of Q is

- (a) P_k is true,
- (b) each P_j in X is false if $j \neq k$,
- (c) there is at least one true proposition A that bears the relation R to the ordered pair $<P_k, X>$

where (a) and (b) taken together constitute the *central presupposition*. The canonical form of a *direct answer* to Q is

> (*) P_k in contrast to the rest of X because A.

The proposition A is known as the *core of an answer* to Q, because the direct answer would normally be abbreviated, "Because A." The following conditions must be met if (*) is to qualify as a direct answer to Q:

- (i) A is true.
- (ii) P_k is true.
- (iii) No member of X other than P_k is true.
- (iv) A bears relation R to $<P_k, X>$.

The context in which a question is posed involves a body of background knowledge K. According to van Fraassen, two of the biggest problems faced by other theories of scientific explanation are *rejections of requests for explanation* and *the asymmetries of explanation*. Rejections are handled in van Fraassen's theory in terms of the presupposition. Unless K entails the truth of the central presupposition—namely, that the topic is true and that every member of the contrast class other than the topic is false—the question does not *arise* in that context. For example, the question "Why was Jimmy Hoffa murdered?" does not arise in my current knowledge situation, for to the best of my knowledge it has not been established that he was murdered—only that he disappeared. In addition, if K entails the falsity of (c)—that is, if K entails that there is no answer—the question does not arise. If, for instance, one asks why a given unstable nucleus decayed at some particular moment, we might answer that there is no reason why. Our best theories tell us that such things just happen. In other words, it is appropriate to raise the question Q if we know that P_k is the one and only member of the contrast class that is true, and we do not know that Q has no answer.

If the question Q does not arise in a given context, we should reject it rather than trying to provide a direct answer. This can be done, as we saw in the preceding paragraph, by providing a corrective answer to the effect that some part of the central presupposition is not entailed by the body of knowledge K, or that K entails that there is no direct answer. If the question does arise, but we find out that (c) is false, then a corrective answer to that effect is appropriate. If the presupposition is completely satisfied, the request for a direct answer is legitimate, and if one is found, it constitutes an explanation of P_k.

There is, I believe, a profound difficulty with van Fraassen's theory centering on the relevance relation R. It can be put very simply, namely, that he imposes no restriction whatever on the nature of the relevance relation. He says explicitly that A is relevant to P_k if A bears relation R to P_k (1980, 143). But if R is not a bona fide relevance relation, then A is 'relevant' to P_k only in a Pickwickian sense. The difficulty can be posed in extremely simple terms. Formally speaking, a relation consists of a set of ordered pairs. Suppose we want to explain any fact P_k. Pick any arbitrary true proposition A. Let the relation R be the unit set of ordered pairs $\{<A, <P_k, X>>\}$ – the set that has $<A, <P_k, X>>$ as its only member. "Because A" is an explanation of P_k.

Although I had studied van Fraassen's theory with some care between 1980 (when it was published) and 1985 (when the Minnesota NEH Institute was held), I had not noticed this difficulty. It emerged as a result of a discussion with Philip Kitcher, and we published a joint paper in which it was exhibited (1987). The problem was masked by a number of van Fraassen's informal remarks. For instance, at the outset of his exposition of his theory of why-questions, he says, "This evaluation [of answers] proceeds with reference to the part of science accepted as 'background theory' in that context" (1980, 141). Earlier, he had remarked that, "To ask that . . . explanations be scientific is only to ask that they rely on scientific theories and experimentation, not on old wives' tales" (1980, 129) and "To sum up: no factor is explanatorily relevant unless it is scientifically relevant; and among the scientifically relevant factors, context determines explanatorily relevant ones" (1980, 126). In conclusion, he says, "To call an explanation scientific, is to say nothing about its form or the sort of information adduced, but only that the explanation draws on science to get this information (at least to some extent) and, more importantly, that the criteria of evaluation of how good an explanation it is, are being applied using a scientific theory" (1980, 155–56). But in the formal account no restriction is imposed on the relation R.

To see the consequences of this lacuna, consider a concrete example. Suppose someone asks why John F. Kennedy died on 22 November 1963; this is the question Q = $<P_k, X, R>$, where

P_k = JFK died 11/22/63   (topic)
X = {JFK died 1/1/63,
JFK died 1/2/63,

.

.

.

JFK died 11/22/63,

.

.

.

JFK died 12/31/63,
JFK survived 1963}   (contrast class)
R = astral influence   (relevance relation)

Suppose that the direct answer is

P_k in contrast to the rest of X because A,

where A (the core of the answer) consists of a *true* description of the configuration of the planets, sun, moon, and stars at the time of Kennedy's birth. Suppose further that the person who supplies this answer has an astrological theory from which it follows that, given A, it was certain, or highly probable, that Kennedy would die on that day. We now have a why-question and an answer; the answer is an explanation. We must ask how good it is.

Van Fraassen's theory does not stop at the definition of an answer to a why-question; obviously, it must offer grounds for evaluating answers. We need to be able to grade explanations as better or worse. And he does provide criteria for this purpose; there are three. First, we must ask how probable the answer is in light of our background knowledge K. Second, we must ask to what extent the answer *favors* the topic vis-à-vis the other members of the contrast class. Third, we must ask how this answer compares with other available answers: (i) are any other answers more probable? (ii) do any other answers more strongly favor the topic? or (iii) do any other answers render this one wholly or partially irrelevant? On each of these criteria the astrological explanation gets highest marks. In the first place, since we have excellent astronomical records we have practical certainty regarding the celestial configuration at the time of Kennedy's birth; indeed, we have stipulated that A is true. In the second place, we must suppose that the astrologer can derive from A, by means of astrological theory, that Kennedy was sure to die on that day—or, at least, that the probability for his death on that day was much greater than for any other day of the year, and also much greater than the probability that he would live to see the beginning of 1964. Since this explanation, like any explanation, is given ex post facto, we must credit the astrologer with sufficient ingenuity to produce such a derivation. In the third place, no other answer is better than A: (i) A is true; hence, no other answer could be more probable. Moreover, since, astrologically speaking, the heavenly configuration at the time of one's birth is the primary determinant of one's fate (ii) no other answer could favor the topic more strongly and (iii) no other facts could supersede answer A or render it irrelevant.

Consideration of the astrology example makes vivid, I hope, the fundamental problem van Fraassen's theory encounters with respect to the relevance relation. In our critique of that theory, Kitcher and I show formally that *any* true proposition A can be an indispensable part of an explanation of *any* topic P_k (with respect to a contrast class that contains P_k and *any* assortment of false propositions), and, indeed, that it gets highest marks as an explanation of P_k (1987, 319–22). Thus, it is natural to suggest that van Fraassen add one item to the list of presuppositions of Q, namely,

(d) R is a relevance relation.

But when we attempt to impose such a condition, we find ourselves in a mare's nest of difficulties. The problem is to characterize in a general way what constitutes a suitable relevance relation, recognizing, of course, that there may be more than one.

Consider, for example, the relation of logical deducibility. That this relation is not satisfactory has been known since antiquity; Aristotle pointed out that some demonstrations provide understanding while others do not. Hempel recognized from the outset that demonstrations like

Horace is a member of the Greenbury School Board.
All members of the Greenbury School Board are bald.

Horace is bald.

cannot qualify as a bona fide explanation of Horace's baldness. Since the beginning of the first decade in 1948 he has insisted that the demonstration must contain essentially at least one law-statement among its premises. But we have seen what a vexed question it is to distinguish lawlike from nonlawlike statements.

We have also seen that, even if the problem of lawlikeness can be handled, a problem about asymmetries arises. Recalling one of our standard counterexamples, we have two valid deductive arguments with suitable lawful premises, one of which seems clearly to provide an explanation, the other of which seems to most of us not to. From the height of the flagpole in conjunction with the elevation of the sun and laws of propagation of light, we can deduce the length of the shadow it casts. This sounds like a good explanation. From the length of the shadow and other premises of the aforementioned sort (excluding, of course, the height of the flagpole), we can deduce the height of the flagpole. This does not sound like a good explanation. Hence, the problem of asymmetries.

Van Fraassen has maintained, as we have already noted, that the two chief problems inherited from the traditional accounts of explanation that he wants to solve are the problem of rejections and the problem of asymmetries. We have already discussed his solution to the problem of rejection of the explanatory questions, and found no fault with it. His solution to the problem of asymmetries is another story – indeed, it is the fable of "The Tower and the Shadow" (1980, 132–34).

Van Fraassen's treatment of the asymmetries is to show that certain why-questions usually arise in certain typical contexts in which a standard sort of answer is satisfactory. According to van Fraassen's story, when first he asks why the shadow of the tower is so long, he is told that it is cast by a tower of a certain height; in addition, his host, the Chevalier, adds that the tower was built to that height on that particular spot for certain historical reasons. That is *his* explanation, but later in the tale we learn that it is false. The servants have a different, and more accurate, explanation of the position and height of the tower. Carefully taking the contextual factors into account, we discover that the correct answer to the question, "Why is the tower that tall and located in that place," is that it casts a shadow long enough to cover a certain spot on the terrace, where the Chevalier had murdered a servant girl in a fit of jealous rage, at a certain time of day. Here we have an admissible why-question and a suitable direct answer. We are tempted to remonstrate, as I did (1984, 95), that it was the antecedent desire of the Chevalier to have a shadow that long that explains the height of the tower. But van Fraassen seems to be maintaining that the answer given is legitimate. The topic P_k of the question is that the tower stands at a particular place and is 175 feet tall. The contrast class X consists of a series of statements about towers of different heights located at various locations in the vicinity. And the relation R is a relation of intentional relevance. Our remonstrance was based on the belief that this relation is not a suitable explanatory relevance relation. As an advocate of a causal/mechanical conception of scientific explanation, I am not prepared to admit that effects explain their causes, even where conscious purposeful behavior is involved.

It is clear from the whole tenor of van Fraassen's account that he is not proposing an 'anything goes' sort of theory. He is *not* suggesting that any answer to a why-question that happens to satisfy a questioner is a satisfactory explanation. Consequently, it seemed to Kitcher and me, van Fraassen's pragmatic theory cannot escape precisely the kinds of problems concerning objective explanatory relevance relations with which other more traditional theories – such as the received view or the statistical-relevance approach – had to struggle. Hence, when I proposed above the addition of presupposition

(d) R is a relevance relation

I was opening up the whole question of what constitutes a satisfactory explanatory relevance relation.

Consider, for example, the relation of statistical relevance, which I once regarded as the key explanatory relation. We suppose that the person who raises a given why-question has a prior probability distribution over the members of the contrast class. When an appropriate answer is given, it results in a different probability distribution over the same set of alternatives. In the S-R model I required only that the probabilities change; van Fraassen's theory requires that the topic be favored—that is, that it be elevated relative to its rivals in the contrast class. What kinds of probabilities are these? Coffa and I took propensities and frequencies, respectively, as the appropriate interpretations. In a pragmatic theory of explanation, because of its emphasis upon the knowledge situation and the context, it is natural to think of epistemic or personal probabilities. Given the well-known difficulties with the former, van Fraassen seems to prefer the latter.

Whatever sort of probability is involved, a basic problem—one we have confronted previously—arises. When an explanation is sought we already know that the explanandum-phenomenon has occurred. For van Fraassen, the truth of the topic is the first presupposition of a why-question. Moreover, the second presupposition of the why-question is that all of the other members of the contrast class are false. The why-question does not arise unless these presuppositions are fulfilled. But if this presupposition is fulfilled, the prior probability distribution with respect to the body of knowledge K is one for the topic and zero for all other members of the contrast class. No change in this probability distribution could possibly favor the topic. So van Fraassen proposes that we must cut back our body of knowledge K to some proper part K(Q) that is appropriate for question Q (1980, 147). The problem of deciding what information K(Q) should contain is precisely the problem Hempel faced in connection with the requirement of maximal specificity. It is also the problem with which I had to deal in characterizing objective homogeneity. It turns out, then, that this cluster of traditional problems is not evaded by van Fraassen's pragmatic account.

The appeal to personal probabilities in this context gives rise to another serious difficulty. Return to the example of John F. Kennedy's assassination. To the sincere believer in astrology the configuration of heavenly bodies at the time of Kennedy's birth is highly relevant to his death on that particular fateful day in 1963. Acquiring that information will produce a redistribution of personal probabilities strongly favoring the topic. Believing, as we do, that there is no objective relevance between the celestial configuration at the time of Kennedy's birth and the occurrence of his death on a particular day, we need to block such explanations. Unless we can impose the demand for objective relevance relations, we cannot arrive at a satisfactory characterization of scientific explanation. As many philosophers have insisted, we need to appeal to objective nomic relations, causal relations, or other sorts of physical mechanisms if we are to provide adequate scientific explanations.

Philosophers have long recognized that scientific explanation has pragmatic dimensions. It is obvious that various features of actual explanations depend upon context. Hempel was aware from the beginning that the individual who answers an explanation-seeking why-question standardly omits parts of the explanation that are already well known to the questioner. I have emphasized the fact that one frequently has to clarify the question in order to ascertain what explanation is being sought. As we saw in the second decade, the ordinary language philosophers placed great emphasis upon pragmatic considerations. The theory of the pragmatics of explanation given by van Fraassen in *The Scientific Image* is highly illuminating and is, I believe, the best that has been given to date. It must be emphasized, however, that he has not succeeded in showing that all the traditional problems of explanation can be solved by appealing to pragmatics. In that sense he has *not* provided a pragmatic theory of explanation. The problems concerning the nature of laws, and those concerning the nature of causality, have not been circumvented by pragmatic considerations.

Another important representative of the erotetic version of the epistemic conception is Peter Achinstein, whose view is articulated in great detail in *The Nature of Explanation* (1983). His theory differs significantly from those of both Bromberger and van Fraassen, but it is much closer in spirit to that of Bromberger. This can be seen most clearly in the emphasis that both Achinstein and Bromberger place on the linguistic analysis of English usage — something van Fraassen does hardly at all. Whereas van Fraassen simply announces that explanations are answers to why-questions, Achinstein invests considerable time and effort in clarifying the usage of "explanation" and closely related terms. Along with Bromberger, he denies that all explanations are answers to why-questions. Nevertheless, as we shall see, questions — not just why-questions — and their answers play a fundamental role in this theory.

Achinstein points out that "explanation" may refer either to a *process or a product*. The process is a linguistic performance; someone explains something to someone by uttering or writing statements. The product is the content of the linguistic performance. The linguistic performance itself involves an intention on the part of person producing the explanation: "Explaining is what Austin calls an illocutionary act. Like warning and promising, it is typically performed by uttering words in certain contexts with appropriate intentions" (1983, 16). According to Achinstein, this process concept of explanation is primary. When we try to characterize the product, we must take account of the intention (or illocutionary force) of the explanation, for the same set of words can be used either to explain or do other sorts of things. A physician might explain John's malaise this morning by saying, "He drank too much last night." John's wife might use the same words to criticize his behavior. His wife's speech act is not an explanation; consequently, what she produced is not an explanation. This same pragmatic consideration arose, incidentally, in the 1948 Hempel-Oppenheim essay for, given the explanation/prediction symmetry thesis, an argument may function in one context as an explanation and in another as a prediction. To deal explicitly with this aspect of explanation, Achinstein adopts what he calls *an ordered pair view* of explanation: an explanation in the product sense is an ordered pair $<x, y>$ in which x is a specified type of proposition and y is a type of speech act, namely, explaining. On this view, y retains the intention involved in the process of explanation (1983, 85–94). Achinstein refers to his account as *the illocutionary theory*.

It should be noted that the foregoing considerations are designed to clarify the notion of explanation without qualification. Up to this point, Achinstein makes no attempt to characterize correct explanations, good explanations, or scientific explanations. It is not until page 117, almost one-third of the way through a fairly large book, that scientific explanation comes up for serious consideration. The philosopher of science who is impatient to get to that topic cannot begin there, however, for the preliminaries are used extensively in the subsequent discussions.

In his preliminary formulations, Achinstein presents two aspects of explanation:

(1) If S explains q by uttering u, then S utters u with the intention that his utterance of u render q understandable (1983, 16);

(2) If S explains q by uttering u, then S believes that u expresses a proposition that is a correct answer to Q (1983, 17).

If Q is a why-question (which is, I repeat, for Achinstein only one among many sorts of explanation-seeking questions), then q is what van Fraassen called the *topic* of the question. To use one of Achinstein's examples, if Q is "Why did Nero fiddle?" then q is "Nero fiddled." According to (1), whatever answer is given is intended to make q understandable, but since explanation and understanding are such closely related concepts, it is necessary for Achinstein to say something about what constitutes understanding. He offers the following necessary condition:

A understands q only if there exists a proposition p such that A knows of p that it is a correct answer to Q, and p is a complete content-giving proposition with respect to Q. (Here p is a proposition expressed by a sentence u uttered by A.) (1983, 42)

He later suggests that it is a sufficient condition as well (1983, 57). Space does not permit a full statement of what constitutes a complete content-giving proposition with respect to a question; the details are spelled out in Achinstein's book. But the crucial point can be raised by considering a special case – one of his examples. A straightforward complete content-giving proposition with respect to the question, "Why did Nero fiddle?" is "The reason Nero fiddled is that he was happy." Given that this is a complete content-giving proposition with respect to that question, A understands Nero's fiddling iff A knows that "The reason Nero fiddled is that he was happy" is a *correct answer* to the question, "Why did Nero fiddle?"

This view of explanation seems seriously question-begging. We may raise essentially the same question with regard to Achinstein's theory as we did concerning van Fraassen's: what objective relationship must obtain between the fact that Nero was happy and the fact that he fiddled to make "The reason Nero fiddled is that he was happy" a correct answer? How must that relationship differ from the relationship between the fact that Caesar was assassinated on the Ides of March and the fact that Nero fiddled? These questions have fundamental importance; to see this it will be useful to make a direct comparison with Hempel's theory of deductive-nomological explanation.

According to Hempel's theory, if "Nero was happy" is part of the explanans of "Nero fiddled," then "Nero was happy" must be true. Hempel's theory of explanation specified the relationship that must exist between the facts if one is to be (part of) an explanation of the other. According to the D-N model, the statement "Nero was happy" must be a premise of a valid deductive argument having "Nero fiddled" as its conclusion, and including essentially at least one other premise stating a lawful regularity. This argument must fulfill the Hempel-Oppenheim empirical condition of adequacy, namely, that all of its premises be true. Given the fulfillment of these conditions, we are then authorized to accept the claim that (at least part of) the reason Nero fiddled is that he was happy. It was *not* part of the empirical condition of adequacy to determine that "The reason Nero fiddled is that he was happy" is true. The whole idea of the Hempel-Oppenheim theory was to provide conditions under which it is correct to make such claims as "The reason Nero fiddled is that he was happy." They do not require us to assess the truth of such a statement to ascertain whether or not we have a correct explanation. Achinstein is clearly aware that this line of argument may be brought against his theory, and he attempts to rebut it (1983, 71–72), but I am not convinced that his defense is successful. I think he ends up—like Bromberger and van Fraassen—lacking an adequate characterization of the kinds of objective relevance relations required in sound scientific explanations.

Despite my skepticism regarding the illocutionary theory of explanation, there are, it seems to me, several especially illuminating features of Achinstein's treatment of scientific explanation. First, he distinguishes carefully between *correct* explanations and *good* explanations. An explanation < x, y > is a correct explanation if the first member of that ordered pair is a true statement. Of course, the fact that the ordered pair is an explanation imposes other qualifications on x. However, for any number of pragmatic reasons, a correct explanation may not be a good explanation. It may be unsuitable to the knowledge and abilities of the listeners, or lacking in salience with respect to their interests. To deal with evaluations of explanations over and above correctness, Achinstein introduces the idea of a set of instructions to be followed in constructing explanations (1983, 53–56).

Such instructions could be of a wide variety of kinds. One might be the Hempelian instruction that the explanation must include essentially at least one law. Another might be an instruction to give a microphysical explanation. Still another might be the instruction to give a causal explanation.

Another important feature is Achinstein's contention that there is no single set of universal instructions that will suffice to judge the merits of scientific explanations at all times and in all contexts. Instructions that are suitable in one context may be quite unsuitable in another. He offers a characterization of appropriate instructions as follows:

I is a set of appropriate instructions for an explainer to follow in explaining q to an audience iff *either*

a. The audience does not understand q in a way that satisfies I, and
b. There is answer to Q (the question that elicits an explanation of q), that satisfies I, the citing of which will enable the audience to understand q in a way that satisfies I, and
c. The audience is interested in understanding q in a way that satisfies I, and
d. Understanding q in a way that satisfies I, if it could be achieved, would be valuable for the audience;

or

It is reasonable for the explainer to believe that a-d are satisfied. (113, slightly paraphrased)

Employing several important historical examples, Achinstein argues that there is no set of instructions that is universally appropriate for science (1983, 119–56).

The discussion of universal instructions leads naturally into another main feature of Achinstein's theory, namely, a consideration of the possibility of formal models of explanation such as the D-N, I-S, S-R. Again the conclusion is negative. Achinstein states two requirements which, he believes, motivate the "modelists." The first is the "No-Entailment-By-Singular-Sentence" (or NES) requirement. According to this requirement, no correct explanation of a particular occurrence can contain, in the explanans, any singular sentence or finite set of singular sentences that entail the explanandum (1983, 159). The second requirement he calls "the a priori requirement." According to this requirement, "the only *empirical* consideration in determining whether the explanans correctly explains the explanandum is the truth of the explanans; all other considerations are a priori" (1983, 162). His strategy in arguing that there cannot be models of explanation, in the traditional sense, is that any model that satisfies one of these requirements will violate the other (1983, 164–92). Even though I have spent a great deal of effort in elaborating the S-R model, this view of Achinstein's is one with which I happen to agree. I now think that an adequate scientific explanation identifies the mechanisms by which the explanandum came about. Consequently, what constitutes a suitable scientific explanation depends on the kinds of mechanisms —

causal or noncausal – that are operative in our world. This is an issue that cannot be settled a priori. Achinstein's arguments against the possibilities of models of explanation are far more precise and detailed than mine.

4.5 Empiricism and Realism

Van Fraassen's seminal book, *The Scientific Image*, has spawned a great deal of discussion, most of which has been directed toward his rejection of scientific realism. Churchland and Hooker, *Images of Science* (1985), is an important collection of critical essays, accompanied by van Fraassen's detailed replies. None of these essays focuses primarily on his treatment of scientific explanation, though many are tangentially relevant to it.[21] Since, however, the realism issue has direct bearing on the nature of theoretical explanation, we must devote some attention to it. To carry out this discussion I will accept van Fraassen's claim – which I believe to be sound – that there is a viable distinction between the observable and the unobservable. Although the dividing line may not be sharp, there are clear cases of observables (e.g., sticks and stones) and clear cases of unobservables (e.g., atoms, electrons, and simple molecules).

The first sentence in the first chapter of *The Scientific Image* is striking: "The opposition between empiricism and realism is old, and can be introduced by illustrations from many episodes in the history of philosophy" (1980, 1). It formulates an assumption that goes unquestioned throughout the rest of the book – namely, that it is impossible to have empirical evidence that supports or undermines statements about objects, events, and properties that are not directly observable by the unaided normal human senses. This assumption should not, I think, go completely unchallenged. Indeed, I believe it is false. However that may be, it raises what I take to be the *key question* for scientific empiricism (W. Salmon 1985).

Let me illustrate the point by means of a simple example. I own a copy of the Compact Edition of the Oxford English Dictionary. It contains print of various sizes. I can read the largest print on the title page with my naked eye, but everything else is blurry without my eyeglasses. When I put on the spectacles I can easily read some of the larger print within the books. The use of corrective lenses does not take us beyond the realm of the directly observable; their effect is to restore normal vision, not to extend it beyond the normal range. The spectacles enable me to see the things I could have seen in my early teens without their aid. But even with their aid much of the print is blurry. With the aid of a magnifying glass (which comes with the set) I can read even the entries in smallest type. Also, with the aid of the magnifying glass, I can see marks of punctuation that were completely invisible to me without it. I claim that I have established the existence of an entity that is not directly observable; I have established a statement to the effect that there is, at a particular place on a given page, an ink spot too small to be detected with the normal unaided human sense of sight. Moreover, I cannot feel it, smell it, hear it, or taste it.

It is important to note that, when I view through the magnifying glass print that I *can* read without it, the letters, words, and marks of punctuation that I see are the same. They simply appear clearer and larger. When I view smaller print with the magnifying glass, the forms I see—letters, words, marks of punctuation—make sense. I see bona fide words, and I read appropriate definitions of them. The words I read appear in their correct places. The same is true of the dot that I could not see at all without the magnifying glass. When it is made visible by the glass, it appears in a syntactically correct place. Moreover, although I confess that I have not performed the experiment, I have complete confidence that a comparison of the entries in the compact edition with those of the unreduced editions would reveal an identity between what is seen without the magnifying glass in the larger edition with what is seen with the aid of the magnifying glass in the compact edition.

Evidently, many different experiments of the type just described can be conducted with a variety of lenses, and on their basis we can establish a theory of geometrical optics. A number of fundamental facts of geometrical optics were known in antiquity and medieval times, and Snell's law was proposed in 1621. This theory is completely empirical even in the narrow sense van Fraassen adopts. With geometrical optics we can develop the theories of the telescope and the microscope. These theories are readily confirmable by means of experiment. Telescopes, for example, can be used to view from a distance terrestrial objects that can be approached for viewing at close range. It is interesting that van Fraassen regards as observable celestial objects—such as the moons of Jupiter—that can be seen from earth only with the aid of a telescope. We can, in principle, travel closer to them and see them with the naked eye. More interesting still is the fact that he takes objects remote from us in time—such as dinosaurs—to be observables. Time-travel into the past is something NASA has not achieved even in a small way. In contrast, he considers objects that can be seen only with the aid of a microscope as unobservables. There is no other place to go to get a better vantage point.

To substantiate my own claim that we can have empirical knowledge of unobservables, I attempted to spell out the sort of inference that underlies the transition from observation to conclusions about unobservables (1984, 231–34). Shortly thereafter I put the matter in these terms:

> When David Hume stated that all of our reasonings about unobserved matters of fact are based upon the relation of cause and effect he was, I suspect, almost completely correct. One notable exception is induction by simple enumeration (or some other very primitive form of induction). As remarked above, I am assuming for purposes of this discussion that some sort of primi-

tive induction is available; I shall not attempt to characterize it or justify it in this context. The type of argument required in connection with microscopic observation is, I think, causal. It is a rather special variety of causal inference that is also analogical. This particular sort of causal/analogical argument is, in my view, quite powerful. To employ it, we must assume that we already have knowledge of cause-effect relations among observables — e.g., that hitting one's thumb with a hammer causes pain, that drinking water quenches thirst, and that flipping a switch turns off the light. Such relations can be established by Mill's methods and controlled experiments. Neither the instrumentalist's nor the constructive empiricist's account of science can get along without admitting knowledge of such relations. Using relations of this sort, we can now schematize what I suspect is the basic argument enabling us to bridge the gap between the observable and the unobservable. It goes something like this:

It is observed that:

An effect of type E_1 is produced by a cause of type C_1.
An effect of type E_2 is produced by a cause of type C_2.

.

.

.

An effect of type E_k occurred.

We conclude (inductively) that:

A cause of type C_k produced this effect of type E_k.

The particular application of this argument that interests us is the case in which $C_1, C_2, \ldots, C_k$ are similar in most respects except size. Under these circumstances we conclude that they are similar in causal efficacy. (1985, 10–11)

The foregoing argument connects quite directly with an analysis of microscopic observation presented by Ian Hacking just after the publication of van Fraassen's *The Scientific Image*. To learn something about the scientific use of microscopes Hacking did something quite extraordinary for a philosopher. He actually went to a laboratory where the use of microscopes is essential to the research being conducted, and he learned how to use a variety of types. The research involved observation of dense bodies in red blood cells, and it employed a standard device known as a *microscopic grid*. "Slices of a red blood cell are fixed upon a microscopic grid. This is literally a grid: when seen through a microscope one sees a grid each of whose squares is labelled with a capital letter" (1981, 315). Making reference to the microscopic grid, he then addresses the issues raised by van Fraassen:

I now venture a philosopher's aside on the topic of scientific realism. Van Fraassen says we can see through a telescope because although we need the telescope to see the moons of Jupiter when we are positioned on earth, we could go out there and look at the moons with the naked eye. Perhaps that fantasy is close to fulfillment, but it is still science fiction. The microscopist avoids fantasy. Instead of flying to Jupiter he shrinks the visible world. Consider the grid that we used for re-identifying dense bodies. The tiny grids are made of metal: they are barely visible to the naked eye. They are made by drawing a very large grid with pen and ink. Letters are neatly inscribed by a draftsman at the corner of each square on the grid. Then the grid is reduced photographically. Using what are now standard techniques, metal is deposited on the resulting micrograph. . . . The procedures for making such grids are entirely well understood, and as reliable as any other high quality mass production system.

In short, rather than disporting ourselves to Jupiter in an imaginary space ship, we are routinely shrinking a grid. Then we look at the tiny disk and see exactly the same shapes and letters as were drawn in the large by the first draftsman. It is impossible seriously to entertain the thought that the minute disk, which I am holding by a pair of tweezers, does not in fact have the structure of a labelled grid. I know that what I see through the microscope is veridical because we *made* the grid to be just that way. I know that the process of manufacture is reliable, because we can check the results with any kind of microscope, using any of a dozen unrelated physical processes to produce an image. Can we entertain the possibility that, all the same, this is some kind of gigantic coincidence[?] Is it false that the disk is, in fine, in the shape of a labelled grid? Is it a gigantic conspiracy of 13 totally unrelated physical processes that the large scale grid was shrunk into some non-grid which when viewed using 12 different kinds of microscopes still looks like a grid? (1981, 316–17)

To avoid possible misinterpretation I should report that Hacking does not mean to use his argument to support the kind of wholesale realism I argued for in "Why Ask, 'Why?'?" but he does offer a strong argument for the conclusion that we can, with the aid of microscopes, have knowledge of objects and properties not visible to the naked eye.[22]

It seems to me that we can usefully distinguish between direct and indirect observation, where direct observation is accomplished by the use of unaided normal human senses, and indirect observation is accomplished by the use of instruments, such as the microscope and the telescope, that extend the range of the senses. I consider my simple argument, based in part on the development of geometrical optics, and Hacking's argument, based on sophisticated microscopy, strong arguments to the effect that we can have indirect observational knowledge of objects and properties that are not directly observable. In addition, I want to claim, our knowledge of unobservables can be extended even further by appealing to appropriate theoretical considerations. That was the upshot of the discussion of Perrin's argument concerning the ascertainment of Avogadro's number N and the issue of molecular reality in §4.2 above. Perrin studied the behavior of Brownian particles that could be observed using microscopes available in his day. He used his *indirect observations* of these entities as a basis for *inferring* the existence of simple molecules that are much too small to be viewed microscopically. It is mildly amusing that Perrin cited 13 independent methods of ascertaining N and that Hacking refers to 13 independent physical processes in dealing with the microscopic grid. The number 13 appears to be especially unlucky for antirealists.

One problem that has traditionally been associated with scientific realism is the problem of the meaning of theoretical terms. That cannot be the crucial problem, for we have been entertaining claims about unobservable entities without using any esoteric theoretical vocabulary. Since we do successfully describe the things we directly observe, our language must contain an observational vocabulary. However we might characterize it in general, "ink spot," "page," and "smaller than" are surely terms within it. Recall, in this connection, Perrin's work with tiny spheres of gamboge — *much* too small to be observed directly — which can be described entirely within our observational vocabulary.

I realize, of course, that there exists a broad range of philosophical opinion on the matter of scientific realism. I have recapitulated the argument that I find compelling. Other philosophers — e.g., Arthur Fine and Larry Laudan — join van Fraassen in rejecting realism, but for reasons quite different from van Fraassen's. Still others — e.g., Richard Boyd and Ernan McMullin — embrace realism, but appeal to different kinds of arguments to support their views.[23] We cannot escape the realism issue, for it is crucial to the debate between Coffa's highly realistic ontic conception of scientific explanation and van Fraassen's highly pragmatic erotetic approach.

4.6 Railton's Nomothetic/Mechanistic Account

In the first year of the fourth decade Peter Railton's first published article on scientific explanation appeared (1978). It is addressed chiefly to Hempel's I-S model, and it embodies an attempt to provide an account of probabilistic explanation that avoids what Railton regards as the two most troublesome aspects of Hempel's model — namely, epistemic relativization and the requirement of maximal specificity. This article can be viewed, in part, as a further development of some of the ideas expressed in Jeffrey (1969). With Jeffrey, Railton rejects the thesis of the received view that all explanations are arguments, but he goes further than Jeffrey in this regard. Whereas Jeffrey had admitted that in some "beautiful cases" (where the difference between the actual probability and unity is so small as to "make no odds") statistical explanations can be arguments, Railton argues that (practical considerations aside) there is no theoretical difference between the beautiful cases and the unbeautiful. In none of the kinds of cases Hempel treated as inductive-statistical, he claims, should the explanation be construed as an argument. Railton also agrees with Jeffrey in maintaining that, for ineluctably statistical phenomena, the key to explanation lies in understanding the stochastic mechanism by which the occurrence came to pass, not in finding some way to render the event nomically to be expected. In this connection, both Jeffrey and Railton agree that, where some results of a given stochastic process are probable and others improbable, we understand the improbable just as well as we understand those that are highly probable. With Jeffrey, consequently, Railton rejects the high-probability requirement. Railton goes far beyond Jeffrey, however, in spelling out the details of a model of scientific explanation that embodies all of these features.

Railton chooses to elaborate what he calls a *deductive-nomological model of probabilistic explanation* – or D-N-P *model* – in terms of an example of an event that has an extremely small probability, namely, the alpha decay of a nucleus of uranium^{238}. Since the mean-life of this radionuclide is 6.5×10^9 years, the probability that such an atom would decay within a specific short period of time is almost vanishingly small – but not quite vanishing, for such decays do occur. Our theory enables us to calculate the probability p that such a decay will occur within a short time interval Δt. Suppose u, a particular nucleus of this sort, has emitted an alpha-particle during such an interval. Then, we can set up the following deductive argument:

(2) (a) All nuclei of U^{238} have probability p of emitting an alpha-particle during any interval of length Δt, unless subjected to environmental radiation.

   (b) Nucleus u was a nucleus of U^{238} at time t and was subjected to no environmental radiation during the interval $[t, t + \Delta t]$.

   (c) Nucleus u had a probability p of emitting an alpha-particle during the interval $[t, t + \Delta t]$. (Railton 1978, 214, slightly paraphrased)

In this argument Railton, like Coffa, construes the probabilities as single-case propensities, and he regards premise (a) as a lawful generalization. He assumes, moreover, that (a) is an irreducibly probabilistic law, and that it incorporates all probabilistically relevant factors. He recognizes that (2) appears to be an explanation of the fact that u had a probability p of decaying within the specified time interval; he maintains, however, that it can be supplemented in a way that transforms it into an explanation of the fact that u actually decayed. The first addition is "a derivation of (2a) from our theoretical account of the mechanism at work in alpha-decay" (1978, 214) – from the theory of quantum-mechanical tunneling. This is, I take it, another deductive argument employing the Schrödinger wave equation and such facts as the atomic weight and atomic number of $_{92}U^{238}$ as its premises. It would qualify as one of Hempel's D-S explanations.

Railton's approach is strongly mechanistic, as can be seen clearly by contrasting his attitude toward these two arguments with Hempel's attitude. For Hempel, we recall, to explain a particular fact it is sufficient to subsume it under a law. Such explanations are complete. If one wants an explanation of a law that entered into the first explanation, it can be supplied by deriving that law from more general laws or theories. The result is another explanation. The fact that a second explanation of this sort can be given does nothing to impugn the credentials of the first explanation. Railton's view is quite different. According to him—and in this I heartily agree—explanation involves revealing the mechanisms at work in the world. Mere subsumption of phenomena under generalizations does not constitute explanation. Explanation involves understanding *how the world works*. For Railton, then, the quantum mechanical explanation of the probabilistic decay law is an integral and indispensable part of the explanation of the decay that occurred.

To transform this pair of arguments, which still appear to explain the probability of the decay of u, into an explanation of the actual decay of u, we are asked to supply a "parenthetic addendum to the effect that u did alpha-decay during the interval" in question (1978, 214). These three components, taken together, though they do not constitute an argument or a sequence of arguments, do constitute an *explanatory account*:

(3) A derivation of (2a) from our theoretical account of the mechanism of
    alpha decay.
    The D-N inference (2).
    The parenthetic addendum.

A D-N explanation and a D-S explanation (or two D-N explanations, if we continue to consider D-S a subtype of D-N) serve as the core of the explanatory account, but they do not comprise the whole explanation. The parenthetic addendum is also required. If the parenthetic addendum were taken as an additional premise for an argument, the explanation would be vitiated by becoming trivially circular.

If the explanatory account is not an argument, Railton realizes, many readers are going to wonder about its explanatory status:

Still, does (3) explain why the decay took place? It does not explain why the decay *had to* take place, nor does it explain why the decay *could be expected to* take place. And a good thing, too: there is no *had to* or *could be expected to* about decay to explain—it is not only a chance event, but a very improbable one. (3) does explain why the the decay *improbably* took place, which is how it did. (3) accomplishes this by demonstrating that there existed at the time a small but definite physical possibility of decay, and noting that, by chance, this possibility was realized. The derivation of (2a) that begins (3) shows, by assimilating alpha-decay to the chance process of potential barrier tunneling, how this possibility comes to exist. If alpha-decays are chance phenomena of the sort described, then once our theory has achieved all that (3) involves, it has explained them to the hilt, however unsettling this may be to *a priori* intuitions. To insist upon stricter subsumption of the explanandum is not merely to demand what (alas) cannot be, but what decidedly should not be: sufficient reason that one probability rather than another be realized, that is, chances without chance. (1978, 216)

Railton argues carefully that his characterization of probabilistic explanation does escape epistemic relativization, and that it has no need for a requirement of maximal specificity. The basic reason is that, by construing (2a) as a (putative) law, he claims that it is simply false if it fails to be maximally specific. If there are further factors relevant to the occurrence of the decay, which have not been included in the alleged law, it is false. Moreover, the kind of maximal specificity to which he refers is fully objective; it is just as objective as my objectively homogeneous reference classes. Inasmuch as Railton adopts a *single case propensity interpretation* of probability, he has no need for reference classes, but the maximal specificity of his laws is strictly analogous to objective homogeneity for a frequentist (see the discussion of propensities in §3.3 above).

In the introductory sections of his 1978 paper, Railton offers some extremely compact remarks about his approach to explanation in general, not just probabilistic explanations of particular facts. Regarding his mechanistic orientation, he says,

The goal of understanding the world is a theoretical goal, and if the world is a machine — a vast arrangement of nomic connections — then our theory ought to give us some insight into the structure and workings of the mechanism, above and beyond the capability of predicting and controlling its outcomes. . . . Knowing enough to subsume an event under the right kind of laws is not, therefore, tantamount to knowing the *how* and *why* of it. As the explanatory inadequacies of successful practical disciplines remind us: explanation must be more than potentially-predictive inferences or law-invoking recipes. (1978, 208)

Some of the mechanisms are, of course, indeterministic.

The D-N probabilistic explanations to be given below do not explain by giving a deductive argument terminating in the explanandum, for it will be a matter of chance, resisting all but *ex post facto* demonstration. Rather, these explanations subsume a fact in the sense of giving a D-N account of the chance mechanism responsible for it, and showing that our theory implies the existence of some physical possibility, however small, that this mechanism will produce the explanandum in the circumstances given. I hope the remarks just made about the importance of revealing mechanisms have eased the way for an account of probabilistic explanation that focuses on the indeterministic mechanisms at work, rather than the "nomic expectability" of the explanandum. (1978, 209)

The views expressed by Railton in this early article are ones I find highly congenial. There may, however, be a difference in our attitudes toward causal explanation. Like Jeffrey before him, Railton discusses the old chestnut of the falling barometric reading and the storm:

(S) The glass is falling.
    Whenever the glass falls the weather turns bad.

The weather will turn bad.

and considers repairing it by adding a causal premise:

(C) The glass is falling.
    Whenever the glass is falling the atmospheric pressure is falling.
    Whenever the atmospheric pressure is falling the weather turns bad.

The weather will turn bad.

Now Railton considers (C) a causal explanation and its third premise a causal law. He points out that, until we understand the mechanism behind that causal relationship, we do not have an adequate explanation of the turn in the weather. He therefore concludes that we need something over and above the causal infusion that transformed (S) into (C). I agree that we need something more, but we also need something less.

Paul Humphreys has noted (in a personal communication) that there is something odd about the second premise of (C) as a component in a causal explanation, for its presence licenses an inference from effect to cause. And once that point is clear, we see that the first premise of that argument has no place in a causal explanation (or any other, I should think). As we noted in our earlier analysis of this example, the storm and the falling barometer are effects of a common cause, the drop in atmospheric pressure, and the cause screens the common effects off from one another. Therefore (C) should be replaced by

(C') The atmospheric pressure is falling.
     Whenever the atmospheric pressure is falling the weather turns bad.

The weather will turn bad.

The first premise of (C) is no part of the explanation; it is our evidence for the truth of the first premise of (C').

Still, I would not call (C') causal, for my conception of causality includes mechanisms of propagation and interaction. But I would regard the explanation that results from supplementation of (C')—in the way we both consider appropriate—a causal explanation. So far, then, the main disagreement is ter-minological. But Railton also makes mention of such structural laws as the Pauli exclusion principle, which he rightly holds to be noncausal (1978, 207). He characterizes explanations based upon such laws as noncausal *structural explana-tions*. In *Scientific Explanation and the Causal Structure of the World* I had taken the attitude that structural laws have explanatory force only if they themselves can be explained causally. In that context I was thinking of such structural laws as the ideal gas law, and I placed great weight upon its explanation on the basis of kinetic theory. Since I did not endeavor to offer an account of quantum mechani-cal explanation, I did not confront such laws as the Pauli exclusion principle.[24]

At the time Railton sent this article off for publication he must have been work-ing on his dissertation (1980). The brief suggestive remarks he offers in the arti-cle, in addition to his articulation of the D-N-P model, are elaborated clearly and at length in the dissertation—a monumental two-volume work that comes to 851 pages in toto.

When first I read the 1978 essay, I must confess, I failed to understand it. Only after seeing a later paper did I begin to appreciate the beauty of Railton's work on explanation. In 1981 a landmark symposium on probabilistic explanation (Fet-zer 1981) was published; it included, among many other valuable contributions, Railton's "Probability, Explanation, and Information." In this essay he offers fur-ther discussion of the D-N-P model, addressing certain objections that might be brought against it. Of particular importance is the objection based on the ac-knowledged fact that many proffered explanations that are widely accepted as cor-rect omit items, such as laws, that form an integral part of any explanation that conforms to the D-N-P model.

To deal with this problem, Railton introduces a distinction—one that turns out to be extraordinarily fruitful—between an *ideal explanatory text* and *explanatory information* (1981, 240). Given an event we wish to explain, the ideal explanatory text would spell out all of the causal and nomic connections that are relevant to its occurrence. For probabilistic explanation, the D-N-P model furnishes the schema for the ideal explanatory text. The ideal explanatory text can be expected—in most, if not all, cases—to be brutally large and complicated. When one considers the myriad molecules, atoms, subatomic particles, and interactions involved in everyday events, it is easy to see that the ideal explanatory text is an *ideal* that might never be realized. That does not matter.[25] The scientist, in seek-ing scientific understanding of aspects of our world, is searching for *explanatory information* that enables us to fill out parts of the ideal explanatory text. The ideal explanatory text constitutes a framework that provides guidance for those who endeavor to achieve understanding of various aspects of the world, and that is its primary function.

Railton acknowledges the possibility that, in addition to the probabilistic explanations characterized by the D-N-P model, there may be nonprobabilistic explanations that are closely related to Hempel's D-N model. For Railton, however, D-N explanations are far more robust than they are for Hempel:

> I would argue that the D-N schema instead provides the skeletal form for ideal explanatory texts of non-probabilistic phenomena, where these ideal texts in turn afford a yardstick against which to measure the explanatoriness of proffered explanations in precisely the same way that ideal D-N-P texts afford a yardstick for proffered explanations of chance phenomena. Thus, proffered explanations of non-probabilistic phenomena may take various forms and still be successful in virtue of communicating information about the relevant ideal text. For example, an ideal text for the explanation of the outcome of a causal process would look something like this: an inter-connected series of law-based accounts of all the nodes and links in the causal network culminating in the explanandum, complete with a fully detailed description of the causal mechanisms involved and theoretical derivations of all of the covering laws involved. This full-blown causal account would extend, via various relations of reduction and supervenience, to all levels of analysis, i.e., the ideal text would be closed under relations of causal dependence, reduction, and supervenience. It would be the whole story concerning why the explanandum occurred, relative to a correct theory of the lawful dependencies of the world. (1981, 246–47)

Does the conception of an ideal explanatory text have any utility? Railton replies,

> [Is it] preposterous to suggest that any such ideal could exist for scientific explanation and understanding? Has anyone ever attempted or even wanted to construct an ideal causal or probabilistic text? It is not preposterous if we recognize that the actual ideal is not to *produce* such texts, but to have the ability (in principle) to produce arbitrary parts of them. It is thus irrelevant whether individual scientists ever set out to fill in ideal texts as wholes, since within the division of labor among scientists it is possible to find someone (or, more precisely, some group) interested in developing the ability to fill in virtually any aspect of ideal texts—macro or micro, fundamental or "phenomenological," stretching over experimental or historical or geological or cosmological time. A chemist may be uninterested in how the reagents he handles came into being; a cosmologist may be interested in just that; a geologist may be interested in how those substances came to be distributed over the surface of the earth; an evolutionary biologist may be interested in how chemists (and the rest of us) came into being; an anthropologist or historian may be interested in how

man and material came into contact with one another. To the extent that there are links and nodes, at whatever level of analysis, which we could not even in principle fill in, we may say that we do not completely understand the phenomenon under study. (1981, 247–48)[26]

The distinction between the ideal explanatory text and explanatory information can go a long way, I think, in reconciling the views of the pragmatists and the realists.[27] In his work on deductive-nomological explanation, Hempel steadfastly maintained an attitude of objectivity. D-N explanations were never epistemically relativized, and they always fulfill the requirement of maximal specificity trivially. As we have seen, Hempel found it necessary to relinquish this attitude when he confronted inductive-statistical explanations, but Coffa, Railton, and I, among others, made serious efforts to restore full objectivity to the domain of probabilistic explanation. At the same time, those philosophers who have emphasized the pragmatics of explanation, from Hanson and Scriven to van Fraassen, have criticized objectivist accounts for demanding inclusion of too much material to achieve what they regard as legitimate explanations. This opposition has, for example, led to the rejection of the covering law conception of explanation by pragmatists, and to insistence upon it by the objectivists.

The issue of how 'fat' or 'thin' explanations should be was raised in *Scientific Explanation and the Causal Structure of the World* (1984, 131–34). and it resurfaced in the 1985 American Philosophical Association Symposium on that book (Kitcher, van Fraassen, and W. Salmon, 1985). One useful way to think about this conflict, I believe, is to regard the objectivists—the advocates of the ontic conception—as focusing on the ideal explanatory text. We all hoped, I suspect, that the ideal explanatory text would be a good deal simpler than Railton had conceived it to be, but it did not work out that way. It was clear to me long ago, for example, in elaborating the statistical relevance model, that an objectively homogeneous partition, based on all objectively relevant factors, would usually have a horrendous number of cells, and, consequently, that a complete S-R explanation would be frightfully complex. However, if it is only the ideal text that is so forbidding, the situation is not so hopeless. The ideal explanatory text contains all of the objective aspects of the explanation; it is not affected by pragmatic considerations. It contains all *relevant* considerations.

When we turn to explanatory information, pragmatic considerations immediately loom large. What part of the ideal explanatory text should we try to illuminate? Whatever is *salient* in the context under consideration. That depends upon the interests and the background knowledge of whoever seeks the explanation. It depends upon the explanation-seeking why-question that is posed. Pragmatic considerations must not be seen as total determinants of what constitutes an adequate explanation, for whatever explanation is offered must contain explanatory information that coincides with something or other in the ideal explanatory text. The ideal explanatory text determines what constitutes explanatory information, and distinguishes it from explanatory misinformation. *Relevance* is a matter of objective fact; *salience* is a matter of personal or social interest. Thus, I should be inclined to say, the putative astrological explanation of President John F. Kennedy's death—discussed in §4.4—can be ruled out because it fails to coincide with any part of the ideal explanatory text. The configuration of stars and planets and satellites at the time of Kennedy's birth is (I firmly believe) irrelevant to the date of his assassination.

Looking at the pragmatics of explanation in this way, we can account for the rejections and asymmetries of explanation, which van Fraassen takes to be crucial (see Railton 1981, 248; fn 15 explicitly refers to van Fraassen). We can evaluate the presuppositions of why-questions to see if they are objectively satisfied. We can take necessary steps to determine—on the basis of contextual factors—just exactly what why-question is being posed. To do so we must specify van Fraassen's contrast class. We can consider the background knowledge of the individual who poses the question to ascertain what is missing—what knowledge gaps need to be filled if that person is to achieve scientific understanding—what aspect of the ideal text needs to be exhibited. We would also take into account the capacity of the questioner to assimilate scientific information in order to determine the depth and detail appropriate for that individual.

Given Railton's conception of explanatory information and the ideal explanatory text, it seems to me that much of the longstanding battle—going back to the early skirmishes between the logical empiricists and the ordinary language philosophers—can be resolved. I see this achievement as one foundation upon which a new consensus might be erected.

Railton refers to his general conception of scientific explanation as the *nomothetic account*; he compares it to the received view in the following way:

> Where the orthodox covering-law account of explanation propounded by Hempel and others was right has been in claiming that explanatory practice in the sciences is in a central way *law-seeking* or *nomothetic*. Where it went wrong was in interpreting this fact as grounds for saying that any successful explanation must succeed either in virtue of explicitly invoking covering laws or by implicitly asserting the existence of such laws. It is difficult to dispute the claim that scientific explanatory practice—whether engaged in causal, probabilistic, reductive, or functional explanation—*aims* ultimately (though not exclusively) at uncovering laws. This aim is reflected in the account offered here in the structure of ideal explanatory texts: their backbone is a series of law-based deductions. But it is equally difficult to dispute the claim that many proffered explanations succeed in doing some genuine explaining *without* either using laws explicitly or (somehow) tacitly asserting their existence. This fact is reflected here in the analysis offered of explanatoriness, which is

treated as a matter of providing accurate information about the relevant ideal explanatory text, where this information may concern features of that text other than laws. (1981, 248–49)

In choosing to call his account *nomothetic* Railton emphasizes the role of laws. I think it equally deserves to be called *mechanistic*. We have already noted the crucial role of mechanisms in connection with the D-N-P model. In his 1981 article he remarks more generally on the standards for judging accounts of explanation:

The place to look for guidance is plainly scientific explanatory practice itself. If one inspects the best-developed explanations in physics or chemistry textbooks and monographs, one will observe that these accounts typically include not only derivations of lower-level laws and generalizations from higher-level theory and facts, but also attempts to *elucidate the mechanisms* at work. Thus an account of alpha-decay ordinarily does more than solve the wave-equation for given radionuclei and their alpha-particles; it also provides a model of the nucleus as a potential well, shows how alpha-decay is an example of the general phenomenon of potential-barrier penetration ("tunnelling"), discusses decay products and sequences, and so on. Some simplifying assumptions are invariably made, along with an expression of hope that as we learn more about the nucleus and the forces involved we will be able to give a more realistic physical model. It seems to me implausible to follow the old empiricist line and treat all these remarks on mechanisms, models, and so on as mere *marginalia*, incidental to the "real explanation", the law-based inference to the explanandum. I do not have anything very definite to say about what would count as "elucidating the mechanisms at work" — probabilistic or otherwise — but it seems clear enough that an account of scientific explanation seeking fidelity to scientific explanatory practice should recognize that part of scientific ideals of explanation and understanding is a description of the mechanisms at work, where this includes, but is not merely, an invocation of the relevant laws. Theories broadly conceived, complete with fundamental notions about how nature works — corpuscularianism, action-at-a-distance theory, ether theory, atomic theory, elementary particle theory, the hoped-for unified field theory, etc. — not laws alone, are the touchstone in explanation. (1981, 242)

In view of these comments, and many others like them, I am inclined to consider Railton's account primarily mechanistic and secondarily nomothetic.

Some time after I had seen his 1981 article, I happened to fall into conversation with Hempel — who was a colleague in Pittsburgh at the time — about Railton's work. He mentioned that he had a copy of Railton's dissertation (1980) and offered to lend it to me. After looking at parts of it rather quickly, I realized that I had found a treasure. Obtaining a copy of my own, I studied it at length, and made considerable use of it in conducting the Minnesota Institute on Scientific Explanation in 1985. In a general survey article such as this I cannot attempt a full summary of its contents, but a little more should be said to indicate the scope of the theory he develops.

In his dissertation Railton distinguishes three kinds of explanations of particular facts. In response to various difficulties with D-N explanations of particular facts as treated by the received view, Railton maintains that an appeal to causality is required in many cases. He characterizes the *ideal causal nomothetic explanatory text for particular facts* and concludes that it "includes an account of the causal mechanisms involved at all levels, and of all the strands in the causal network that terminates in the explanandum" (1980, 725). However, he resists the notion that all particular-fact explanations are causal. There are, as we have seen, probabilistic explanations of particular facts, and we have discussed the appropriate type of ideal explanatory text. As an advocate of the propensity interpretation of probability, he regards probabilistic explanations as dispositional. But he is willing to take the concept of disposition in a sense broad enough to include dispositions of strengths zero and unity; consequently, he introduces, in connection with his second type of particular-fact explanation, the notion of the *ideal explanatory text for dispositional particular-fact explanations*, of which the D-N-P model provides a special case (1980, 735–36). The third type of particular-fact explanation, for Railton, is structural. Such explanations appeal typically to such laws of coexistence as the Pauli exclusion principle, Archimedes' principle, laws of conservation, and so on. Maintaining that such explanations cannot plausibly be construed as disguised causal explanations, he also introduces the notion of the *ideal explanatory text for structural particular-fact explanations* (1980, 736–39). Explanations of this type—like causal and dispositional explanations—involve essential reference to the underlying mechanisms.

Recognizing that actual explanations of particular occurrences may involve elements of more than one of the foregoing, Railton also gives us the notion of a *nomothetic ideal encyclopedic text* (1980, 739). "Harmonious co-operation among the elements of an encyclopedic text is possible because all three forms reflect a conception of explanation that we may intuitively describe as this: an explanation should invoke the factors, laws, etc., that actually *bring about* or *are responsible for* the explanandum (once these relations are broadened beyond the purely causal), i.e., the explanations should show what features of the world the explanandum is *due to*" (1980, 739–40).

Because of certain differences in terminology among the three of us, perhaps it would be helpful to say something about the relationships among Coffa, Railton, and me. First, it should be noted, I am an advocate of probabilistic causality, so when I speak of causal explanation it is explicitly intended to include probabilistic explanation. Both Coffa and Railton use the term "cause" in the narrower deterministic sense. Furthermore, Coffa and Railton advocate the propensity interpretation of probability, while I reject the notion that propensities satisfy the axioms of probability, and hence that there is any such thing as a propensity interpretation. However, I do believe that the concept of a probabilistic propensity is an extremely useful notion; I would identify it with probabilistic cause. Hence, not to put too fine a point on it, Coffa, Railton, and I all agree on the general idea of a causal or dispositional type of explanation that coincides with Railton's causal and dispositional types. Up to this point, the differences are mainly terminological. A major difference arises, however, between Coffa on the one hand and Railton and me on the other, for both of us place far more importance upon an appeal to mechanisms than does Coffa. On another point Railton differs from Coffa and me; neither of us accords a distinctive place to Railton's structural explanations. My *constitutive explanations* bear some resemblance, but I regard them as fundamentally causal (1984, 270–76), whereas Railton explicitly denies that his structural explanations are. While I am still inclined to think that many of Railton's structural explanations can be analyzed causally, I am far from confident that all can. That point seems particularly clear with regard to such quantum mechanical examples as the Pauli exclusion principle, but, as we know, the nature of quantum mechanical explanation is deeply perplexing (W. Salmon 1984, 242–59). Nevertheless, had I read Railton's dissertation before the publication of my book, I would have devoted considerably more attention to structural explanation.

Railton's overall account is not confined to particular-fact explanations; he also offers an account of theoretical explanation. In summarizing his approach he remarks,

> It was argued that by beginning with particular-fact explanation, we had not prejudiced the account against theoretical explanation, for, in fact, theoretical explanation had been involved all along in the full development of ideal explanatory texts. As might have been expected, the nomothetic account recognizes that regularities and laws may have causal, dispositional, or structural elements in their explanations, and these elements have virtually the same ideal forms as in particular-fact explanation. Thus the same requirements of true ingredients, basic covering-law structure, an account of mechanisms, thorough-going theoretical derivation, asymmetries, etc., apply in theoretical explanation. The nomothetic account thereby preserves the estimable unity of theoretical and particular-fact explanation that is characteristic of covering-law approaches to explanation. (1981, 746–47)[28]

Railton's dissertation offers a deep and sensitive treatment of a wide range of problems, issues, and views regarding the nature of scientific explanation. Although it does not attempt to resolve some of the most fundamental problems we have encountered in our discussions of explanations – such problems as the nature of laws, the analysis of causality, the nature of mechanisms, the notion of a purely qualitative predicate—it is a rich source of philosophical insight on the nature of scientific explanation. While I do not by any means agree with all of his views, I do believe that anyone who is seriously interested in philosophical work on scientific explanation should study his dissertation with care.

4.7 Aleatory Explanation: Statistical vs. Causal Relevance

In the early part of the third decade—when I was busily expounding the statistical-relevance model—I was aware that explanation involves causality, but I hoped that the required causal relations could be fully explicated by means of such statistical concepts as screening off and the conjunctive fork. A decade later, I was quite thoroughly convinced that this hope could not be fulfilled (W. Salmon 1980; see also 1984, chap. 7). Along with this realization came the recognition that statistical relevance relations, in and of themselves, have no explanatory force. They have significance for scientific explanation only insofar as they provide evidence for causal relations. By 1984 (34–47) they had been relegated to *the S-R basis* upon which causal explanations can be founded. Causal explanation, I argued, must appeal to such mechanisms as causal propagation and causal interactions, which are not explicated in statistical terms.

The question arises of what to do with the S-R basis. Given the fact that it will often be dreadfully complex, we might consign it to Railton's ideal explanatory text, recognizing that it is the sort of thing that will not often be spelled out explicitly. We will refer to parts of it when we need to substantiate causal claims.

Another sensible approach to this problem is, in the words of Frank Lloyd Wright, "Abandon it!"[29] This is the tack taken by Paul Humphreys in articulating his theory of *aleatory explanation* (1981, 1983). A basic difference between Humphreys's model and other models of probabilistic or statistical explanation extant at the time is that all of the latter require one or more probability values to appear explicitly in the completed explanation. Although knowledge of probabilities is used in constructing aleatory explanations, values of probabilities are absent from the explanation itself.

According to Humphreys, factors that are causally relevant are also statistically relevant. Just as there are two kinds of statistical relevance—positive and negative—so also are there two kinds of causes. Causes that tend to bring about a given effect are *contributing causes*; those that tend to prevent the effect are *counteracting causes*. The canonical form for an aleatory explanation is " 'A because Φ, despite Ψ', where Φ is a non-empty set of contributing causes, Ψ is a set, possibly empty, of counteracting causes, and A is a sentence describing what is to be explained." It is assumed that Φ and Ψ include all known causally relevant factors. The set of contributing causes must not be empty, for we have no explanation at all if only counteracting causes are present.

Consider a modified version of one of Humphreys's examples. Suppose that a car has gone off a concrete road at a curve and that the only known conditions that are causally relevant are the fact that the driver was fully alert and the car was traveling at an excessive speed. The first is obviously a counteracting cause of the accident; the latter a contributing cause. We might then say that the car went off the road because it was traveling too fast, despite the fact that the driver was alert. There are, of course, many other statistically relevant factors, and if they are taken into account the probability of the explanandum will change. But as long as there are no other factors that screen off excessive speed, or render it a counteracting cause, its status as a contributing cause is unchanged. Similarly, if there are no other factors that screen off driver alertness or transform it into a contributing cause, its status as a counteracting cause holds.

Suppose now we find out that, in addition, visibility was clear, but there was sand on the road at the curve. The first of these is a counteracting cause of the accident, and the second is a contributing cause. We may add both to the original explanation, transforming it into the following: The car went off the road, despite the fact that the driver was alert and visibility was clear, because the car was traveling too fast and there was sand on the road at the curve. The result is that the first explanation, though incomplete, is correct; we have not been forced to retract any part of the original. If, in contrast, we had been trying to construct an S-R explanation of the same phenomenon, we would have been required to retract the original statistical relevance relations and replace them with others.

In constructing aleatory explanations, we must be aware of the possibility that the introduction of an additional causally relevant factor may change a contributing cause into a counteracting cause or vice-versa, or it may render a cause of either type irrelevant. Suppose we learn that there was ice on the road at this curve. This would change the presence of sand on the road from a contributing cause to a counteracting cause, for if the road were icy the sand would tend to prevent the car from skidding. In this case, the additional factor would force us to reject the former explanation, even as a correct partial explanation.

When constructing aleatory explanations, we are aware of the danger of a defeating condition—i.e., a condition, such as ice on the road in the preceding example, that transforms a contributing cause into a counteracting cause or vice-versa. By careful investigation we try to assure ourselves that none are present. Establishing such a result would be analogous to verifying one of Coffa's extremal clauses (§3.3). Just as we might check for the absence of compressing forces before applying the law of thermal expansion to a heated iron bar, so also would we check for road surface conditions that could transform sand on the road from a contributing to a counteracting cause before offering an explanation of the car leaving the road.

Humphreys's theory of aleatory explanation falls clearly within the ontic conception; it constitutes a valuable contribution to our understanding of causal explanation, where probabilistic causes are included. Inasmuch as its fullest articulation is contained in "Scientific Explanation: The Causes, Some of the Causes, and Nothing but the Causes," his contribution to this volume (Kitcher and Salmon 1989), I shall resist the temptation to discuss his theory at greater length and let him speak for himself.

4.8 Probabilistic Causality

In the third decade, as we saw, Coffa's theory placed the notion of a probabilistic disposition (what I would call a probabilistic cause) in a central position with respect to scientific explanation. During the same decade, a similar line of thought was pursued by Fetzer, whose theory was fully articulated early in the fourth decade. By the beginning of the fourth decade I was thinking seriously about how the S-R (statistical-relevance) model could be augmented by suitable causal considerations. At that time I was aware of only three theories of probabilistic causality that had been reasonably well worked out, namely, those of Reichenbach (1956), I. J. Good (1961–62), and Patrick Suppes (1970). Neither Reichenbach nor Suppes drew any connection between his theory of probabilistic causality and scientific explanation; Good used his mathematical definition of degree of explicativity (1977) in a key role in his probabilistic causal calculus, but without much philosophical elaboration. I surveyed these theories (1980) and pointed to severe difficulties in each; soon thereafter Richard Otte (1981) exhibited still greater problems in Suppes's theory. The primary moral I drew was that causal concepts cannot be fully explicated in terms of statistical relationships; in addition, I concluded, we need to appeal to causal processes and causal interactions. (For details see W. Salmon 1984, chaps. 5–7.)

Early in the fourth decade Fetzer and Nute (1979) published a new theory of probabilistic causality that was intended to play a crucial role in the explication of scientific explanation.[30] Fetzer's theory of scientific explanation is based on two cardinal principles: first, the interpretation of probability as a single-case propensity, and second, the inadequacy of extensional logic for the explication of such fundamental concepts as lawlikeness and causality. Single-case propensities are understood as probabilistic dispositions, indeed, as probabilistic causes. To elaborate a theory of scientific explanation that embodies these ideas, Fetzer and Nute construct a modal logic in which conditional statements embodying three kinds of special connectives are introduced, namely, subjunctive conditionals, causal conditionals (involving dispositions of universal strength), and probabilistic conditionals (involving dispositions having numerical degrees of strength). Those of the last type are statements of probabilistic causality.

The formal system they construct has 26 axiom schemas in all. The first six form the basis for the logic of subjunctive conditionals; the symbol for the connective is a fork ($\looparrowright$). This logic is not different in kind from various well-known systems of modal logic. The next eight axiom schemas form the basis for the logic of universal causal conditionals; the symbol for this connective is the u-fork ($\ni_u$). There is nothing wildly nonstandard in this part of the system either. The final dozen axiom schemas form the basis for the logic of probabilistic causality. At this stage, however, instead of introducing one new connective symbol, the authors introduce a *nondenumerable infinity* of symbols ($\ni_n$), the n-forks, where n assumes the value of each real number in the unit interval. They then talk about establishing the well-formed formulas of this calculus "in the usual way."

What they have done is, on the contrary, most unusual. In the first place, the standard representation of real numbers is by means of sequences of digits, almost all of which are infinite sequences. It appears, then, that the vast majority of symbols of the form "$\ni_n$" are not single symbols, or finite strings of symbols, but infinitely long strings. Standard logical and mathematical languages, though they usually admit countable infinities of symbols, generally limit the well-formed formulas to finite length. In standard logical and mathematical languages there can exist names for no more than a denumerable subset of the real numbers. The use of a nondenumerable infinity of symbols and infinitely long well-formed formulas signal fundamental difficulties for the Fetzer-Nute proposed system.

In several of the axiom schemas, for example, the n-fork appears twice with the same subscript. If we look at a given formula containing two n-forks with their numerical subscripts — one that coincides with an axiom schema in all other respects than the equality or inequality of the two subscripts — we must determine whether the subscripts are identical to ascertain whether the formula is an axiom. This is done by comparing the two sequences digit by digit. If the two sequences are different, we will discover a discrepancy within a finite number of comparisons; but if they are the same we will never be able to establish that fact. There is no effective way of deciding, in general, whether two representations designate the same real number.

One of the basic virtues of a good axiomatic system is that it provides us with the capability of recognizing an axiom when we see one. If we have a finite number of finite formulas that capability is evident. In addition, given a finite number of *suitable* axiom schemas, each of which admits an infinite set of axioms, it is possible to provide an effective method for recognizing an axiom when we meet one. The Fetzer-Nute system, as presented, does not give us any such method. Unless the formation rules are spelled out in full detail, and a way is found to circumvent this problem, the probabilistic part of the calculus cannot get off the ground. Fetzer and Nute acknowledge that their nondenumerable infinity of well-formed formulas presents a difficulty when it comes to an attempt to establish the completeness of this part of their calculus (Fetzer and Nute 1979, 473; Fetzer 1981, 67), but I think the problem lies much deeper.

Another conspicuous feature of the probabilistic part of the calculus is that from twelve rather complicated axioms only four utterly trivial theorems are offered:

If p probabilistically implies q, then

    (a) p does not necessarily imply q, and p does not necessarily imply not-q;

    (b) p does not have a universal disposition to produce q, and p does not have a universal disposition to produce not-q;[31]

    (c) p is possible and not-q is possible;[32]

    (d) p and q are jointly possible.

Such a large tree should bear more fruit.

The foregoing criticisms are not meant to imply that all attempts to construct a probabilistic causal calculus within a nonextensional logic are bound to be futile or fruitless, but only that the Fetzer-Nute version needs further work if it is to succeed.

In the less formal portion of his work on scientific explanation, Fetzer offers searching discussions of Hempel's requirement of maximal specificity and of the homogeneity requirement I imposed on the S-R model. As we recall, in 1968 Hempel replaced his 1965 RMS with a revised version RMS*. Fetzer offers further revisions and offers RMS** which, he believes, deals adequately with the problems concerning relevance that I had raised (1981, 99). In addition, he offers a revised explication of homogeneity designed to escape certain difficulties he alleged to be present in mine.

Two main points of his critique are worth at least brief comment. The most serious, I think, involves "the mistaken identification of statistical relevance with explanatory relevance" (1981, 93), since "*statistically relevant* properties are not necessarily *causally relevant* (or *nomically relevant*) properties, and conversely . . . " (1981, 92). This is a criticism whose validity I have completely endorsed, as I have remarked repeatedly in foregoing sections. When the S-R model was first published I believed that causal relevance could be explicated entirely in terms of statistical relevance relations, making great use of the screening off relation. In that place, however, I did not attempt to carry out any such analysis. During the next several years I became increasingly skeptical about the viability of any such approach, arriving finally at the fairly strong conviction that it could not be done (1980). However, the details of the mechanisms of causality and their role in scientific explanation were not spelled out in full detail until (1984), three years after the publication of Fetzer's book.

Another one of Fetzer's basic arguments is that, according to my definition, the only objectively homogeneous reference classes are those that have only one member, and, consequently, the relative frequency of any attribute in such a class is, of logical necessity, either zero or one (1981, 86–94). Therefore, he claims, whereas I had accused Hempel's theory of an implicit commitment to determinism, in fact my account is at least equally guilty. As I had long been aware, the reference class problem is extremely serious for any theory of statistical explanation. In Hempel's account, RMS (later RMS*) was designed to deal with this problem; in mine, objective homogeneity was intended to handle it. In my initial presentations of the S-R model, early in the third decade, I handled the problem quite cavalierly, making passing reference to Richard von Mises's concept of a place selection. By the end of the third decade, I realized that it had to be taken very seriously. My first attempt at a detailed explication of objective homogeneity was badly flawed (1977b); an improved treatment, which I hope is more success-ful, was not published until 1984 (chap. 3). However that may be, I am still con-vinced that the concept of an *objectively homogeneous reference class* is legiti-mate and important. I maintain, for example, that the class of carbon-14 atoms is objectively homogeneous with respect to the attribute of spontaneous radioac-tive decay within 5730 years. If our concept does not fit cases of that sort, our explication must be at fault.

Obviously I cannot accept Fetzer's view that problems with objective homogeneity are merely unfortunate consequences of adopting of the *extensional* frequency interpretation of probability; they are problems that can be solved and deserve to be solved. Moreover, as I have argued in detail (1979a), the single-case propensity interpretation does not escape what amounts to the same type of problem, namely, the specification of the chance set-up that is supposed to pos-sess the probabilistic disposition, for on any propensity interpretation it is neces-sary to specify what counts as repeating the experiment. Fetzer addresses this problem by imposing *the requirement of strict maximal specificity*:

> An explanation of why an explanandum event . . . occurs is adequate *only if* every property described by the antecedent condition(s) is nomically rele-vant to the occurrence of its attribute property . . . . (1981, 125–26)

Nomic relevance is relevance by virtue of a causal or noncausal law. Nomic rele-vance is emphatically not to be identified with statistical relevance.

Fetzer uses this requirement in the formulation of his characterization of causal explanation:

> A set of sentences S, known as the "explanans," provides *an adequate nomi-cally significant causal explanation of the occurrence of a singular event de-scribed by another sentence E, known as its explanandum, relative to [a given] language framework*, if and only if:
>
> (a) the explanandum is either a deductive or a probabilistic consequence of its explanans;
> (b) the explanans contains at least one lawlikè sentence (of universal or statisti-cal) 'causal' form that is actually required for the deduction or probabilistic derivation of the explanandum from its explanans;
> (c) the explanans satisfies the requirement of strict maximal specificity (RSMS) with respect to its lawlike premise(s); and

(d) the sentences constituting the explanation—both the explanans and the explanandum—are true, relative to the [given] language framework. (1981,126-27)

As a gloss on this formulation, Fetzer remarks, "If the law(s) invoked in the explanans are essentially universal, the logical properties of the relationship between the sentences constituting the explanans and its explanandum will be those of complete (deductive) entailment; while if they are essentially statistical, this relationship will be that of only partial (deductive) entailment. The logical relation, in either case, is *strictly deductive*" (1981, 127). In the following chapter, Fetzer offers an entirely parallel analysis of *nomically significant theoretical explanation*, a noncausal form of explanation.

Fetzer has offered a nonextensional explication which, by virtue of its appeal to partial entailment, bears striking resemblance to Mellor's nondeterministic version of the modal conception of scientific explanation. In commenting (above) on Mellor's views I expressed my strong doubts about the viability of the concept of partial entailment; these qualms apply equally to Fetzer's employment of it. Given the intensional analysis, however, Fetzer has a straightforward answer to the issue we raised in §4.3 concerning the relationship between descriptive and explanatory knowledge. He suggests that the distinction between *description and prediction*, on the one hand, and *explanation*, on the other, is that the former can proceed in an extensional language framework, while the latter demands an intensional language framework. It remains to be seen whether the intensional logic can be satisfactorily formulated.

Fetzer has offered an account of explanation that is, in an extended sense, deductive. His book is dedicated to Karl Popper. In the next section we shall consider a more orthodox Popperian approach to statistical explanation.

4.9 Deductivism

The thesis that all legitimate scientific explanations are deductive arguments has a long and proud history, going back at least to Aristotle. It has been reiterated by many philosophers, including John Stuart Mill in the nineteenth century, and Karl R. Popper in the early part of the twentieth century. During our four decades it has been advocated by Brodbeck, Stegmüller, and von Wright, as well as many others, including Popper and his followers. It is *not*, however, a view that Hempel ever explicitly held, for he steadfastly maintained that there are, in addition, explanations of the inductive-statistical type.[33]

Before the twentieth century, during the reign of classical physics and Laplacian determinism, deductivism with respect to scientific explanation was a natural and appealing view. One might have tolerated something like explanations of the inductive-statistical variety as long as it was clearly understood that they were incomplete, the appeal to probability or induction being merely a result of our ignorance of the full explanatory laws or facts. With the advent of quantum mechanics it became necessary to admit that determinism might well be false, and that deductive-nomological explanations of some important phenomena may be impossible in principle.

One possible response to these new developments in physics is simply to hold stubbornly to determinism; this reaction seems utterly anachronistic. New–as yet undiscovered–physical theories may eventually convince us that determinism is true; even so, we have no *a priori* guarantee that determinism will be reinstated. At this point in the history of physics our philosophical theories of explanation must leave open the possibility that the world is indeterministic.

Another possible response is to admit that quantum mechanics is an indeterministic theory, but then to deny that quantum mechanics furnishes explanations of any physical phenomena. This reaction also seems unwarranted. Quantum mechanics (including quantum electrodynamics and quantum chromodynamics) has had more explanatory success than any other theory in the history of science. Classical physics could not explain the distribution of energy in the spectrum of blackbody radiation; quantum mechanics provided the explanation. Classical physics could not explain the photoelectric effect; quantum mechanics could. Classical physics could not explain the stability of atoms; quantum mechanics could. Classical physics could not explain the discrete spectrum of hydrogen; quantum mechanics could. Unless one wants to retreat to the old position that science never explains anything, it seems implausible in the extreme to deny that quantum mechanics has enormous explanatory power.

Another avenue is, however, open to the deductivist. Without relinquishing the deductivist position, one can concede that quantum mechanics has great explanatory power, but that all of its explanations are of the type Hempel characterized as deductive-statistical. This means, of course, that quantum mechanics can furnish no explanations of particular facts; the only things it can explain are statistical laws. To take just one example, the deductivist who adopts this stance must maintain that, while quantum mechanics can explain how, in general, electrons are scattered by crystals, it cannot explain the particular patterns actually obtained by Davisson and Germer in their famous experiment.[34] I find it difficult to accept this conclusion. Given the fact that quantum mechanics governs the microstructure of all macroscopic phenomena, it would appear to lead to the conclusion that science cannot explain any particular occurrences whatever. As I argue in a paper that will appear in the first year of the fifth decade (1988), even if theoretical science could be shown to have no need of explanations of particular facts, it seems impossible to make the same claim for applied science. When, for example, questions of legal liability arise, we seek causal explanations (which may have probabilistic components) for such particular occurrences as the collapse of a tank holding vast quantities of oil or the contracting of lung cancer by a long-term heavy smoker of cigarettes. Philosophy of science that confines its attention to pure science, to the complete neglect of applied science, is, I suggest, severely biased.

Another maneuver available to the deductivist is to maintain that, although we cannot have a scientific explanation of a particular chance event, we can have an explanation of the fact that such an event has some particular probability. An easy way to see how it would work is to reconsider Railton's D-N-P model (discussed in section 4.6). In focusing on Railton's model, it is essential to keep in mind that he is not a deductivist. According to Railton, we can have explanations of chance events; such explanations consist of two parts: (1) a deductive argument, whose conclusion is that the event in question has a certain probability, and (2) a parenthetic addendum, which states that the event in fact occurred. If we were to accept the first part, while rejecting the second, we would be left with a deductive explanation of the fact that the explanandum-event has a particular probability.[35] Such a view leaves us in the position of having to say that science cannot explain what happens in the world; it can only explain why those things that do happen have certain probabilities. It can also explain why things that do not happen have certain probabilities of occurring.

When Richard Jeffrey challenged Hempel's claim that statistical explanations are arguments, he excepted certain "beautiful cases" in which the probability of occurrence is so great that there seems no point in giving any weight at all to their non-occurrence. One of his examples is the failure of a flat tire to reinflate spontaneously as a result of a jet of air formed by chance from the random motion of molecules in the surrounding air. Another example would be the melting of an ice-cube placed in a glass of tepid water. Although he allowed that such explanations may be construed as arguments, he did not, of course, claim that they are deductive. Railton maintained, on the contrary, that the mere fact that some probabilities are high and some not so high does not make a difference in principle between the "beautiful" and less attractive cases. If the latter are not arguments then neither are the former. My own view is that Railton is correct in this observation, but my opposition to the inferential conception of explanation is so deep that I may simply be prejudiced.

One staunch deductivist who has recently considered the problem of Jeffrey's "beautiful cases" is John Watkins, whose neo-Popperian *Science and Scepticism* appeared in 1984. Although he is a deductivist, he maintains that, while science cannot provide explanations of individual micro-events, it can furnish explanations of macro-events that consist of large aggregates of chance micro-events:

> There is a far-reaching analogy between the explanatory power of the deterministic theories of classical physics and the indeterministic theories of modern micro-physics . . . Both explain empirical regularities by appealing to higher level structural laws that are taken as immutable and absolute. In the

case of classical physics, such a law says that, given only that such-and-such conditions are satisfied, nothing whatever can prevent a certain outcome from following. In the case of microphysics, it says that, given only that such-and-such conditions are satisfied, nothing whatever can alter *the chance* that a certain outcome will follow. One could as well call the latter an "iron law" of chance as the former an "iron law" of nomic necessity. Both kinds of law, in conjunction with appropriate initial conditions, can explain empirical regularities and, indeed, singular macro-events (provided that the macro-event in question is the resultant of a huge aggregate of micro-events . . .). The analogy breaks down, however, when we come down to individual events at the micro-level. . . . I have argued that it is a mistake to call upon microphysics to *explain* an individual micro-event, such as the disintegration of a radon atom or the reflection of a photon; for if it really was a matter of chance which way it went, then the fact that it chanced to go this way rather than that simply defies explanation. What we *can* explain with the help of an appropriate indeterministic microphysical theory is why there was a precise objective probability that it would go this way. (1984, 246)

To support his claim that particular macro-events can be explained, Watkins discusses one of Hempel's familiar examples. Suppose we have a sample consisting of 10 milligrams of radon. We find, after 7.64 days (two half-lives of radon), that the sample contains 2.5 $\pm$ 0.1 milligrams of radon. Given the law of spontaneous radioactive decay and the fact that the original sample contained more than 10^{19} atoms, we can deduce the probability that at the end of 7.64 days 2.5 $\pm$ 0.1 milligrams of radon will remain. This probability is extremely close to unity; indeed, it differs from one by less than $10^{-(10^{15})}$

At this point, anyone, whether a deductivist, inductivist, or whatever, concerned with the nature of microphysics, faces a crucial question: what *physical* meaning should be given to vanishingly small values [of probability]? Should we interpret our disintegration law as allowing that the macro-outcome *might* fall outside the [given] interval? Should we interpret the laws of thermodynamics as allowing that patches of ice and wisps of steam *might* form spontaneously in an ordinary bathtub because of hugely improbable distributions of molecules? (1984, 243)

Following an approach to statistical laws that had been advocated by Popper (1959, 198–205), Watkins argues that in cases like Hempel's radon decay example, the statistical law is *physically equivalent* to a universal law, namely, given *any* 10 mg. sample of radon, it will contain 2.5 $\pm$ 0.1 mg. of radon after two half lives have transpired. If the statistical decay law is replaced by its universal surrogate, we can, of course, furnish a deductive explanation of the phenomenon. This is the foundation of Watkins's claim that, although we cannot explain individual micro-occurrences, we can provide explanations of macro-events that involve the behavior of extremely large numbers of micro-entities.[36]

Whether one accepts or rejects the Popper-Watkins thesis about deductive explanations of large aggregates of micro-events, the deductivist position – which rejects all such models of explanation as the I-S or S-R – is appealing because of its avoidance of the problems associated with maximal specificity, epistemic ambiguity, and explanations of improbable events. According to deductivism, given an indeterministic world, we must forego explanations of *what actually happens* (in many cases, at least); however, we can, by revealing the stochastic mechanisms, understand how the world works. I shall not go into greater detail about the merits of deductivism here, since Philip Kitcher's contribution to this volume (Kitcher & Salmon 1989) contains an extensive elaboration of that position.

There is, however, a residual problem with deductivism that merits attention. It has to do with the relationship of causality to explanation. In his celebrated work *The Cement of the Universe*, J. L. Mackie invites consideration of three machines that dispense candy bars (1974, 40–43). One of the machines is deterministic; the other two are indeterministic. Leaving aside some details that are inessential to our discussion, we can say that the deterministic machine gives you a candy bar if and only if you insert a shilling.

Among the indeterministic machines, the first gives a candy bar only if a shilling is inserted, but it sometimes fails to give one when the coin is put in. There is no deterministic explanation of these failures; they simply happen occasionally by chance. Suppose a shilling is inserted and the machine yields a candy bar. In this case, according to Mackie, putting the coin in the slot *causes* the candy bar to come out, for without the coin there would have been no candy bar. This is the *sine qua non* conception of causality; a cause is a necessary condition. I agree with Mackie about the causal relation here, and I would add that putting the coin in the slot *explains* the ejection of the candy bar.

The second indeterministic machine is the converse of the first. Whenever a shilling is inserted a candy bar is forthcoming, but occasionally, by chance, the machine ejects a candy bar when no coin is put in. The insertion of the coin is a sufficient, but not a necessary, condition of getting a candy bar. Suppose someone puts a shilling in this machine and receives a candy bar. According to Mackie it would be wrong to say that putting in the coin causes the candy bar to come out, for a candy bar might have been forthcoming even if no coin had been inserted. Again, I think that Mackie is right, and I would go on to say that the insertion of the coin does *not* furnish a (nonstatistical) explanation of the appearance of the candy bar.

If the intuition about these cases – shared by Mackie and me – is correct, it leaves the deductivist in an awkward position regarding scientific explanation. In the case of the first machine, we have identified the cause of the explanandum-event, but we cannot provide a D-N explanation, for the insertion of the coin is *not* a sufficient condition of getting a candy bar. In the case of the second machine, we can provide a D-N *explanation* of the explanandum-event, but there is no (nonprobabilistic) cause of it.

In my discussion of this point, to escape the blatant artificiality of Mackie's candy machines, I introduced two photon-detectors. The first never produces a click unless a photon impinges on it, but it fails to click in a small percentage of cases in which a photon impinges. The second detector never fails to click when a photon impinges, but occasionally it gives a spurious click when no photon is present. These detectors are strictly analogous to the candy-dispensing machines. With the first detector we have the cause of a click, but (according to the deductivist) no explanation. With the second detector we have (according to the deductivist) an explanation, but no cause. The deductivist, it seems to me, needs to come to terms with examples of this sort (W. Salmon 1988).

4.10 Explanations of Laws Again

There is one further point in Watkins's book that has important bearing on one of the recalcitrant problems we have encountered regarding deductive explanation. It will be recalled that Hempel and Oppenheim—in their notorious footnote 33—explained why they offered no account of deductive explanation of laws. Their difficulty was that Kepler's laws K could be deduced from the conjunction of those very laws with Boyle's law B, but this would surely fail to qualify as a bona fide explanation. The problem is to characterize precisely the distinction between deductions of that sort, which do not constitute explanations, and those derivations of regularities from more general laws that do constitute legitimate explanations.

According to Popperians, the fundamental aim of science is to produce and test bold explanatory theories. Watkins regards scientific theories as finite sets of axioms. The axioms must be logically compatible with each other and they must be mutually independent. He remarks, "It is rather remarkable that, although scientific theories are taken as the basic units by many philosophies and nearly all histories of science, there is no extant criterion, so far as I am aware, for distinguishing between a theory and an assemblage of propositions which, while it may have much testable content, remains a rag-bag collection" (1984, 204). As an answer to this problem he offers what he calls *the organic fertility requirement* (1984, 205). If a theory T contains more than one axiom, then it fulfills this requirement if it is impossible to partition the axiom set into two mutually exclusive and exhaustive nonempty subsets T′ and T″, such that the testable content of T is equal to the sum of the testable contents of T′ and T″. In other words, the axioms must work together to yield testable consequences that they cannot generate separately. With reasonable restrictions on what qualifies as testable content, it seems clear that the testable content of Kepler's laws and Boyle's law is no greater than the set theoretical union of the testable content of Kepler's laws and the testable content of Boyle's law. Kepler's three laws — K_1, K_2, K_3 — presumably satisfy the organic fertility requirement, while the set consisting of those together with B would not.

The situation is somewhat complicated by the fact that any given theory can be axiomatized in many different ways. We could, for instance, form the conjunction of all four laws to give us a theory with just one axiom. To deal with this kind of move, Watkins provides five rules for "natural" axiom sets. Among them is one he designates as "Wajsberg's requirement"; it says, in part, "An axiom is impermissible if it contains a (proper) component that is a theorem of the axiom set" (1984, 208) This clearly disposes of the Hempel-Oppenheim example. Whether Watkins's requirements block all counterexamples is a question I shall not try to answer. If they do, then they can be used to solve the problem with which Michael Friedman was concerned. Theories that satisfy the organic fertility requirement and are axiomatized "naturally" serve to unify our scientific knowledge.

4.11 A Fundamental Principle Challenged

There is a principle that has long been considered a cornerstone in the theory of scientific explanation, namely, if a set of circumstances of type C on one occasion explains the occurrence of an event of type E, then circumstances of the same type C cannot, on another occasion, explain the nonoccurrence of an event of type E (or the occurrence of an event of a type E' that is incompatible with E). Since it has so often been taken as a first principle, let us call it "Principle I." If this principle were relinquished, it has been thought, the floodgates would be open to all sorts of pseudo-explanations that are scientifically unacceptable. Nevertheless, careful consideration of probabilistic or statistical explanation has led some authors to reject that principle.

The D-N model of scientific explanation clearly satisfies this principle, for it is impossible validly to deduce two incompatible propositions from any consistent set of premises. Hempel's I-S model also satisfies it, as long as the high-probability requirement is enforced; since the sum of the probabilities of two incompatible statements with respect to any given consistent body of evidence cannot exceed one, it is impossible for both to have high probabilities. Adherents of the modal conception are committed to Principle I, for to show that an occurrence is necessary obviously implies that any incompatible occurrence is impossible. In van Fraassen's pragmatic theory Principle I is embodied in the view that an explanation shows why the topic *rather than* any other member of the contrast class is true.

At various junctures in our discussion of statistical explanation, I have made reference to a symmetry principle, namely, if a given stochastic process gives rise to some outcomes that are highly probable and to others that are improbable, then we understand the improbable ones just as well (or as poorly) as the probable ones. If this symmetry principle is correct—and I believe it is—it puts us in a dilemma. It forces us to choose between abandoning Principle I or forgoing explanations of probabilistic outcomes. For if circumstances C explain the probable outcome E on many occasions, then the same circumstances C explain the improbable outcome E′ on some other occasions. For closely related reasons, both Stegmüller and von Wright rejected the claim that probabilistic explanations of particular occurrences are possible. Since I am not inclined to give up statistical explanations of single events, and since I am strongly opposed to any high probability requirement, I have (in company with Achinstein and Railton) rejected Principle I. Although there may be non-deductivists—such as Hempel in "Aspects of Scientific Explanation"—who reject the symmetry principle, thus embracing Principle I *and* statistical explanations of particulars, Principle I does seem to be one of the chief bludgeons of the deductivists (see, e.g., Watkins 1984, 246).

What are the hazards involved in the rejection of Principle I? We have often been warned—and rightly so—that science has no place for theological or metaphysical 'theories' that explain whatever happens. Suppose someone is critically ill. If the person dies, the loved ones explain it as owing to "God's will." If the person recovers, they explain it as owing to "God's will." Whatever happens is explained in terms of the will of the Almighty. However comforting such 'explanations' might be, they are vacuous because there is no independent way of determining just what God wills. To rule out 'explanations' of this sort we do not need to appeal to Principle I; it is sufficient to insist that scientific explanations invoke *scientific* laws and facts. Scientific assertions, including those employed for purposes of explanation, should be supported by evidence. If we are dealing with what Hempel and Oppenheim called "potential explanation," then the laws or theories involved must be capable of independent support.

At the outset of our story we mentioned the attitude of scientific philosophers toward Driesch's attempts to explain biological phenomena in terms of entelechies and vital forces. During the same era, Popper was severely critical of Freudian psychoanalytic theory and Marxian economics. In all of these cases the criticisms were directed against the empirical vacuousness of the theories involved. To the extent that the theories in question are, indeed, empirically vacuous, to that extent they *are* devoid of scientific explanatory import. Such theories differ radically from the basic statistical theories of contemporary physics. These statistical theories offer a range of possible outcomes from a given indeterministic situation, attaching a definite probability value to each. They are far from vacuous, and they are supported by a vast amount of empirical evidence. The fact that such theories are not deterministic does not rob them of explanatory power. Nowadays most philosophers would agree that they have the capacity to explain statistical regularities; the question is whether they can be used to explain individual occurrences. If Principle I—which is *not* needed to block vacuous explanations—is relinquished we can give an affirmative answer.[37]

Conclusion
Peaceful Coexistence?

We have arrived, finally, at the conclusion of the saga of four decades. It has been more the story of a personal odyssey than an unbiased history. Inasmuch as I was a graduate student in philosophy in 1948, my career as a philosopher spans the entire period. I do not recall when I first read the Hempel-Oppenheim essay, but I did discuss it in my classes in the early 1950s. My specific research on scientific explanation began in 1963, and I have been an active participant in the discussions and debates during the past quarter-century. Full objectivity can hardly be expected.

Certain large areas have been left virtually untouched. I have deliberately neglected such topics as explanation in biology, explanation in psychology, explanation in history, and explanation of human action largely because I am not well enough informed about the substantive subject matter. Certain related issues, especially causality, have been treated only in passing, because the literature is so vast, and because this essay is already far too long. I know that there are other important pieces of work, especially during the latter part of the four decades, that have not been mentioned. As the decades passed the volume of literature increased exponentially, and judgments about the recent past are probably more subjective than those pertaining to more remote times. My decisions about what to discuss and what to omit are, without a doubt, idiosyncratic, and I apologize to the authors of such works for my neglect. But I have made an honest effort to discuss the material that has mainly influenced the present state of the subject. That said, let me offer my personal appraisal of the situation.

5.1 Consensus or Rapprochement?

Is there a new consensus in philosophy of science regarding the nature of scientific explanation? Not to any very noticeable extent. There are, however, a few basic points on which there seems to be widespread agreement among those who are contributing actively to the philosophical discussion of the subject.

(1) At the beginning of the four decades, the view was rather widely held that it is no part of the business of science to provide explanations – that explanations can be found, if at all, only in the realms of theology and transcendental metaphysics. At present, virtually all philosophers of science of widely diverse persuasions agree that science can teach us, not only *that*, but also *why*. This is an important piece of progress. It is now generally acknowledged that one of the most important fruits of modern science is understanding of the world. We do not have to go outside of science to find it.

(2) There seems to be general agreement – but by no means as unanimous as opinion on the preceding point – that the 'received view' of the mid-1960s is not viable. Contemporary deductivism may retain some of the core of the 'received view', but it rejects the expansive attitude toward probabilistic explanations of particular facts, and recognizes the need for an improved view of the nature of deductive explanation.

(3) It is noteworthy that the Hempel-Oppenheim paper undertook to furnish a formal explication of at least one type of scientific explanation, providing the syntax and semantics of a formal language and offering precise logical definitions. Most subsequent treatments, including those of Hempel himself, have not aimed for that degree of formal rigor. There is, I think, a general tacit recognition that the kinds of tools employed by Hempel and Oppenheim are not especially fruitful for handling the problems encountered in this area. For my own part – but I emphasize that this is *not* part of any general consensus – I believe that what constitutes a satisfactory scientific explanation depends upon certain contingent facts about the universe, for example, what kinds of mechanisms actually operate in the physical world.

(4) There appears to be fairly wide agreement on the importance of the pragmatics of explanation, and on the recognition that this aspect was not accorded sufficient emphasis in the 'received view'.

Beyond these four points, I cannot think of any other areas in which consensus actually obtains. Nevertheless, another question should be raised, namely, is a new consensus *emerging* in philosophy of science? This question calls for a risky prediction, but I shall hazard a guess. It seems to me that there are at least three powerful schools of thought at present – the pragmatists, the deductivists, and the mechanists – and that they are not likely to reach substantial agreement in the near future.

Still another question should be raised. Even if there is no general consensus at present, and no bright prospects for one to emerge soon, is there a basis for a substantial degree of rapprochement among the differing viewpoints? Here I think an affirmative answer can be given.

Around the beginning of the second decade, when serious controversy about the nature of scientific explanation erupted, two points of view—those associated primarily with the names of Hempel and Scriven—appeared in opposition to each other. The Hempelian view, which became the received view, emphasized deductive subsumption, the covering law thesis, and nomic expectability. We have chronicled many of the vicissitudes it has suffered; the net result, I think, is that it emerges at the close of the fourth decade in the form of the unification thesis. This view, whose chief proponents are Friedman and Kitcher, holds that scientific understanding increases as we decrease the number of independent assumptions that are required to explain what goes on in the world. It seeks laws and principles of the utmost generality and depth. This is a view to which I believe Popperians also subscribe. As I have said above (§4.3), the explanatory goal of this approach is the construction of a coherent world picture, and the fitting of particular facts within this framework. On this conception, explanatory knowledge is not of some additional kind that transcends descriptive knowledge. Explanations serve to organize and systematize our knowledge in the most efficient and coherent possible fashion. Understanding, on this view, involves having a world-picture—a *scientific Weltanschauung*—and seeing how various aspects of the world and our experience of it fit into that picture. The world-picture need not be a deterministic one; nothing in this view precludes basic laws that are irreducibly statistical.

The unification approach, as defended recently by Watkins and pursued in this book (Kitcher & Salmon 1989) by Kitcher, is, I believe, viable. In Kitcher's terminology, it is a "top-down" approach. It fits admirably with the intuitions that have guided the proponents of the received view, as well as those that inspire scientists to find unifying theories. This is the form in which the epistemic conception of scientific explanation can flourish today.

From the beginning, the most prominent critics of Hempel—especially Scriven—stressed two themes, causality and pragmatics. Often the two were closely related, for many of them took causality to be inescapably context-dependent. As things developed, however, the emphasis upon causality and upon objective statistical relevance relations issued in the ontic conception of scientific explanation, in which the objective relations among events could be considered quite apart from pragmatic considerations. As this approach had developed by the close of the fourth decade, it became the causal/mechanical view that is advocated by—among others—Humphreys, Railton, and me.

Although it would be unfair for me to assume that Humphreys and Railton would agree, as I see it, this version of the ontic conception has developed into a view that makes explanatory knowledge into knowledge of the hidden mechanisms by which nature works.[1] It goes beyond phenomenal descriptive knowledge into knowledge of things that are not open to immediate inspection. Explanatory knowledge opens up the black boxes of nature to reveal their inner workings. It exhibits the ways in which the things we want to explain come about. This way of understanding the world differs fundamentally from that achieved by way of the unification approach. Whereas the unification approach is "top-down," the causal/mechanical is "bottom-up."

When we pause to consider and compare these two ways of looking at scientific explanation, an astonishing point emerges. These two ways of regarding explanation are *not incompatible* with one another; each one offers a reasonable way of construing explanations. Indeed, they may be taken as representing two different, but compatible, aspects of scientific explanation.[2] Scientific understanding is, after all, a complex matter; there is every reason to suppose that it has various different facets.

Let me illustrate this point by recounting an actual incident.[3] Several years ago, a friend and colleague—whom I shall call *the friendly physicist*—was sitting on a jet airplane awaiting takeoff. Directly across the aisle was a young boy holding a helium-filled balloon by a string. In an effort to pique the child's curiosity, the friendly physicist asked him what he thought the balloon would do when the plane accelerated for takeoff. After a moment's thought the boy said that it would move toward the back of the plane. The friendly physicist replied that *he* thought it would move toward the front of the cabin. Several adults in the vicinity became interested in the conversation, and they insisted that the friendly physicist was wrong. A flight attendant offered to wager a miniature bottle of Scotch that he was mistaken—a bet he was quite willing to accept. Soon thereafter the plane accelerated, the balloon moved forward, and the friendly physicist enjoyed a free drink.

Why did the balloon move toward the front of the cabin? Two explanations can be offered, both of which are correct. First, one can tell a story about the behavior of the molecules that made up the air in the cabin, explaining how the rear wall collided with nearby molecules when it began its forward motion, thus creating a pressure gradient from back to front of the cabin.[4] This pressure gradient imposed an unbalanced force on the back side of the balloon, causing it to move forward with respect to the walls of the cabin. Second, one can cite an extremely general physical principle, Einstein's *principle of equivalence*, according to which an acceleration is physically equivalent to a gravitational field. Since helium-filled balloons tend to rise in the atmosphere in the earth's gravitational field, they will move forward when the airplane accelerates, reacting just as they would if a gravitational field were suddenly placed behind the rear wall.

The first of these explanations is causal/mechanical. It appeals to unobservable entities, describing the causal processes and causal interactions involved in the explanandum phenomenon. When we are made aware of these explanatory facts we understand how the phenomenon came about. The second explanation illustrates the unification approach. By appealing to an extremely general physical principle, it shows how this odd little occurrence fits into the universal scheme of things. It does not refer to the detailed mechanisms. This explanation provides a different kind of understanding of the same fact. It is my present conviction that both of these explanations are legitimate and that each is illuminating in its own way.

If this assessment of the situation is correct, we have grounds for a substantial degree of rapprochement between two approaches to scientific explanation that have been in conflict for at least three decades. In so saying, I do *not* intend to suggest that there was no real opposition between these views from the beginning, and that all of the controversy has been beside the point. On the contrary, at the beginning of the second decade there was genuine disagreement between Hempel and Scriven, and there was subsequently genuine disagreement between Hempel and me on the relative merits of I-S and S-R explanations. Over the intervening years, however, both viewpoints have evolved and become more mature. This evolution has, I believe, removed the sources of earlier conflict and made possible an era of peaceful coexistence.

At this point there is a strong temptation to take a page from Carnap's *Logical Foundations of Probability* and announce that there are two concepts of scientific explanation – explanation$_1$ and explanation$_2$ – both of which are perfectly legitimate, and which must not be confused with one another. The question is how to characterize them and their relations to each other. Let us identify explanation$_1$ with causal/mechanistic explanation. It could fairly be said, I believe, that mechanistic explanations tell us how the world works. These explanations are local in the sense that they show us how particular occurrences come about; they explain particular phenomena in terms of collections of particular causal processes and interactions – or, perhaps, in terms of noncausal mechanisms, if there are such things. This does not mean that general laws play no role in explanations of this kind, for the mechanisms involved operate in accordance with general laws of nature. Furthermore, it does not mean that explanations$_1$ can only be explanations of particular occurrences, for causal/mechanical explanations can be provided for general regularities. The causal/mechanical explanation offered in the case of the helium-filled balloon can be said to apply quite generally to lighter-than-air entities in similar circumstances. The molecular kinetic theory of gases provides a causal/mechanical explanation of such regularities as Boyle's law. Explanations$_1$ are *bottom-up* explanations, in Kitcher's terminology, because they appeal to the underlying microstructure of what they endeavor to explain.

Explanation$_2$ then becomes explanation by unification. Explanation in this sense is, as Friedman emphasized, global; it relates to the structure of the whole universe. Explanation$_2$ is *top-down* explanation. To reinforce the global character of explanation$_2$ we might restrict its applicability, as Friedman did, to explanations of regularities, but I do not think anything of much importance hinges on such a limitation.

If the foregoing suggestions are correct, we can reconcile the currently viable versions of the epistemic and ontic conceptions. That leaves the pragmatic view.

If the pragmatic approach is construed as the claim that scientific explanation can be explicated entirely in pragmatic terms, then I think our examination of Achinstein's, Bromberger's, and van Fraassen's work seriously undermines it. To give a correct characterization of scientific explanation we need to identify the kinds objective relevance relations that make an explanation *scientifically correct*. This task falls outside of pragmatics.

If, however, we see the pragmatic approach as illuminating *extremely important features* of explanation, without doing the whole job, we can fit it nicely into the foregoing account. As I remarked in §4.6, if we adopt Railton's concepts of *ideal explanatory text* and *explanatory information*, pragmatic considerations can be taken as determining what aspects of the ideal text are *salient* in a given context. The ideal text contains the objective *relevance relations* upon which correct explanations must be founded.

When we make reference to the ideal explanatory text, the question naturally arises whether it conforms to the unification conception or the causal/mechanical conception. The answer is, I think, both. If one looks at the main example Railton offered to illustrate his notion of a D-N-P ideal text — namely, alpha-decay of a uranium nucleus — we see that it contains elements of both. It appeals to the most general theory of quantum phenomena, but it also details the specific mechanism involved in the decay of one particular nucleus. Indeed, looking at that example, one is tempted to say that the top-down and bottom-up approaches are just two different ways of 'reading' the ideal explanatory text. Pragmatic considerations determine which way of 'reading' is appropriate in any given explanatory context. In the case of the friendly physicist, for example, an appeal to Einstein's equivalence principle would have been totally inappropriate; however, the causal/mechanical explanation might have been made intelligible to the boy and the other interested adults.

My remarks about the relationships among (what I see as) the three currently viable approaches to explanation have necessarily been brief and sketchy. They are reflections that have grown directly out of the writing of this essay. It remains to be seen whether they can be filled out in detail and applied more generally to a wide variety of examples of scientific explanation. If so, a new consensus can emerge from our present understanding of the topic.

5.2 Agenda for the Fifth Decade

Nothing could be more pleasing than the emergence of a new consensus along the lines just sketched, but whether that occurs or not, there are certain remaining problems that demand further serious consideration.

First among them, I believe, is a problem — or pair of problems — that has been with us from the beginning. This is the problem — raised, but not adequately answered, in the 1948 Hempel-Oppenheim paper — of lawlike statements and purely qualitative (or projectable) predicates. On this issue dissensus reigns. It is obviously crucial to anyone who adopts any sort of covering law view, but it is equally crucial to those who reject that conception if, like Scriven, they admit that laws have a role-justifying function.

Second, the problem of causality is still with us. It becomes especially critical for those who find an explanatory role for probabilistic causes. As my brief remarks have indicated, considerably more work is needed to clarify this concept. Recent interchanges with I. J. Good (W. Salmon 1988a), in addition to my critique of the Fetzer-Nute system, have convinced me that conflicting fundamental intuitions are rampant.

The foregoing two problems are hoary philosophical chestnuts on which much ink has been spilled. That situation will, no doubt, continue. The third problem is rather different, I believe. It has to do with quantum mechanical explanation. The chief source of the problem is the famous Einstein-Podolsky-Rosen paper.[5] It has, of course, been widely discussed, but not often explicitly in the context of scientific explanation. That paper, in effect, describes a thought-experiment and predicts its result. The issue has become more urgent recently as a result of Bell's theorem and the Aspect experiment on remote correlations (see, e.g., Mermin 1985 or Shimony 1988). Now there is an actual experimental outcome to explain. Opinions vary on the significance of these results. Some say that they are of little consequence, showing only that the results conform to the theoretical prediction of quantum mechanics. To the discredit of the received view, it would support that position and claim that, because of subsumption under a well-confirmed theory, the experimental result is explained. At the opposite end of the spectrum of opinion are some who say that Bell's theorem and Aspect's confirmation of the quantum mechanical prediction constitute the most important development in the history of physics. I cannot accept that assessment either. With N. David Mermin, I accept the *moderate view* that "anyone who isn't worried about this problem has rocks in their head."[6]

The situation, basically, is this. There are impressive remote correlations in the spin-states of photons that cannot be explained by local causal principles; action-at-a-distance appears to be manifested. The results can be derived from spin conservation, but it is nonlocal conservation, and we have no mechanism by which to explain it. I have no idea what an appropriate explanation would look like; we may need to know more about the microcosm before any explanation can be forthcoming. But I do have a profound sense that *something* that has not been explained needs to be explained. As I said at the close of *Scientific Explanation and the Causal Structure of the World*, "to provide a satisfactory treatment of microphysical explanation constitutes a premier challenge to contemporary philosophy of science" (279). That still strikes me as correct.

Notes

Introduction

1. See Rudolf Carnap (1966, reissued 1974, pp. 12–17) for interesting comments on the transition from the denial to the acceptance of the view that science can furnish explanations. This passage includes an interesting discussion of Hans Driesch's vitalism (including the appeal to entelechies), and of reactions to it by the logical positivists of the Vienna Circle.

2. It is possible, of course, to adopt a hypothetical or 'suspend the truth' attitude in which one asks how a particular event could be explained *if it were to occur*. This is *not* Velikovsky's attitude.

3. For a brief and accessible introduction see Gale (1981). The heading of this rather flamboyant article reads, "Certain conditions, such as temperature, were favorable to the emergence of life on earth. The anthropic principle argues the reverse: the presence of life may 'explain' the conditions." In the article Gale adds, "It is fair to say, however, that not all cosmologists and philosophers of science assent to the utility of the anthropic principle, or even to its legitimacy. Here I shall describe some of the ways in which the principle has been applied and let the reader judge its validity" (p. 154). For a more thorough and technical, as well as more recent, treatment see Barrow and Tipler (1986).

4. That feature, in itself, should give us pause, for it is an elementary logical fallacy to infer the truth of the premises from the truth of the conclusion of a valid deductive argument. As a matter of fact, neither Hempel nor I considers the traditional hypothetico-deductive schema an adequate characterization of scientific confirmation, but it seems to be so regarded by many people; see, for example, Braithwaite, *Scientific Explanation*, (1953 p. 9). Discussions of the shortcomings of the hypothetico-deductive method can be found in W. Salmon (1984, §30) or W. Salmon (1967, Chap. VII).

5. Samuel E. Gluck (1955) made a brief stab at the task, but it was insufficiently general and failed to take notice of such basic difficulties as the ambiguity of statistical explanation. One interesting feature of his article is the claim that, because of the inherently probabilistic character of physical measurement, even what we take to be D-N explanations in quantitative sciences are actually statistical explanations. This point has been a longstanding source of worry for deductivists.

6. There are, of course, precursors of the theory set forth in this article, including Aristotle and John Stuart Mill. An important twentieth-century precursor is Karl R. Popper, whose 1935 work sketches a version of D-N explanation, though not in the depth of detail of Hempel-Oppenheim. Moreover, Popper's book was not highly influential until the English translation (1959) was published.

The First Decade

1. For one explicit comment, see 1953, p. 347.

2. For a recent detailed and sophisticated discussion of the nature of laws see John Earman (1986, chap. 5).

3. For purposes of this discussion it is not necessary to draw a distinction between counterfactual and subjunctive conditionals. In the example, "If this table salt were placed in water it would dissolve," it does not matter whether or not the particular sample is at some future time placed in water.

4. I once made a very crude estimate of the amount of gold in the earth's seas, and it came out to be more than 1,000,000 kg; if sufficient resources were devoted to the project, somewhat more than 100,000 kg could be extracted from sea water and fashioned into a sphere.

5. It has been speculated that tachyons—particles that travel faster than light—exist, but on pain of contradiction with special relativity, they cannot be used to send messages.

6. In these remarks about the coextensiveness of laws and statements that support counterfactuals I am, of course, excluding counterfactuals based on logical truths and/or on definitions.

7. This may represent Braithwaite's fundamental view of the problem.

8. This point is closely related to the pragmatic view of John Stuart Mill, Charles Saunders Peirce, and David Lewis, according to which the laws are those generalizations that end up as fundamental principles in the ideal limit of scientific investigation. One crucial question with respect to this approach is whether ideal science must eventuate in a unique final form, or whether alternatives are possible. If alternatives are possible, this pragmatic resolution of the problem would not pick out a unique set of laws, for the future course of science is unpredictable. Consequently, the distinction between laws and nonlaws would be basically epistemic; the distinction would not be objective.

9. See the concept of resiliency in Brian Skyrms (1980).

10. A great deal of light will be shed on this issue by Bas van Fraassen's forthcoming book, *Laws and Symmetries*.

11. At this point in his discussion he is advancing this view for consideration, rather than asserting it, but this is precisely the conclusion he does draw at the end of this chapter, pp. 317-18.

12. Partly because of its somewhat opaque title and partly because of its formidable complexity this book received little attention. It was later reprinted under the more descriptive title, *Laws, Modalities, and Counterfactuals* (Berkeley & Los Angeles: University of California Press, 1976). In my foreword to this volume I attempted to survey the issues as they stood in 1976, and to provide a more accessible account of Reichenbach's major ideas.

13. Ernest Nagel, *The Structure of Science* (New York: Harcourt, Brace and World, 1961), Chap. 4.

14. This discussion was continued in (Goodman, 1947) and (Carnap, 1947a).

15. I am paraphrasing the Hempel–Openheim definitions, but I shall preserve their original numbering.

16. See W. Salmon, foreword, in Reichenbach (1947), pp. xxxii–xxxiii.

17. The expression "iff" is a standard abbreviation for "if and only if."

18. For the sake of a concrete interpretation of this counterexample, it will do no harm to restrict the range of our variables to humans.

19. In *no* sense is this remark intended as a historical comment; I do *not* mean to suggest that they were on the verge of sending it off for publication and spotted the problem just in the nick of time. I am merely drawing a parallel between this technical problem and that pointed out by Eberle, Kaplan, and Montague.

20. By the *conjunctive normal form* of a formula we mean an equivalent formula (containing just those sentential variables occurring essentially in the original) which is a conjunction each of whose terms is a disjunction of sentential variables or their negations.

21. It is a serious terminological error, I believe, to refer to the D-N model of scientific explanation as "*the* covering law model," for although it is one model that conforms to the general idea that explanation always involves subsumption under laws, it is by no means the only such model. Indeed, Hempel's I-S model is also a covering law model.

22. In conversation, when he visited the workshop at the Minnesota Center for Philosophy of Science.

23. It appears that Hempel published nothing on scientific explanation, beyond the Hempel-Openheim article, during this first decade of our chronicle.

24. For purposes of historical accuracy it should be noted that this chapter is based upon Braithwaite's 1946 presidential address to the Aristotelian Society. At about the same time Ernest Nagel (1956) produced an important study of functional explanation. This paper was prepared in 1953 as a reserach report at Columbia University. Neither Braithwaite nor Nagel makes any reference to the Hempel-Openheim paper in their discussions of teleological or functional explanation.

25. Like Braithwaite and Nagel, Scheffler also makes no mention of the Hempel-Oppenheim article. It might be remarked facetiously that if Scheffler had been sufficiently attuned to the historical situation he would have published this paper on teleology in 1957, and the 1957 critique of Hempel and Oppenheim in 1958.

26. Both Hempel and Nagel rely heavily on Merton's extensive discussion of *functional analysis*. Nagel's work (1956) contains a detailed analysis of Merton's treatment of this topic.

27. Although the first published version of the I-S model did not appear until three years later, it seems evident that Hempel had a pretty clear idea of what it would be like when he wrote the 1959 article.

28. Scriven (1959) would, I believe, be one example.

29. See, however, Scheffler (1957) for a discussion of the bearing of the referential opacity of intentional contexts in this connection.

30. Nagel (1961, chap. 1) lists four types of explanation: deductive, probabilistic, functional or teleological, and genetic, leaving open at that stage whether the latter two can be reduced to one or both of the former two.

31. John Canfield (1966) covers the discussion of functional explanation during a good deal of the first two decades of our story. It contains, in addition to classic papers by important authors, a clear and perceptive introduction by Canfield.

32. Wright (1973) contains important anticipations of the theory set out in his book.

The Second Decade

1. As we noted in §0.2, the most serious attempt before 1962 seems to be (Gluck, 1955), but it is much too sketchy to present a clearly articulated theory, and it fails to notice the serious difficulties involved in statistical explanation.

2. His book *The Philosophy of Science* (1953) contains many references to scientific explanation, but does not provide an explicit account of that concept. His *Foresight and Understanding* (1961) deos offer an explicit account.

3. In *Scientific Explanation* (1984, p. 130) I incorrectly attributed to Bromberger the view that all explanations can be appropriately requested by means of why-questions. I regret this error.

4. [Bromberger's footnote] 'Can think of no expression' and 'can think of no answer' as short for 'can imagine nothing, conjure up nothing, invent nothing, remember nothing, conceive nothing' . . . does not cover the familiar states of momentary amnesia during which one has the answer 'on the tip of one's tongue' but cannot utter it.

5. Broberger's footnote: "As an achievement term 'to explain' is also often used to credit people with certain scientific discoveries. 'Newton explained why the tides vary with the phases of the moon' may serve to mark the fact that Newton was the one who solved the riddle, who found the answer to a question with regard to which everybody had been in either a p-predicament or a b-predicament. To have explained something in this sense is to be one of the first to have explained it in the [previous] sense . . . to be one of the first to have been in a position to explain . . . it to a tutee."

6. See the charming dedication to Hempel under "NOTES" on p. 107.

7. These restrictions are contained in the definition of *general abnormic law* given in Bromberger (1966, p. 98).

8. For technical reasons an additional restriction—(4) the general rule completed by L has the property that if one of the conjuncts in the antecedent is dropped the new general rule cannot be completed by an abnormic law—is needed, but the main thrust if this account can be appreciated without it.

9. As Broberger notes explicitly, "unless" has to be contrued as the exclusive disjunction.

10. However, see Grünbaum (1963), Fetzer (1974), and Rescher (1963).

11. It should be recalled, from our discussion of functional or teleological explanation in §1.3, that Hempel did not base his qualms about explanations of these sorts on the problem of the temporal relation between the function and its goal.

12. I do not believe Bromberger ever published this precise example; his actual examples, which have the same import, are the height of a tower and the height of a utility pole to which a guy wire is attached.

13. Bas van Fraassen is an exception. In his 1980 work (pp. 132–34) he suggests that there are possible contexts in which such an explanation would be legitimate. We shall discuss his theory of explanation in §4.4 below.

14. This 'law' does, of course, make reference to particular entitites—earth, sun, and moon—but that, in itself, is not too damaging to the example. After all, in this respect it is just like Kepler's laws of planetary motion and Galileo's law of falling bodies. Like these, it qualifies as a derivative law, though not a fundamental law.

15. Scheffler (1957) subjected the symmetry thesis to searching criticism. To the best of my knowledge, this article is the first significant published critique of the Hempel-Oppenheim article.

16. See Grünbaum (1963) for a fuller discussion of this point.

17. As Philip Kitcher pointed out to me in a personal communication, Hempel could be defended against this example by arguing that "a *natural selection* explanation of the presence of a trait is really a deduction that the probability that the trait becomes fixed in a finite population is high."

18. This example is due to Henry Kyburg (1965). A variant of this example (due to Noretta Koertge) has unhexed table salt placed in holy water; and the 'explanation' of the dissolving is that the water was blessed and whenever table salt is placed in holy water it dissolves.

19. I offered one such method, based on Reichenbach's treatment of laws (Salmon, 1979).

20. In the introduction to this essay I noted that an earlier paper on statistical explanation by Samuel E. Gluck may qualify as a precursor.

21. I shall us the term "statistical law" to refer to factual generalizations, such as the chance of getting 6 if a standard die is tossed, or the probability that a carbon-24 nucleus will decay in 5730 years. One could say that statistical laws are empirical generalizations, provided it is clearly understood that this does *not* mean that only directly observable properties are involved. I shall use the term "law of probability" to refer to axioms and theorems of the mathematical calculus of probability. Such laws are not empirical and they do not have factual content.

22. I am not objecting to the use of idealizing simplifications in dealing with philosophical problems, but their limitations must be kept in mind.

23. Recalling the fact that, for the limiting frequency interpretation, the individuals have to be taken in some specific order, I must confess to a bit of queasiness here.

24. We shall find reasons below for doubting this assessment of the relative importance of D-S and I-S explanations.

25. Whether this statistical law asserts that r is the actual fraction of Fs within a finite class F, or the limiting frequency of Fs within an infinite class F, is an issue that need not concern us here.

26. If we are thinking of probabilities as limiting frequencies in infinite sequences, $P(G|F)$ may equal 1 and $P(G|F.H)$ may equal 0.

27. Hempel (1962) succumbed to this temptation, but he soon realized the essential difficulty, as he explains in (1965, p. 401, note 21).

28. I have slightly modified Hempel's notation, but not in any way that affects our discussion.

29. As we noted in discussing the precise Hempel-Oppenheim explication, the actual requirement for D-N explanation is that the explanans contain essentially at least one theory, where a theory may contain existential quantifiers and need not contain any universal quantifier. In the technical explication of a potential explanans, the explanatory theory must be true (since all theories, in the special sense of Hempel and Oppenheim, are, by definition, true). It would seem more reasonable to require only that the general premise be essentially generalized, for certainly we want to consider the explanatory power of theories (in the usual sense) that need not be true.

30. As we noted above, p. 16, although Braithwaite's *Scientific Explanation* contains a chapter devoted to explanations of laws, he does not come to grips with the fundamental problem noted by Hempel and Oppenheim.

31. Hempel published a revised version of RMS (1968), pp. 116–33, which he designated "RMS*." It embodies some technical revisions designed to overcome an objection by Richard Grandy, but these do not affect any of the philosophical issues we are discussing.

32. Linus Pauling's claims in this regard were receiving a good deal of publicity in the early 1960s (see Pauling, 1970).

33. I owe this example to the eminent geneticist Tracy Sonneborn. In a social conversation about hormones and sex, I asked him what would happen to a man who took oral contraceptives. Without a monent's hesitation he replied, "Well, he wouldn't get pregnant."

I am indebted to the eminent archaeologist William A. Longacre for furnishing further information regarding this question. He kindly sent me a copy of the 2 June 1987 issue of the tabloid *Sun* (vol. 5, no. 22) which carried the front page headline "WIFE FEEDS HUBBY BIRTH CONTROL PILLS TO STOP HIM CHEATING." According to the story (p. 35), "Determined to stop her husband from fooling around, a betrayed housewife came up with the idea of feeding him birth control pills—and the dumbfounded hubby was frightened out of his wits when the pills made him impotent." When she confessed that she had been concealing them in his food he became furious and "stormed out of the house and has since filed for divorce . . . on the grounds of mental and physical cruelty."

The Third Decade

1. It is worth noting that Hempel introduces the term "statistically relevant" in this article, but he does not use it to refer to a statistical relevance relation (as that concept is generally understood).

2. As the issue was addressed here, it was buried rather deep in a paper whose title gave hardly any hint that it dealt with this topic.

3. This paper is a much expanded and highly revised version of one presented at a University of Pittsburgh workshop in 1965.

4. I discuss these three conceptions—the epistemic, the modal, and the ontic—rather briefly in (1982, 1986); and also in greater detail in (1984, chaps. 1 and 4). We shall return to them in this essay when we get to the fourth decade.

5. The answer is obvious in the sense of Christopher Columbus. The story is told of a dinner attended by Columbus sometime after his first voyage to the new world. Some of the guests were belittling his accomplishment by suggesting that it was not all that difficult. Columbus requested an egg from the kitchen, which he passed around asking each guest to try to stand it on end. When all had failed and it came back to him, he gave it a sharp tap on the table, breaking in the shell at one end, whereupon it remained upright. "You see," he said, "it is easy after you have been shown how."

6. I have in mind chiefly Peter Railton and myself.

7. Suppose, for example, that we have a class of tosses consisting of tosses of many different coins. Suppose further that many of the coins are biased, some toward heads, others toward tails. Some of the coins are fair. Now if each of the biased coins is biased in the same degree toward heads or tails (as the case may be), if as many are biased toward heads as toward tails, and all of the coins are tossed equally often, then the probability of heads in the entire class will be ½. This class may be relevantly partitioned into three cells—tosses with coins biased toward heads, tosses with coins biased toward tails, and tosses of fair coins. The probability of heads in the third cell is equal to the probability of heads in the original class, but that does not mean that the partition is not relevant.

8. I am modifying the example somewhat for purposes of the present discussion.

9. These arguments were also contained in his doctoral dissertation, *Foundations of Inductive Explanation* (1973).

10. The treatment of the topic in this article was seriously flawed, and it is completely superseded by my 1984 work, (chap. 3).

11. We might roughly define violet light as that having wave-lengths in the range of approximately 3600–3900 Å and red light as that having wave-lengths in the range of approximately 6500–7100 Å. It is a *fact* that the normal human range of vision extends from about 3600 Å to about 7100 Å.

12. The addition to Hempel's RMS is italicized.

13. In developing the statistical-relevance model of scientific explanation, I employ a non-epistemically-relativized counterpart of RMS. It is the requirement that reference classes used in explanations of this sort be *objectively homogeneous*. A number of the considerations that enter into the foregoing revision of Hempel's RMS were developed in the attempt to provide an adequate characterization of objective homogeneity. See W. Salmon (1984, chap. 3).

14. This attempt was seriously flawed in several ways. It is completely superseded by my 1984 work (chap. 3).

15. The members of F must be taken in some order. If F is finite the order makes no difference; if F is infinite, choose some natural order, such as the temporal order in which the events occur or are discovered.

16. This requirement is analogous to Richard von Mises's principle of insensitivity to place selections in his definition of the *collective*.

17. Philip Kitcher criticizes this proposal as follows, "I suspect that the omnipresence of correlations will make for inhomogeneity everywhere. This [proposal] will not get around the spurious correlation problem. There's bound to be some (possibly hokey) property that all correlates of the Bs have." If he is right, there is more work to be done on the concept of objective homogeneity. One suggestion, essentially adopted by Coffa, is to restrict consideration to nomically relevant (not just statistically relevant) factors.

18. This 'law' needs more careful formulation. Innumerable automotive and plumbing problems have been occasioned by the fact that water expands when it is cooled.

19. I have treated these issues in some detail (1979).

20. This is a highly contentious issue. In a personal communication Philip Kitcher comments, "Here I think you are describing a different chance setup. It depends how far you go back in the causal chain. If the setup involves the producing machine, then the Laplacian description seems right. If we start with some already selected coin of unknown constitution, then it seems that [the other] is right. We don't have an ambiguous description *of the same situation.*" My answer to this ploy is that we have two different descriptions of two different *types* of chance setups. The event in question actually belongs to both. If we want to characterize *this chance setup,* we have to decide which type to assign it to.

In another personal communication Paul Humphreys writes, "Your argument here applies only to long-run propensity interpretations, and not to single case propensities. Under the single case account, the continued operation of the machine is irrelevant to each specific trial, and under either experiment you describe, the propensity will be either less than or greater than ½, although in the first experiment it will remain fixed, whereas in the second experiment it will change depending on which kind of disk is picked on each trial." My answer to Humphreys is that he has done what any good limiting frequency or long-run propensity theorist would do. He has picked the appropriate type of chance setup to characterize *this trial.*

21. Because Coffa did not develop the idea or spell out any details, I had completely missed the point and forgotten all about it until I very recently reread his disseration. Every important philosophical discovery has to have a precursor; Coffa plays that role for Humphreys.

22. Hempel's response to Rescher can be found in (Hempel, 1965, pp. 403–6).

23. This book contains a comprehensive bibliography of works on scientific explanation up to its date of publication.

24. Indeed, any but the simplest putative Hempelian explanation of an individual event, because it involves subsumption under one or more laws, can be construed as an explanation of a regularity. The explanation of the bursting radiator, for example, is an explanation of why *any* radiator of a similar type, subjected to similar conditions, will burst.

25. Kicher responds, of course, that the Hempel-Oppenheim criteria are too weak. Where they required merely that the primitive predicates be purely qualitative, Kicher seems to want to require that they signifiy natural kinds.

26. Kitcher will argue, of course that typically not all of the Hs will represent predicates that are projectable from their instances.

27. Kitcher responds, "I'm not sure you could project either generalization without having a background theory about organisms that would give you both at once—or that would tell you that one is true if and only if the other is."

28. In a personal communication Kitcher also rejects this example: "I'm not sure we could accept any of these without accepting them all. The reason is that the projection from a finite sample seems to depend on believing that the relevant predicates pick out crucial classes in nature. There's no basis for thinking of the large-small cases, for example, as a privileged class for projection without thinking that the class of *all* two-body systems is projectable. Sometimes, I believe, you can't make a *restricted* projection. You either go all the way or nowhere."

29. Philip Kitcher, "Explanation, Conjunction, and Unification," *Journal of Philosophy* LXXIII (1976), p. 209,

30. In a personal communication Kitcher reponds: "Not necessarily, if the conditions on acceptability concern the projectability of predicates." Kitcher's remarks about projectability in response to my criticism of Friedman are interesting and important. If, however, this is the avenue to salvation for Friedman's program, he clearly needs to add a theory of projectability to the proposal he has offered.

31. A.A. Few, "Thunder," *Scientific American* (July, 1975), pp. 80–90. [Kitcher's reference.]

32. The first two dogmas were elaborated in W. V. Quine's classic essay "Two Dogmas of Empiricism," (1951).

33. This is the position Jeffrey took (1969); however, I do not agree with his way of making the distinction between those that are and those that are not.

34. If these temporal relations are not obvious, think of a given photon approaching the vicinity of the flagpole. If it passes by the flagpole without being absorbed, it reaches the ground a little later than it passed by the flagpole. A companion photon, traveling alongside of the above-mentioned one in a parallel path that intersects with the flagpole, will be absorbed a little before the other reaches the ground. As a very rough rule of thmb, the speed of light is a billion ft/sec; the photons travel about one foot per nanosecond.

35. I offered one suggestion on how this could be done in "Postscript: Laws in Deductive-Nomological Explanation—An Application of the Theory of Nomological Statements" (1979).

36. Moves of this sort were made in Fetzer (1981).

37. Carnap argues in careful detail for his denial of rules of acceptance (1950, §44b, 50–51). We need not go into his reasons for this approach here, for we are concerned only with its consequences in this context.

38. It has been noted that the special theory of relativity, while prohibiting the acceleration of ordinary material particles from speeds below that of light to superluminal speeds, may not preclude the existence of particles—called *tachyons*—that always travel faster than light. It is generally agreed, I believe, that there is, at present, no empirical evidence for their existence. Moreover, the presumption is that, if they should exist, they would be incapable of serving as signals or transmitting information. See my 1975 work (2nd ed., 1980, pp. 105, 122–24) for a discussion of the problems that would arise if tachyons actually existed.

39. Many philosophers would analyze processes as continuous series of events. While this can surely be done, there is, in my opinion, no particular reason for doing so,and there are some significant disadvantages. See my 1984 work (pp. 140–41, 156–57) for a discussion of this issue. See also John Venn's nice remark quoted on p. 183.

40. A material object that is at rest in one frame of reference will, of course, be in motion in other frames of reference.

41. In speaking of interventions that produce marks, I do *not* mean to suggest that such occurrences must be a result of human agency. For example, light from a distant star may be marked as a result of passage through a cloud of interstellar gas that selectively absorbs light of certain frequencies while letting others pass through unaffected.

42. The pseudo-process consisting of the moving spot of light can be changed from white to red from some point onward by putting a red lens on the beacon, but that does not qualify as an intervention *in the pseudo-process* because it is done elsewhere.

43. I often put so-called scare-quotes around the word "process" when it is being used to refer to pseudo-processes, for many people might be inclined to withhold that term when they realize that the process in question is pseudo. Reichenbach called them "unreal sequences."

44. The Lone Ranger's horse Silver was described as "A fiery horse with the speed of light." For the image of the horse on the screen that is possible in principle.

45. In the infinitesimal calculus we do, to be sure, define the concept of *instantaneous velocity*, but that definition requires consideration of the position of the object at neighboring instants of time. See my 1970 work, which contains Russell's article on Zeno's Paradoxes, and my 1975 work (chap. 2) for a fuller account.

46. Because of my own disciplinary limitations I have not participated in this field.

47. Perhaps I should add, or a bit of teleological non-action, such as not removing a tree that just happened to grow where it did, because it provides some wanted shade.

48. I do not find Nagel's criticisms of Wright particularly weighty.

49. These include Andrew Woodfield (1976) and Michael Ruse (1973).

50. It should be recalled, as previously mentioned, that Hempel addressed the problem of functional explanation in such areas as anthropology, sociology, and psychology, as well as biology. Nagel confines his attention, in these lectures, to biology. It is *much* harder to argue against functional equivalents in these other fields than it is in biology.

The Fourth Decade

1. I have argued this issue briefly (1967, pp. 729–32), and in greater detail (1969).

2. Coffa had previously provided searching critiques of both of these models in his doctoral dissertation (1973).

3. This theory was also presented in his doctoral dissertation; see §3.3 above.

4. Before his 1977 publication at any rate. As I remarked above, it is not clear how the I-S model can survive abandonment of the high-probability requirement.

5. I discussed these general conceptions at length (1984, chaps. 1 and 4), and more briefly (1982, 1985).

6. I have no qualms about considering "micrometer" a term of the observational vocabulary, for it is defined in terms of "meter," which in view of the International Prototype and zillions of meter sticks in homes, stores, laboratories, and shops throughout the world, is surely an observational term.

7. Perrin actually studied the Brownian movements of particles suspended in liquids, but that fact does not affect the final results as applied to gases.

8. In offering this response on behalf of the modal conception I am, of course, assuming that knowledge of physical necessity and impossibility are not fully contained within our descriptive knowledge of the world. I shall return to this issue shortly.

9. I take it that predictive knowledge qualifies directly as part of descriptive knowledge, for knowledge of what is going to happen in the future is surely included in descriptive knowledge of the world.

10. See J. L. Mackie (1974) for detailed discussion of this approach. Mackie, himself, does not adopt it.

11. See, for example, Carnap (1966, 1974) for an exposition of this view of causality in the context of scientific explanation.

12. See David Lewis, *Counterfactuals* (1973). As I mentioned above, Lewis himself attemps to break out of the circle by adopting the view that laws are the basic principles in an ideally complete science. In my foregoing remarks on this topic, it will be recalled, I raised the crucial question of the uniqueness of characterization—a problem Lewis acknowledges.

13. Some physicists have proposed the "many-worlds" interpretation of quantum mechanics to resolve the problem of measurement (see De Witt and Graham, 1973). I am not favorably incluined toward this interpretation for much the same reasons.

14. It should be recalled that the term "theory," as it occures in the Hempel-Oppenheim formal explication, does not involve any appeal to unobservables. It simply refers to general statements that may contain existential as well as universal quantifiers.

15. I shall dicuss the legitimacy of this claim to empirical knowledge of unobservables in §4.5.

16. To those who suggested that such feelings are observable, at least by the subject, they responded that introspection is not a scientifically acceptable kind of observation.

17. I have argued this point most explicitly in my 1985 publication; it is also discussed in my 1984 work (pp. 229–38).

18. See my *Scientific Explanation* (pp. 13–15) for a striking historical example.

19. There are, of course, unsatisfactory answers to questions, but I would suppose they should be called unsatisfactory explanations. Not all explanations are good explanations.

20. See Frolich (1980), especially the photographs on p. 154.

21. Indeed, the only detailed critiques of van Fraassen's treatment of explanation so far published of which I am aware are in Salmon (1984) and Kitcher and Salmon (1987). The criticisms expressed in my work are entirely different from those given in Kitcher and Salmon. Achinstein (1983) gives brief attention to van Fraassen's theory of explanation.

22. For van Fraassen's response to this argument, as well as to mine, see Churchland and Hooker (1985, pp. 297–300).

23. A rather good sample of the range of opinion can be found in (Churchland and Hooker, 1985) and in (Leplin, 1984). Essays by Fine, Laudan, Boyd, and McMullin can be found in the Leplin volume.

24. In his useful discussion, "Structural Explanation" (1978, pp. 139–47), Ernan McMullin stresses the causal foundations of structural explanations, but he too steers clear of quantum phenomena.

25. It is ideal in the sense not of something we should necessarily strive to realize, but, rather, of (a Platonic?) something that may not exist in the physical world. It is the sort of thing Laplace's demon might be able to realize, if it concerned itself with explanations.

26. Ironically, the published article contains a misprint, which I have corrected, in the first two words of this passage. Instead of "Is it" the original reads "It is." Has Freud struck yet again?

27. I suspect Railton might agree (see Railton, 1981, pp. 243–44).

28. At the close of §3 of his paper in this volume (Kitcher & Salmon 1989), Kitcher offers his critique of Railton and me on the issue of the role of mechanisms in explanations.

29. This was reportedly the terse recommendation of this famous architect when, many years ago, he was brought to Pittsburgh by the city fathers to give advice on what could be done to improve the city.

30. This material is incorporated, to a large degree, in Fetzer (1981). Nute was Fetzer's collaborator on the formal aspects of probabilistic causality, but not on the general theory of scientific explanation.

31. A probability of one is not equated with a universal disposition and a probability of zero is not equated with a universal negative disposition.

32. One would expect the stronger result, that p and not-q are jointly possible, to be derivable.

33. As we saw in §3.2, however, Hempel's doctrine of essential epistemic relativization of inductive-statistical explanation brought him dangerously close to the brink of deductivism.

34. I discuss this example and others in some detail (1984, pp. 111–20).

35. Railton remarks, incidentally, that he is not strongly opposed to discarding the parenthetic addendum (1981, p. 236). He remians, however, strongly opposed to the view that explanations are always arguments.

36. Popper (1959, pp. 198–205) offers a criterion for deciding when the number of micro-events is large enough to justify this kind of replacement; Watkins employs the same criterion in his account.

37. I have discussed the status of Principle I, and other closely related principles (1984, especially pp. 111–20).

Conclusion

1. It should not be supposed that all mechanical explanations appeal to unobservable mechanisms. One might explain the workings of some gadget solely in terms of observables to someone who had not noticed the mechanical relationships among them—e.g., the way in which squeezing the handbrake on a bicycle brings the bicycle to a stop. However, deeper scientific explanations do seem usually to invoke unobservables.

2. Gregory Cooper, who was an active participant in the Minnesota Workshop, independently recognized the compatability of these two approaches earlier than I did. His thought on this matter is contained in his doctoral dissertation. I am happy to acknowledge his priority with respect to this point.

3. This little story was previously published (W. Salmon, 1981 pp. 115–25). I did not offer an explanation of the phenomenon in that article.

4. Objects denser than air do not move toward the front of the cabin because the pressure difference is insufficient to overcome their inertia.

5. Albert Einstein et al., "Can Quantum-Mechanical Description of Physical Reality Be Considered Complete?" *Physical Review* 47 (1935), pp. 777–80. In *The Book of Revelations* 13:18 it is said that the number of The Beast is 666; I believe it is 777.

6. N. David Mermin, "Is the Moon Really There When No One Is Looking? Quantum Theory and Reality," *Physics Today* (1985).

Chronological Bibliography

• Designates item on scientific explanation of special significance and/or receiving more than passing mention in this essay.

☐ Designates an item not directly on scientific explanation cited in this essay.

Standard textbooks on philosophy of science—many of which contain sections on scientific explanation, without contributing to research on the subject—are not included in this bibliography.

Antiquity

• Aristotle, *Posterior Analytics*.

Modern Prehistory

1843

• Mill, John Stuart, *A System of Logic*, first edition. London: John W. Parker.

1906

• Duhem, Pierre Maurice Marie, *La Théorie physique, son objet et sa structure*, first edition. Paris. [Second edition, 1914; English translation of second edition, *The Aim and Structure of Physical Theory*. Princeton: Princeton University Press, 1954.]

1913

☐ Perrin, Jean, *Les Atomes*. Paris: Alcan. [English translation New York: Van Nostrand, 1923.]

1919

Broad, C. D., "Mechanical Explanation and Its Alternatives," *Proceedings of the Aristotelian Society* 19, pp. 85–124.

1926

Ducasse, C. J., "Explanation, Mechanism, and Teleology," *Journal of Philosophy* 23, pp. 150–55.

1935

Broad, C. D., C. A. Mace, G. F. Stout, and A. C. Ewing, "Symposium: Mechanical and Teleological Causation," *Proceedings of the Aristotelian Society*, supplementary vol. 14, pp. 83–112.

Cornforth, K., "Symposium: Explanation in History," *Proceedings of the Aristotelian Society*, supplementary vol. 14, pp. 123–41.

☐ Einstein, Albert, B. Podolsky, and N. Rosen, "Can Quantum-Mechanical Description of Reality Be Considered Complete?" *Physical Review* 47, pp. 777–80.

• Popper, Karl R., *Logik der Forschung*. Vienna: Springer. [Imprint 1935, actually published in 1934]

1942

• Hempel, Carl G., "The Function of General Laws in History," *Journal of Philosophy* 39, pp. 35–48. [Reprinted in Hempel (1965); reprint pagination used for references in this book.]

1943

☐ Goodman, Nelson, "A Query on Confirmation," *Journal of Philosophy* 40, pp. 383–85.

• Rosenblueth, Arturo, Norbert Wiener, and Julian Bigelow, "Behavior, Purpose, and Teleology," *Philosophy of Science* 10, pp. 18–24.

White, Morton, "Historical Explanation," *Mind* 52, pp. 212–29.

1945

Feigl, Herbert, "Some Remarks on the Meaning of Scientific Explanation," *Psychological Review* 52, pp. 250–59. [Reprinted in Herbert Feigl and Wilfrid Sellars, eds., *Readings in Philosophical Analysis* (New York: Appleton-Century-Crofts).]

1946

Braithwaite, R. B., "Teleological Explanations: The Presidential Address," *Proceedings of the Aristotelian Society* 47, pp. i–xx.

Hospers, John, "On Explanation," *Journal of Philosophy* 43, pp. 337–46.

Miller, D. L., "The Meaning of Explanation," *Psychological Review* 53, pp. 241–46.

1947

☐ Carnap, Rudolf, "On the Application of Inductive Logic," *Philosophy and Phenomenological Research* 8, pp. 143–47.

☐ —— (a), "Reply to Nelson Goodman," *Philosophy and Phenomenological Research* 8, pp. 461–62.

Ginsberg, Morris, "The Character of Historical Explanation," *Proceedings of the Aristotelian Society*, supplementary vol. 21, pp. 69–77.

☐ Goodman, Nelson, "The Infirmities of Confirmation Theory," *Philosophy and Phenomenological Research* 8, pp. 149–51.

☐ —— (a), "The Problem of Counterfactual Conditionals," *Journal of Philosophy* 44, pp. 113–28. [Reprinted as chap. I of Goodman's *Fact, Fiction, and Forecast* (Cambridge, MA: Harvard University Press, first edition 1955).]

MacIver, A. M., "The Character of Historical Explanation," *Proceedings of the Aristotelian Society*, supplementary vol. 21, pp. 33–50.

Miller, D. L., "Explanation vs. Description," *Philosophical Review* 56, pp. 306–12.

☐ Reichenbach, Hans, *Elements of Symbolic Logic*. New York: Macmillan Co.

Strong, E. W., "Fact and Understanding in History," *Journal of Philosophy* 44, pp. 617–25.

Walsh, W. H., "The Character of Historical Explanation," *Proceedings of the Aristotelian Society*, supplementary vol. 21, pp. 51–68.

The First Decade

1948

• Hempel, Carl G., and Paul Oppenheim, "Studies in the Logic of Explanation," *Philosophy of Science* 15, pp. 135–75. [Reprinted in Hempel (1965); reprint pagination used for references in this book.]

☐ Quine, Willard van Orman, "Two Dogmas of Empiricism," *Philosophical Review* 60, pp. 20–43.

1949

Feigl, Herbert, "Some Remarks on the Meaning of Scientific Explanation," in Herbert Feigl and Wilfrid Sellars, eds., *Readings in Philosophical Analysis* (New York: Appleton-Century-Crofts), pp. 510–14.

Flew, Antony, "Psychoanalytic Explanation," *Analysis* 10, pp. 8–15.

Kneale, William, *Probability and Induction*. Oxford: Clarendon Press, pp. 92–110. [Reprinted, under the title "Induction, Explanation, and Transcendent Hypotheses," in Herbert Feigl and May Brodbeck, eds., *Readings in the Philosophy of Science* (New York: Appleton-Century-Crofts, 1953), pp. 353–67.]

☐ Merton, R. K., *Social Theory and Social Structure*. New York: Free Press. [Second edition, revised and enlarged, 1957.]

1950

☐ Carnap, Rudolf, *Logical Foundations of Probability*. Chicago: University of Chicago Press. [Second edition 1962.]

Cohen, Jonathan, "Teleological Explanation," *Proceedings of the Aristotelian Society* 51, pp. 225–92.

Nagel, Ernest, "Mechanistic Explanation and Organismic Biology," *Philosophy and Phenomenological Research* 11, pp. 327–38.

Taylor, Richard, "Comments on the Mechanistic Conception of Purposefulness," *Philosophy of Science* 17, pp. 310–17.

☐ Velikovsky, Immanuel, *Worlds in Collision*. Garden City, N. Y.: Doubleday & Co.

Wilkie, J. S., "Causation and Explanation in Theoretical Biology," *British Journal for the Philosophy of Science* 1, pp. 273–90.

1951

Deutsch, Karl W., "Mechanism, Teleology and Mind," *Philosophy and Phenomenological Research* 12, pp. 185–223.

Nilson, S. S., "Mechanics and Historical Laws," *Journal of Philosophy* 48, pp. 201–11.

1952

☐ Carnap, Rudolf, *The Continuum of Inductive Methods*. Chicago: University of Chicago Press.

• Gardiner, Patrick, *The Nature of Historical Explanation*. Oxford: Clarendon Press.

☐ Radcliffe-Brown, A. R., *Structure and Function in Primitive Society*. London: Cohen and West Ltd.

Strong, E. W., "Criteria of Explanation in History," *Journal of Philosophy* 49, pp. 57–67.

Watkins, J. W. N., "Ideal Types and Historical Explanation," *British Journal for the Philosophy of Science* 3, pp. 22–43.

1953

• Braithwaite, Richard Bevin, *Scientific Explanation*. Cambridge: Cambridge University Press.

Gregory, R. L., "On Physical Model Expanations in Psychology," *British Journal for the Philosophy of Science* 4, pp. 192–97.

Hofstadter, Albert, "Universality, Explanation, and Scientific Law," *Journal of Philosophy* 50, pp. 101–15.

Nagel, Ernest, "Teleological Explanation and Teleological Systems," in Sidney Ratner, ed., *Vision and Action* (Rutgers University Press: New Brunswick, NJ), pp. 192–223.

Skinner, B. F., "The Scheme of Behavior Explanations," in B. F. Skinner, *Science and Human Behavior* (New York: Free Press).

☐ Toulmin, Stephen, *The Philosophy of Science*. London: Hutchinson's University Library.

☐ Wittgenstein, Ludwig, *Philosophical Investigations*. New York: Macmillan.

1954

Brown, Robert, "Explanation by Laws in Social Science," *Philosophy of Science* 21, pp. 25–32.

Dray, William H., "Explanatory Narrative in History," *Philosophical Quarterly* 4, pp. 15–27.

☐ Malinowski, B., *Magic, Science and Religion, and Other Essays*. Garden City, N. Y.: Doubleday Anchor Books.

☐ Reichenbach, Hans, *Nomological Statements and Admissible Operations*. Amsterdam: North-Holland Publishing Co.

1955

Gallie, W. B., "Explanations in History and the Genetic Sciences," *Mind* 64, pp. 161–67. [Reprinted in Gardiner (1959), pp. 386–402.]

Gluck, Samuel E., "Do Statistical Laws Have Explanatory Efficacy?" *Philosophy of Science* 22, pp. 34–38.

□ Goodman, Nelson, *Fact, Fiction, and Forecast*. Cambridge: Harvard University Press.

Hayek, F. A., "Degrees of Explanation," *British Journal for the Philosophy of Science* 6, pp. 209–25.

1956

Danto, Arthur C., "On Explanations in History," *Philosophy of Science* 23, pp. 15–30.

Ellis, Brian, "The Relation of Explanation to Description," *Mind* 65, pp. 498–506.

Feigl, Herbert, and Michael Scriven, eds., *Minnesota Studies in the Philosophy of Science*, vol. I. Minneapolis: University of Minnesota Press.

Hospers, John, "What is Explanation?" in Antony Flew, ed., *Essays in Conceptual Analysis* (London: Macmillan & Co., Ltd), pp. 94–119.

Hutten, E. H., "On Explanation in Psychology and in Physics," *British Journal for the Philosophy of Science* 7, pp. 73–85.

• Nagel, Ernest, "A Formalization of Functionalism (With Special Reference to its Application in the Social Sciences)," in Ernest Nagel, Logic Without Metaphysics (Glencoe, Ill.: The Free Press), pp. 247–83.

□ Reichenbach, Hans, *The Direction of Time*. Berkeley & Los Angeles: University of California Press.

1957

Argyle, Michael, "Explanation of Social Behaviour," in Michael Argyle, *The Scientific Study of Social Behaviour* (London: Methuen), chap. 3.

• Donagan, Alan, "Explanation in History," *Mind* 66, pp. 145–64. [Reprinted in Gardiner (1959).]

• Dray, William, *Laws and Explanation in History*. London: Oxford University Press.

Frankel, Charles, "Explanation and Interpretation in History," *Philosophy of Science* 24, pp. 137–55. [Reprinted in Gardiner (1959), pp. 408–27.]

Goldstein, L. J., "The Logic of Explanation in Malinowskian Anthropology," *Philosophy of Science* 24, pp. 156–66.

Madden, Edward H., "The Nature of Psychological Explanation," *Methodos* 9, pp. 53–63.

□ Merton, R. K., *Social Theory and Social Structure*, 2nd ed., revised and enlarged. New York: Free Press.

• Scheffler, Israel, "Explanation, Prediction, and Abstraction," *British Journal for the Philosophy of Science* 7, pp. 293–309.

Watkins, J. W. N., "Historical Explanation in the Social Sciences," *British Journal for the Philosophy of Science* 8, pp. 104–17.

The Second Decade

1958

Feigl, Herbert, Michael Scriven, and Grover Maxwell, *Minnesota Studies in the Philosophy of Science* II. Minneapolis: University of Minnesota Press.

Goudge, T. A., "Causal Explanations in Natural History," *British Journal for the Philosophy of Science* 9, pp. 194–202.

□ Hanson, Norwood Russell, *Patterns of Discovery*. Cambridge: Cambridge University Press.

• Hempel, Carl G., "The Theoretician's Dilemma," in Feigl et al. (1958), pp. 37–98. [Reprinted in Hempel (1965); reprint pagination used for references in this book.]

Papandreou, Andreas G., "Explanation and Prediction in Economics," *Science* 129, pp. 1096–1100.

Passmore, John, "Law and Explanation in History," *Australian Journal of Politics and History* 4, pp. 269–76.

Rescher, Nicholas, "On Prediction and Explanation," *British Journal for the Philosophy of Science* 8, pp. 281–90.

□ ——(a), "A Theory of Evidence," *Philosophy of Science* 25, pp. 83–94.

• Scheffler, Israel, "Thoughts on Teleology," *British Journal for the Philosophy of Science* 9, pp. 265–84.

• Scriven, Michael, "Definitions, Explanations, and Theories," in Feigl, Scriven, and Maxwell (1958), pp. 99–195.

Skarsgard, Lars, "Some Remarks on the Logic of Explanation," *Philosophy of Science* 25, pp. 199–207.

Yolton, J. W., "Philosophical and Scientific Explanation," *Journal of Philosophy"* 55, pp. 133–43.

1959

Beattie, J. H. M., "Understanding and Explanation in Social Anthropology," *British Journal of Sociology* 10, pp. 45–60.

Dray, William H., " 'Explaining What' in History," in Gardiner (1959), pp. 403–8.

Ebersole, F. B., and M. M. Shrewsbury, "Origin Explanations and the Origin of Life," *British Journal for the Philosophy of Science* 10, pp. 103–19.

• Gardiner, Patrick, ed., *Theories of History.* New York: The Free Press.

Hanson, Norwood Russell, "On the Symmetry Between Explanation and Prediction," *Philosophical Review* 68, pp. 349–58.

Harris, E. E., "Teleology and Teleological Explanation," *Journal of Philosophy* 56, pp. 5–25.

• Hempel, Carl G., "The Logic of Functional Analysis," in Llewellyn Gross, ed., *Symposium on Sociological Theory* (New York: Harper & Row). [Reprinted in Hempel (1965); reprint pagination used for references in this book.]

Joynt, C. B., and Nicholas Rescher, "On Explanation in History," *Mind* 68, pp. 383–88.

Pitt, Jack, "Generalizations in Historical Explanation," *Journal of Philosophy* 56, pp. 578–86.

• Popper, Karl R., *The Logic of Scientific Discovery.* New York: Basic Books. [English translation of Popper (1935).]

• Scriven, Michael, "Explanation and Prediction in Evolutionary Theory," *Science* 30, pp. 477–82.

• ——(a), "Truisms as the Grounds for Historical Explanation," in Patrick Gardiner, ed., *Theories of History* (Glencoe: The Free Press), pp. 443–75.

Stannard, Jerry, "The Role of Categories in Historical Explanation," *Journal of Philosophy"* 56, pp. 429–47.

Sutherland, N. S., "Motives as Explanations," *Mind* 68, pp. 145–59.

Yolton, J. W., "Explanation," *British Journal for the Philosophy of Science* 10, pp. 194–208.

1960

Dodwell, P. C., "Causes of Behaviour and Explanation in Psychology," *Mind* 69, pp. 1–13.

□ Hempel, Carl G., "Inductive Inconsistencies," *Synthese* 12, pp. 439–69. [Reprinted in Hempel (1965); reprint pagination used for references in this book.]

1961

Barker, Stephen, "The Role of Simplicity in Explanation," in Herbert Feigl and Grover Maxwell, eds., *Current Issues in the Philosophy of Science* (New York: Holt, Rinehart, and Winston), pp. 265–73. [See also comments by Wesley C. Salmon, Paul Feyerabend, and Richard Rudner, and Barker's replies.]

Bromberger, Sylvain, "The Concept of Explanation," Ph. D. dissertation, Princeton University.

• Eberle, R., D. Kaplan, and R. Montague, "Hempel and Oppenheim on Explanation," *Philosophy of Science* 28, pp. 418–28.

Feigl, Herbert, and Grover Maxwell, eds., *Current Issues in the Philosophy of Science*. New York: Holt, Rinehart, and Winston.

Good, I. J., "A Causal Calculus (I-II)," *British Journal for the Philosophy of Science* 11, pp. 305–18; 12, pp. 43–51. See also "Corrigenda," 13, p. 88.

• Kaplan, David, "Explanation Revisited," *Philosophy of Science* 28, pp. 429–36.

Mandelbaum, Maurice, "Historical Explanation: The Problem of 'Covering Laws'," *History and Theory* 1, pp. 229–42.

• Nagel, Ernest, *The Structure of Science: Problems in the Logic of Scientific Explanation*. New York: Harcourt, Brace & World.

Rescher, Nicholas, and Carey B. Joynt, "The Problem of Uniqueness in History," *History and Theory* 1, pp. 150–62.

Scriven, Michael, "Discussion: Comments on Weingartner," *Philosophy of Science* 28, p. 306.

Sellars, Wilfrid, "The Language of Theories," in Herbert Feigl and Grover Maxwell, eds., *Current Issues in the Philosophy of Science* (New York: Holt, Rinehart, and Winston), pp. 57–77.

Stover, R. C., "Dray on Historical Explanation," *Mind* 70, pp. 540–43.

• Toulmin, Stephen, *Foresight and Understanding*. Bloomington: Indiana University Press.

Weingartner, Rudolph H., "Discussion: Explanations and Their Justifications," *Philosophy of Science* 28, pp. 300–5.

——(a), "The Quarrel about Historical Explanation," *Journal of Philosophy* 58, pp. 29–45.

1962

Alexander, Peter, "Rational Behaviour and Psychoanalytic Explanation," *Mind* 71, pp. 326–41.

Bartley, W. W. III, "Achilles, the Tortoise, and Explanation in Science and History," *British Journal for the Philosophy of Science* 13, pp. 15–33.

Brodbeck, May, "Explanation, Prediction, and 'Imperfect' Knowledge," in Feigl and Maxwell (1962), pp. 231–72.

• Bromberger, Sylvain, "An Approach to Explanation," in R. S. Butler, ed., *Analytical Philosophy— Second Series* (Oxford: Basil Blackwell), pp. 72–105. [American edition, New York: Barnes and Noble, 1965.]

Chomsky, Noam, "Explanatory Models in Linguistics," in Ernest Nagel, Patrick Suppes, and Alfred Tarski, eds., *Logic, Methodology and Philosophy of Science*, Proceedings of the 1960 International Congress (Stanford: Stanford University Press), pp. 528–50.

Feigl, Herbert, and Grover Maxwell, eds., *Minnesota Studies in the Philosophy of Science* III. Minneapolis: University of Minnesota Press.

Feyerabend, Paul, "Explanation, Reduction, and Empiricism," in Feigl and Maxwell (1962), pp. 231–72.

• Grünbaum, Adolf, "Temporally-Asymmetric Principles, Parity Between Explanation and Prediction, and Mechanism vs. Teleology," *Philosophy of Science* 29, pp. 146–70.

• Hempel, Carl G., "Deductive-Nomological vs. Statistical Explanation," in Feigl and Maxwell (1962), pp. 98–169.

——(a), "Explanation in Science and in History," in Robert G. Colodny, ed., *Frontiers of Science and Philosophy* (Pittsburgh: University of Pittsburgh Press), pp. 7–34.

Kim, Jaegwon, "Explanation, Prediction, and Retrodiction: Some Logical and Pragmatic Considerations," Ph. D. dissertation, Princeton University.

Margenau, Henry, "Is the Mathematical Explanation of Physical Data Unique?" in Ernest Nagel, Patrick Suppes, and Alfred Tarski, eds., *Logic, Methodology and Philosophy of Science*, Proceedings of the 1960 International Congress (Stanford: Stanford University Press), pp. 348–55.

Passmore, John, "Explanation in Everyday Life, in Science, and in History," *History and Theory* 2, pp. 105–23.

Rescher, Nicholas, "*The Stochastic Revolution and the Nature of Scientific Explanation*," *Synthese* 14, pp. 200–15.

• Scriven, Michael, "Explanations, Predictions, and Laws," in Feigl and Maxwell (1962), pp. 170–230.

—— (a), "Discussion: Comments on Professor Grünbaum's Remarks at the Wesleyan Meeting," *Philosophy of Science* 29, pp. 171–74.

Treisman, Michel, "Psychological Explanation: The 'Private Data' Hypothesis," *British Journal for the Philosophy of Science* 13, pp. 130–43.

1963

Alexander, Peter, *Sensationalism and Scientific Explanation*. New York: Humanities Press.

Baker, A. J., "Historical Explanation and Universal Propositions," *Australasian Journal of Philosophy* 41, pp. 317–35.

Baumrin, Bernard H., ed., *Philosophy of Science: The Delaware Seminar*, vols. 1–2. New York: John Wiley & Sons.

Brandt, Richard, and Jaegwon Kim, "Wants as Explanations of Actions," *Journal of Philosophy* 60, pp. 425–35.

• Bromberger, Sylvain, "A Theory about the Theory of Theory and about the Theory of Theories," in Baumrin (1963, vol. 2), pp. 79–105.

Brown, Robert, *Explanation in Social Science*. Chicago: Aldine Publishing Co.

Dray, William H., "Historical Explanation of Actions Reconsidered," in Hook (1963), pp. 105–35.

Fain, Haskell, "Some Problems of Causal Explanation," *Mind* 72, pp. 519–32.

Grünbaum, Adolf, "Temporally Asymmetric Principles, Parity Between Explanation and Prediction, and Mechanism versus Teleology," in Baumrin (1963, vol. 1), pp. 57–96. [Substantially the same as Grünbaum (1962).]

Hempel, Carl G., "Explanation and Prediction by Covering Laws," in Baumrin (1963, vol. 1), pp. 107–33.

Henson, R. B., "Mr. Hanson on the Symmetry of Explanation and Prediction," *Philosophy of Science* 30, pp. 60–61.

Hesse, Mary, "A New Look at Scientific Explanation," *Review of Metaphysics* 17, pp. 98–108.

Hook, Sidney, ed., *Philosophy and History*. New York: New York University Press.

• Kim, Jaegwon, "On the Logical Conditions of Deductive Explanation," *Philosophy of Science* 30, pp. 286–91.

Morgenbesser, Sidney, "The Explanatory-Predictive Approach to Science," in Baumrin (1963, vol. 1), pp. 41–55.

Rescher, Nicholas, "Discrete State Systems, Markov Chains, and Problems in the Theory of Scientific Explanation and Prediction," *Philosophy of Science* 30, pp. 325–45.

—— (a), "Fundamental Problems in the Study of Scientific Explanation," in Baumrin (1963, vol. 2), pp. 41–60.

☐ Salmon, Wesley C., "On Vindicating Induction," *Philosophy of Science* 30, pp. 252–61.

• Scheffler, Israel, *The Anatomy of Inquiry*. New York: Alfred A. Knopf.

• Scriven, Michael, "The Temporal Asymmetry Between Explanations and Predictions," in Baumrin (1963, vol. 1), pp. 97–105.

—— (a), "The Limits of Physical Explanation," in Baumrin (1963, vol. 2), pp. 107–35.

—— (b), "New Issues in the Logic of Explanation," in Sidney Hook, ed., *Philosophy and History* (New York: New York University Press), pp. 339–61.

—— (c), "Review of Ernest Nagel, *The Structure of Science*," *Review of Metaphysics* 17, pp. 403–24.

Sellars, Wilfrid, "Theoretical Explanation," in Baumrin (1963, vol. 2), pp. 61–78.

1964

Canfield, John, "Teleological Explanations in Biology," *British Journal for the Philosophy of Science* 14, pp. 285–95.

Donagan, Alan, "Historical Explanations: The Popper-Hempel Theory Reconsidered," *History and Theory* 4, pp. 3–26. [Reprinted in Dray (1966).]

Gallie, W. B., *Philosophy and the Historical Understanding*. London: Chatto & Windus.

Gustafson, Don F., "Explanation in Psychology," *Mind* 73, pp. 280–81.

Jarvie, Ian C., "Explanation in Social Science," *Philosophy of Science* 15, pp. 62–72. [Review of Robert Brown, *Explanation in Social Sciences*.]

Kim, Jaegwon, "Inference, Explanation, and Prediction," *Journal of Philosophy* 61, pp. 360–68.

Martin, Michael, "The Explanatory Value of the Unconscious," *Philosophy of Science* 31, pp. 122–32.

Sorabji, Richard, "Function," *Philosophical Quarterly* 14, pp. 289–302.

• Taylor, Charles, *The Explanation of Behaviour*. London: Routledge and Kegan Paul.

Workman, Rollin W., "What Makes an Explanation," *Philosophy of Science* 31, pp. 241–54.

1965

Ackermann, Robert, "Discussion: Deductive Scientific Explanation," *Philosophy of Science* 32, pp. 155–67.

Balmuth, Jerome, "Psychoanalytic Explanation," *Mind* 74, pp. 229–35.

Canfield, John, "Teleological Explanation in Biology: A Reply," *British Journal for the Philosophy of Science* 15, pp. 327-31.

Feyerabend, Paul, "Reply to Criticism," in Robert S. Cohen and Marx Wartofsky, eds., *Boston Studies in the Philosophy of Science* 2 (Dordrecht: D. Reidel Publishing Co.), pp. 223–61.

Gorovitz, Samuel, "Causal Judgments and Causal Explanations," *Journal of Philosophy* 62, pp. 695–711.

☐ Harman, Gilbert, "Inference to the Best Explanation," *Philosophical Review* 74, pp. 88–95.

• Hempel, Carl G., *Aspects of Scientific Explanation and Other Essays in the Philosophy of Science*. New York: The Free Press.

• —— (a), "Aspects of Scientific Explanation," in *Aspects*.

Hesse, Mary B., "The Explanatory Function of Metaphor," in Yehoshua Bar-Hillel, ed., *Logic, Methodology and Philosophy of Science*, Proceedings of the 1964 International Congress (Amsterdam: North-Holland Publishing Co.), pp. 249–59.

• Kyburg, Henry E., Jr., "Comment," *Philosophy of Science* 32, pp. 147–51.

Lehman, Hugh, "Functional Explanation in Biology," *Philosophy of Science* 32, pp. 1–19.

—— (a), "Teleological Explanation in Biology," *British Journal for the Philosophy of Science* 15, p. 327.

Mischel, Theodore, "Concerning Rational Behaviour and Psychoanalytic Explanation," *Mind* 74, pp. 71–78.

Newman, Fred, "Discussion: Explanation Sketches," *Philosophy of Science* 32, pp. 168–72.

Putnam, Hilary, "How Not to Talk about Meaning," in Robert S. Cohen and Marx Wartofsky, eds., *Boston Studies in the Philosophy of Science* 2 (Dordrecht: D. Reidel Publishing Co.), pp. 205–22.

• Salmon, Wesley C., "The Status of Prior Probabilities in Statistical Explanation," *Philosophy of Science* 32, pp. 137–46.

Sellars, Wilfrid, "Scientific Realism or Irenic Realism," in Robert S. Cohen and Marx Wartofsky, eds., *Boston Studies in the Philosophy of Science* 2 (Dordrecht: D. Reidel Publishing Co.), pp. 171–204.

Smart, J. J. C., "Conflicting Views About Explanation," in Robert S. Cohen and Marx Wartofsky, eds., *Boston Studies in the Philosophy of Science* 2 (Dordrecht: D. Reidel Publishing Co.), pp. 151–69.

1966

Ackermann, Robert, and Alfred Stenner, "A Corrected Model of Explanation," *Philosophy of Science* 33, pp. 168–71.

• Bromberger, Sylvain, "Why-Questions," in Robert G. Colodny, ed., *Mind and Cosmos* (Pittsburgh: University of Pittsburgh Press), pp. 86–111.

☐ Canfield, John, ed., *Purpose in Nature*. Englewood Cliffs, NJ: Prentice-Hall, Inc.

Canfield, John, and Keith Lehrer, "Discussion: A Note on Prediction and Deduction," *Philosophy of Science* 33, pp. 165–67.

☐ Carnap, Rudolf, *Philosophical Foundations of Physics* (edited by Martin Gardner). New York: Basic Books. [Reissued in 1974 as *An Introduction to the Philosophy of Science*.]

Collins, Arthur W., "The Use of Statistics in Explanation," *British Journal for the Philosophy of Science* 17, pp. 127–40.

——(a), "Explanation and Causality," *Mind* 75, pp. 482–85.

Dray, William H., *Philosophical Analysis and History*. New York: Harper & Row.

Feyerabend, Paul, "Article-Review: *The Structure of Science* by Ernest Nagel," *British Journal for the Philosophy of Science* 17, pp. 237–48.

Fine, Arthur I., "Explaining the Behavior of Entities," *Philosophical Review* 75, pp. 496–509.

Gruner, Rolf, "Teleological and Functional Explanation," *Mind* 75, pp. 516–26.

Hempel, Carl G., *Philosophy of Natural Science*. Englewood Cliffs, NJ: Prentice-Hall, Inc. [Unlike other introductory texts in philosophy of science, this book has profoundly influenced the views of scientists, especially behavioral scientists, on the nature of scientific explanation.]

Leach, James, "Discussion: Dray on Rational Explanation," *Philosophy of Science* 33, pp. 61–69.

Lehman, Hugh, "R. E. Merton's Concepts of Function and Functionalism," *Inquiry* 9, pp. 274–83.

Ling, J. F., "Explanation in History," *Mind* 75, pp. 589–91.

Madden, Edward H., "Explanation in Psychoanalysis and History," *Philosophy of Science* 33, pp. 278–86.

Mellor, D. H., "Inexactness and Explanation," *Philosophy of Science* 33, pp. 345–59.

Mischel, Theodore, "Pragmatic Aspects of Explanation," *Philosophy of Science* 33, pp. 40–60.

Scriven, Michael, "Causes, Connections, and Conditions in History," in Dray (1966), pp. 238–64.

Simon, Herbert A., and Nicholas Rescher, "Cause and Counterfactual," *Philosophy of Science* 33, pp. 323–40.

Stegmüller, W., "Explanation, Prediction, Scientific Systematization, and Non-Explanatory Information," *Ratio* 8, pp. 1–24.

1967

Alston, William, "Wants, Actions, and Causal Explanation," in H. Casteneda, ed., *Intentionality, Minds, and Perceptions* (Detroit: Wayne State University Press).

Angel, R. B., "Discussion: Explanation and Prediction: A Plea for Reason," *Philosophy of Science* 34, pp. 276–82.

Coffa, J. Alberto, "Feyerabend on Explanation and Reduction," *Journal of Philosophy* 64, pp. 500–8.

Cunningham, Frank, "More on Understanding in the Social Sciences," *Inquiry* 10, pp. 321–26.

Dietl, Paul, "Paresis and the Alleged Asymmetry Between Explanation and Prediction," *British Journal for the Philosophy of Science* 17, pp. 313–18.

Goh, S. T., "Discussion: Newman and Explanation Sketches," *Philosophy of Science* 34, pp. 273–75.

Gruner, Rolf, "Understanding in the Social Sciences and History," *Inquiry* 10, pp. 151–63.

Kim, Jaegwon, "Explanation in Science," in Paul Edwards, ed., *The Encyclopedia of Philosophy* (New York: Macmillan Publishing Co. and Free Press), vol. 3, pp. 159–63.

Madell, Geoffrey, "Action and Causal Explanation," *Mind* 76, pp. 34–48.

Malcolm, Norman, "Explaining Behavior," *Philosophical Review* 76, pp. 97–104.

Noble, Denis, "Charles Taylor on Teleological Explanation," *Analysis* 27, pp. 96–103.

☐ Salmon, Wesley C., *The Foundations of Scientific Inference*. Pittsburgh: University of Pittsburgh Press.

Shope, Robert K., "Explanation in Terms of 'the Cause'," *Journal of Philosophy* 64, pp. 312–20.

Suchting, W. A., "Deductive Explanation and Prediction Revisited," *Philosophy of Science* 34, pp. 41–52.

Taylor, Charles, "Teleological Explanation—a Reply to Denis Noble," *Analysis* 27, pp. 141–43.

The Third Decade

1968

•Beckner, Morton, *The Biological Way of Thought*. Berkeley & Los Angeles: University of California Press.

Brody, B. A., "Confirmation and Explanation," *Journal of Philosophy* 65, pp. 282–99.

Coffa, J. A., "Discussion: Deductive Predictions," *Philosophy of Science* 35, pp. 279-83.

Finn, D. R., "Categories of Psychological Explanation," *Mind* 77, pp. 550-55.

• Fodor, Jerry A., *Psychological Explanation*. New York: Random House.

——(a), "The Appeal to Tacit Knowledge in Psychological Explanation," *Journal of Philosophy* 65, pp. 627-40.

Good, I. J., "Corroboration, Explanation, Evolving Probability, Simplicity and a Sharpened Razor," *British Journal for the Philosophy of Science* 19, pp. 123-43.

Harman, Gilbert, "Knowledge, Inference, and Explanation," *American Philosophical Quarterly* 5, pp. 161-73.

• Hempel, Carl G., "Maximal Specificity and Lawlikeness in Probabilistic Explanation," *Philosophy of Science* 35, pp. 116-33.

Humphreys, Willard C., "Discussion: Statistical Ambiguity and Maximal Specificity," *Philosophy of Science* 35, pp. 112-15.

Leach, James, "Explanation and Value Neutrality," *British Journal for the Philosophy of Science* 19, pp. 93-108.

Martin, Michael, "Situational Logic and Covering Law Explanation in History," *Inquiry* 11, pp. 388-99.

Massey, Gerald J., "Hempel's Criterion of Maximal Specificity," *Philosophical Studies* 19, pp. 43-47.

Mellor, D. H., "Two Fallacies in Charles Taylor's *Explanation of Behaviour*," *Mind* 77, pp. 124-26.

Paluch, Stanley, "The Covering Law Model of Historical Explanation," *Inquiry* 11, pp. 368-87.

Wright, Larry, "The Case Against Teleological Reductionism," *British Journal for the Philosophy of Science* 19, pp. 211-23.

1969

Aronson, Jerrold L., "Explanations without Laws," *Journal of Philosophy* 66, pp. 541-47.

Boden, Margaret A., "Miracles and Scientific Explanation," *Ratio* 11, pp. 137-44.

Goodfield, June, "Theories and Hypotheses in Biology: Theoretical Entities and Functional Explanation," in Robert S. Cohen and Marx Wartofsky, eds., *Boston Studies in the Philosophy of Science* 5, pp. 421-49.

Gruner, Rolf, "The Notion of Understanding: Replies to Cunningham and Van Evra," *Inquiry* 12, pp. 349-56.

Hanna, Joseph, "Explanation, Prediction, Description, and Information," *Synthese* 20, pp. 308-44.

Hein, Hilde, "Molecular Biology vs. Organicism: The Enduring Dispute between Mechanism and Vitalism," *Synthese* 20, pp. 238-53.

Iseminger, Gary, "Malcolm on Explanations and Causes," *Philosophical Studies* 20, pp. 73-77.

• Jeffrey, Richard C., "Statistical Explanation vs. Statistical Inference," in Nicholas Rescher, ed., *Essays in Honor of Carl G. Hempel* (Dordrecht: D. Reidel Publishing Co.), pp. 104-13. [Reprinted in Salmon, et al. (1971).]

Levi, Isaac, "Are Statistical Hypotheses Covering Laws?" *Synthese* 20, pp. 297-307.

Levison, A. B., and I. Thalberg, "Essential and Causal Explanations," *Mind* 78, pp. 91-101.

Macklin, Ruth, "Explanation and Action: Recent Issues and Controversies," *Synthese* 20, pp. 388-415.

Manier, Edward, "'Fitness' and Some Explanatory Patterns in Biology," *Synthese* 20, pp. 185-205.

McCullagh, C. B., "Narrative and Explanation in History," *Mind* 78, pp. 256-61.

Rescher, Nicholas, "Lawfulness as Mind-Dependent," in Nicholas Rescher, ed., *Essays in Honor of Carl G. Hempel* (Dordrecht: D. Reidel Publishing Co.), pp. 178-97.

Rosenkrantz, Roger D., "On Explanation," *Synthese* 20, pp. 335-70.

□ Salmon, Wesley C., "Partial Entailment as a Basis for Inductive Logic," in Nicholas Rescher, ed., *Essays in Honor of Carl G. Hempel* (Dordrecht: D. Reidel), pp. 47-82.

Stegmüller, Wolfgang, *Wissenschaftliche Erklärung und Begründung*. Berlin, Heidelberg, & New York: Springer-Verlag.

Van Evra, James W., "Understanding in the Social Sciences Revisited," *Inquiry* 12, pp. 347-49.

Wilson, Fred, "Explanation in Aristotle, Newton, and Toulmin," *Philosophy of Science* 36, part I, pp. 291–310; part II, pp. 400–28.

1970

Ayala, Francisco J., "Teleological Explanations in Evolutionary Biology," *Philosophy of Science* 37, pp. 1–15.

Borger, R., and F. Cioffi, *Explanation in the Behavioural Sciences*. Cambridge: The University Press.

Brittan, Gordon, Jr., "Explanation and Reduction," *Journal of Philosophy* 67, pp. 446–57.

Churchland, Paul, "The Logical Character of Action-Explanations," *Philosophical Review* 79, pp. 214–36.

Ellis, B. D., "Explanation and the Logic of Support," *Australasian Journal of Philosophy* 48, pp. 177–89.

Goh, S. T., "Some Observations on the Deductive-Nomological Theory," *Mind* 79, pp. 408–14.

——(a), "The Logic of Explanation in Anthropology," *Inquiry* 13, pp. 339–59.

• Greeno, James G., "Evaluation of Statistical Hypotheses Using Information Transmitted," *Philosophy of Science* 37, pp. 279–93. [Reprinted under the title "Explanation and Information" in Salmon, et al. (1971).]

Hedman, C. G., "Gustafson on Explanation in Psychology," *Mind* 79, pp. 272–74.

• Hesse, Mary, *Models and Analogies in Science*. Notre Dame: University of Notre Dame Press.

Kyburg, Henry E., Jr., "Discussion: More on Maximal Specificity," *Philosophy of Science* 37, pp. 295–300.

Margolis, Joseph, "Puzzles regarding Explanation by Reasons and Explanation by Causes," *Journal of Philosophy* 67, pp. 187–95.

Morgan, Charles G., "Discussion: Kim on Deductive Explanation," *Philosophy of Science* 37, pp. 434–39.

Nissen, Lowell, "Canfield's Functional Translation Schema," *British Journal for the Philosophy of Science* 21, pp. 193–95.

Omer, I. A., "On the D-N Model of Scientific Explanation," *Philosophy of Science* 37, pp. 417–33.

☐ Pauling, Linus, *Vitamin C and the Common Cold*. San Francisco: W. H. Freeman and Co.

• Rescher, Nicholas, *Scientific Explanation*. New York: Free Press.

• Salmon, Wesley C., "Statistical Explanation," in Robert G. Colodny, ed., *The Nature and Function of Scientific Theories* (Pittsburgh: University of Pittsburgh Press), pp. 173–231. [Reprinted in Salmon et al. (1971).]

☐ ——(a), ed., *Zeno's Paradoxes*. Indianapolis: Bobbs-Merrill.

Schlegel, Richard, "Statistical Explanation in Physics: The Copenhagen Interpretation," *Synthese* 21, pp. 65–82.

Suppes, Patrick, *A Probabilistic Theory of Causality*. Amsterdam: North-Holland.

Taylor, Charles, "Explaining Action," *Inquiry* 13, pp. 54–89.

1971

Achinstein, Peter, *Law and Explanation*. Oxford: Clarendon Press.

Alexander, Peter, "Psychoanalysis and the Explanation of Behaviour," *Mind* 80, pp. 391–408.

Alston, William P., "The Place of Explanation of Particular Facts in Science," *Philosophy of Science* 38, pp. 13–34.

Economos, John James, "Explanation: What's It All About?" *Australasian Journal of Philosophy* 49, pp. 139–45.

Fetzer, James H., "Dispositional Probabilities," in Roger C. Buck and Robert S. Cohen, eds., *PSA 1970* (Dordrecht: D. Reidel Publishing Co.), pp. 473–82.

Gaukroger, Stephen, *Explanatory Structures*. Brighton: Harvester Press.

Greeno, James G., "Theoretical Entities in Statistical Explanation," in Roger C. Buck and Robert S. Cohen, eds., *PSA 1970* (Dordrecht: D. Reidel Publishing Co.), pp. 3–26.

Hovard, Richard B., "Theoretical Reduction: The Limits and Alternatives to Reductive Methods in Scientific Explanation," *Philosophy of the Social Sciences* 1, pp. 83–100.

Jeffrey, Richard C., "Remarks on Explanatory Power," in Roger C. Buck and Robert S. Cohen, eds., *PSA 1970* (Dordrecht: D. Reidel Publishing Co.), pp. 40–46.

Martin, Michael, "Neurophysiological Reduction and Psychological Explanation," *Philosophy of the Social Sciences* 1, pp. 161-70.

Mullane, Harvey, "Psychoanalytic Explanation and Rationality," *Journal of Philosophy* 68, pp. 413-26.

Nickles, Thomas, "Covering Law Explanation," *Philosophy of Science* 38, pp. 542-61.

Nilson, Sten Sparre, "Covering Laws in Historical Practice," *Inquiry* 14, pp. 445-63.

Ruse, Michael, "Discussion: Functional Statements in Biology," *Philosophy of Science* 38, pp. 87-95.

Salmon, Wesley C., "Explanation and Relevance," in Roger C. Buck and Robert S. Cohen, eds., *PSA 1970* (Dordrecht: D. Reidel Publishing Co.), pp. 27-39.

• Salmon, Wesley C., et al., *Statistical Explanation and Statistical Relevance*. Pittsburgh: University of Pittsburgh Press.

• von Wright, G. H., *Explanation and Understanding*. Ithaca, NY: Cornell University Press.

Watson, Patty Jo, Steven A. LeBlanc, and Charles Redman, *Explanation in Archaeology: An Explicitly Scientific Approach*. New York: Columbia University Press. [Second edition, Watson et al., (1984).]

White, J. E., "Avowed Reasons and Causal Explanation," *Mind* 80, pp. 238-45.

Zaffron, Richard, "Identity, Subsumption, and Scientific Explanation," *Journal of Philosophy* 68, pp. 849-60.

1972

Ball, Terrence, "On 'Historical' Explanation," *Philosophy of the Social Sciences* 2, pp. 181-92.

☐ Binford, Lewis, *An Archaeological Perspective*. New York: Harcourt.

Boden, Margaret A., *Purposive Explanation in Psychology*. Cambridge, MA: Harvard University Press.

Brody, Baruch, "Towards an Aristotelian Theory of Scientific Explanation," *Philosophy of Science* 39, pp. 20-31.

Fabian, Robert G., "Human Behavior in Deductive Social Theory: The Example of Economics," *Inquiry* 15, pp. 411-33.

Harré, Rom, and P. F. Secord, *The Explanation of Social Behaviour*. Totowa, N.J.: Rowman and Littlefield.

Lehman, Hugh, "Statistical Explanation," *Philosophy of Science* 39, pp. 500-6.

Martin, Michael, "Confirmation and Explanation," *Analysis* 32, pp. 167-69.

—— (a), "Explanation in Social Science: Some Recent Work," *Philosophy of the Social Sciences* 2, pp. 61-81.

Miettinen, Seppo K., "Discussion: On Omer's Model of Explanation," *Philosophy of Science* 39, pp. 249-51.

Morgan, Charles G., "Discussion: On Two Proposed Models of Explanation," *Philosophy of Science* 39, pp. 74-81.

☐ Nye, Mary Jo, *Molecular Reality*. London: Macdonald.

Rapoport, Anatol, "Explanatory Power and Explanatory Appeal of Theories," *Synthese* 24, pp. 321-42.

Tuomela, Raimo, "Deductive Explanation of Scientific Laws," *Journal of Philosophical Logic* 1, pp. 369-92.

Varela, Francisco G., and Humberto R. Maturana, "Discussion: Mechanism and Biological Explanation," *Philosophy of Science* 39, pp. 378-82.

Wallace, William A., *Causality and Scientific Explanation*. Ann Arbor: University of Michigan Press.

White, J. E., "Hedman on Explanation," *Mind* 81, pp. 595-96.

Wimsatt, William, "Teleology and the Logical Structure of Function Statements," *Studies in the History and Philosophy of Science* 3, pp. 1-80.

Wright, Larry, "Explanation and Teleology," *Philosophy of Science* 39, pp. 204-18.

—— (a), "Discussion: A Comment on Ruse's Analysis of Function Statements," *Philosophy of Science* 39, pp. 512-14.

1973

Bennett, P. W., "Avowed Reasons and the Covering Law Model," *Mind* 82, pp. 606–7.
• Coffa, J. Alberto, *The Foundations of Inductive Explanation*. Doctoral dissertation, University of Pittsburgh.
  ☐ DeWitt, Bryce S., and Neill Graham, *The Many-Worlds Interpretation of Quantum Mechanics*. Princeton: Princeton University Press.
Finocchiaro, Maurice A., *History of Science as Explanation*. Detroit: Wayne State University Press.
Gower, Barry, "Martin on Confirmation and Explanation," *Analysis* 33, pp. 107–9.
Hull, David, "A Belated Reply to Gruner," *Mind* 82, pp. 437–38.
Krausser, Peter, "A Cybernetic Systemstheoretical Approach to Rational Understanding and Explanation, especially Scientific Revolutions with Radical Meaning Change," *Ratio* 15, pp. 221–46.
  ☐ Lewis, David, *Counterfactuals*. Cambridge, MA.: Harvard University Press.
Morgan, Charles G., "Discussion: Omer on Scientific Explanation," *Philosophy of Science* 40, pp. 110–17.
Nickles, Thomas, "Discussion: Explanation and Description Relativity," *Philosophy of Science* 40, pp. 408–14.
Ruse, Michael, *Philosophy of Biology*. London: Hutchinson University Library.
——(a), "Discussion: A Reply to Wright's Analysis of Functional Statements," *Philosophy of Science* 40, pp. 277–80.
——(b), "Teleological Explanations and the Animal World," *Mind* 82, pp. 433–36.
Salmon, Wesley C., "Reply to Lehman," *Philosophy of Science* 40, pp. 397–402.
Shelanski, V. B., "Nagel's Translations of Teleological Statements: A Critique," *British Journal for the Philosophy of Science* 24, pp. 397–401.
Sklar, Lawrence, "Statistical Explanation and Ergodic Theory," *Philosophy of Science* 40, pp. 194–212.
Srzednicki, Jan, "Statistical Indeterminism and Scientific Explanation," *Synthese* 26, pp. 197–204.
• Stegmüller, Wolfgang, *Probleme und Resultate der Wissenschaftstheorie und analytischen Philosophie*, Band 4, Studienausgabe Teil E. Berlin/New York: Springer-Verlag.
Steinberg, Danny, "Discussion: Nickles on Intensionality and the Covering Law Model," *Philosophy of Science* 40, pp. 403–7.
Stemmer, Nathan, "Discussion: Brody's Defense of Essentialism," *Philosophy of Science* 40, pp. 393–96.
Tondl, Ladislav, *Scientific Procedures, Boston Studies in the Philosophy of Science* 10, chap. V.
Wright, Larry, "Function," *Philosophical Review* 82, pp. 139–68.

1974

Bogen, James, "Moravcsik on Explanation," *Synthese* 28, pp. 19–26.
Brody, Baruch, "More on Confirmation and Explanation," *Philosophical Studies* 26, pp 73–75.
Buroker, Jill Vance, "Kant, the Dynamical Tradition, and the Role of Matter in Explanation," in Kenneth F. Schaffner and Robert S. Cohen, eds., *PSA 1972* (Dordrecht: D. Reidel Publishing Co.), pp. 153–64.
• Coffa, J. Alberto, "Hempel's Ambiguity," *Synthese* 28, pp. 141–63.
Collins, Paul W., "The Present Status of Anthropology as an Explanatory Science," in Raymond J. Seeger and Robert S. Cohen, eds., *Philosophical Foundations of Physics, Boston Studies in the Philosophy of Science* 11, pp. 337–48.
• Fetzer, James H., "Grünbaum's 'Defense' of the Symmetry Thesis," *Philosophical Studies* 25, pp. 173–87.
• ——(a), "Statistical Explanations," in Kenneth Schaffner and Robert S. Cohen, eds., *PSA 1972* (Dordrecht: D. Reidel Publishing Co.), pp. 337–47.
• ——(b), "A Single Case Propensity Theory of Explanation," *Synthese* 28, pp. 87–97.
• Friedman, Michael, "Explanation and Scientific Understanding," *Journal of Philosophy* 71, pp. 5–19.
• Grene, Marjorie, *The Understanding of Nature: Essays in the Philosophy of Biology*. Dordrecht: D. Reidel Publishing Co.

Hopson, Ronald C., "The Objects of Acceptance: Competing Scientific Explanations," in Kenneth Schaffner and Robert S. Cohen, eds., *PSA 1972* (Dordrecht: D. Reidel Publishing Co.), pp. 349-63.

Klein, Martin J., "Boltzmann, Monocycles and Mechanical Explanation," in Raymond J. Seeger and Robert S. Cohen, eds., *Philosophical Foundations of Physics, Boston Studies in the Philosophy of Science* 11, pp. 155-75.

☐ Mackie, J. L., *The Cement of the Universe*. Oxford: Clarendon Press.

Mayr, Ernst, "Teleologica and Teleonomic, a New Analysis," in Robert S. Cohen and Marx Wartofsky, eds., *Boston Studies in the Philosophy of Science* 14, pp. 91-118.

Moravcsik, Julius M. E., "Aristotle on Adequate Explanations," *Synthese* 28, pp. 3-18.

Salmon, Wesley C., "Comments on 'Hempel's Ambiguity' by J. Alberto Coffa," *Synthese* 28, pp. 165-69.

Singleton, J., "The Explanatory Power of Chomsky's Transformational Generative Grammar," *Mind* 83, pp. 429-31.

• Teller, Paul, "On Why-Questions," *Nous* 8, pp. 371-80.

Thorpe, Dale A., "Discussion: The Quartercentenary Model of D-N Explanation," *Philosophy of Science* 41, pp. 188-95.

Wigner, Eugene P., "Physics and the Explanation of Life," in Raymond J. Seeger and Robert S. Cohen, eds., *Philosophical Foundations of Physics, Boston Studies in the Philosophy of Science* 11, pp. 119-32.

Wright, Larry, "Mechanisms and Purposive Behavior," *Philosophy of Science* 41, pp. 345-60.

1975

Achinstein, Peter, "The Object of Explanation," in Körner (1975), pp. 1-45. [See also comments by Rom Harré and Mary Hesse, and Achinstein's reply.]

Baublys, Kenneth K., "Discussion: Comments on Some Recent Analyses of Functional Statements in Biology," *Philosophy of Science* 42, pp. 469-86.

• Brody, Baruch, "The Reduction of Teleological Sciences," *American Philosophical Quarterly* 12, pp. 69-76.

Burian, Richard M., "Conceptual Change, Cross-Theoretical Explanation, and the Unity of Science," *Synthese* 32, pp. 1-28.

Crowell, E., "Causal Explanation and Human Action," *Mind* 84, pp. 440-42.

Cummins, Robert, "Functional Analysis," *Journal of Philosophy* 72, 741-65.

Eglin, Peter, "What Should Sociology Explain—Regularities, Rules or Interpretations?" *Philosophy of the Social Sciences* 5, pp. 277-92.

Englehardt, H. T., Jr., and S. F. Spicker, eds., *Evaluation and Explanation in the Biomedical Sciences*. Dordrecht: D. Reidel Publishing Co.

Fetzer, James H., "On the Historical Explanation of Unique Events," *Theory and Decision* 6, pp. 87-97.

——(a), "Discussion Review: Achinstein's *Law and Explanation*," *Philosophy of Science* 42, pp. 320-33.

☐ Few, A. A., "Thunder," *Scientific American* 233, no. 1 (July), pp. 80-90.

Good, I. J., "Explicativity, Corroboration, and the Relative Odds of Hypotheses," *Synthese* 30, pp. 39-74.

Koertge, Noretta, "An Exploration of Salmon's S-R Model of Explanation," *Philosophy of Science* 42, pp. 270-74.

——(a), "Popper's Metaphysical Research Program for the Human Sciences," *Inquiry* 18, pp. 437-62.

• Körner, Stephan, ed., *Explanation*. Oxford: Basil Blackwell.

Martin, Michael, "Explanation and Confirmation Again," *Analysis* 36, pp. 41-42.

Minton, Arthur J., "Discussion: Wright and Taylor: Empiricist Teleology," *Philosophy of Science* 42, pp. 299-306.

Mucciolo, Laurence F., "Neurophysiological Reduction, Psychological Explanation, and Neuropsychology," *Philosophy of the Social Sciences* 5, pp. 451-62.

Salmon, Wesley C., "Theoretical Explanation," in Körner (1975), pp. 118–45. [See also comments by L. J. Cohen and D. H. Mellor, and Salmon's reply.]

☐ ——(a), *Space, Time, and Motion: A Philosophical Introduction.* Encino, CA: Dickenson Publishing Co. [Second edition, Minneapolis: University of Minnesota Press, 1980.]

Sanders, Robert E., and Larry W. Martin, "Grammatical Rules and Explanation of Behavior," *Inquiry* 18, pp. 65–82.

• Scriven, Michael, "Causation as Explanation," *Nous* 9, pp. 3–16.

Studdert-Kennedy, Gerald, *Evidence and Explanation in Social Science.* London: Routledge & Kegan Paul.

1976

Addis, Laird, "On Defending the Covering Law Model," in R. S. Cohen, C. A. Hooker, A. C. Michalos, and J. W. Van Evra, eds., *PSA 1974* (Dordrecht: D. Reidel Publishing Co.), pp. 361–68.

Beckner, Morton, "Functions and Teleology," in Grene and Mendelsohn (1976), pp. 197–212.

☐ Belnap, Nuel D., Jr., and J. B. Steel, Jr., *The Logic of Questions and Answers.* New Haven: Yale University Press.

Boorse, Christopher, "Wright on Functions," *Philosophical Review* 85, pp. 70–86.

Cherry, Christopher, "Explanation and Explanation by Hypothesis," *Synthese* 33, pp. 315–40.

Davidson, Donald, "Hempel on Explaining Action," *Erkenntnis* 10, pp. 239–54.

Downes, Chauncey, "Functional Explanations and Intentions," *Philosophy of the Social Sciences* 6, pp. 215–25.

Fetzer, James H., "The Likeness of Lawlikeness," in Robert S. Cohen et al., eds., *Boston Studies in the Philosophy of Science* 32 (Dordrecht: D. Reidel Publishing Co.), pp. 337–91.

Gardenfors, Peter, "Relevance and Redundancy in Deductive Explanations," *Philosophy of Science* 43, pp. 420–31.

Girill, T. R., "The Problem of Micro-Explanation," in Frederick Suppe and Peter D. Asquith, eds., *PSA 1976*, vol. 1 (East Lansing, MI: Philosophy of Science Assn.), pp. 47–55.

——(a), "Evaluating Micro-Explanations," *Erkenntnis* 10, pp. 387–406.

• Grene, Marjorie, and Everett Mendelsohn, eds., *Topics in the Philosophy of Biology, Boston Studies in the Philosophy of Science* 27, part III, Problems of Explanation in Biology, pp. 145–263.

Grobstein, Clifford, "Organizational Levels and Explanation," in Grene and Mendelsohn (1976), pp. 145–52.

Hempel, C. G., "Dispositional Explanation and the Covering Law Model: Response to Laird Addis," in R. S. Cohen, C. A. Hooker, A. C. Michalos, and J. W. Van Evra, eds., *PSA 1974* (Dordrecht: D. Reidel Publishing Co.), pp. 369–76.

Hesslow, Germund, "Two Notes on the Probabilistic Approach to Causality," *Philosophy of Science* 43, pp. 290–92.

Jobe, Evan, "A Puzzle Concerning D-N Explanation," *Philosophy of Science* 43, pp. 542–49.

Kauffman, Stuart, "Articulation of Parts Explanation in Biology and the Rational Search for Them," in Grene and Mendelsohn (1976), pp. 245–63.

King, John L., "Statistical Relevance and Explanatory Classification," *Philosophical Studies* 30, pp. 313–21.

• Kitcher, Philip, "Explanation, Conjunction, and Unification," *Journal of Philosophy* 73, pp. 207–12.

Krüger, Lorenz, "Discussion: Are Statistical Explanations Possible?" *Philosophy of Science* 43, pp. 129–46.

Levin, Michael D., "The Extensionality of Causation and Causal-Explanatory Contexts," *Philosophy of Science* 43, pp. 266–77.

——(a), "On the Ascription of Functions to Objects, with Special Reference to Inference in Archaeology," *Philosophy of the Social Sciences* 6, pp. 227–34.

• Manninen, J., and R. Tuomela, eds., *Essays on Explanation and Understanding.* Dordrecht: D. Reidel Publishing Co. [A collection of essays pertaining to von Wright (1971).]

Martin, Rex, "Explanation and Understanding in History," in Manninen and Tuomela (1976), pp. 305–34.

McLachlan, Hugh V., "Functionalism, Causation and Explanation," *Philosophy of the Social Sciences* 6, pp. 235–40.

• Mellor, D. H., "Probable Explanation," *Australasian Journal of Philosophy* 54, pp. 231–41.

Morgan, Charles G., "Tuomela on Deductive Explanation," *Journal of Philosophical Logic* 5, pp. 511–25.

Munch, Peter A., "The Concept of Function and Functional Analysis in Sociology," *Philosophy of the Social Sciences* 6, pp. 193–214.

Nathan, N. M. L., "On the Non-Causal Explanation of Human Action," *Philosophy of the Social Sciences* 6, pp. 241–43.

Niiniluoto, Ilkka, "Inductive Explanation, Propensity, and Action," in Manninen and Tuomela (1976), pp. 335–68.

Pattee, H. H., "Physical Theories of Biological Co-ordination," in Grene and Mendelsohn (1976), pp. 153–73.

☐ Reichenbach, Hans, *Laws, Modalities, and Counterfactuals.* Berkeley & Los Angeles: University of California Press. [Reprint of Reichenbach (1954) with a new foreword by Wesley C. Salmon.]

Riedel, Manfred, "Causal and Historical Explanation," in Manninen and Tuomela (1976), pp. 3–26.

Ringen, Jon D., "Explanation, Teleology, and Operant Behaviorism: A Study of the Experimental Analysis of Purposive Behavior," *Philosophy of Science* 43, pp. 223–53.

☐ Salmon, Wesley C., foreword to Reichenbach (1954).

Simon, Thomas W., "A Cybernetic Analysis of Goal-Directedness," in Frederick Suppe and Peter D. Asquith, eds., *PSA 1976*, vol. 1 (East Lansing, MI: Philosophy of Science Assn.), pp. 56–67.

Tuomela, Raimo, "Causes and Deductive Explanation," in R. S. Cohen, C. A. Hooker, A. C. Michalos, and J. W. Van Evra, eds., *PSA 1974* (Dordrecht: D. Reidel Publishing Co.), pp. 325–60.

——(a), "Explanation and Understanding of Human Behavior," in Manninen and Tuomela (1976), pp. 183–205.

——(b), "Morgan on Deductive Explanation: A Rejoinder," *Journal of Philosophical Logic* 5, pp. 527–43.

van Fraassen, Bas C., "To Save the Phenomena," *Journal of Philosophy* 73, pp. 623–32.

Williams, Mary B., "The Logical Structure of Functional Explanations in Biology," in Frederick Suppe and Peter D. Asquith, eds., *PSA 1976*, vol. 1 (East Lansing, MI: Philosophy of Science Assn.), pp. 37–46.

Wimsatt, William C., "Complexity and Organization," in Grene and Mendelsohn (1976), pp. 174–93.

• Woodfield, Andrew, *Teleology.* New York: Cambridge University Press.

• Wright, Larry, *Teleological Explanations.* Berkeley/Los Angeles/London: University of California Press.

——(a), "Functions," in Grene and Mendelsohn (1976), pp. 213–42.

1977

Achinstein, Peter, "What Is an Explanation?" *American Philosophical Quarterly* 14, pp. 1–16.

——(a), "Function Statements," *Philosophy of Science* 44, pp. 341–67.

Causey, Robert L., *Unity of Science.* Dordrecht: D. Reidel Publishing Co.

• Coffa, J. Alberto, "Probabilities: Reasonable or True?" *Philosophy of Science* 44, pp. 186–98.

Cummins, Robert, "Programs in the Explanation of Behavior," *Philosophy of Science* 44, pp. 269–87.

Cupples, B., "Three Types of Explanation," *Philosophy of Science* 44, pp. 387–408.

Fetzer, James H., "A World of Dispositions," *Synthese* 34, pp. 397–421.

Gluck, Peter, and Michael Schmid, "The Rationality Principle and Action Explanations: Koertge's Reconstruction of Popper's Logic of Action Explanations," *Inquiry* 20, pp. 72–81.

Good, I. J., "Explicativity: A Mathematical Theory of Explanation with Statistical Applications," *Proceedings of the Royal Society of London* 354, pp. 303–30.

Goode, Terry M., "Explanation, Expansion, and the Aims of Historians: Toward an Alternative Account of Historical Explanation," *Philosophy of the Social Sciences* 7, pp. 367–84.

• Hempel, Carl G., "Nachwort 1976: Neuere Ideen zu den Problemen der statistischen Erklärung," in Carl G. Hempel, *Aspekte wissenschaftlicher Erklärung* (Berlin/New York: Walter de Gruyter), pp. 98–123.

Kung, Joan, "Aristotle on Essence and Explanation," *Philosophical Studies* 31, pp. 361–83.

Martin, Michael, "Neurophysiological Reduction and Type Identity," *Philosophy of the Social Sciences* 7, pp. 91–94.

McCarthy, Tim, "On an Aristotelian Model of Scientific Explanation," *Philosophy of Science* 44, pp. 159–66.

Nagel, Ernest, "Teleology Revisited," *Journal of Philosophy* 74, pp. 261–301. [Reprinted in Nagel, 1979.]

Nickles, Thomas, "On the Independence of Singular Causal Explanation in Social Science: Archaeology," *Philosophy of the Social Sciences* 7, pp. 163–87.

—— (a), "Davidson on Explanation," *Philosophical Studies* 31, pp. 141–45.

Rosenberg, Alexander, "Concrete Occurrences vs. Explanatory Facts," *Philosophical Studies* 31, pp. 133–40.

• Salmon, Wesley C., "An 'At-At' Theory of Causal Influence," *Philosophy of Science* 44, pp. 215–24.

—— (a), "Hempel's Conception of Inductive Inference in Inductive-Statistical Explanation," *Philosophy of Science* 44, pp. 180–85.

• —— (b), "Objectively Homogeneous Reference Classes," *Synthese* 36, pp. 399–414.

• —— (c), "A Third Dogma of Empiricism," in Robert Butts and Jaakko Hintikka, eds., *Basic Problems in Methodology and Linguistics* (Dordrecht: D. Reidel Publishing Co.), pp. 149–66.

—— (d), "Indeterminism and Epistemic Relativization," *Philosophy of Science* 44, pp. 199–202.

Sayre, Kenneth, "Statistical Models of Causal Relations," *Philosophy of Science* 44, pp. 203–14.

Shrader, Douglas W., Jr., "Discussion: Causation, Explanation, and Statistical Relevance," *Philosophy of Science* 44, pp. 136–45.

Tuomela, Raimo, *Human Action and its Explanation*. Dordrecht: D. Reidel Publishing Co.

—— (a), "Dispositions, Realism, and Explanation," *Synthese* 34, pp. 457–78.

Utz, Stephen, "Discussion: On Teleology and Organisms," *Philosophy of Science* 44, pp. 313–20.

• van Fraassen, Bas C., "The Pragmatics of Explanation," *American Philosophical Quarterly* 14, pp. 143–50.

von Bretzel, Philip, "Concerning a Probabilistic Theory of Causation Adequate for the Causal Theory of Time," *Synthese* 35, pp. 173–90.

Wright, Larry, "Rejoinder to Utz," *Philosophy of Science* 44, pp. 321–25.

The Fourth Decade

1978

Aronovitch, Hilliard, "Social Explanation and Rational Motivation," *American Philosophical Quarterly* 15, pp. 197–204.

Bridgstock, Martin, and Michael Hyland, "The Nature of Individualistic Explanation: A Further Analysis of Reduction," *Philosophy of the Social Sciences* 8, pp. 265–70.

Cummins, Robert, "Explanation and Subsumption," in Peter D. Asquith and Ian Hacking, eds., *PSA 1978*, vol. 1 (East Lansing, MI: Philosophy of Science Assn.), pp. 163–75.

Derden, J. K., "Reasons, Causes, and Empathetic Understanding," in Peter D. Asquith and Ian Hacking, eds., *PSA 1978*, vol. 1 (East Lansing, MI: Philosophy of Science Assn.), pp. 176–85.

Dorling, Jon, "Discussion: On Explanations in Physics: Sketch of an Alternative to Hempel's Account of the Explanation of Laws," *Philosophy of Science* 45, pp. 136–40.

Essler, Wilhelm K., "A Note on Functional Explanation," *Erkenntnis* 13, pp. 371–76.

Girill, T. R., "Approximative Explanation," in Peter D. Asquith and Ian Hacking, eds., *PSA 1978*, vol. 1 (East Lansing, MI: Philosophy of Science Assn.), pp. 186–96.

Glymour, Clark, "Two Flagpoles Are More Paradoxical Than One," *Philosophy of Science* 45, pp. 118–19.

• Hanna, Joseph, "On Transmitted Information as a Measure of Explanatory Power," *Philosophy of Science* 45, pp. 531–62.

Hempel, Carl G., "Dispositional Explanation," in Raimo Tuomela, ed., *Dispositions* (Dordrecht: D. Reidel Publishing Co.), pp. 137–46. [An extensively revised version of Hempel (1965), pp. 457–63.]

Levin, Michael E., and Margarita Rosa Levin, "Flagpoles, Shadows and Deductive Explanation," *Philosophical Studies* 32, pp. 293–99.

• McMullin, Ernan, "Structural Explanation," *American Philosophical Quarterly* 15, pp. 139–47.

Moor, James H., "Explaining Computer Behavior," *Philosophical Studies* 34, pp. 325–27.

Pinkard, Terry, "Historical Explanation and the Grammar of Theories," *Philosophy of the Social Sciences* 8, pp. 227–40.

Quay, Paul M., "A Philosophical Explanation of the Explanatory Functions of Ergodic Theory," *Philosophy of Science* 45, pp. 47–59.

• Railton, Peter, "A Deductive-Nomological Model of Probabilistic Explanation," *Philosophy of Science* 45, pp. 206–26.

Rosen, Deborah, "In Defense of a Probabilistic Theory of Causality," *Philosophy of Science* 45, pp. 604–13.

Saliers, Don E., "Explanation and Understanding in the Social Sciences," *Philosophy of the Social Sciences* 8, pp. 367–71.

• Salmon, Wesley C., "Why Ask, 'Why?'? – An Inquiry Concerning Scientific Explanation," *Proceedings and Addresses of the American Philosophical Association* 51, pp. 683–705.

Sayre, Kenneth, "Discussion: Masking and Causal Relatedness: An Elucidation," *Philosophy of Science* 45, pp. 633–37.

Shrader, Douglas, Jr., "Discussion: Sayre's Statistical Model of Causal Relations," *Philosophy of Science* 45, pp. 630–32.

Stern, Cindy, "Discussion: On the Alleged Extensionality of 'Causal-Explanatory Contexts'," *Philosophy of Science* 45, pp. 614–25.

Stiner, Mark, "Mathematical Explanation," *Philosophical Studies* 34, pp. 135–51.

Thagard, Paul R., "The Best Explanation: Criteria for Theory Choice," *Journal of Philosophy* 75, pp. 76–92.

1979

Chopra, Y. N., " 'Explaining and Characterizing Human Action," *Mind* 88, pp. 321–33.

Cupples, Brian, "Discussion: Moor and Schlesinger on Explanation," *Philosophy of Science* 46, pp. 645–50.

Enc, Berent, "Function Attributions and Functional Explanations," *Philosophy of Science* 46, pp. 343–65.

Fair, David, "Causation and the Flow of Energy," *Erkenntnis* 14, pp. 219–50.

• Fetzer, James H., and Donald E. Nute, "Syntax, Semantics, and Ontology: A Probabilistic Causal Calculus," *Synthese* 40, pp. 453–95.

☐ Gardner, Michael, "Realism and Instrumentalism in the 19th Century," *Philosophy of Science* 46, pp. 1–34.

Hanna, Joseph F., "An Interpretive Survey of Recent Research on Scientific Explanation," in Peter D. Asquith and Henry E. Kyburg, Jr., *Current Research in Philosophy of Science* (East Lansing, MI: Philosophy of Science Assn.), pp. 291–316.

Martin, Michael, "Reduction and Typical Individuals Again," *Philosophy of the Social Sciences* 9, pp. 77–79.

Meixner, John B., "Homogeneity and Explanatory Depth," *Philosophy of Science* 46, pp. 366–81.

Nagel, Ernest, "Teleology Revisited," in *Teleology Revisited and Other Essays in the Philosophy and History of Science* (New York: Columbia University Press), pp. 275–316. [First published in 1977.]

Nerlich, Graham, "What Can Geometry Explain," *British Journal for the Philosophy of Science* 30, pp. 69–83.

Salmon, Wesley C., "Postscript: Laws in Deductive-Nomological Explanation – An Application of the Theory of Nomological Statements," in Wesley C. Salmon, ed., *Hans Reichenbach: Logical Empiricist* (Dordrecht: D. Reidel Publishing Co.), pp. 691–94.

☐ (a), "Propensities: A Discussion Review," *Erkenntnis* 14, pp. 183–216.

Thomas, J/P, "Homogeneity Conditions on the Statistical Relevance Model of Explanation," *Philosophical Studies* 36, pp. 101–6.

Van Parijs, Philippe, "Functional Explanation and the Linguistic Analogy," *Philosophy of the Social Sciences* 9, pp. 425–44.

Wilson, Ian, "Explanatory and Inferential Conditionals," *Philosophical Studies* 35, pp. 269–78.

Wolfson, Paul, and James Woodward, "Scientific Explanation and Sklar's Views of Space and Time," *Philosophy of Science* 46, pp. 287–94.

Woodward, James, "Scientific Explanation," *British Journal for the Philosophy of Science* 30, pp. 41–67.

1980

Cooke, Roger, "Discussion: A Trivialization of Nagel's Definition of Explanation for Statistical Laws," *Philosophy of Science* 47, pp. 644–45.

Cupples, B., "Four Types of Explanation," *Philosophy of Science* 47, pp. 626–29.

Currie, Gregory, "The Role of Normative Assumptions in Historical Explanation," *Philosophy of Science* 47, pp. 456–73.

Fetzer, James H., and Donald E. Nute, "A Probabilistic Causal Calculus: Conflicting Conceptions," *Synthese* 44, pp. 241–46.

Forge, John, "The Structure of Physical Explanation," *Philosophy of Science* 47, pp. 203–26.

☐ Frohlich, Cliff, "The Physics of Somersaulting and Twisting," *Scientific American* 242 (March), pp. 154–64.

Gardenfors, Peter, "A Pragmatic Approach to Explanations," *Philosophy of Science* 47, pp. 404–23.

Good, I. J., "Some Comments on Probabilistic Causality," *Pacific Philosophical Quarterly* 61, pp. 301–4.

—— (a), "A Further Note on Probabilistic Causality: Mending the Chain," *Pacific Philosophical Quarterly* 61, pp. 452–54.

Gruender, David, "Scientific Explanation and Norms in Science," in Peter D. Asquith and Ronald N. Giere, eds., *PSA 1980*, vol. 1 (East Lansing, MI: Philosophy of Science Assn.), pp. 329–35.

Hooker, C. A., "Explanation, Generality, and Understanding," *Australasian Journal of Philosophy* 58, pp. 284–90.

Humphreys, Paul, "Cutting the Causal Chain," *Pacific Philosophical Quarterly* 61, pp. 305–14.

Klein, Barbara V. E., "What Should We Expect of a Theory of Explanation," in Peter D. Asquith and Ronald N. Giere, eds., *PSA 1980*, vol. 1 (East Lansing, MI: Philosophy of Science Assn.), pp. 319–28.

Kline, A. David, "Screening-off and the Temporal Asymmetry of Explanation," *Analysis* 40, pp. 139–43.

Lambert, Karel, "Explanation and Understanding: An Open Question?" in Risto Hilpinin, ed., *Rationality in Science* (Dordrecht: D. Reidel Publishing Co.), pp. 29–34.

Laymon, Ronald, "Idealization, Explanation, and Confirmation," in Peter D. Asquith and Ronald N. Giere, eds., *PSA 1980*, vol. 1 (East Lansing, MI: Philosophy of Science Assn.), pp. 336–50.

Omer, I. A., "Minimal Law Explanation," *Ratio* 22, pp. 155–66.

Post, John F., "Infinite Regresses of Justification and of Explanation," *Philosophical Studies* 38, pp. 31–52.

• Railton, Peter, *Explaining Explanation*. Ph.D. dissertation, Princeton University.

☐ Salmon, Wesley C., "Probabilistic Causality," *Pacific Philosophical Quarterly* 61, pp. 50–74.

☐ Skyrms, Brian, *Causal Necessity*. New Haven: Yale University Press.
• van Fraassen, Bas C., *The Scientific Image*. Oxford: Clarendon Press.
Woodward, James, "Developmental Explanation," *Synthese* 44, pp. 443–66.

1981

Addis, Laird, "Dispositions, Explanation, and Behavior," *Inquiry* 24, pp. 205–27.
Cooke, Roger M., "Discussion: A Paradox in Hempel's Criterion of Maximal Specificity," *Philosophy of Science* 48, pp. 327–28.
Creary, Lewis G., "Causal Explanation and the Reality of Natural Component Forces," *Pacific Philosophical Quarterly* 62, pp. 148–57.
• Fetzer, James H., *Scientific Knowledge*. Dordrecht: D. Reidel Publishing Co.
——(a), "Probability and Explanation," *Synthese* 48, pp. 371–408.
Gale, George, "The Anthropic Principle," *Scientific American* 245, no. 6 (Dec.), pp. 154–71.
• Garfinkel, Alan, *Forms of Explanation*. New Haven and London: Yale University Press.
☐ Hacking, Ian, "Do We See Through a Microscope?" *Pacific Philosophical Quarterly* 62, pp. 305–22.
Hanna, Joseph, "Single Case Propensities and the Explanation of Particular Events," *Synthese* 48, pp. 409–36.
Huff, Douglas, and Stephen Turner, "Rationalizations and the Application of Causal Explanations of Human Action," *American Philosophical Quarterly* 18, pp. 213–20.
• Humphreys, Paul, "Aleatory Explanation," *Synthese* 48, pp. 225–32.
• Kitcher, Philip, "Explanatory Unification," *Philosophy of Science* 48, pp. 507–31.
Matthews, Robert J., "Explaining Explanation," *American Philosophical Quarterly* 18, pp. 71–77.
Niiniluoto, Ilkka, "Statistical Explanation," in G. Floistad, ed., *Contemporary Philosophy 1966–1978* (The Hague: Martinus Nijhoff), pp. 157–87.
——(a), "Statistical Explanation Reconsidered," *Synthese* 48, pp. 437–72.
Otte, Richard, "A Critique of Suppes' Theory of Probabilistic Causality," *Synthese* 48, pp. 167–90.
Pitt, Joseph C., *Pictures, Images, and Conceptual Change*. Dordrecht: D. Reidel Publishing Co.
• Railton, Peter, "Probability, Explanation, and Information," *Synthese* 48, pp. 233–56.
• Rogers, Ben, "Probabilistic Causality, Explanation, and Detection," *Synthese* 48, pp. 201–23.
Salmon, Merrilee H., "Ascribing Functions to Archaeological Objects," *Philosophy of the Social Sciences* 11, pp. 19–26.
Salmon, Wesley C., "Causality: Production and Propagation," in Peter D. Asquith and Ronald N. Giere, eds., *PSA 1980*, vol. 2 (East Lansing, MI: Philosophy of Science Assn.), pp. 49–69.
☐ ——(a), "Rational Prediction," *British Journal for the Philosophy of Science* 32, pp. 115–25.
Suppes, Patrick, and Mario Zanotti, "When Are Probabilistic Explanations Possible?" *Synthese* 48, pp. 191–200.
Tuomela, Raimo, "Inductive Explanation," *Synthese* 48, pp. 257–94.
Wicken, Jeffrey S., "Causal Explanations in Classical and Statistical Thermodynamics," *Philosophy of Science* 48, pp. 65–77.

1982

Abel, Ruben, "What is an Explanandum?" *Pacific Philosophical Quarterly* 63, pp. 86–92.
Brown, James Robert, "Realism, Miracles, and the Common Cause," in Peter D. Asquith and Thomas Nickles, eds., *PSA 1982*, vol. 1 (East Lansing, MI: Philosophy of Science Assn.), pp. 98–105.
Cohen, G. A., "Functional Explanation, Consequence Explanation, and Marxism," *Inquiry* 25, pp. 27–56.
Ellett, Frederick S., Jr., and David P. Ericson, "On Reichenbach's Principle of the Common Cause," *Pacific Philosophical Quarterly* 64, pp. 330–40.
Farr, James, "Humean Explanations in the Moral Sciences," *Inquiry* 25, pp. 57–80.
Forge, John, "Physical Explanation: With Reference to the Theories of Hempel and Salmon," in Robert

McLaughlin, ed., *What? Where? When? Why?* (Dordrecht: D. Reidel Publishing Co.), pp. 211–29.

Glymour, Clark, "Causal Inference and Causal Explanation," in Robert McLaughlin, ed., *What? Where? When? Why?* (Dordrecht: D. Reidel Publishing Co.), pp. 179–91.

Hutchison, Keith, "What Happened to Occult Qualities in the Scientific Revolution?" *Isis* 73, pp. 233–53.

Leslie, John, "Anthropic Principle, World Ensemble, Design," *American Philosophical Quarterly* 19, pp. 141–52.

Levy, Edwin, "Critical Discussion: Causal-Relevance Explanations: Salmon's Theory and its Relation to Reichenbach," *Synthese* 50, pp. 423–45.

Lipton, Peter, "Nagel Revisited" (Review of Ernest Nagel: *Teleology Revisited*), *British Journal for the Philosophy of Science* 33, pp. 186–94.

MacKinnon, Edward M., *Scientific Explanation and Atomic Physics*. Chicago: University of Chicago Press.

Mandelbaum, Maurice, "G. A. Cohen's Defense of Functional Explanation," *Philosophy of the Social Sciences* 12, pp. 285–88.

Meixner, John, "Are Statistical Explanations Really Explanatory?" *Philosophical Studies* 42, pp. 201–7.

Pork, Andrus, "A Note on Schemes of Historical Explanation: Problem of Status," *Philosophy of the Social Sciences* 12, pp. 409–14.

• Salmon, Merrilee H., *Philosophy and Archaeology*. New York: Academic Press.

Salmon, Wesley C., "Comets, Pollen, and Dreams: Some Reflections on Scientific Explanation," in Robert McLaughlin, ed., *What? Where? When? Why?* (Dordrecht: D. Reidel Publishing Co.), pp. 155–78.

——(a), "Further Reflections," ibid., pp. 231–80.

Taylor, Denise Meyerson, "Actions, Reasons and Causal Explanation," *Analysis* 42, pp. 216–19.

van Fraassen, Bas C., "Rational Belief and the Common Cause Principle," in Robert McLaughlin, ed., *What? Where? When? Why?* (Dordrecht: D. Reidel Publishing Co.), pp. 193–209.

1983

• Achinstein, Peter, *The Nature of Explanation*. New York: Oxford University Press.

Cartwright, Nancy, *How the Laws of Physics Lie*. New York: Oxford University Press.

• Cummins, Robert, *The Nature of Psychological Explanation*. Cambridge, MA: MIT Press.

Fetzer, James H., "Probabilistic Explanations," *PSA 1982*, vol. 2, pp. 194–207.

• Hanna, Joseph, "Probabilistic Explanation and Probabilistic Causality," in Peter D. Asquith and Thomas Nickles, eds., *PSA 1982*, vol. 2, pp. 181–93.

Hausman, Daniel M., "Causal and Explanatory Asymmetry," in Peter D. Asquith and Thomas Nickles, eds., *PSA 1982*, vol. 2, pp. 43–54.

Howson, Colin, "Statistical Explanation and Statistical Support," *Erkenntnis* 20, pp. 61–78.

• Humphreys, Paul, "Aleatory Explanation Expanded," in Peter Asquith and Thomas Nickles, eds., *PSA 1982*, vol. 2, pp. 208–23.

Perry, Clifton, "The Explanatory Efficacy of Individualism," *Philosophy of the Social Sciences* 13, pp. 65–68.

Platts, Mark, "Explanatory Kinds," *British Journal for the Philosophy of Science* 34, pp. 113–48.

Short, T. L., "Teleology in Nature," *American Philosophical Quarterly* 20, pp. 311–20.

Sober, Elliott, "Equilibrium Explanation," *Philosophical Studies* 43, pp. 201–10.

Walt, Stephen, "A Note on Mandelbaum's 'G. A. Cohen's Defense of Functional Explanation'," *Philosophy of the Social Sciences* 13, pp. 483–86.

1984

Achinstein, Peter, "A Type of Non-Causal Explanation," in Peter A. French, Theodore E. Uehling, Jr., and Howard K. Wetterstein, eds., *Causation and Causal Theories, Midwest Studies in Philosophy* 9, (Minneapolis: University of Minnesota Press) pp. 221–44.

Amundson, Ron, and Laurence Smith, "Clark Hull, Robert Cummins, and Functional Analysis," *Philosophy of Science* 51, pp. 657–60.

Collins, Arthur, "Action, Causality, and Teleological Explanation," in Peter A. French, Theodore E. Uehling, Jr., and Howard K. Wetterstein, eds., *Causation and Causal Theories, Midwest Studies in Philosophy* 9, pp. 345–70.

F̯etzer, James H., "Review of Peter Achinstein, *The Nature of Explanation*," *Philosophy of Science* 51, pp. 516–19.

☐ Leplin, Jerrett, ed., *Scientific Realism*. Berkeley/Los Angeles/London: University of California Press.

McMullin, Ernan, "Two Ideals of Explanation in Natural Science," in Peter A. French, Theodore E. Uehling, Jr., and Howard K. Wetterstein, eds., *Causation and Causal Theories, Midwest Studies in Philosophy* 9, pp. 205–20.

Nissen, Lowell, "Discussion: Woodfield's Analysis of Teleology," *Philosophy of Science* 51, pp. 488–94.

Papineau, David, "Representation and Explanation," *Philosophy of Science* 51, pp. 550–72.

• Salmon, Wesley C., *Scientific Explanation and the Causal Structure of the World*. Princeton: Princeton University Press.

☐ ——(a), *Logic*, third edition. Englewood Cliffs, NJ: Prentice-Hall.

• Sintonen, Matti, *The Pragmatics of Scientific Explanation*. Acta Philosophica Fennica 37. Helsinki: Societas Philosophica Fennica.

——(a), "On the Logic of Why-Questions," in Peter D. Asquith and Philip Kitcher, eds., *PSA 1984*, vol. 1 (East Lansing, MI: Philosophy of Science Assn.), pp. 168–76.

• Sober, Elliott, "Common Cause Explanation," *Philosophy of Science* 51, pp. 212–41.

Tuomela, Raimo, "Social Action-Functions," *Philosophy of the Social Sciences* 14, pp. 133–48.

• Watkins, John, *Science and Scepticism*. Princeton: Princeton University Press.

Watson, Patty Jo, Steven A. LeBlanc, and Charles Redman, *Archaeological Explanation: The Scientific Method in Archaeology*. New York: Columbia University Press.

Woodward, James, "Explanatory Asymmetries," *Philosophy of Science* 51, pp. 421–42.

——(a), "A Theory of Singular Causal Explanation," *Erkenntnis* 21, pp. 231–62.

1985

Apel, Karl-Otto, *Understanding and Explanation: A Transcendental-Pragmatic Perspective*. Cambridge, MA: MIT Press.

Carleton, Lawrence R., "Levels in Discription and Explanation," *Philosophy Research Archives* 9, pp. 89–110.

☐ Churchland, Paul M., and Clifford A. Hooker, *Images of Science*. Chicago: University of Chicago Press.

Doppelt, Gerald, "Finocchiaro on Rational Explanation," *Synthese* 62, pp. 455–58.

Ehring, Douglas, "Dispositions and Functions: Cummins on Functional Analysis," *Erkenntnis* 23, pp. 243–50.

Finocchiaro, Maurice A., "Aspects of the Logic of History-of-Science Explanation," *Synthese* 62, pp. 429–54.

Grene, Marjorie, "Explanation and Evolution," in Robert S. Cohen and Marx Wartofsky, eds. *A Portrait of Twenty-five Years* (Dordrecht: D. Reidel Publishing Co.), pp. 177–97.

Hausman, David B., "The Explanation of Goal-Directed Behavior," *Synthese* 65, pp. 327–46.

Heil, John, "Rationality and Psychological Explanation," *Inquiry* 28, pp. 359–71.

☐ Humphreys, Paul, "Why Propensities Cannot Be Probabilities," *Philosophical Review* 94, pp. 557–70.

• Jobe, Evan K., "Explanation, Causality, and Counterfactuals," *Philosophy of Science* 52, pp. 357–89.

• Kitcher, Philip, "Two Approaches to Explanation," *Journal of Philosophy* 82, pp. 632–39.

——, van Fraassen, B., and Salmon, W. "Symposium on Wesley Salmon's *Scientific Explanation and the Causal Structure of the World," Journal of Philosophy* 82, pp. 632–54.

☐ Mermin, N. David, "Is the Moon There When Nobody Looks? Reality and the Quantum Theory," *Physics Today* 38 (no. 4, April), pp. 38–47.

Olding, A., "Short on Teleology," *Analysis* 45, pp. 158–61.

Papineau, David, "Probabilities and Causes," *Journal of Philosophy* 82, pp. 57–74.

Pearce, David, and Veikko Rantala, "Approximate Explanation is Deductive Nomological," *Philosophy of Science* 52, pp. 126–40.

Prior, Elizabeth W., "What is Wrong with Etiological Accounts of Biological Function?" *Pacific Philosophical Quarterly* 66, pp. 310–28.

• Salmon, Wesley C., "Empiricism: The Key Question," in Nicholas Rescher, ed., *The Heritage of Logical Positivism* (Lanham, MD: University Press of America), pp. 1–21.

——(a), "Conflicting Conceptions of Scientific Explanation," *Journal of Philosophy* 82, pp. 651–54.

• Sober, Elliott, "A Plea for Pseudo-Processes," *Pacific Philosophical Quarterly* 66, pp. 303–9.

• van Fraassen, Bas C., "Salmon on Explanation," *Journal of Philosophy* 82, pp. 639–51.

Wilson, Fred, *Explanation, Causation, and Deduction*. Dordrecht: D. Reidel Publishing Co.

1986

Achinstein, Peter, "The Pragmatic Character of Explanation," in Peter D. Asquith and Philip Kitcher, eds., *PSA 1984*, vol. 2 (East Lansing, MI: Philosophy of Science Assn.), pp. 275–92.

☐ Barrow, John D., and Frank J. Tipler, *The Anthropic Cosmological Principle*. Oxford: Clarendon Press.

Bromberger, Sylvain, "On Pragmatic and Scientific Explanation: Comments on Achinstein's and Salmon's Papers," in Peter D. Asquith and Philip Kitcher, eds., *PSA 1984*, vol. 2 (East Lansing, MI: Philosophy of Science Assn.), pp. 306–25.

☐ Earman, John, *A Primer of Determinism*. Dordrecht: D. Reidel Publishing Co.

Forge, John, "The Instance Theory of Explanation," *Australasian Journal of Philosophy* 64, pp. 127–42.

Humphreys, Paul, "Review of *Scientific Explanation and the Causal Structure of the World* by Wesley C. Salmon," *Foundations of Physics* 16, 1211–16.

Kincaid, Harold, "Reduction, Explanation, and Individualism," *Philosophy of Science* 53, pp. 492–513.

Robinson, Joseph D., "Reduction, Explanation, and the Quests of Biological Research," *Philosophy of Science* 53, pp. 333–53.

Salmon, Wesley C., "Scientific Explanation: Three General Conceptions," in Peter D. Asquith and Philip Kitcher, eds., *PSA 1984*, vol. 2 (East Lansing, MI: Philosophy of Science Assn.), pp. 293–305.

Skarda, Christine A., "Explaining Behavior: Bringing the Brain Back in," *Inquiry* 29, pp. 187–202.

Smith, Quentin, "World Ensemble Explanations," *Pacific Philosophical Quarterly* 67, pp. 73–86.

Sober, Elliott, 'Explanatory Presupposition,' *Australasian Journal of Philosophy* 64, pp. 143–49.

1987

Bigelow, John, and Robert Pargetter, "Functions," *Journal of Philosophy* 84, pp. 181–96.

Fetzer, James H., "Critical Notice: Wesley Salmon's *Scientific Explanation and the Causal Structure of the World*," *Philosophy of Science* 54, pp. 597–610.

Grimes, Thomas R., "Explanation and the Poverty of Pragmatics," *Erkenntnis* 27, pp. 79–92.

Irzik, Gurol, and Eric Meyer, "Causal Modeling: New Directions for Statistical Explanation," *Philosophy of Science* 54, pp. 495–514.

• Kitcher, Philip, and Wesley C. Salmon, "Van Fraassen on Explanation," *Journal of Philosophy* 84, pp. 315–30.

Levine, Joseph, "*The Nature of Psychological Explanation* by Robert Cummins," *Philosophical Review* 96, pp. 249–74.

Seager, William, "Credibility, Confirmation and Explanation," *British Journal for the Philosophy of Science* 38, pp. 301–17.

Sober, Elliott, "Explanation and Causation" (Review of Wesley Salmon, *Scientific Explanation and the Causal Structure of the World*), *British Journal for the Philosophy of Science* 38, pp. 243–57.

Torretti, Roberto, "Do Conjunctive Forks Always Point to a Common Cause?" *British Journal for the Philosophy of Science* 38, pp. 186–94.

The Fifth Decade

1988

Grünbaum, Adolf, and Wesley C. Salmon, eds., *The Limitations of Deductivism*. Berkeley & Los Angeles: University of California Press.

Koura, Antti, "An Approach to Why-Questions," *Synthese* 74, pp. 191–206.

Pitt, Joseph C., ed., *Theories of Explanation*. New York: Oxford University Press. [An anthology on scientific explanation.]

——(a), "Galileo, Rationality and Explanation," *Philosophy of Science* 55, pp. 87–103.

• Salmon, Wesley C., "Deductivism Visited and Revisited," in Grünbaum and Salmon, (1988).

——(a), "Intuitions: Good and Not-So-Good," in William Harper and Brian Skyrms, eds., *Causation, Cause, and Credence*, vol. I (Dordrecht: Kluwer Academic Publishers), pp. 51–71.

☐ Shimony, Abner, "The Reality of the Quantum World," *Scientific American* 258, no. 1 (Jan. 1988), pp. 46–53.

Temple, Dennis, "Discussion: The Contrast Theory of Why-Questions," *Philosophy of Science* 55, pp. 141–51.

1989

• Kitcher, Philip, and Wesley C. Salmon, eds., *Scientific Explanation*. Minneapolis: University of Minnesota Press.

Date Unknown

Humphreys, Paul, *The Chances of Explanation*. Princeton University Press.

van Fraassen, Bas C., *Laws and Symmetries*.

Explanation and Metaphysical Controversy

I

A seachange has occurred in the study of explanation.[1] As recently as a decade ago, students of explanation had a fairly standard way of proceeding. They had before them a dominant theory of explanation, C. G. Hempel's covering-law account. They would begin by producing counterexamples to it and could go on either to construct arguments and epicycles to escape these counterexamples or to propose abandonment of all or part of Hempel's account, perhaps also advancing some replacement of their own devising that better fit the examples given. This all had the air of a well-defined activity, and it had gone on for a long time. Part of what gave the activity its apparent definiteness was the existence of a large and carefully drawn target at which to aim—Hempel's account.[2] Part, too, was a tacit understanding of the rules of the game: one would adduce intuitive judgments of sample explanations, intuitions one expected one's readers to share, and then one would use these intuitions to test analyses. An acceptable analysis, like an acceptable grammar, should get the intuitions of native informants right.

These rules functioned at the same time as limits, for they defined a way of proceeding in the analysis of explanation that prevented philosophers from wandering off into other areas, such as metaphysics. Two philosophers of explanation could draw up to the same table, lay out for inspection their examples and their analyses, produce at appropriate times their favorite counterexamples to each other and their various strategies for handling them, try on these grounds to convince each other, and then depart, without having breathed a word about metaphysical disputes in the philosophy of science, such as the growing debate between realists and irrealists. Except in a few polemical places, theories of explanation were described by their formal features—"covering-law," "why-question," "speech-act," "statistical-relevance"—and did not come prefixed with such metaphysical codes as "empiricist," "pragmatist," or "realist." Yet at the table sat empiricists, pragmatists, and realists.

This activity began to lose its sense of purpose once dissatisfaction with the

Hempelian account became widespread and no new orthodoxy emerged to take its place as a focus of discussion. Many philosophers grew weary and turned to other questions, including questions about realism and irrealism. Interestingly, within *that* dispute the concept of explanation proved to be indispensable: one side often claimed that a realist interpretation of scientific theory was justified by inference to the best explanation; the other side often responded that the realist's posits could do no explanatory work, since they yielded no empirical predictions beyond those already afforded by the observational reduction of the theory.

For a while, it seemed as though explanation were able to play so central a role in this debate precisely because there had ceased to be any widely accepted idea of what explanation is. Each side could use the notion for its purposes because, in the absence of any agreed-upon account, nothing prevented them from doing so. Dissatisfaction with such a state of affairs may in some measure have contributed to the seachange in studies of explanation. For, after a low ebb, the flow of writing on explanation has once again quickened, but in the opposite direction. Bas van Fraassen's *The Scientific Image*[3] and Wesley Salmon's *Scientific Explanation and the Causal Structure of the World*,[4] to mention only two prominent examples, are entirely explicit about placing analysis of explanation in a metaphysical setting, the one irrealist, the other, realist. The accounts they give of explanation, while of course making use of familiar strategies of exampling and counterexampling, are defended on grounds that do not purport to be innocent of metaphysics, and that involve metaphysically driven reinterpretations of intuitions.[5]

What else should one expect? It is inconceivable that a notion such as explanation could fail to depend crucially upon one's most general picture of the world and its ways. A pair of examples may suffice.

First, it seems in retrospect that the empiricist view that the subject matter of science is in the first and last resort experiential, that no individual experience ever intrinsically points to another, and that therefore neither science nor the philosophy of science can admit such a notion as physical necessity, accounts in part for the lasting appeal of the Hempelian doctrine that explanations must be deductive inferences from covering laws.[6] At least since Aristotle, explanation has been thought to involve demonstrating the *necessity* of the phenomenon to be explained. Thus, Hempel wrote in his first major paper on explanation that "explanation . . . aims at showing that the event in question was not a 'matter of chance', but was to be expected in view of certain antecedent or simultaneous conditions."[7] But if individual experiences never on their own necessitate other experiences, how is explanation via demonstration from "antecedent or simultaneous conditions" to proceed? The answer is that a statement of a general character is needed to establish the linkage: the fact that a given piece of copper wire was heated cannot by itself necessitate the fact that this wire's electrical resistance increased, but it could do so in the presence of a generalization stating that the resistance of a copper wire always increases as a function of temperature. Necessity of the explanandum given "simultaneous conditions" is established, but the relation is among descriptions of these events, not the events themselves, and the necessity involved is logical, not physical — the Aristotelian thought has found expression without exceeding the bounds of sense as delimited by empiricism.

If one did not confine oneself to the empiricist's world of individual experiences, or if one otherwise made room for the idea of necessary connections among events "in the world," then to show the necessity of a particular phenomenon one would not have to resort to law-based deduction. Indeed, the order of explanation might be reversed: one could explain the generalization as holding in virtue of the necessary features of concrete physical systems, much as some now think that the relative frequencies displayed by physically indeterministic systems are to be explained as accumulated outcomes of individual chance events.

This mention of indeterminism brings me to the second example. According to the Hempelian account of probabilistic explanation, probabilistic explanations are analogous to more familiar nonprobabilistic explanations in two important ways: they involve empirical laws essentially and they show that the explanandum phenomenon was to have been expected. The degree of law-based expectation is however weakened to accommodate the statistical character of probabilistic laws — outright demonstration is not required for explanation, only inference with "practical certainty." Yet, according to Hempel, probabilistic explanations also show three crucial disanalogies with nonprobabilistic explanations: they are inductive rather than deductive in form, they are relativized to an epistemic context, and they must satisfy a principle of maximal specificity relative to that context.[8] This has some rather odd results. Many of the explanations of chance phenomena in contemporary physics and biology are of events with low probability given the initial conditions, and these would be ruled out by Hempel's criteria. Moreover, if our evidence currently supports a lawlike statistical generalization assigning high probability to an explanandum, then even if this evidence is limited, a valid explanation can be given; if our evidence improves, but in a way that lowers the probability attributed to the explanandum, then we must say that explanation now is precluded, yet all the while continuing to hold that the old explanation was strictly correct. Isn't the implication of the new evidence instead that the old explanation was not strictly correct? Hempel is adamant in the case of nonprobabilistic explanation that changes in evidence can do no more than make a potential explanation *seem* right or wrong, just as he would think it a confusion to say that a statement about, say, the price of a slave on a given date in colonial Havana fluctuates in truth value as material on deposit in tropical archives deteriorates.[9]

Why should probabilistic explanation be so different from nonprobabilistic explanation, in Hempel's eyes, and why different in just these ways? The answer may reflect a metaphysical assumption. Suppose that one were a determinist at heart, and thought that in principle all indeterminism could be removed from our theories by the discovery of hidden variables. What view might one have of the probabilistic explanations offered by contemporary science? One might think that they are part of an inductive process, steps along the road to genuine, deductive explanation. They could serve to summarize the current state of knowledge about possible causal factors, and to approximate deduction as best that permits. Of course, if the factors known at a given time failed to determine a high expectation value for an outcome, then science clearly would be missing some very crucial variable or variables, so that an inductive inference based upon these factors would not be very acceptable as a proxy for explanation. On the other hand, if the known factors conferred a high probability upon the explanandum, even when compared to all extant sample populations, then we might think that the induction comes closer to approximating explanation. There would be no need to be specific—as, indeed, Hempel rather uncharacteristically is not—about exactly where the line demarcating "high probability" is to be drawn, since we would be certain that we were not in possession of a true explanation until all hidden variables were revealed, at which time we should be able in principle to bring the probability of the explanandum arbitrarily close to one. Thus, if one were a closet determinist, or if one at least had not extricated oneself from determinist ways of thought, one naturally would gravitate toward a Hempelian model of probabilistic explanation: its peculiarities would be virtues.[10]

Suppose by contrast that one did not have such a picture, but thought instead that the world's physical and social processes might be irreducibly indeterministic, albeit in an orderly way, with stable, law-abiding probabilities that manifest themselves in highly predictable relative frequencies. Then why should it matter for explanation whether the probability of the phenomenon to be explained is low (e.g., radioactive decay by long-lived isotopes of uranium) or high (e.g., radioactive decay by unstable isotopes of actinium)? And why should probabilistic explanation be epistemically relativized if nonprobabilistic explanation is not? If we discover that our current theory omitted some variables relevant to determining the probabilistic state of a system, then our current theory would be to that extent wrong, now and always, about why the system behaves as it does.[11]

In these two cases there is an evident coincidence between the requirements laid down by a model of explanation, on the one hand, and substantive metaphysical assumptions, on the other. Coincidence is not explanation, and I do not purport to have given explanations. Rather, I want simply to enter the remark that these coincidences went largely unnoticed in the vast debate over the Hempelian models of explanation.[12]

Things have changed, and overtly metaphysical discussion has now become much in evidence in the theory of explanation. Welcome as this development is, it does very much change the rules for the philosophical study of explanation. To say that one's background picture of the world is involved in one's *conception* of explanation is to suggest that one's intuitions about particular kinds or instances of purported explanation may not constitute a body of neutral data for testing theories of explanation—exampling and counterexampling will never be quite what they used to be. Equally, it is to suggest that there may not be a unitary, substantial *concept* of explanation to analyze, or, perhaps more accurately, that the concept of explanation is rather thin, too slight, perhaps, to be asked to resolve deep philosophical disputes.[13]

II

If we come to believe that the analysis of explanation is not a metaphysically neutral activity, then we must appreciate how this affects the debate between realists and irrealists. I feel this particularly keenly as a realist interested in the philosophical study of explanation who finds himself uncomfortable with certain uses of "inference to the best explanation" to defend realism.

Typically, a realist about the external world will say that the existence of such a world affords the best explanation of the stability, coherence, and so on, of our sensations. A realist about science will say that the approximate truth of current theory—where this means something like its approximate correspondence, under a literal interpretation, to the world—is the best explanation of the predictive and manipulative successes of science. In such claims it appears that we are being given a special sort of confirmation of the existence of the realist's posits, confirmation stemming from their contribution to explanation. Now I do not for a moment doubt that near the heart of realism lies a concern about explanation, but I am inclined to think that if the issue is to be put in a way that does not beg the question against the irrealist, then what is at stake must be considerations that, while making some contribution to confirmation, make their primary contribution to the scientific rationale and epistemic warrant of realism in other ways. It will take me most of this essay to say what that might mean.[14]

Consider contemporary science, with its theoretical vocabulary of 'cells', 'genes', 'atoms', 'energy', and the like. This theory is the result of a protracted history of experimentation and innovation, and has been vastly successful in practice. We are inclined to say that our evidence for this theory is observational: the manifest experimental and practical successes. Yet for familiar reasons, we can at best give only a partial interpretation of such terms as 'cell' and 'energy' in the observational vocabulary.[15] We therefore face the question: What stance should we adopt toward these theoretical terms and the sentences containing them?

Two answers to this question are realism and irrealism, but these terms have become desperately nonspecific. Let us try to fix ideas for the purposes of this paper, without worrying about the comprehensiveness of our category scheme. The realist with whom I will be concerned interprets at least some theoretical terms literally, and holds that we sometimes have good scientific reasons for believing the sentences containing them to be literally true.[16] The irrealists I will have in mind are characterized by the part of realism they deny. The first sort of irrealist does not interpret theoretical terms literally and therefore holds that sentences containing them do not have literal truth-values. The second sort interprets theoretical terms literally but holds that we never have good scientific reasons for believing the sentences containing them to be literally true.

Irrealism of the first sort directs itself against semantic realism, usually on the grounds of an empiricist theory of meaning: only statements with implications for the actual or possible course of experience have genuine cognitive content, and they have it only insofar as they have such implications.[17] This claim finds its defense in the argument that for language to be learnable and to be a medium of communication, it must be conditioned to publicly observable states of affairs, not unobservable ones. Once certain metalinguistic vapors have evanesced, according to this view, the cognitive content of any statement or system of statements is seen to be the difference to the observable world that the truth of the statement would make. Where there is no such difference, there is no genuine statement, but only, at best, noncognitive sorts of linguistic function. Let us call this sort of irrealism *observationalism*, avoiding the term 'verificationism' and its association with certain rather definite views about criteria of observability.

Irrealism of the second sort directs itself against the realist's claim that we sometimes have good scientific reasons for believing the literal content of theoretical statements to be true. Here empiricism is wielded in the first instance as an epistemic rather than semantic doctrine: all our evidence is observational; theoretical statements, interpreted literally, involve claims not only about the course of experience, but also about things in principle unobservable; therefore the evidence we have for our theories is at most evidence only that they are correct in their claims about what is observable, i.e., that they are empirically adequate.[18] Let us call this sort of irrealism *agnosticism*.[19] I will mostly be concerned with the agnostic in what follows, although the argumentative strategy employed can readily be extended to the observationalist.

Both sorts of irrealist are concerned to show that all genuinely scientific functions of theory can be carried out without (what they view as) the extreme resort of realism. For example, an irrealist may appeal to Craig's theorem to demonstrate that we can in principle derive from scientific theory (in conjunction with its interpretive system) an axiomatizable reduced theory, couched entirely in the observational vocabulary, which effects all the same deductive connections among observables as the original theory. Call this the *observational theory*.[20]

Now the realist typically counters that there are scientific functions served by the unreduced theory that cannot be served as well by the observational theory alone. In particular, the unreduced theory affords *good explanations* of the observational theory's regularities. Thus, the observed ratios of combination in chemical compounding can be explained by the microtheory of atomic structure and bonding. If we did not accept the reality of atoms, electrons, etc., we would be left only with a large number of brute observational regularities. Therefore, the argument goes, by inference to the best explanation we can say that the observational evidence supports going beyond the observational theory and believing as well the unreduced theory.

The task of assessing this argument is not made easier by the fact that so little has been said explicitly about what the criteria of "better explanation" might be.[21] We will have to consider several possible notions of "better explanation" and ask of each whether it could do the job.

Sometimes "better explanation" seems to mean something like this: one explanation is better than another if it allows us to see the explanandum as more likely.[22] The obvious cases spring to mind. Suppose you are staying at a midtown hotel and the large office building across the street, which had been ablaze with light, suddenly goes altogether dark. Simultaneously, the noisy air-conditioning towers on its roof fall silent. What would explain this? Well, the ballasts in fluorescent lamps, though highly dependable, fail at random every so often. Similarly for the electric motors that drive air-conditioning compressors. And similarly for the large step-down transformers that bring power to modern office buildings. One explanation of the sudden changes observed in the office building is that the ballasts in all the fluorescent lamps independently failed at once, and at the same moment for unrelated reasons the motors in the air-conditioning towers broke down. This explanation makes the event observed a very remarkable coincidence. An alternative explanation is that the power supply to the entire building has been interrupted because the transformer in the basement shorted out. This explanation makes the event observed one which still may not be terribly common, but one which is not so grotesquely unlikely. Hence, we deem it the better explanation of the two considered, and take it to receive more support from the evidence than its competitor. In a similar way, some realists have said that the approximate truth of scientific theory, under a realist interpretation, is the only hypothesis that does not make the success of science a coincidence of cosmic scale.[23]

Stated as a principle of evidence, we might put this sort of argument as follows. Since noncoincidental explanations show their explananda to have been less unlikely, they therefore receive more support from the occurrence of the explananda than their competitors. This is akin to a maximum likelihood theory of confirmation.[24]

Now let us see how this argument might be used to compare the support for our original theory as opposed to its observational reduction. Suppose we have observed certain regularities in chemical combination. The observational theory will contain as a proper part a large conjunction of observational regularities regarding chemical combination, but the original theory will also contain an account of underlying atomic structures and mechanisms that implies these regularities, and makes their conjunction likely. So, if these regularities are observed, the argument goes, then the original theory should be better confirmed than the observational theory.

The irrealists, however, have a ready response. Since the observational theory implies this conjunction of observed regularities, it naturally assigns that conjunction a high likelihood. Admittedly, it does so by a rather quick logical deduction rather than by a complex derivation from underlying structures and mechanisms. But still, if the original theory implies the conjunction, then the observational theory will do so as well.[25] The mere conjunction of these observed regularities may have low *a priori* probability in the absence of an underlying theory, but our comparison is not between belief in the underlying theory and belief in no theory; instead, it is between belief in the underlying theory and belief only in the observational theory.[26]

In reply the realist may complain that the observational theory does not give us a good explanation of the conjunction, but now 'better explanation' must mean something more than "likelihood-enhancing hypothesis."

III

One important recent suggestion is that explanation proceeds by the reduction of independent phenomena; theories are explanatory in virtue of the unification they effect among diverse phenomena. This suggestion, which has taken various forms,[27] has seemed to some especially congenial to realism, for it gives a role to the postulation of underlying structures and mechanisms beyond the mere entailment of observations. Thus, when physical theory treats of such *prima facie* disparate phenomena as fixed compounding ratios, emission and absorption spectra, the conductivity of metals, and the periodic table of the elements, by providing a unifying physical model of the atom, it achieves a substantial reduction in the number of phenomena that must independently be taken as basic. At issue is not whether the theory supplies a higher likelihood for these diverse phenomena than does its observational reduction, but whether it renders these phenomena more comprehensible or better understood[28] in virtue of tracing them to a common structural and causal basis. This common basis is characterized theoretically, and if we were to eliminate scientific commitment to atoms, electrons, etc., we would lose the physical model by means of which unification is achieved. This affords a sense in which the unreduced theory provides a better explanation of observations than the observational theory, and if in science we infer to the best explanation of the data, then, it seems, we will infer to the unreduced theory, not merely to its observational reduction.

Irrealists are entitled to find this argument somewhat puzzling—at least, in the mouth of a realist. For the realist wishes to give a literal construal of physical theory. If inference to the best explanation is to be an account of confirmation and if confirmation is a matter of (something like) epistemic probability of truth relative to evidence, then this, plus literal construal, requires that the realist take the criteria of better explanation to be indicators of increasing epistemic probability of literal truth. But why should the fact that a theory renders phenomena more comprehensible to us constitute evidence of that theory's truth? Some contemporary philosophers have sought to build criteria of comprehensibility into truth, as for example in coherence theories of truth. It is open to realists to take such a view, and to make explanatory unification part of coherence and therefore part of truth. But at least one mark of the sort of realism with which we are here concerned has been resistance to the idea that considerations of comprehensibility have a hand in determining truth. Realists of this stripe have held that the world and its explanatory structure could be very unlike what we think them to be, or could comprehend them being. It would be a major revision of such realism to accept synthetic *a priori* constraints of comprehensibility upon the nature of reality.

The realist may instead insist that, in speaking of the reduction of independent phenomena as yielding comprehensibility, he is not employing a notion tied to the character of human thought. Core elements of quantum mechanics may be much harder for us to understand than any finite fragment of the complex array of independent observational regularities it might entail. Nonetheless, there is a structural sense in which quantum mechanics renders this complex array more comprehensible, by showing it to be the multifaceted manifestation of a limited number of elementary entities satisfying a small set of quite general laws.

It begins to look as if possessing "comprehension" in this structural sense is nothing more than possessing unifying explanatory accounts, so perhaps it is best to state this defense of realism directly in terms of a unification criterion of *explanation* without invoking any notion of comprehensibility. And once more, the irrealist is entitled to puzzlement. If unification provides a criterion of explanation, and if explanation is evidence of truth, then unification is evidence of truth. Yet how does the realist know *a priori* that the world we inhabit is a unified one? In the case of the office building blacking out, we find the unifying explanation afforded by a power failure quite plausible, but it might for all that be false. The simultaneous blackout might have been the coincidental result of multiple independent causal chains, and, if so, then the single-power-failure explanation certainly would not be "better," though it would be more unifying. Similarly, the realist cannot rule out *a priori* the possibility that such chemical phenomena as combination ratios, spectral lines, conductivity, and so on, are in fact due to a host of independent underlying entities and mechanisms. But then unification cannot be the *criterion* which separates explanations from nonexplanations, or better from worse. This conclusion appears strengthened by the difficulty of seeing how a realist could settle a dispute between competing conceptions of unification.

Nonetheless, it might be argued, unification may be *evidence* for explanation, which then is evidence for truth. After all, in the past, unified physical theories have been sought, and, perhaps surprisingly, found. These theories have subsequently been replaced by other, still more unified theories. And these more unified theories have survived the most rigorous and extensive testing. Do we not then have *a posteriori* evidence that nature is unified? Indeed, do we not have an *a posteriori* means of settling disputes about what sorts of unification count for scientific explanation?

This argument manages to beg the question the realist originally turned to explanation to answer. For it simply assumes that the observational success of these theories is strong evidence that their theoretical claims are literally true—that nature really is unified in the ways they have claimed.[29] But suppose that the history of science had been slightly different: instead of the theories actually adopted, other theories had been proposed and accepted, theories with the same observational content, but which achieved different degrees of unification, or achieved unification in some different sense. We would, under this supposition, now find ourselves saying of *these* theories that their observational success entitles us to take these quite different degrees or kinds of unification as confirmation of the truth of those explanations characterized by such degrees or kinds of unification. Experience, it seems, cannot by itself single out one degree or kind of unification as privileged to confirm literal truth when an indefinite number of theories—exhibiting an indefinite array of degrees or kinds of unification—are all compatible with the evidence.

This objection might receive a familiar response: theory testing is always a matter of deciding among actual rival hypotheses, not of comparing an indefinite number of possible hypotheses. Historically, the number of rivals has always been small, and, within this limited competition, those hypotheses that have emerged triumphant have exhibited certain definite degrees and kinds of unification. This is sufficient to justify us in appealing to such degrees and kinds of unification when assessing potential explanations.

But again, the realist is making use of a consideration with no obvious connection to literal truth. Have we the miraculous gift that the mere fact that we entertain certain hypotheses but not others (others, that is, that would also be compatible with the data but that would exhibit different degrees or kinds of unification) counts as a reliable sign of their literal truth?

The irrealists may at this point re-enter the argument. If the realist is saying that unification is a pragmatic desideratum in theories, and that it is impractical to consider all possible hypotheses, so that we are well advised to seek theories that are simple in ways similar to successful theories in the past, then they can heartily agree, and welcome the realist's return to his senses. Moreover, if the realist is saying that there is some sort of linkage between such unification and explanation, it is open to the irrealists to concur. Where the irrealists will of course balk is at any attempt by the realist to market such straightforwardly pragmatic considerations as if they were indicators of truth in some nonepistemic sense. Such advertising is surely false, and it begins to seem possible that the real-ist will find himself in the embarrassing position that his antagonists are better able than he to give an account of inference to the best explanation. For irrealists have no difficulty in treating unification as at once a merely practical virtue of a theory and a criterion of explanation—they can treat explanation itself as a prag-matic matter. The realist, by contrast, can establish a criterial connection between explanation and unification only at the expense of throwing into doubt the exis-tence of an intrinsic linkage between explanation and truth.

IV

This is more than the realist can bear. How, he asks, can irrealists draw upon the unification a theory effects among observational phenomena when they have replaced belief in the theory with belief only in its observational reduction?

Our irrealists, however, are entitled to hang on to the original theory so long as they make no use of it inconsistent with the fact that they abjure commitment to its literal truth. They are as free to use it instrumentally—without compromis-ing their irrealism—as they are to use any other instrument of science. They may, for example, employ it as a mechanism for generating predictions and retrodic-tions, or otherwise assisting in the gathering, sorting, and ordering of observa-tional data. One of the useful features of the theory, from this standpoint, is that it provides an especially economical way of organizing experience. If, as Mach seems to have thought, explanation *is* economical organization of experience by mental constructs not assumed to have any literal reflection in reality, then even a Machian irrealist is able to appeal to the apparatus of the unreduced theory to generate unifying explanations.

In response, the realist might pull *Aspects of Scientific Explanation* down from the bookshelf and point out that Hempel requires that the explanans of a success-ful explanation be true. Or he might argue that the intuitive notion of explanation-by-unification involves commitment to the existence of the reducing entities or properties, or to the truth of the underlying principles. No genuinely explanatory unification is achieved, he could claim, when one says that a wide array of obser-vational regularities crop up *as if* there were some unifying underlying structures and mechanisms, any more than genuinely extrasensory perception is achieved when one realizes that it is *as if* an attentive friend has read one's mind.

I find this appeal to intuition congenial, but I cannot expect the irrealist to share my feeling. Here we have come to one of those points, prefigured in section I, where issues about the nature of explanation depend upon, and thus are incapable of resolving, substantive metaphysical issues. Whatever one's view of the form of theoretical explanation, Hempelian or unificationist, if one includes among the material criteria of explanation a condition of literal truth of theoretical premises, one risks begging the question against irrealism. Of course, realist explanations are "better explanations" if theoretical explanations must be interpreted *ontically*;[30] but, by the same token, irrealist explanations are "better explanations" if an ontic interpretation of theory is incoherent or indefensible. Our irrealists may help themselves to any of a variety of theories of explanation – whether something as simple as the idea that explanations are no more than economical symbolizations, or something as intricate as recent work on the pragmatics of answers to why-questions, or something as familiar as Hempel's own deductive-nomological account (suitably reinterpreted on the matter of truth). On such accounts, it may be possible to view the derivations afforded by a partially interpreted unreduced theory, or by a theory held to be no more than a model useful for scientific purposes, as legitimate explanations.

V

Is the debate over explanation between realist and irrealists thus a standoff, with each side able to do no more than beg the question against the other? To push beyond this point, we must locate some considerations that both sides find to some degree compelling, but that nonetheless tell one way rather than the other. The irrealists might think that there are such considerations, and that they tell against realism. For if realist arguments involving inference to the best explanation are seen as question-begging, then the agnostic's argument – to the premises of which our irrealists happily assent and our realist seems, despite some grumbling, obliged to listen – once again threatens to strip realism of any epistemic respectability.

The agnostic argues as follows. First, all the (internally available) evidence we will ever have for our theories is observational evidence.[31] Second, it is possible in principle to generate a theory – the observational theory – that contains all and only the observational implications of our original theory.[32] Third, since the original theory implies the observational theory but not conversely, the original theory could not be more likely to be true than observational theory. Fourth, since the original theory and the observational theory assign the same likelihood to any collection of potential observations, it could not be the case that the original theory is better confirmed by actual observations than the observational theory. Fifth, since the observational theory is thus weaker than the unreduced theory, and since it, like our evidence, does not extend beyond the observable, we will always have more reason to believe the observational theory than the unreduced theory. Sixth and finally, since there will always be an indefinite number of unreduced theories compatible with any set of observations, however large, we will never have sufficient evidence to accept any one of these unreduced theories as true, or nearly so.

Although this is the agnostic's argument, it may be borrowed by the observationalist, who can, with its help, pose a dilemma for realists: to the extent that the observationalist's semantic argument (section II) succeeds in showing that statements about unobservables lack full cognitive content, then *a fortiori* we will never have good reason to accept them as literally true; to the extent that the argument fails, then the agnostic's argument can be brought in to show that nonetheless we still will never have good reason to accept such statements as literally true.[33] Other irrealists may make other arguments and draw other conclusions, but this argument is available to our irrealists. Against them, what has our realist to say?

Recently, realists have expended some ingenuity trying to construct an argument to the effect that it is wrong to suppose that the stronger, unreduced theory must always fare worse in confirmation.[34] Consider again the conjunction of observable regularities about chemical combination, spectral lines, conductivity, etc. contained in the observational theory. If a new observation is made about spectral lines, this may help confirm the conjuncts concerning spectral lines, but it will not be relevant to the conjuncts concerning chemical combination. Similarly, observations of chemical combination will not help confirm conjuncts concerning spectral lines. By contrast, in the presence of the unreduced theory, observed regularities of spectral behavior can help confirm underlying hypotheses about the atom, which in turn imply regularities in chemical combination. If these regularities are then observed to obtain, that adds confirmation to the underlying theory, which then can confer its enhanced confirmation upon the other observational regularities it implies, e.g., those concerning spectral behavior. So, given the same evidence, the presence of the unreduced theory appears to permit more confirmation than is possible for the observational theory alone: its very strength—in postulating unobservable phenomena that connect observable regularities—allows it to be better tested by the same data.

However, since the unreduced theory is equivalent to the conjunction of its observational component—i.e., the observational theory—and the unreduced remainder, we should find this conclusion difficult to accept. How could A & B be better confirmed by a given body of evidence than A alone?

Recall that our irrealists are allowed to employ the unreduced theory in any way that does not require that it be believed to be literally true. In particular, they are allowed to use it to generate inferences from one set of observations to another, inferences *as if* the unreduced theory were true.[35] Whatever confirmation the unreduced theory might pick up of course also accrues to the hypothesis that observation will be *as if* the unreduced theory were true. And the *as if* hypothesis will do as much by way of logically relating various sorts of evidence as does the unreduced theory itself.[36] Any evidence we have that observation will be *as if* the unreduced theory were true is evidence that, for example, if we observe certain spectral behavior, we should expect certain behavior in chemical combination. If such chemical behavior is observed, that will further support the *as if* hypothesis, and this hypothesis can then confer its additional support upon its deductive implications, such as the existence of certain regularities in spectral behavior. But now note that the *as if* hypothesis is weaker than the unreduced theory. Relative to any given body of positive evidence, it should be better confirmed than the unreduced theory, and so able to supply that much more additional support to the observational theory. The irrealists, then, have built a better bootstrap.

This same conclusion can be reached in a way that may be more intuitive, albeit less careful. The observational theory does not involve the claim that there *are no* underlying mechanisms; rather, it is silent on this question. We may thus think of the unreduced theory as *one* way the unobservable world might be that would make the observational theory come out true: there might be atoms, they might behave as the unreduced theory says, and so on.[37] Since the truth of the unreduced theory is one way that the observational theory might be true, then any evidence for the unreduced theory is automatically evidence for the observational theory. But notice that there are in addition many other ways in which the observational theory might be true, corresponding to all the possible ways unobservable reality might be and still yield the regularities of the observational theory. Of course, we may not find these other theories as plausible as familiar atomic theory, but it hardly seems possible to say that the evidence rules decisively against their truth, since in fact their truth is compatible with all our evidence. So however slight our confidence that any one of these competitors might be true, still, our fallibilism requires us to give their disjunction some nonzero probability. But then we must conclude that the observational theory is better supported than the unreduced theory, for it will inherit not only the support of the unreduced theory, but also the support of all its empirically equivalent competitors.[38]

Perhaps the realist should give up tugging at his bootstraps, and accept the fact that the unreduced theory could not be more likely to be true than the observational theory. Still, he might hope that the unreduced theory is nonetheless sufficiently well confirmed by the evidence that it has passed some threshold of acceptance. If, however, we take confirmation to be a measure of epistemic probability of literal truth relative to the evidence, this seems a forlorn hope. For there simply will be too many theories that fit all existing evidence to allow the unreduced theory to achieve high confirmation. And as soon as we introduce criteria such as simplicity, unification, entrenchment, etc. to limit the field of alternatives, we break the connection between degree of confirmation and probability of truth — unless we are prepared to make *a priori* assumptions about the simplicity and unity of the world, or about the aptness of our concepts to it.[39]

A final realist strategy fails for similar reasons. The realist may insist that if science has been spectacularly successful in establishing observational theories, then it is an epistemic desideratum that science give a noncoincidental explanation of this success, i.e., we have epistemic grounds for preferring a theory that yields such an explanation over one that does not.[40] This strategy differs from inference to the best explanation since it sets forth an internal explanatory desideratum, which could be interpreted according to realist or irrealist theories of explanation.

Suppose then that we compare the observational theory with the unreduced theory. Let us say that unreduced, current science gives a noncoincidental explanation of the success of scientific practices at yielding a theory of observables by showing how the experimental procedures and apparatus of current science are able to detect the unobservable phenomena causally responsible for the regularities of experience, and how the inferential patterns of current scientists enable them to translate these inputs into a fairly accurate map of the world around them. We have already seen how irrealists might defend the *general* claim that the observational theory is able to offer explanations, though depending upon the sort of analysis of explanation an irrealist provides, this general claim may not suffice to show that *particular* explanations of the unreduced theory will be preserved. So it is possible that the unreduced theory's explanation of the success of science will not be preserved, and thus possible that we will have at least this reason for preferring realism about the unreduced theory on epistemic grounds.

There are, however, irrealist analyses that would preserve all the explanations of the unreduced theory (suitably interpreted, of course), some of which have been mentioned above. And in any event the realist is able to see that there will exist many distinct yet observationally equivalent theories capable of giving noncoincidental explanations of the observational success of science. Thus both a theory that space is Euclidean and universal forces are present and a theory that space-time is curved could afford noncoincidental explanations of the success of current physics. So the desideratum seems unlikely to narrow the field of competing theories sufficiently to allow the unreduced theory to acquire confirmation at or near a reasonable threshold of acceptance.[41]

Our realist, then, appears to suffer by the comparison with our irrealists. He is in some difficulty explaining the nature or role of inference to the best explanation, and he is committed to a theory that it seems he must concede to be both less well confirmed than the observational theory and far from a threshold of confirmation-driven acceptance.

VI

However, the realist may wonder whether the arguments made thus far really are on the side of our irrealists, and whether, if not, the debate might be a closer call than it now appears to be.

Suppose that we have an effective specification in observational terms of the range of circumstances in which an *observation* takes place.[42] We then can work our way through the recursively generated observational theory culling out all statements concerning relations among observable states of affairs where the description of the states of affairs does not imply that an observation occurs. This leaves as a residue a theory, call it the *manifest theory*, which says all that the observational theory says about observed observables—past, present, or future—but which is altogether silent about unobserved observables. Whereas the observational theory says things (something like) 'All swans are white' as well as 'All swan observations are white swan observations,' the manifest theory says only the latter.[43]

The manifest theory bears the same relation to the observational theory that the observational theory bears to the original, unreduced theory. First, all the evidence we ever will have for our theories is observed evidence. Second, it is possible in principle to generate a theory—the manifest theory—that contains all and only the implications for actual observation of the observational theory.[44] Third, since the observational theory implies the manifest theory, but not conversely, it could not be more probable that the observational theory is true than that the manifest theory is true. Fourth, since the observational theory and the manifest theory assign the same likelihood to any collection of actual observations, it could not be the case that the observational theory is better confirmed than the manifest theory. Fifth, since the manifest theory is thus weaker than the observational theory, and since it, like our evidence, does not extend beyond the observed, we will always have more reason to believe the manifest theory than the observational theory. Sixth and finally, since there will always be an indefinite number of ways in which the unobserved portions of the observable world might be that nonetheless are compatible with any particular manifest portion, we can hardly expect any one of these observational theories to be sufficiently probable relative to the evidence to warrant belief in its literal truth.

The observational theory, despite its modesty in comparison to the unreduced theory, makes sweeping claims about undetected observable states of affairs, including states of affairs that would be so distant in time or space as not to be epistemically accessible to us. If we never have a reason for believing a theory that goes beyond whatever evidence we now have or ever will have, a theory facing myriad competitors that will never be ruled out by the evidence, then we never have a reason for believing the observational theory. That is, we never have reason for believing our theory to be empirically adequate, rather than merely manifestly adequate.

VII

Although they must recognize the force of these arguments in favor of the manifest theory, scientific irrealists are perhaps unlikely to deflate their commitments and embrace manifestationalism. Why not? For our irrealists, it is a fortuitous matter rather than a deep fact which points in the history and expanse of the universe are occupied by observers, depending as it does upon such trivial matters as who happens to be where and looking at what when. To confine scientific theory to claims about experiences and refrain from generalization to unobserved (yet "in principle observable") states of affairs would be to carve out one part of the world and ignore the rest, imposing a boundary of little inherent scientific interest despite its connection with greater confirmation. Our irrealists see it as part of the ambition of science to give a theory adequate not only to that which is experienced, but that which could be experienced.[45]

Perhaps an irrealist could defend belief in the observational theory on the ground that it is the most reasonable projection outward from the boundary of the manifest theory, since it merely involves lifting the restriction to observations. Such a response would only show that the irrealist had forgotten his own arguments against the realist who would defend the unreduced theory as the most reasonable projection of the observational theory, involving as it does no more than lifting the restriction to observables. The difficulty in both cases lies in saying what 'reasonable projection' might mean. It cannot mean "the projection that best fits the evidence," for there are indefinitely many projections from observed to unobserved, or observable to unobservable, compatible with all the evidence. And it cannot mean "the simplest projection," for we have no rationally favored account of simplicity – indeed, were simplicity in some chosen sense held to be *a priori* evidence for the literal truth of a theory, we would in effect be claiming to have *a priori* knowledge that the world is simple in that sense. This would hardly be consistent with the empiricist scruples of our irrealists or the naturalist scruples of our realist. Similarly, our irrealists would fall victims to their own arguments if they were to attempt an *a posteriori* validation of a criterion of simplicity, or if they were to appeal to inference to the best explanation as a way of undercutting the manifest theory. For example, all of the arguments made earlier to show that irrealists are able to preserve explanation (under an appropriate interpretation, to be sure) can now be made on behalf of the manifestationalist.

Where might an irrealist turn to justify his conception of the proper scope of scientific theory? Recall that we began by considering two irrealists, one observationalist, the other agnostic. Thus far, the manifestationalist has used the agnostic's epistemic argument to challenge irrealism. Perhaps irrealism might find support in the observationalist's semantic argument. Unfortunately, the manifestationalist can with equal success turn the observationalist's argument against *him*. Only actual experiences – and not "in principle observables" – have any real effect shaping speakers' acquisition and use of language. In the teaching of language one can of course *say* that the word 'pippin' is meant to pick out unobserved as well as observed apples, but the only tests of competence with 'pippin' involve actual, experienced circumstances. After a certain number of tests our pupil may say "now I can go on . . . ," but the dispositional competence this reflects is notoriously underdetermined by any actual history of instruction, and the only constraint upon future use will be his manifest experience of success or failure in communicating or meeting expectations. Therefore, according to this argument, only those statements whose conditions of assertion are exhausted by manifest experience can have determinate cognitive content.

Scientific irrealists thus face a problem of their own making. If they accept the agnostic's argument, then it seems they must conclude that the observational theory could not be epistemically warranted. If they accept the observationalist's argument, then it seems they must conclude that the observational theory exceeds the bounds of sense. So the irrealists' positions would appear to be self-destabilizing: the logic of their criticisms of realism should by rights carry them past observational reduction and at least as far as manifest reduction. Arguably, it should carry them further still: the logical terminus may be phenomenalism, perhaps even phenomenalism of the present moment.

Yet familiar difficulties in accommodating the content and practice of modern science have led those irrealists with whom we have been concerned, namely *scientific* irrealists, away from phenomenalism. When Vienna discovered that scientific laws are not finitely verifiable, it ultimately decided to give up the criterion of finite verifiability rather than abandon the idea that propositions expressing laws are cognitively meaningful. When Vienna discovered that statements about physical objects could not be translated without loss into the language of sense-contents, it ultimately decided to shift the basic vocabulary from the language of sense-contents to the physical-thing language. And when Vienna discovered that explicit definitions or sentence-by-sentence translations of the theoretical expressions of science were not possible within the language of observable properties of physical things, it ultimately decided to give up the requirement of full interpretability in favor of partial interpretability. At each of these points a trade-off was made against confirmation in an effort to preserve the scope of science — scope adequate to encompass whatever the truth might be about the lawful features of the observable properties of the physical world, the rendering of which was taken to be the aim of science.

These eminently sensible moves inched scientific irrealism off its argumentative foundations, and care was not taken to ensure restabilization. We have noticed two points of possible collapse. First, without a theory of meaning that goes beyond the constraints of manifest conditions, it is unclear how the observational theory's unrestricted quantification over observables could have cognitive content. Second, without a theory of epistemic warrant or rationality in belief-formation that goes beyond confirmation, it is unclear how we could ever be warranted or rational in believing that the observational theory is literally true.

Let us make our job more manageable by setting aside the first, or semantic, sort of instability for the remainder of this paper. What might be done by a scientific irrealist to contend with the second, the epistemic?

Here follows one proposal, not the only one and perhaps not the best, but one that has the advantage of incorporating a perspective on epistemology and belief-formation that places the issues in a different light from discussions based upon confirmation.

There are at least two ingredients in knowledge: truth and warrant. The equation of epistemic warrant with degree of confirmation (as understood here) has the unnerving implication that even if our current scientific theory is correct—or very nearly so—in its claim about observables, still, belief in these claims could not be knowledge. We might have true belief—but not warrant, for the epistemic probability of the highly informative observational theory relative to our evidence will always be low.

Why is warrant important for knowledge? In part because true opinion might be reached by arbitrary, unreliable means. But now suppose that the experimental and inferential procedures followed by scientists in coming to believe the observational theory are in fact such as to be highly reliable in the production of true belief about observables. On this supposition, scientists are excellent detectors of the observable world, and it is not accidental that they have come to believe a true theory of the observable. If that is the way things are, then one of the worries about warrant has been addressed even though the theory arrived at is not well confirmed.

But warrant involves other worries as well. We can imagine epistemic agents following reliable belief-forming practices "in the dark," without being able to see themselves as reliable, or to understand the nature of this reliability and its basis. So let us further suppose—a supposition of slight difficulty, as we noted—that the irrealist is able to make nonliteral use of the unreduced theory, or literal use of the observational theory, to give good explanations of our reliability at detecting observables. A second worry about warrant can then be addressed even by a theory that is not well confirmed.

Still, a third worry remains. We could be reliable, able to see and understand our reliability, and yet unable to demonstrate our reliability *with high epistemic probability on the basis of internally available evidence*. Confirmation is just the thing to address this worry, for it is a measure of epistemic probability relative to evidence. Yet we have seen how little would be left of scientific theory were we to insist that it be well confirmed. The irrealist may be well advised to consider whether this third worry is one worry too many.

If the irrealist were to accept a reliabilist account of epistemic warrant, then the manifestationalist's challenge might be blocked, for the irrealist would then be able to say that although the observational theory is below a threshold of confirmation-driven theory acceptance—as virtually any interesting general theory must be—it might nonetheless be the product of a practice that is above a threshold of reliability in belief-formation. It is perhaps unusual to combine scientific irrealism with epistemic externalism, but then the usual sort of scientific irrealism has just been seen to have some difficulty in saying how it might ever be possible that (the observational reductions of) scientific theories could be objects of knowledge. (Recall that we have set aside irrealisms that are revisionist about truth.)

One source of resistance to epistemic externalism is the sense that it can establish no more than the bare possibility of knowledge, and thus may be unable to provide a rationale for any particular set of belief-forming practices. Reasons for belief seem bound to be internal even if warrant for belief is not, and where is the externalist to look for scientific reasons for belief? He must, it would appear, look to the goals of science. Securing confirmation for what one believes is certainly a goal of science, but if one simply wanted highly confirmed beliefs, one could achieve this by the expedient of restricting one's beliefs to tautologies and observation reports. Irrealists have tended to say that among the goals of science is attainment of an empirically adequate theory, but that claim is bound to be controverted by the manifestationalist. If irrealists are to have a convincing defense of their conception of the aim of science, and of their corresponding view of legitimate scientific belief-forming practices, they must do so by reference to uncontroversial scientific goals.

It would seem uncontroversial that *scientists seek theories that are predictively, manipulatively, and explanatorily successful over as wide a range of experience as possible, and moreover seek genuine knowledge of the natural world, that is, beliefs that are (at least) both true and warranted.* We need not attempt to decide which among these ends are intrinsic, which instrumental, so long as none of the parties to our debate wishes to dispute them. The manifestationalist can accept them, for although he confines belief to the realm of experience, he is prepared to assign theories literal truth over that realm, and to provide a suitable interpretation of such notions as explanation and warrant. Nor need the realist dispute them, for although his first loves may be truth and explanation, he has no doubts about the centrality to scientific practice of the aims of successful prediction and control.

Let us say, then, that a belief-forming practice is scientifically rational to the extent that it advances these ends. However, while reasons may be internal, we may not always be able to assess them internally. We are, for example, unable to generate internally ratings of belief-forming practices in terms of their truth-conduciveness. Moreover, uncontroversial ends are subject to controversial interpretation: manifestationalists, irrealists, and realists will differ over such notions as warrant and explanation, so no agreed upon rating in these terms is possible. Let us begin, then, with this goal: *success at prediction and manipulation over a wide range of experience.* If one belief-forming practice could be shown to be superior to others in this regard, that would furnish it a rationale in terms of the ends of science. Because this is but one of the goals of science, only a *prima facie* rationale is possible, but because predictive and manipulative success may be the most important of those scientific goals amenable to uncontroversial assessment, such a rationale could carry considerable weight.

To see what is at stake here, and how this dispute might be carried out, we need to say what a belief-forming practice involves. It is a matter of which beliefs, or what degrees of belief, are formed in response to new information. For example, how much, if at all, does one allow one's observation that all Fs thus far sampled are G to strengthen one's belief that all Fs are Gs? does one—or does one not—restrict this generalization to manifest or observable Fs and Gs? does one invest significant belief in all those hypotheses that are consistent with present observations, or only in some? how does one select certain properties or classes of properties for projection rather than others (e.g., green vs. grue)? does one come to accept beliefs as true or approximately true, and if so, when? and so on.

The belief-forming practices endorsed by the manifestationalist require that we remain agnostic among all hypotheses about observables consistent with observations, while irrealists think such suspension of belief about unobserved observables is excessive, and that the sort of evidence scientists now possess can justify investing a high degree of belief in one rather than another manifestly equivalent theory of the observable.

What sort of difference would it make to science to adopt irrealist rather than manifestationalist belief-forming practices? At first glance, no difference, for observables come within the grasp of science only during observation. Moreover, a manifestationalist is able make instrumental use of the observational theory. He can without compromising his manifestationalism believe that experience is *as if* the observational theory held, and thereby can take advantage of whatever inductive and deductive systematization that theory effects among the observed.

Nonetheless, if we think of scientists situated at a certain point in the development of theory, then which sorts of proposals they will make, or which experiments they will undertake, and therefore which conclusions they will in fact reach, will be affected by the character of their individual and, as it were, collective credence functions at that time, and by the dynamical character of these credence functions over time—that is, by their belief-forming practices. If a community of scientists is agnostic about what goes on in unobserved states of affairs, and resists accepting substantive theories about such states, then its behavior in the formulation and testing of hypotheses can be expected to differ from a community that believes and revises theories about unobserved observables on the strength of observations. For example, a community of manifestationalists would not infer from a series of experiments that they should believe a principle of conservation requiring unobserved magnitudes to have determinate values. Nor would it seem appropriate to them to invest a very substantial portion of research resources in an experimental program the design of which presupposes determinate values for unobserved magnitudes. They are, after all, rational decision-makers, and their willingness to commit resources will reflect their estimation of the expected values of outcomes, which in turn will reflect their credence functions. But their credence functions are agnostic over the whole range of unfalsified theories of the unobserved, and this would hardly yield a convincing reason for concentrating resources in any particular venture whose prospective worth is highly dependent upon how things are with the unobserved.

Of course, nothing would prevent manifestationalists from thinking that actual observations show regularities *as if* (for example) certain conservation principles extending to the unobserved held. But the question is whether, if they had no significant tendency to believe any principle of this kind, they would be inclined to consider or develop the idea that observation behaves *as if* any particular conservation principle were true, and so be less likely to discover those regularities of actual experience that one would expect were things *as if* it were true.

This concern will afford a scientific rationale if, historically, beliefs in theoretical principles involving unobserved magnitudes – and commitments to experimental procedures motivated by such beliefs – have played a large role in the development of predictively and manipulatively successful theories. It is impossible to know how scientific practice would have differed in its character or accomplishment had scientists allocated their degrees of belief in strict accord with degrees of confirmation instead of accepting stronger, less agnostic theories. Yet it does seem evident that belief in powerful theories about observables has been pervasively involved in the development and testing of successful scientific theories.

Scientists, like the rest of us, have a remarkable thirst for underconfirmed, overambitious theories, theories that strive to be comprehensive. Scarcely is a fraction of possible data points in hand than smooth curves are drawn connecting the points and racing outward in both directions. This is *not* an application of a well-confirmed second-order theory according to which predicates apply uniformly over time. For there is a much better confirmed second-order theory which tells us only that those predicates science in fact keeps track of apply uniformly over observed time. Any perfectly general second-order theory of uniformity would just be another underconfirmed, overambitious theory.

It is better to admit straight out that we are beings who are strongly inclined to hold opinions about how the world is, even when it is not observed, and who view agnosticism about what we take to be meaningful questions as a kind of cost. Scientists, especially, seem so inclined. We do not know exactly what role this inclination has played in the development of science, but we do know that this development has been highly successful. To abstain from belief about all states of affairs not observed would not only frustrate the ambition of obtaining a theory with wide scope, but would also have unknown effects upon the future progress of science.

If one is to have beliefs about what is literally true where no observation occurs, as our irrealists do, then one will have to sacrifice confirmation. Such a trade-off will seem especially worth making if one also believes that theory development might thereby be enhanced. And it does seem that the development of existing theories—theories which have manifestly done well—has been tightly bound up with the prevalence of these more ambitious beliefs. If so, then a rationale in terms of uncontroversial scientific goals is at hand for irrealist rather than manifestationalist belief-forming practices—why abandon something that is working in favor of something untried?

VIII

The manifestationalist may concede that a non-question-begging rationale of this kind for irrealist belief-forming practices is available, but insist that the rationale is of an entirely practical rather than epistemic character. Even if it does succeed in giving a scientific reason for believing the observational theory over and above the manifest theory, it does not do so by showing that this stronger belief is *epistemically warranted*, only that it is useful.

The irrealist may reply that the predictive and manipulative success of general theories about observables does provide epistemic warrant of a kind the manifestationalist must accept, since it affords some *confirmation* of observational theory. Moreover, it affords some confirmation that the belief-forming practices that led to the observational theory are reliable. But the manifestationalist will quickly point out that, given the range of manifestly equivalent competitors, the confirmation is weak.

Yet it would be question-begging against our (revisionist, externalist) irrealists to confine epistemic warrant to confirmation. Thus the manifestationalist cannot say in advance that the rationale provided for irrealist belief-forming practices is without epistemic force, for some of the very features of scientific belief-forming practices picked out by the rationale may contribute to its reliability, and thus be warrant-conferring. This gives the irrealist an answer to the manifestationalist's charge that, given the sort of evidence available to science, the observational theory *could not* be an object of knowledge. It could be, and, if it is not terribly wrong in its claims, it is.

Has the irrealist thus restabilized his position? He may have a way of resisting collapse into manifestationalism, but he has strengthened his position only by weakening the argument he used against realism. If he is to resist realist pressures, he must show that scientific irrealism is both an appropriate equilibrium point in the trade-off between confirmation and scope and the natural stopping place for belief-forming practices given the rationale afforded by the aims of science. It is not clear that he can do either.

As far as confirmation is concerned, the observational theory is of course more probable relative to the evidence than the unreduced theory, but neither is remotely well confirmed. What could make unobservables seem the right stopping place for extending the scope of scientific belief? After all, we could make substantial gains in confirmation by restricting scope still further.

This particular stopping place would seem appropriate if it corresponded to a point of special epistemic significance. Traditionally, irrealists have appealed to the notion of *epistemic access* to explain the significance of in-principle observability. We have, it is said, more direct access to the observable. But unobservable states of affairs in our neighborhood may be more directly accessible than in-principle observable states of affairs elsewhere. It can be easier for us to detect proximate photons, or changes in inertial force, or collisions of air molecules, than it is for us to detect observable states of affairs in distant galaxies, the remote past, or the uncertain future. An eye accustomed to darkness can register the arrival of a photon, an inner ear in equilibrium can sense inertial accelerations, and a child can hear molecular collisions against the walls of a conch shell. Such detection is of course indirect and inferential by the time it reaches the level of self-conscious perception, but it would be a very naïve picture of the operation of the human perceptual system to think that we see proximate middle-sized objects noninferentially. And, of course, all judgments about unobserved observable events will be inferential, often highly so. The observable thus does not uniformly enjoy an inherent advantage in directness of access over the unobservable. Similar remarks apply to *reliability* of detection.

Moreover, the irrealist's claim that only statements about what is in-principle observable are *testable* cannot establish a convincing epistemic distinction. Van Fraassen argues that an unreduced theory's "vulnerability to future experience consists *only* in that the claim of its empirical adequacy is thus vulnerable," but of course in this sense the vulnerability of a theory resides only in its claim to manifest, not empirical, adequacy. Thus van Fraassen's barb that realism is "but empty strutting and posturing" rather than "courage under fire" would apply to irrealism as well.[46] Happily, however, the barb pricks neither, since theories about observables and unobservables alike can face the barrage of experience – as corporate bodies if need be – whenever we perform an experiment for the outcome of which they have implications.

The distinction between observables and unobservables does not, it seems, mark off a divide of special epistemic interest with regard to directness of access, potential reliability of detection, or liability to test. Nor does this distinction appear to be of special scientific interest. Just as the irrealist pointed out that the realm of the observed is essentially fortuitous from the standpoint of large-scale scientific theorizing, so may the realist point out that it is a fortuitous matter rather than a deep fact that certain objects are of a size or nature to be observable to humans without technical assistance. No doubt within that part of evolutionary psychology that treats of the phylogeny of human sensory receptors there will be interesting things to be said about the nature of our thresholds of discrimination, but these thresholds are unlikely to define a category with far-reaching significance in natural science as a whole.

The irrealist was offered an externalist response to the question of how perfectly general statements about observables might be objects of knowledge. Yet without an argument that assigns statements about the observable a special status with regard to reliability, it will be difficult for the irrealist who would make use of this response to explain why he draws the line of epistemic tolerance to include unobserved observables but to exclude all other unobservables. If the observational theory, despite its low degree of confirmation, could nonetheless be warranted because it was arrived at by reliable means, then the unreduced theory might for the same reason be warranted as well: scientific belief-forming practices might be quite reliable with regard to viruses, cells, molecules, and so on. It will be a contingent matter whether warrant gets as far as the observational theory, and whether, if it does, it goes on to further parts of the unreduced theory. No quick, *a priori* argument, such as the agnostic's argument promised to be, will be able to secure the epistemic respectability of irrealism, or the epistemic indecency of realism.

Nonetheless, the irrealist might argue, belief that the unreduced theory is true involves substantially greater exposure to the possibility of error than belief only in the observational theory. This much the agnostic's argument usefully shows. Should we not, then, see whether we can do without it? After all, even though a rationale has been suggested for belief-forming practices strong enough to yield the observational theory, much stronger practices would be required to reach the unreduced theory. Since avoidance of error counts among the uncontroversial values of science, a rationale might be constructed for eschewing such strong belief-forming practices. The cost seems small for a large reduction in epistemic risk: we need only give up (fallible) belief that the unreduced theory is (approximately) literally true, and we may go on (fallibly) believing that it is (approximately) empirically adequate.

Yet the costs of altogether abandoning these strong belief-forming practices may in fact be quite high. In the first place, error has disvalue for science thanks in part to the value of truth. If scientific methods are reasonably reliable with regard to unobservables of various kinds, then we could lose more in truth than we gain in reduction of falsehood. Indeed, we might do better with regard to avoidance of error by giving up projection to certain remote or exotic observables and keeping projection to humdrum, local unobservables like the micro-organisms on this planet.

In the second place, even if jettisoning some particular part of our theory at a given time would reduce epistemic risk at that time, it might have a different effect over time, for—as the irrealist argued against the manifestationalist— scientific practices in the formulation and testing of hypotheses are highly dependent upon what scientists believe to be the case. Altering scientific credence functions from belief to agnosticism on questions about the nature and behavior of unobservables might dramatically alter theory testing and development, with unpredictable effects upon the reliability of scientific practices.

If one believed the unreduced theory's claims, literally interpreted, one would believe that certain entities really exist and certain mechanisms really are at work. This inevitably would influence one's behavior as a scientist. For example, if one thought we had reason to believe that there are such things as viruses, that they are made of protein, that they interact with antibodies in certain ways, and so on, one would have a reason to design and carry out certain experiments (but not others), to draw certain conclusions from the outcomes of these experiments (but not others), and to attempt theoretical innovations consistent with the existence and functioning of the mechanism in question (but not others).

If by contrast one allowed oneself to believe nothing stronger than that the regularities among observables are *as if* viruses existed, had certain properties, and so on, this would be only a weak reason for such a sharp focus in experimental design and theoretical innovation, for there are infinitely many ways the world might be and still be *as if* certain mechanisms were at work. Analogies are cheap, and there would be no inconsistency or even incoherence in deploying different analogies at different points in the testing or development of a theory, or in restricting the scope of an analogy that failed to fit certain aspects of the phenomenon under study. Which principles are likely to occur to an investigator, or to attract sufficient commitment to bring about much investigation or development, will depend upon what he believes to be true. To adapt the irrealist's own example, a conservation principle that requires unobservable entities to take on certain magnitudes would not come to be believed by an irrealist, and yet, were this principle not to be believed, a large-scale regularity in what *is* observable might fail to be noticed.

A realist would be able to extend the response offered to irrealists without begging questions because the features of theory testing and development that involve literalism about unobservables are part of a scientific tradition that has been successful in terms recognizable to the irrealist, that is, successful at extending the scope and accuracy of prediction and control. It would be difficult to show that these achievements could have been reached had the community of scientists not invested substantial belief in the existence of viruses, cells, molecules, atoms, and the like.[47]

It is perhaps a sense of the difficulty of establishing such a counterfactual that leads van Fraassen, an advocate of an agnostic form of irrealism, to recommend to scientists "total immersion" in the unreduced theory. He seeks, however, to distinguish total immersion from outright belief; according to him, the former relation to a theory involves " 'bracketing' its ontological implications"—although not when these concern the existence of observables, whether observed or not.[48] What is this attitude? "Total immersion" sounds like—and seems to be taken to be[49]—acting exactly as if the theory were literally true when it comes to doing science. Now doing science is not just going through motions; it is inextricable from one's credence function and belief-forming practices, for it involves designing experiments, making inferences, deciding which hypotheses are worth pursuing, deciding when to withhold judgment and when not. Of course, scientists who *act on the belief that observation behaves as if the unreduced theory were literally true* are able to bracket ontological commitments, but for that very reason their practice can be expected to differ from that of scientists who *act on the belief that the unreduced theory is literally true.* Yet the achievements of science may owe something – perhaps a good deal – to the fact that brackets have long been removed from such things as viruses, cells, molecules, and atoms.

Van Fraassen makes the intriguing suggestion that a scientist expressing his "epistemic commitment" is "stepping back" from his theory.[50] This retreat is reminiscent of a now unfashionable image of epistemology as prior to, and on a different footing from, on-going science. But fashions come and go. So one must ask: if such a retreat were to take place, and if a scientist were to conclude upon reflection that he had no business being "epistemically committed" to unobservables, what would his credence function and his belief-forming practices look like when he returns to science? He must make decisions about large, expensive, long-lived experimental programs whose selection from the realm of all possible experimental programs depends for its rationale upon quite strong views about the properties unobservables actually have. Yet he is "epistemically committed" to full-fledged agnosticism about the accuracy of these views. Mustn't he regard any such program as an irrational stab in the dark? And mustn't he regard the success of such a program, should it arrive, as no reason to believe that the views upon which it is based are accurate – beyond their empirical (really, manifest) adequacy? Would that sort of attitude be compatible with the total immersion van Fraassen recommends? And is it the sort of attitude that has characterized contemporary scientific beliefs about viruses, cells, molecules, and the like?[51] Perhaps the natural description of the attitude van Fraassen seeks to encourage is not 'immersion plus agnosticism' but 'realism plus fallibilism.' The latter is self-critical, the former, it would seem, self-undermining.

If literalism about various unobservables has been a central feature of theory testing and development in modern science, then there may be a rationale in terms of the goals of science for adopting belief-forming practices strong enough to support holding significant degrees of belief in detailed hypotheses about the nature and behavior of unobservables. And the rationale is the same the irrealist offered: why fix what is not broken?

As with the matter of epistemic warrant and reliability, the dispute between realist and irrealist over the rationality of belief-forming practices would become an empirical question: how much of a difference has it made to the success of science that realist interpretations of unobservables have been accepted so widely and pushed so hard? Here, too, there is no short, *a priori* argument (this time, about rationality in belief-formation) to settle the matter.

I have sketched a response to manifestationalism that is available to an irrealist who will countenance epistemic externalism, and have argued that this same response could in turn be adapted by realists to provide them a response to irrealism. My aim has not been to advocate this response to either realists or irrealists, but rather to indicate its existence and point out that it does allow us to see these two views as occupying points along a continuum. Moreover, it enables us to see the debate between them as not merely question-begging, as the debate over explanation has often been. For, as the two views have been understood here, both embrace the idea that even when the beliefs of scientists extend to that which is not actually given in experience, this may yet be rational (in light of the aims of science) and warranted. In particular, it may be rational and warranted when these ambitious beliefs (and the ambitious belief-forming mechanisms from which they result) play an important role in a successful tradition of theory testing and development. No questions are begged when we pose the dispute between realist and irrealist in terms of where along the path of forming beliefs about unobserved phenomena they think – or better, predict – either rationale or warrant will give out. .

Both camps recognize that the scientific ambition is not merely to be able to answer lots of questions about the world, but to answer them well. So both realist and irrealist want good reasons for making any sort of trade-off against confirmability. Religious and natural teleologies answer many questions that contemporary science leaves hanging, and thus are attractive for their potential informativeness. The difficulty of such teleologies, in light of the present discussion, is that over the last few centuries theoretical traditions *un*constrained by literalism about teleological commitments have done vastly better in promoting the development of theories that refine and extend our success at prediction and control. Even though it is a goal of science to achieve a comprehensive system of beliefs, advancement in this regard is always viewed as having attendant costs, assessed in terms of increased risk of error and all that goes with it. Those costs can be seen as scientifically rational only so long as an appropriately large gain in theory development has accompanied them.

IX

I do not consider leaps of faith or belief in things unseen, arrived at for whatever reason, necessarily irrational – only the pretense that we are rationally compelled (e.g., through arguments concerning explanatory value) to embrace more than strict empiricism prescribes.

B. C. van Fraassen[52]

Let us end with a story. You are visiting a friend who lives in a city spread along the sea at the foot of a sierra. A chain of small, rocky islands extends out into the ocean, linked to each other and to the shore by a series of picturesque bridges.[53] Each island commands a fine vista of the city and its setting, although the view becomes more spectacular the farther out one walks. Your friend, however, does not venture beyond one of the middle islands.

"I always stop *here*," he says with cheery firmness.

"Why?" you ask, "Do you prefer a less sweeping view?"

"Oh, no," he replies, "it isn't that. It's just that I worry about support, and so don't think you should cross any more bridges than, according to reason, you strictly have to."

"Well, we didn't strictly *have to* come out this far," you note. "As far as support is concerned, we could have stopped at one of the inner islands, or stayed on the shore in the first place—that's better supported than any bridge."

"But if you don't come at least this far, you will never see the lovely, wide view of the city."

You are puzzled. "If you think a lovely, wide view is worth crossing a few bridges, why isn't a lovelier, wider view ever worth crossing a few more? After all," you add, "if you always stop here, you will never see the mountains behind the city."

The story is left for the reader to finish.

Notes

1. I would like to thank Richard Boyd, Risto Hilpinen, Alan Goldman, David Lewis, Richard Miller, and Tim Maudlin for helpful comments on an earlier draft. I am grateful to Lawrence Sklar and Nicholas White for numerous helpful conversations.

2. See especially, C. G. Hempel, *Aspects of Scientific Explanation* (New York: Free Press, 1965).

3. Oxford: Clarendon, 1980.

4. Princeton: Princeton University Press, 1984.

5. Revisionism with regard to intuition is certainly not altogether new, but, characteristically, it previously did not find expression in overtly metaphysical language. Thus, Hempel wrote in response to complaints that his D-N model allowed unintuitive explanations in which the phenomena mentioned in the explanans do not cause or "bring about" the explanandum phenomenon: "while such considerations may well make our earlier examples of explanation, and all causal explanations, seem more natural or plausible, it is not clear what precise construal could be given to the notion of factors 'bringing about' a given event." *Aspects of Scientific Explanation*, p. 353. The point is put as a matter of analysis, but 'precise construal' in the context is in fact elliptical for "precise construal in terms of the categories of analysis permissible under empiricism—observables, observable patterns of observables, extensional logic, and so on."

6. There are, of course, other ways of accounting for this appeal. Cf. Hempel's remarks in *Aspects of Scientific Explanation* about fitting explananda "into a nomic nexus" (p. 488). Along a similar line, see also the remarks on the "nomothetic" character of scientific explanatory practice in P. Railton, "Probability, Explanation, and Information," *Synthese* 48 (1981): 233–56.

7. "The Function of General Laws in History," first printed in 1942, reprinted in *Aspects of Scientific Explanation*, p. 235.

8. See Hempel, *Aspects of Scientific Explanation*, pp. 376–410.

9. For Hempel's rejection of epistemic relativity in the case of nonprobabilistic explanation, see Hempel and P. Oppenheim, "Studies in the Logic of Explanation," reprinted in *Aspects of Scientific Explanation*, where it is said of a nonprobabilistic case similar to the one discussed in the text that "the ampler evidence now available makes it highly probable that the explanans is not true, and hence

that the account in question is not – and never was – a correct explanation" (pp. 248–49). See also *Aspects of Scientific Explanation*, pp. 388 and, especially, 488: "the concept of deductive-nomological explanation requires no such relativization."

10. For further discussion, see P. Railton, "Taking Physical Probability Seriously," in M. Salmon (ed.), *The Philosophy of Logical Mechanism* (Dordrecht: D. Reidel, forthcoming).

11. See P. Railton, "A Deductive-Nomological Model of Probabilistic Explanation," *Philosophy of Science* 48 (1978): 206–26.

12. An interesting exception with regard to the first example is to be found in a discussion of the Humean background of the covering-law model by Israel Scheffler, *The Anatomy of Inquiry: Philosophical Studies in the Theory of Science* (New York: Alfred A. Knopf, 1963), pp. 19–20 . Counterexamples to the Hempelian account based upon "singular causal statements" might have furnished raw material for substantive metaphysical discussion, but instead both Hempel and his critics tended to keep discussion of singular causal statements within the realms of linguistic analysis and epistemology.

13. For the distinction between a concept and a conception, as applied to justice, see John Rawls, *A Theory of Justice* (Cambridge: Harvard University Press, 1971), pp. 5ff.

14. Roughly, what I have in mind by *confirmation* is a measure of the epistemic probability that a theory – perhaps taken as an interpreted whole – is true relative to a given body of evidence. The clearest example of a theory of confirmation in this sense is a theory of logical probability, which gives what is in effect a measure of the semantic information of a hypothesis and the semantic information of a given body of evidence and an explication of the logical relation between the two. The evidence will confirm the hypothesis to the extent that it agrees with the information content of the hypothesis.

15. I am assuming here and elsewhere that it is possible to give an interesting account of the theory/observation distinction, and that this account will take the form of distinguishing observable properties – or predicates – from theoretical ones. There are reasons for worrying about just what this might involve, reasons that do, I think, play an important role in the argument for realism. But I would like at this point to concentrate on a set of issues that can perhaps most readily be seen if, for the sake of argument, a theory/observation distinction of the sort just mentioned is allowed.

16. It is, of course, a vexed question how one should formulate the idea of a literal interpretation. It is not enough to say that under the interpretation the phrase " 'virus' refers to viruses" comes out true, since the phrase 'refers to viruses' must – so to speak – itself be understood literally, for example along the lines characteristic of (nonvacuous) correspondence theories.

17. Again, we should set aside difficulties of characterization. For the sake of argument, we will assume some definite sense can be attached to the idea of the observational implications of a statement or set of statements. See also note 20, below.

18. Cf. the "constructive empiricism" of van Fraassen in *The Scientific Image*.

19. The controversy between realists and irrealists often involves a controversy over the nature of truth, but I will assume that our realist and our irrealists share a (nonvacuous) correspondence theory of truth, although the observationalist thinks that sentences containing theoretical vocabulary may, owing to partial interpretation, fail to have truth-value. I will only consider in passing irrealists who propose nonstandard theories of truth.

20. Use of Craig's theorem presupposes that scientific theory can be put into axiomatized form and that its interpretive system can be effectively specified. Moreover, for the *observationalist* to make use of Craig's construction he must employ a purely syntactic notion of logical consequence. Once again, for the sake of argument, let us set aside some doubts that might be raised about the irrealists' assumptions. Here, too, we ignore the worry that, except relative to a *ceteris paribus* clause we do not know how to complete, theories may not have determinate observational consequences. Let us also ignore Hempel's influential argument that the observational theory will not preserve the *inductive* systematization affected by the unreduced theory. We will return to the question of inductive systematization, below. For Hempel's argument, see his essay "The Theoretician's Dilemma," reprinted in *Aspects of Scientific Explanation*, especially pp. 214–17.

Those who employ a model-theoretic conception of theories, e.g., van Fraassen, may prefer to speak not of the observational theory (constructed in a syntactic, Craigian way) but of the observational *substructures* of a theory. What I say of the observational theory will, however, be applicable *mutatis mutandis* to the unreduced theory's observational substructures. For example, I will make nothing of the unwieldy character of a Craigian reduction or of the question whether a Craigian observational theory might in fact contain statements describing or otherwise referring to unobservables, albeit using an observational vocabulary to do so.

21. An early paper of Harman's is perhaps typical. He writes, "There is, of course, a problem about how one is to judge that one hypothesis is sufficiently better than another. Presumably such a judgment will be based on considerations such as which hypothesis is simpler, which is more plausible, which explains more, which is less *ad hoc*, and so forth. I do not wish to deny that there is a problem about explaining the exact nature of these considerations; I will not, however, say anything more about this problem." From Gilbert Harman, "The Inference to the Best Explanation," *The Philosophical Review* 74 (1965): 88–95. For an interesting discussion of possible criteria, see Paul Thagard, "The Best Explanation: Criteria for Theory Choice," *The Journal of Philosophy* 75 (February 1978): 76–92.

22. This construal is suggested by many of the examples in the literature, as well as by Harman's remarks concerning statistical explanation in *Thought* (Princeton: Princeton University Press, 1973), pp. 137ff: "the greater the statistical probability an observed outcome has in a particular chance set-up, the better that set-up explains that outcome." In the example that follows, the various explanations might be thought of as involving chance mechanisms.

23. See J. J. C. Smart, *Between Science and Philosophy* (New York: Random House, 1968), pp. 150–51.

24. See, for example, Jerrold Aronson, *A Realist Philosophy of Science* (New York: St. Martin's Press, 1984). One might further complicate this account by the introduction of prior probabilities. Some of the difficulties the realist would create for his argument by introducing prior probabilities *in this connection* will be discussed below.

25. Things get more complicated when we consider relations weaker than outright logical implication. Let us, again, avoid both complication and attendant worries.

26. It should be emphasized that this latter belief does not involve commitment to the *non*existence of underlying mechanisms. The former belief, of course, involves the latter, since the underlying theory implies the observational theory.

27. See for example Michael Friedman, "Explanation and Scientific Understanding," *The Journal of Philosophy* 71 (1974): 250–61, and Philip Kitcher, "Explanation, Conjunction, and Unification," *The Journal of Philosophy* 73 (1976): 207–12.

28. Friedman characterizes his motivation in part as seeking to give a theory of explanation that appropriately links this notion to a conception of understanding. See his "Explanation and Scientific Understanding," p. 251.

29. It is important to see that this objection is not a form of blanket skepticism about induction. The argument allows one to infer from past observational success to future observational success. It questions whether one can infer from observational success—past, present, or future—to theoretical accuracy.

30. For discussion of an "ontic" conception of explanation and its contrast with, among others, an "epistemic" conception, see Salmon, *Scientific Explanation*, pp. 84–111, 121–24.

31. The (vague) qualification 'internally available' is necessary since a realist may opt for a causal theory of evidence, according to which unobservable entities can enter into our evidence in virtue of their physical effects upon us, even when the causal origin of these effects is not "internally available" to their recipients. But the bald claim that we have reason to believe that unobservable phenomena are in this causal sense part of our evidence would beg the question against irrealism. By contrast, it is not question-begging for the irrealist to *begin* from premises involving internally available evidence alone, with the aim of reaching as a conclusion the claim that we never have reason to believe that a theory of unobservables is literally true. For although the realist may hold that evidence *need not* be internally available, it appears that if he is to provide any guidance for actual epistemic agents, he will have to provide an account of scientific theory from an internal point of view. Such an account necessarily would start out from what is "available." We will see whether it must also end there. (The qualification about internal availability will often be taken to be understood below, and omitted.)

32. The realist's suspensions of disbelief remain in effect.

33. The observationalist's primary argument will be addressed below only in passing.

34. The argument that follows is in the spirit of a number of recent proposals, especially that of Clark Glymour, in *Theory and Evidence* (Princeton: Princeton University Press, 1980), pp. 162–63. Although the argument as given by Glymour is not in the form of an inference to the best explanation, he does suggest a link between his "bootstrap" strategy and explanatoriness: "Informally, the body of evidence may be explained in a uniform way in one theory but have to be explained in several different ways in the second theory. We prefer the first." (p. 353; a typographical error in the original has been corrected)

35. There is an awkwardness here. The observationalist may be unable to give full cognitive meaning to the unreduced theory's claims, and so to that extent must profess that there could be no fact corresponding to the truth or falsity of that theory. To this extent, one supposes, he professes not to know what it would be like for that theory to be true or false. He does, however, admit to understanding this much of what it would be like for the unreduced theory to be true: what would be the case in the realm of the observable. Hence he is able to attach a definite sense to the idea that the world is *as if* the unreduced theory were true—i.e., that the theory implies successful observational predictions—though not perhaps the sense attached to this idea by the realist or by an agnostic who claims to have no principled difficulty in understanding what it would be like were the unreduced theory literally true. Of course, both the realist and the agnostic may claim to be up against psychological limitations in attempting to understand what state of affairs corresponds to, say, a quantum-mechanical superposition.

36. These relations will be effected syntactically according to the observationalist; they may be effected semantically according to the agnostic.

37. The carelessness of this argument, at least, when employed by the observationalist, is conspicuous here. For this is precisely the sort of supposition that the observationalist might deny has full cognitive meaning. See note 35, above. Still, the observationalist might be able to claim that he has some notional understanding of what the unreduced theory says regarding viruses, gas molecules, etc. on the strength of something akin to semantic analogy. (See Lawrence Sklar, "Semantic Analogy," *Philosophical Studies* 38 (1980): 217–34.)

38. David Lewis has suggested to me a much less cumbersome way of putting this point for any irrealist who allows the unreduced theory to take on truth-value. Call the observational theory O and the unreduced theory T. T implies O, and so is equivalent to O & T. Now the observational theory is equivalent to O & $(Tv$ -$T)$, or to $(O$ & $T)$ v $(O$ & -$T)$. But then we can see at once that, given evidence that does not rule decisively against -T, O must be better confirmed than T, since O picks up not only the confirmation of O & T, but also the confirmation of O & -T.

39. At one point, when discussing scientific grounds for preferring one of two empiricially equivalent theories over the other, Glymour claims that this preference "need not be founded on *a priori* conceptions about how the world is or probably is" (p. 353). His alternative explanation is that we prefer "better tested theories," in his sense. Yet he concedes in the concluding paragraph of *Theory and Evidence* that he has given no argument for expecting theories better tested in *his* sense to be more likely to be true (p. 377).

40. Richard Boyd follows a related approach in "Realism, Underdetermination, and a Causal Theory of Evidence," *Noûs* 7 (1973): 1–12, especially pp. 3ff.

41. It should be clear that the realist cannot claim that this desideratum is a *constraint* upon which theories might be true—for the observational success of science might after all be coincidental. Rather, it is a condition which, if satisfied by a theory, would contribute to the likelihood of that theory relative to evidence that includes observational success.

42. It may not be easy to do this, but it should be no harder than (indeed, it is likely to be a part of) giving an effective specification of the notion of potential observability.

43. Just as, for the sake of argument, we extended considerable charity to irrealists in formulating their account of the observational theory—see, e.g., notes 15 and 20—, so will we extend charity to manifestationalists in formulating *their* account of the manifest theory. Much more work is needed before it can be judged whether either theory can satisfactorily be formulated.

44. Just as the observational theory does not say *everything* about the course of experience that the unreduced theory says (since the unreduced theory says, for example, that certain observations indicate the presence of unobservable entities), so the manifest theory does not say *everything* about the course of experience the observation theory says (since the observational theory says, for exam-

ple, that certain observations are instances of generalizations extending to the unobserved). This partly shows that one must be more careful of one's wording than I have been.

45. Cf. van Fraassen, "*Science aims to give theories which are empirically adequate* . . . I must emphasize that this refers to *all* the phenomena; these are not exhausted by those actually observed, nor even by those observed at some time, whether past, present, or future." (*The Scientific Image*, p. 12) He goes on to give a rough characterization of 'observable' as "X is observable if there are circumstances which are such that, if X is present to us under those circumstances, then we observe it." (*Ibid.*, p. 16) Despite his use of the indicative mood, this plainly is a modal rather than a manifest notion.

46. B. C. van Fraassen, "Empiricism in the Philosophy of Science," in P. M. Churchland and C. A. Hooker, *Images of Science: Essays on Realism and Empiricism* (Chicago: The University of Chicago Press, 1985), pp. 254–55.

47. Note that this extension of the irrealist's argument does not make the question-begging assumption that earlier scientific theories were approximately correct in their claims about unobservables, only that they were (in effect) interpreted in scientific practice as making claims about unobservables, and that this feature of practice played a significant role in subsequent theory testing and development.

48. *The Scientific Image*, pp. 81, 12.

49. *The Scientific Image*, p. 82.

50. *The Scientific Image*, p. 82.

51. It is no simple matter to characterize contemporary scientific beliefs about unobservables. It suffices for the realist's argument if there are some areas of current science where ontological brackets have been removed and literalism plays a crucial role in theory testing and development. Few biologists would, even in reflective moments, bracket the existence of cells, and few chemists would bracket molecules. Indeed, many in these disciplines would say that we have successfully observed both. But few physicists would say they are entirely sure what removing the brackets from quantum mechanics would involve. What is striking, however, and what the realist's argument turns on, is a pattern of experimentation and theory development extending across various disciplines, a pattern that would make sense under literalist assumptions, but appear irrational under the assumption of genuine agnosticism. Even in the face of quantum mysteries, it is commoner to hear physicists express doubts about *what* the psi-function describes than *whether* there is some hard-to-conceive but nonetheless real unobservable state of affairs represented by it, and it is commoner still to hear physical hypotheses and experimental designs defended on grounds that involve reference to mechanisms and entities, literally understood. Literalism tends to be pursued whenever there is a robust enough sense of what could be going on in terms deemed "physically realistic."

52. "Empiricism in the Philosophy of Science," p. 286.

53. I am grateful to David Lewis for help in improving this image.

Explanation: In Search of the Rationale

The interrogative construal of explanations is one of the earliest analytical devices in the philosophy of science. The general idea is still with us, in part because Hempel and Oppenheim endorsed it in their classic formal account. But although their work starts with the claim that explaining phenomena amounts to answering why-questions, the interrogative perspective did not find its way to either the analysandum or the analysans. And it is fair to say that despite intensive discussion there still is no consensus over the scope and credentials of the perspective. One conceivable reason for this lack of progress is that we haven't had, until recently, a well-developed erotetic logic to deal with questions and answers. Another consideration has kept the doors closed to influences from another direction. Taking the idea of questions and answers literally invites us to look at explanations as anthropomorphic goods, produced and consumed by real-life people. But this brings metatheory to the vicinity of pragmatics, an area which was simply off limits for Hempel and Oppenheim.

The story of pragmatics is a Cinderella story. For quite some time she has had access to the court. However, there still are a number of advisors in the court who keep whispering that the shoe didn't quite fit to begin with. Apart from the objection that pragmatics brings in an undesired element of subjectivity there is the complaint that it has failed to create sufficient order in some notoriously messy corners in the court. I shall argue, in a largely informal fashion, that Cinderella manages her daily routines with grace. The rather loose interrogative idea ramifies into distinct insights, from the logic of questions to the logic and pragmatics of conversation and inquiry at large. Roughly, I shall try to establish that where the former fails to bring philosophical illumination, the latter provides am-

This paper was written while I enjoyed a Fulbright Research Grant in Boston Center for the Philosophy and History of Science. I want to thank the Finland-U.S. Educational Exchange Commission for the financial support and Boston University for its hospitality during my stay. Some passages of this paper are based on my article "On the Logic of Why-questions," printed in *PSA 1984*, Vol. 1, pp. 168–76. (East Lansing, Michigan: The Philosophy of Science Association).

ple consolation. Let me just briefly outline why I think a pragmatic interrogative perspective is needed to restore the lost rationale of scientific explanation.

First, scientific inquiry is goal-directed and rational activity which aims at the acquisition of certain cognitive objectives or goals, such as information, truth, and explanatory power. The interrogative model looks into the conditions under which an answer, intended to promote such objectives, in fact does so. I may fail to convince the philosophical advisors who think that these conditions have little to contribute toward explicating specifically scientific explanation. But it is worth a try, for a number of general reasons. To start with, it is not clear that there are sweeping differences between good everyday explanations and good scientific ones, although there are between silly everyday explanations and good scientific ones. But equally important, airing questions, answers, and explanations is a primordial mode of action, and we should not think that a philosophical account which draws on this insight is any worse off just for that.

Second, to be successful in promoting cognitive goals questions and answers must be to the point. My claim is that the syntactic and semantic mold in which most previous accounts of explanation have been cast creates difficulties which can be overcome, in a quite natural way, by referring to some pragmatic features of questions and answers. For instance, modeling explanations through questions and answers brings in a needed dynamic aspect to explanations. Another such obstacle has to do with the identity of questions: there usually is more to a question than meets the ear, and for an answer to be relevant it must respect this contextual residue. The point to emphasize here is not that syntactic-cum-semantic explicates won't do — that has been clear for a while — but, rather, that to the extent they do they do because they manage to capture pragmatic restrictions by syntactic and semantic means.

Third, purely formal models are incapable of accounting for the variety of explanation-seeking questions, and they remain silent on how questions arise. It might be thought that this opens an opportunity for the logic of questions which insists that the first question always is if the question in fact arises. I shall argue (in the next section) that the logic of questions is too impoverished to be of much assistance, but sketch (in sections 4–6) a pragmatic account of inquiry which is rich enough to give some compensation. Coupling a weak logic with a rich theory of inquiry brings one virtue of Cinderella to full light: questions provide a passage through which salient parts of our more or less amorphous body of background knowledge surface to our consciousness, to constitute objects of wonder and bewilderment.

1. Why-Questions

Let me turn to the first obstacle in the way of a more rigorous interrogative model of explanation. Now that we do have a logic of questions we could perhaps use it to throw light on the notion of explanation. To explore the proposal we need some terminology and notation. To begin with we distinguish between propositional questions (such as yes-no-questions) which take complete propositions (or sets of propositions) as answers and wh-questions (what-, where-, who-questions) which receive singular terms as answers.

To proceed via examples, let us represent the logical forms of the wh-questions "What is the color of copper?" and "Who authorized the arms deal?" by (1) and (2), respectively,

(1) (?x) (x is the color of copper),

(2) (?x) (x authorized the arms deal),

and call the quantifier expressions in them *interrogative operators* and the open sentences following them the *matrices* of the questions. Now there is a simple relationship between the logical forms of the questions and their presuppositions, i.e., the existential generalizations

(3) (Ex)(x is the color of copper),

(4) (Ex)(x authorized the arms deal),

for questions (1) and (2) have (direct) answers only if (3) and (4) are true. Finally, there is an easy way to tell what a direct answer to a wh-question is: it is simply a substitution instance of the matrix of the question.

Wh-questions have the nice feature that they allow us to use epistemic logic to study conditions under which an answer is conclusive for an inquirer H. An answer is conclusive for H if H is, after hearing it, in a position to say, truly, "I know what (who, where, etc.) x is." A correct direct answer does not automatically fulfill this condition. For instance a (unique) definite description, say "The head of the National Security Council (NSC)," to H's question (2) is not, if H does not know who the head is (or what NSC is). The model provides a tool for the study of answerhood because there is a simple condition under which a direct answer satisfies the questioner. It does so, namely, when and only when the answer together with the inquirer's background knowledge suffices to entail the state of affairs he desires, described by "I know who authorized the arms deal." Clearly, an answer is conclusive only if it does not give rise to such further questions as "But who is the head?" or "But what is NSC?"

There are alternative ways of representing questions, answers and presuppositions, but they all suffice to highlight a difficulty for our project. The idea works where the questioner knows, when putting the question, what counts as an answer. This requirement can be understood in a stronger and a weaker sense. The stronger amounts to an ability to enumerate all possible substitution instances, as in yes-no- (and which-) questions: here the answerer's task boils down to the choice of one of the displayed alternatives. The weaker sense is satisfied when the questioner knows precisely what *types* of substitution instances would count as direct answers, and is (usually) fulfilled by wh-questions. An answer to a what color-question must specify a color, an answer to a who-question a person, and an answer to a where-question a place to count as an answer of the right type. There may of course be other contextual demands which further narrow down possible substitution instances, but the general observation holds: mere knowledge of language and its categories suffices to guarantee that both the questioner and the answerer know precisely what would count as an answer.[1]

Unfortunately this is not true of why-questions or, more exactly, of the contextually open (if not ill-defined) why-questions which cry for explanatory answers. Granted that explanations are answers to why-questions, there is no logic for why-questions to match in precision the logic of wh-questions. To appreciate the difference, let us try to apply the idea to why-questions, such as the much discussed "Why did the radiator of my car crack?" There is no problem in casting the question into the canonical notation. We can represent the question which arises for H by "(?p)E(p)," where the interrogative operator "(?p)" ranges over causes or reasons and E is the presupposition of the question, i.e., "The radiator of my car cracked." The question, then, has the logical form:

(5) (?p) (The radiator of my car cracked because p).

Direct answers to the question are substitution instances of its matrix "E(p)," and they have the form "The radiator of my car cracked because p," or its truncated form "Because p." Now if I am knowledgeable enough I am able to think of a number of singular facts which, together with some known general laws governing temperatures and volumes, would be sufficient to entail the presupposition of (5). Then the availability of a potential explanation for the event would amount to the fact that I am in a position to enumerate a set of potential causes $p_1, p_2, \ldots, p_n$, any one of which would be a filling of the right type. But this would mean that I would have been able to turn a why-question into a which-question, and the burden of the answer would have been reduced to singling out one alternative.

But although some why-questions have this form, the most interesting and troublesome ones do not. Search for such an explanation is not search for evidence to determine which one of a set of well-defined potential answers is the actual one. In the interesting cases the interrogative operator ("Why?") ranges over causes or, more generally, reasons, the identity of which is contextual in the extreme. Whether (5) in fact is a covert which-question cannot be read from the sentence alone (it depends on the background knowledge of the inquirer). When it came to delineating answers to a scientific question like "Why do the equinoxes precess?," raised by the ancients, the strong knowledgeability condition was not fulfilled. And to the extent the weak one was, it was not because the interrogative operator had a type tag glued to it. It hardly requires pointing out that when we come to puzzles in quantum mechanics things get out of hand: we encounter questions in which we are completely at a loss to infer, from the interrogative operator alone, what would count as an answer.

I am not suggesting that all why-questions are open-ended, only that some are. And although I shall qualify the account later by suggesting possible category restrictions on why-questions, the value of a logic of questions to any concrete type of inquiry is in proportion to the severity of the constraints on possible answers in that inquiry. And it is noteworthy also that the following anomalies associate with some why-questions. First, it would be tempting to model their presuppositions along those of wh-questions, so that the presupposition of (5) would be

(6) (Ep) (The radiator of my car cracked because p).

i.e., the existential claim that there is a reason or cause for my car radiator's cracking. The reason why erotetic logicians have tilted toward such a rendering of why-questions is that they generally hold that a question has an answer only if its presupposition is true. However, although I am inclined to think that (6) is true (it certainly would be bad news if it wasn't), I am less prone to think that (7) is the presupposition of (8), that is, that there is a direct answer to (8) only if (7) is true:

(7) (Ep)(This uranium atom split because p).

(8) (?p)(This uranium atom split because p).

Now it might be thought that this result merely indicates that (7) does not really arise, and that it only allows corrective answers such as "Because it just did." Be this as it may, it is clear that the choice hinges on knowledge of physics (and maybe metaphysics), not on knowledge of language and the logical form of why-questions.[2]

Second, the following elementary result for wh-questions fails for some why-questions. One attractive view of knowledge has it that someone knows who a person is only if he can tell that person apart from others (in the jargon of this view, if he can identify this individual in the various possible worlds not ruled out by the rest of his knowledge). But I might be puzzling over a why-question and know for sure that I fail to be in an analogous position. If the presupposition of the why-question (that is, the syntactically chosen counterpart of the question-sentence) flatly contradicts what I believe (or believed!), I know that some of my background knowledge is false. It is inconceivable that I could be so baffled over a who-question.

2. A Thin Logic of Questions

It appears, then, that we are at an impasse: maybe the reason why the interrogative model of explanation has had difficulties getting airborne is that there is no logic of why-questions. Here I want to retract a bit, and ask what happens if we take the pragmatic and intentional idiom seriously. We have an explainee H who addresses a question, and an explainer S, also a member of the scientific community, who intends, by uttering a linguistic expression u, to get the explainee H to understand why E.

For the general idea to work we need to assume that members of the scientific community have a common language in which questions and answers can be phrased, common contextual background beliefs and interests which delineate admissible answers and provide clues for interpretation, as well as common conventions which, if not enable, at least make it easier for speakers and hearers to identify one another's intentions. We could, more specifically, resort to Grice's Cooperation Principle and to the associated more specific maxims, such as the maxims of relevance and truthfulness. All these assumptions can be easily justified, for we have already agreed that scientific inquiry is a cooperative and rational search for such epistemic goods as truth, information content, and explanatory power. So with these pragmatic elements at hand we arrive at a new five-placed analysandum "S explained to H why E by uttering u in a problem context P," instead of Hempel's two-placed "{T, C} is an explanans for E."[3] But to counter the philosophical worries briefly mentioned we must show that these conditions bring in something new. Do they?

The role of u, as intended by S, is to cause in H an epistemic change vis-à-vis the explanandum E. When successful it works along time-honored communicative strategies: S relies on H's knowledge of the meaning of u (if it has conventional meaning) and on whatever means there are for getting H to recognize that u is intended as an answer to H's question. But there are further conditions for the truth of the pragmatic locution which give flesh to the interrogative proposal. First, we saw that there always is more to a question than meets the ear, and for u to bring about the desired epistemic change it must represent a right type of answer to "$(?p)E(p)$." Second, u must bring in something H did not know already. Third, an answerer S who cooperates commits himself to the principle that u is to the point and does not burden H with information which has nothing to do with the topic. Fourth, the answer expressed by u must be factually relevant: E must be true *because* the answer is. Finally, u must have the intended effect, that is, change H's epistemic state.[4] The five types of condition under which u brings about an improvement in H's epistemic state vis-à-vis E formulate distinct requirements of relevance. In the sequel I shall adopt a notational shortcut and talk, instead of the answer expressed by u, simply of the answer u.

More needs to be said about all of these, but I shall begin with the requirements that u must be news to H, and that the news must pertain to the question at hand, i.e., to why E. Let us start with the commonplace that explanation has to do with the improvement of knowledge. Some H may perceive E as problematic because it, say, contradicts (or is improbable relative to) H's background knowledge. This means that there is, first, a (more or less) harmonious *initial epistemic state K* in which H is peacefully unaware of forthcoming epistemic trouble. Second, there is the intermediate *problem state K_E* which characteristically results when the observation that E throws the initial state into disequilibrium. Believing is a cognitive and not conative attitude, so that H is forced to admit that E and forced to admit that he doesn't know why E. Finally there is the happy *end state K_u* in which harmony is restored (at least locally) by bringing in the answer u: H then knows both that E and why E. The three epistemic states reveal in fact a dialogue pattern: the problem state gives rise to H's es-question "(?p)E(p)" and the answer is intended to cause in H the end state K_u.

Now to some specifics. It has proved to be dishearteningly difficult to design a satisfactory formal model of deductive explanation within the static confines of Hempel and Oppenheim's original proposal (hence called H-O). My suggestion is that one reason for this difficulty is that the formal models purport to capture some pragmatic, dynamic and contextual features of questions and answers *by purely syntactic and semantic means*.

Here are some examples. According to H-O a pair {T, C} is an explanans for a (singular) E if and only if (1), T is a theory, (2), C is a true singular sentence, (3), {T, C} $\vdash$ E, and (4), there is a class K of basic sentences such that K is compatible with T, K $\vdash$ C, but not K $\vdash$ E.[5] Here condition (4) is a syntactic restriction which rules out arguments in which the singular premise C entails the explanandum E. Complete self-explanations in which C and E are identical are good examples. But what is wrong with them? They are answers in which an inquirer H is offered a piece of information which he already knows.

Consider next what Hempel and Oppenheim called self-evidencing explanations, such as

(9)   T: (x)Mx
       C: Ma $\supset$ Pa
       _____

       E: Pa

According to Hempel and Oppenheim (9) is unacceptable because, given that T is true, the only way of verifying the singular premise C goes via the verification of the explanandum E. But why the requirement of independent verification? Look at the matter from a dynamic point of view. In the problem state H knows that E but not why. His puzzlement needs for alleviation an answer which gives grounds to believe (expect) that E, but not all answers which, together with H's prior knowledge, entail E will do. No answer which is derivable from what H already knows in K_E can improve his epistemic state with respect to E. The requirement of independent verification points to an important fact: for the explanans to be conclusive for H, it must be such that, when added to H's initial state, it would give independent grounds to expect that E.

One intuition behind H-O was that all scientific explanations involve laws, hence the requirement that E be derivable from {T, C} but not from C alone. But what is the rationale behind this? Let us take another look at (9). Robert Ackerman has used it to demonstrate what he calls the *trivialization principle*: an explanation is unacceptable if the explanans can be used to explain anything.[6] And clearly if (9) is all right, then we could replace Pa by anything (excepting Ma and −Ma) and get an explanation. Now there is an obvious interrogative intuition behind shunning such versatility, for the law in (9) completely fails to address the specific question 'Why E?' Laws, then, are needed to secure what was called above the factual relevance condition, i.e., that E is true *because* the answer is.

Another example is one of the famous counterexamples to H-O produced by Eberle, Kaplan, and Montague.[7] Assuming that we have a true fundamental law (x)Fx and a true singular sentence E, we can derive from (x)Fx another fundamental law T which "explains" E, even if the original law and E have no predicates in common:

(10)   T: $(x)(y)(Fx \lor (Gy \supset Hy))$
        C: $(Fb \lor \sim Ga) \supset Ha$
        ———————————————
        E: Ha

Now (10) satisfies the conditions of H-O but is nevertheless unacceptable as explanation. One way to block it was proposed by Jaegwon Kim: add to H-O the requirement that no conjunct in the conjunctive normal form of C be entailed by E, and (10) becomes unacceptable. Hempel welcomed the device but added: it would be desirable to have a non-ad hoc justification for it, one which is not based on the brute fact that the requirement blocks the derivation but finds support in "the rationale of scientific explanation."[8]

The rationale is to improve H's knowledge vis-à-vis the explanandum. Just recall the basic dynamic feature of the question-answer model. Three epistemic states are involved. In the problem state H already knows that E, and therefore knows everything E entails. In the example at hand E entails all of the conjuncts of the conjunctive normal form of C, so that C cannot bring news to H. More specifically, it cannot bring anything relevant about E, viz., independent grounds to expect that it is true. What Kim's syntactic requirement does is bring in the intermediate problem state into the assessment of an argument.

That there is a simple interrogative intuition behind the syntactic requirement can also be seen from rival proposals. D. A. Thorpe has given one in terms of MECs. A MEC (Minimum Evidence Class) for a singular sentence S is the smallest class of basic sentences which suffices to verify S. All we need to require, then, is that no T-consistent MEC for C can verify E. So what is the rationale? The MEC-requirement is in accordance with the scientific method, writes Thorpe, because "we would want the theory to mediate or else we would have no D-N explanation at all."[9] But of course a theory mediates between C and E when C neither entails nor is entailed by E. The interrogative intuition gives the rationale for both: if C entails E, the law is idle and the inquirer H has no reason to believe that E is true *because* C is. Where E entails C the answer C or any part thereof fails to enlighten H in any way at all, because it is already known in K_E.

The requirement that T should mediate between C and E points to one way in which an answer should be relevant. Another one derives from the answerer S's commitment not to flood H with information which does not bear on E (and hence, from the implicature that E is true because the answer is). Consider Henry Kyburg's counterexample to H-O: The sentence E ("This piece of salt dissolved in water") can be "explained" by referring to the general law T ("All hexed samples of salt dissolve in water") and to the singular sentences C_1 ("This sample of salt was placed in water") and C_2 ("This sample is hexed").[10] But even assuming that the sample indeed was hexed (so that the explanans becomes true), $\{C_1, C_2\}$ and C_2 both fail as answers to the question "Why E?" Although both are true, neither is a true answer to the question: it would be false to claim that E is true because the answer is. Clearly both answers flout the general conversational requirement of pertaining to the point.

An analogous requirement rules out redundant information in the explanandum.[11] But these examples should suffice to show that the insight in H-O—that explanations are answers to why-questions—is not buried in any interesting logic of why-questions, but rather in the admittedly thin conversational logic of question-answer sequences: to the extent the syntactic constraints on arguments work they work because they capture unabashedly pragmatic features of discourse.

3. The Epistemic Conception of Explanation

I have all along taken it for granted that the capacity of an answer u to enlighten H resides in its capacity to make the event described by E nomically expected. It is noteworthy that Hempel and Oppenheim baked this rationale into their reading of (singular) why-questions from the outset. The assumption defines what Salmon has called epistemic and Stegmüller informational concept of explanation (in two books dedicated to Hempel).[12] There are a number of considerations which show that the assumption generates trouble, but no consensus over the remedy. As a modest step toward such remedy I shall generate some more trouble, this time from a dynamic epistemic model which steers around problems of relevance. The hope is that this point of view gives us easier access to the root difficulties.

The model is Peter Gardenfors's knowledge situations model which has it that the purely logical tools employed in H-O (and most other formal models) are not rich enough for an adequate explication of explanation.[13] Instead, Gardenfors insists that singular explanations, deductive and statistical alike, are evaluated with respect to the knowledge situations (or epistemic states) of a rational agent, and, more generally, that determining whether something is an explanation depends on the pragmatic context. Now a knowledge situation is a quintuple of the form

(11) $K = <U, W, \{I_w\}_{w \in W}, \{P_w\}_{w \in W}, B>$.

where U is a domain of individuals, W a set of possible worlds (compatible with what H knows), $\{I_w\}_{w \in W}$ an interpretation function which assigns each individual constant of the language L an element from U and each predicate of L a set of individuals. $\{P_w\}_{w \in W}$ is a probability measure for the subsets of individuals in W. It assigns each singular sentence of L its value in every world $w \in W$. B is H's *belief function* which assigns for each subset V of W H's degree of belief that the actual world is an element of V. Now H is characteristically ignorant of the objective probabilities of different individuals' having various properties. To obtain the probability with which H expects an individual a to have the property G Gardenfors resorts to mixed probabilities: the values of the measure P_w are multiplied by H's subjective probability that w is the actual world. As a result we get the following formula for the expected probability of an individual a to be G (for a finite V):

(12) $P_V(G) = \dfrac{1}{B(V)_{w \in V}} P_w(I_w(G)) \, B(\{w\})$.

Assume then that some singular event described by the sentence E is unexpected for H. This means that H's initial epistemic state assigns the sentence E which describes the event a low belief value. Assume now that H nevertheless observes that E. This observation causes (brings about) a new belief state K_E and gives rise to the why-question "(?p)E(p)." What kind of answer would be conclusive for H? Gardenfors does not require that the answer raises the belief value to the vicinity of 1, nor above .5, but he does insist that it raises it to some degree at any rate. But raises relative to what? H already knows in K_E that E is true, therefore its value is 1 and cannot be raised. Gardenfors opts for another candidate, the initial epistemic state. So an explanans is conclusive for H if and only if it raises the value of E relative to its value in the initial epistemic state?

The proposal gives a precise measure for the capacity of an answer u (or explanans $\{T, C\}$) to make the truth of a singular explanandum E expected. The degree of rational tension created by E is its belief value in the problem state minus its belief value in the initial epistemic state. The relief offered by a given explanans—its local explanatory power with respect to E—is measured as follows: first add, by making necessary subtractions and alterations, the answer to the initial epistemic state. Then determine the expected value of E in the end state and subtract from this value the value of E in the initial one. The result tells you how much relief you got.

The model also recommends a strategy for finding the best explanans: it is the one which, when added to the initial knowledge state, maximizes the belief value of E. It also bars partial and complete self-explanations, for adding to the initial epistemic state general (and singular) sentences which are already included in the intermediate one do not improve H's epistemic state vis-à-vis E. Making the relevance of an answer u dependent on the inquirer's knowledge states K and K_E also explains why the acceptability of the answer "Because he was a heavy smoker" to the question "Why did John get lung cancer?" varies from person to person. It is acceptable only for an explainee who does not already know that John was a heavy smoker (or that smoking causes lung cancer). Knowledge situations and the associated belief functions introduce a fine-grained criterion of identity for why-questions: there simply is no one-one mapping from the (surface) why-questions to the explananda, and an inquirer who knows that John was a heavy smoker expects to hear an answer which tells him why, among heavy smokers, just John got the disease.[14]

But the model is incomplete at best. First, assuming that making things expected, or at least raising belief value, is all there is to explanation, why should H in the intermediate epistemic state bother searching for an explanation, *for the belief value of the explanandum E in that state already is the maximum possible?* Put in slightly different terms, if the belief value of the explanandum E in K_E is one, and if increase in belief value earmarks growth of knowledge and understanding, what could possibly count as improvement? Moreover, in statistical explanation at least the belief value of E is *higher* in the problem state K_E than it is in the allegedly harmonious end state, for the simple reason that statistical explanations do not raise the belief value of the explanandum E to one. Therefore, if high or at least higher belief value is the goal, you should stick to the problem state in which you know that E but not why.

Second, there is an oddity in the epistemic account which raises related questions. In discussing—and dismissing—the view that explanations must reduce or link the puzzling event to something with which the questioner is already familiar, and that the familiar needs no explaining, Hempel cites a number of familiar facts, such as tides, lightnings, and even the fact that it is dark at night, which have been subjected to explanation.[15] The examples are puzzling because the familiar facts and the regularity are highly expected and have high belief values. They do not fit the epistemic model for the simple reason that whatever explanations are offered the belief values of the sentences "There are tides," "There are thunders" and "It is dark at night" are not inched up a bit. Again, improvement in our knowledge cannot be more expectedness.

One way out of this difficulty would be to deny that the explananda in the problem situations do have the maximum value one. The reason for this move is easy to appreciate. E is singled out as an explanandum because the inquirer H's background knowledge develops to the point when, suddenly, the fact that E becomes problematic. From a purely logical point of view one could think that this merely shows that E does not, after all, have belief value one. But this way out – Bradley's way because he blamed facts when they conflicted with his high principles – is not open to an H who wishes to stay sensitive to experience. And it *is* implausible to think that the degree of belief of the proposition "It is dark at night" is decreased just because there are theoretical considerations which show that it should be low. H is far more confident of the truth of this common observation than of the background theory which made it problematic.

Another possible way out seizes on H's epistemic split-mindedness and appeals to H's sense of epistemic integrity. What has happened is that H has been saddled with two belief functions which deliver different verdicts on a salient E, and a respectable way out is to move to a third state in which the discrepancy is removed. This way out is, I agree, the right one. The trouble is that it finds little support in the official epistemic doctrine which views explanations as arguments within an interpreted language: it is a static view, geared to maximizing belief value, and it does not tell why an inquirer H should keep an eye on past states.

The difficulties gain in intensity when we move away from H-O to the explanation of regularities and laws, to an area which was not the concern of Gardenfors's model.[16] There is a sense in which regularities might be unexpected – say, when they conflict with known laws. And one could imagine that surprise elimination could cover some theoretical explanations. But whether or not this would fit the intent of the model, there are why-questions which are more resistant to the treatment. First, there are well-entrenched brute regularities which are not surprising or unexpected in any sense but which still pop up as question marks. Second, it often happens in science that we are able to find similarities and formal analogies between phenomena or types of phenomena, such as the inverse square form of laws found in different but perhaps partly overlapping physical theories. Now there may be no uncertainty about the descriptive adequacy of the empirical laws involved – they have high belief values. But of course such formal analogies cry for explanations, in terms of common physical mechanisms, if possible. Whether or not such mechanisms can be found is of course a matter of empirical inquiry. In any case it is clear that we have here a why-question 'Why do the empirical laws have a similar form?' which does not arise from the low belief value or unexpectedness of the generalizations but from the entirely different desire to bring the two established regularities under one roof.

Nor is this the end of troubles for the epistemic model. It is not reasonable for an agent to worry about all isolated beliefs or sentences which happen to have low belief values, or over all changes of belief values toward the worse. It is essential to the mental health of an agent that he be able to live with such tensions, on pain of being paralyzed. For an E to become a genuine problem it is also necessary that E matters, i.e., that it is cognitively central enough. One more shortcoming in the epistemic model is that it provides no account at all why some events in the ocean of unexpected and unrelated events need explanation and others do not.[17]

4. Theory Nets and Explanatory Commitments

A need for an explanation arises when there is a particular why- (or how-, or sometimes what-) question but no satisfactory answer. We can also agree that a belief state K_E in which you intuitively speaking know that E but don't know why is a state in which something is found to be missing. But to make these common-places more than platitudes we need an account which is able to provide the inquirer H motivation for worrying not just over low belief values but also stranded beliefs and, more generally, over forms of disintegrity weaker than inconsistency. The knowledge states model of Gardenfors contains some of the right ingredients, for there are the three epistemic states K, K_E, and K_u. However, what we need is an additional note on epistemic integrity: how do these knowledge states hang together? What makes an inquirer in K_E responsible for his commitments in $K-$ and what are these commitments, anyway? The lead we shall follow is this: the crucial impetus must reside in the tension created by the claim of a theory (or more generally, a body of knowledge) to handle an area of experience, and its (temporary or chronic) incapability of substantiating this claim.

This is a handsome order, and we should not expect that anyone can deliver all ingredients at once. But in setting out to fill it we do not need to start from scratch, for there is a theory notion which is designed to meet some of these desiderata and which can be augmented to meet the rest. I have in mind the so-called structuralist theory-notion. I shall give an informal account of it, and confine my discussion to those features of the notion that are of importance to our immediate concerns.[18] A theory, or more precisely a *theory-element T* is an ordered pair $< K,I >$ in which the *core K* is a conceptual apparatus, and I a set of (classes of) *intended applications*. The core K is a quadruple $< M_{pp}, M_p, M, C >$ in which M_{pp} stands for the potential partial models of the theory, roughly, non-theoretically described objects or constellations of objects, and M for the laws of the theory. M_p represents those objects which have enough structure of the right sort so that it makes sense to ask whether they can be enriched with theoretical functions so as to satisfy the laws M of the theory. Finally, C is a set of constraints whose role in the core is to guarantee that functions which appear in distinct intended applications receive same values.

The range of intended applications I (a subset of M_{pp}) is in turned delineated intentionally through some paradigmatic exemplars. This means that there is no strict set of necessary and sufficient conditions for an object to fall into the domain of the theory. Rather, there are measures of similarity which require that any object or constellation of objects sufficiently similar to one of the exemplars is in the domain of responsibility of the theory. A theory-core can now be used to make empirical claims, viz., claims that some intended applications are models of T, that is, belong to M. The claim of the theory-element at large is the claim that the entire range of its intended applications are structures which can be supplied with theoretical functions in a way which makes them models of T.

Now theory-elements in science often form larger "maxi-theories" or *theory-nets N*, finite sequences of theory-elements connected with one another by various types of intertheory relations. A theory-net has a fundamental or basic theory-element $<K_o, I_o>$ and a number of specialized theory-elements $<K_i, I_i> \epsilon N$, introduced to make more specific claims about some more limited classes of applications ($I_i \subset I_o$). The fundamental core of the theory-net may then give rise to several branches of specializations, and the result may be a hierarchial tree-structure. To give an example, the fundamental core of classical particle mechanics comprised the three Newtonian laws, but these were immediately specialized to distance-depending forces and then, by adding the gravitation law, into a specialized theory-element for gravitation phenomena. Newton's theory was a branching structure in that the distance-depending theory-element was also specialized to nongravitational phenomena (such as oscillating systems), and there were still other branches.[19]

Reference to the specialized theory-elements discovered by a certain time already contains a pragmatic allusion, and it can be made more explicit still by relativizing theory-elements and nets to scientific communities and time-intervals: then we could pick out a set of nontheoretical structures, a subset of I, to which a scientific community SC intends to apply the core at a time (or during a period). Clearly, such intentions to tackle explanatory and other questions reflect not just pocketed achievements but also beliefs and hopes for future net-expansions. One way to codify a scientific community SC's aspirations is to think that it possesses not just a conceptual cutlery K and food for thought I but also other items which, though perhaps more difficult to access formally, are of almost equal importance. These include the paradigmatic exemplars I_p which generate the set of intended applications, as well as the analogies, explanatory ideals, and cognitive values of the scientific community. We can assume that the latter are bequeathed to members of SC through a paradigmatic theory net N_p which shows how the core K_o has been applied thus far, and where it ought to be expanded.

We have now ended up with what has been called a *Kuhn-theory* $<K_o, I_p, N_p, \mathcal{M}>$, where K_o is the fundamental core, I_p its set of paradigmatic exemplars, and N_p an associated paradigmatic theory-net. Now one trouble with any such prag-matized theory-notion is that there are, at any given time, several alternative (and incompatible) directions in which an extant theory-net could be expanded. Some constraints are therefore needed. I shall follow a previous suggestion by the struc-turalists and represent these contextual beliefs and hopes of SC by a restriction function which weeds out (from among the logically possible ones) expansions which are contrary to the spirit of the theory, and which gestures toward promising ones.

The pragmatized notion of a Kuhn-theory allows us to study the development of a theory in time. When a Kuhn-theory is proposed, only some of the intended applications, namely those in I_p, have been shown to be models of the theory. It is a task of later generations to refine and expand the theory-net to cover the remaining envisioned but so far unexamined or unsuccessfully examined applications. A *theory-evolution* represents such historical development: it is a finite sequence of $< K_o, I_p >$-based theory-nets N^1, N^2, . . . such that each N^{i+1} contains at least one theory-element obtained by specialization from an element in the historically preceding theory-net N^i.

It is no part of my argument that the set-theoretic setting in which the structuralist theory notion was cast provides a philosophically impeccable foundation for either the statics or dynamics of theory formation.[20] But representing a theory as a quadruple of a conceptual apparatus, a set of paradigmatic applications, a paradigmatic theory-net and a set of values, analogies and standards embodied in the net has a number of attractions. To begin with, it removes a difficulty with purely formal views which fail to address the problem of how questions arise. If we think that an inquirer H draws on a pragmatically enriched Kuhn-theory, anchored in a set of intended applications, it becomes obvious that H's questions (explananda) do not fall out of the sky: there is a "disciplinary matrix" which gives its holders title to assume that some questions, viz., those which arise from salient nontheoretical structures, are both sound (the presuppositions are true) and motivated while others are not.

Equally important, a Kuhn-theory displays the explanatory commitments of a scientific community, and therefore explains why its member, inquirer H in K_E, is responsible for his commitments in K. H's primary commitment lies in a Kuhn-theory, and since they have a more than ephemeral span they provide a sense of continuity over the specific claims which come and go—therefore H is not at liberty stop in K_E. Note specifically that such a Kuhn-theory could be instrumental in specifying items needed in the knowledge situations model, viz., the domain of individuals U, set of worlds W considered by H to be seriously possible, and belief function of H in a knowledge situation of the type (11). Clearly analogies and heuristic models furnished by a theory-net reflect on H's subjective degree of belief that the actual world is among a given subset of the possible ones.[21]

The pragmatic account also enables us to overcome other problems in the epistemic model. A Kuhn-theory, or, more precisely, a scientific community SC which subscribes to it, characteristically makes a very strong claim, viz., that all of the intended applications picked out by I_p are amenable to treatment by the conceptual cutlery K_o. However, at any given time only some of the intended applications have been confirmed, i.e., only part of the claim has been substantiated. This means that there is a constant tension between the claim of its holders and the goods delivered thus far. But it also allows us to recover the missing motivation for an attempt to find an answer for an explanandum sentence E, whether singular or general, which already has the maximum belief value 1 in the *initial* epistemic state. For any thus far unexamined or recalcitrant intended application is a challenge of the required form: holders of the theory have known all along that E but haven't known why. Since they are committed to a theory-net which specifically addresses the question by claiming that all intended applications are models, an answer u which fills the lacuna becomes a pressing desideratum.[22]

The proposal also does away with a related anomaly. The epistemic model does not find room for why-questions which arise as queries concerning unaccounted for similarities between two types of phenomena, or as queries over formal analogies between laws and regularities. It is easy to see how such questions arise within a Kuhn-theory, for there are existing special laws which provide analogies (and models) for future net-expansions. A good example of the grip of a formal analogy is provided by the inverse-square form of Newton's law of gravitation, for once it was established it gave rise to well-defined questions in other intended applications, such as electrostatistics. The architectonic beauty of the structuralist theory-notion is now easy to appreciate: already available theory-elements provide heuristic guidance for net-expansions which, if successfully carried out, would have explanatory appeal because the results would fit an already existing pattern. But finding such a law does not always end the inquiry, for there may be a further query about the physical foundations of this common pattern.[23]

Let me next show how this account fits our overall picture. We started with a difficulty with why-questions, viz., with the unavailability of a "logic" which could impose restrictions on the substitution instances of the matrices. And it *is* a living concern for any interrogative account of explanation or inquiry at large that questions have enough structure to rule out some logical possibilities, for otherwise the questioner or inquirer literally does not know what he is looking for. We can now see that there is no reason to despair, for the impoverished logic of why-questions has a rich supporter behind the scene. Although the matrix of a why-question, as it was defined, does not contain strict requirements for admissible fillings (as they do in wh-questions), a disciplinary matrix in the form of a Kuhn-theory gives ample encouragement and reward.

We have now joined the pragmatic idea of questions and answers with a pragmatic account of how questions surface from our more or less amorphous background knowledge. But to the extent a Kuhn-theory provides lacunae of various sizes and shapes we can see that this background knowledge is not all that amorphous, in the mature sciences at least. Normal scientific puzzle-solving presupposes well-defined questions—and that is precisely what a Kuhn-theory gives. Moreover, there is a qualitative change in the types of questions, for the rich theory notion chops the unmanageable why-questions into more manageable wh-questions and even yes-no questions. A query which starts as a loose why-question concerning an area of experience turns (in the "paradigmatic" phase) into a what question: *what* special laws are needed to govern this particular application? It is a further task to find yes-no- and which-questions that portray specific alternatives and still narrow down admissible substitution instances. There is of course no logic for generating questions – but there is a contextual and nontrivial heuristics.

Notice, next, that the pragmatized theory-notion outlines an account of cognitive centrality which goes some way toward alleviating two related troubles. We noted that not all beliefs with low belief values motivate a search for explanantia. On the other hand only a portion of our background knowledge is relevant for assessing answers, once a salient question is fixed. Van Fraassen has raised the latter problem: assuming that we do have a specific question (with a built-in relevance relation and a contrast class which single out requirements for an answer), there still is the problem of determining the portion of background knowledge which is taken to be relevant in the assessment of goodness. "The evaluation," he writes,"uses only that part of the background information which constitues the general theory about these phenomena, plus other 'auxiliary' facts which are known but which do not imply the fact to be explained." But, he continues, neither he nor others have had much to say about how the portion is delineated. And he concludes that this must be a further contextual factor. Now of course a Kuhn-theory not only singles out questions but also provides a first delineation of background information relevant for answers: clearly all laws and singular facts which belong to the branch of the theory-tree which gave rise to the question are relevant. Similarly, special laws from neighboring branches count as relevant, at least to the extent that they furnish analogies.[24]

5. Pruning the Web of Belief

We now have an interrogative account which makes an inquirer in K_E responsible for his commitments in K. But there is another surprise waiting for us. The *rationale* of an answer u (or a scientific explanans {T, C}), to the extent there is one such thing, is not merely or even mainly to raise the belief value of E, but to synchronize the inquirer's belief states K_E and K. This result is in perfect harmony with the view that explanation is not a local matter in which an explanans conveys some privileged epistemic status, such as familiarity, naturalness, or whatever, to an explanandum. Rather, it is a global matter of reducing the total number of (types of) phenomena that must be accepted as ultimate.[25]

In this section I shall draw attention to one further attraction of Cinderella: embedding the interrogative proposal into the pragmatized theory-notion throws light on explanatory unification. But before seeing how, note two features an analysans should have. First, it should cherish the distinction between singular facts and kinds of facts. Any genuinely lawlike sentence generalizes over an infinite domain. But as examples gleaned from the history of science show, the sheer number of facts explained is immaterial if not outright ill-defined—what counts is the variety and the number of independently established laws.[26]

Second, the thin logic required that for an answer to be conclusive for H it must contain enough information, but not more, to entail, together with H's background knowledge, the desideratum of H's questions "I know why E." This means that when H's background knowledge expands the answers become shorter. Moreover, if the answerer faces an array of questions, a good strategy, when possible and financially feasible, is to invest in theory. It just might turn out that a number of answers draw on the same concepts and laws. Conversely, it adds to the appeal of a theory if it yields not just a great number of answers, but does so with the minimum of conceptual machinery.

The next step is to show that the proposal comes some way toward securing these features. To begin with, an inquirer H who subscribes to a Kuhn-theory literally inherits a range of intended applications. But the members of this set are not particular structures but classes of applications (formally I_p is a subset of the power set of M_{pp}) and they fall into clearly distinguishable types. There is a clear sense in which Newton's theory brought about unification, for it showed that despite wide variety all instances in intended applications (gravitation phenomena, oscillations, etc.) satisfied the three fundamental laws. Or rather, since we need to distinguish between confirmed and hoped for claims, for each class of applications either there are special laws derived from the fundamental core or there is a premium for deriving them. One ingredient in the degree of unification of a theory-net (at a certain time or during a period) can therefore be assessed with the help of the number of distinct types of intended applications derived from a basic theory-element.[27]

The second desideratum follows from another built-in feature of the structured theory-notion, because, in it, an inquirer equipped with a Kuhn-theory can only avail himself of a limited number of concepts in an attempt to derive special laws (or in an attempt to obtain answers to other questions). The structuralists have stressed that the fundamental core K_o of a theory-net is a *tool* which is used in the derivation of special laws. Thus a Newtonian working on electromagnetic phenomena finds only some types of special laws admissible and initially plausible. A perspicuous way to put this is to say that an existing theory-net defines a number of asymmetric intertheory relations within its theory-elements, so that each theory-element *presupposes* a number of more fundamental theory-elements.

A recent generalization of intertheory-relations, the notion of a link, goes further still, by extending the prospect of unification beyond theory-nets.[28] On that construal the global structure of science exhibits systematic conceptual connections not just within the elements of a theory-net but also more generally between theories which, however liberally understood, are not offsprings of the same basic theory-core. Just to give the idea, the concept of a force appears both in classical particle mechanics and Lorentz's electrodynamics, and it must be assumed that at least where the two theories have joint domains the values of the force function are identical.[29] Where an intertheoretic link is asymmetric it can with justification be called a presupposition link. Whether or not the two theory-elements have a common core or conceptual structure it may turn out that the very identification of the set of non-theoretical structures picked out as intended applications of a theory-element T (say, thermodynamics) may require the use of another theory-element T' (fluid dynamics), but not vice versa.

To see how intertheoretic links between theory-elements help us to understand unification, consider Philip Kitcher's proposal.[30] Crucial for it is the concept of an explanatory store of arguments relative to a set of accepted sentences. Explanations, according to him, are arguments, and scientific theories bring about unification by providing, not unrelated individual arguments, but arguments which exhibit common patterns, imposed by the extra-logical terms available for explanatory purposes in individual arguments. While logic is only interested in the logical form of arguments, theories impose further restrictions. The general problem, then, is to specify the set of patterns which gives the most parsimonious way of generating accepted truths as conclusions.

Furthermore, Kitcher writes, apart from argument patterns there are also what he calls core patterns: some of the arguments which have accepted sentences as conclusions are based on the conclusions of other arguments. Thus, e.g., in Newtonian dynamics there is a core pattern which is used to compute equations of motion. This core pattern is supplemented by patterns of arguments (problem-reducing patterns) used for deriving further conclusions. It may then turn out that all patterns contain a unique core pattern. Clearly the degree of unification of a theory with a single core pattern is greater than the unifying power of a rival with several core patterns. And this must be reflected in the criteria for the best-unifying theory.

Now one consequence of the existence of presupposition links is that a question phrased in a theory T may presuppose prior answers by T', when T' is needed to establish some existential presuppositions of questions in T (say, classical particle mechanics requires for its application prior answers to wh-questions in kinematics). Furthermore, presupposition links have a methodological role in that answers provided by T derive argument patterns from T'. An example, not quite as rigorous as those in physical sciences, can be found in evolutionary theory. Suppose you have the question why island populations exhibit traits different from but clearly similar to those found on the mainland. The evolutionary explanations characteristically refer to the relative geographic isolation of the islands and to the new selective forces operative in them. But to get a satisfactory answer to the question one needs to supplement these empirical observations with a genetic account of gene distributions, for only such an account can give an answer to the crucial questions of why the advantageous phenotypes survived, and why the selective forces lead to a new species. This shows that answers to questions in systematics presuppose answers by population genetics. Furthermore, population genetics is *the* fundmental core of evolutionary theory, because precisely the same pattern of arguments occurs in paleontology, morphology, and other applications. Whether or not evolutionary studies form a theory-net analogous to those in physical sciences, the concepts and laws of population genetics are needed in all of them.[31]

The account still needs an addendum and a qualification. Both have to do with what it is for an answer to be better. Theory-nets provide a static report on the state of the art, but to explain explanatory progress in evolutionary theory we must resort to theory-evolution as it was defined above. Theory-evolutions can be progressive in three ways at least, and all exhibit increase in unification. First, the claim of a theory-net N is the claim that for all $< K_i, I_i > \epsilon N$, I_i can be turned into structures which satisfy the special laws of K_i. One line of progressive theory evolution, then, consists of the substantiation of such claims (and of the redemption of these hopes) by producing new special laws to fill gaps.

Second, there can be refinement in theory-element cores, that is, replacing of existing special laws by more stringent ones, while leaving the sets of intended applications and potential models untouched, as in the various types of theory reduction. Such replacements result in more stringent answers to particular questions, but there is an even more dramatic road to increased unification: finding a new, more restrictive theory-core close to the fundamental element would increase the coherence of the entire edifice, for it would create a new core pattern and hence have an effect on all answers presupposed by it. Third, there is no need to require that the set of intended applications remain fixed during an evolution. One more type of increase in unification results when new classes of applications are added, i.e., a theory-net N_{i+1} replaces a net N_i such that and that $I_i \subset I_{i+1}$, where the new classes were not previously thought to fall within the applications of the theory at all.[32]

The qualification we need has to do with a distinction between providing an answer and providing a good or conclusive—or better or more conclusive— answer. It would be too harsh to say that there can be no answers or even good answers to questions in, say, systematics without prior answers to questions in population genetics. But the more modest comparative claim does hold. A unified theory in which the former questions enjoy support from the latter provides better (more conclusive) answers to all questions, in that it does not leave (or leaves fewer) unanswered questions. It then appears that two outwardly unrelated answers within distinct applications can warm up each other—if they use a pattern derived from a common core.

This introduces a global or holistic element to explanation which needs some comments. First, to the extent theory-evolution redeems hopes in expansion the fundamental theory-core gains in confirmation. But this gain of course manifests in overall explanatory success, for clearly the fundamental theory-element can be used to provide answers as to why the derived special laws hold. There need not be one type of relationship for all theory-nets for an explanatory relationship to hold. The intertheory-relation which ties two (or more) elements T' and $T(\epsilon$ N) together can range from reduction (in which, say, the intended applications I' of a macrotheory are correlated with those of a microtheory T, and where whatever T' claims of I' is entailed by what T claims of I) to weaker interpretative relations. And the notion of a link is more versatile still, because it can tie together theories with distinct conceptual structures. But however varied in form, links enhance global coherence, and consequently add to the appeal of the particular answers which emanate from the global structure.[33]

Furthermore, we have received some light on the distinction between theoretical and singular explanation. Although scientific questions often arise over particulars, the concern is not with this or that particular, say the whereabouts of Mercury. Rather, the trouble with the anomalous conduct of Mercury was that it was bad news for theory – it radiated epistemic pangs inward in the web of belief (i.e., upward in the theory-net). Similarly, exciting as the particular discoveries of the geographic distribution of finch populations in the Galapagos Islands were, the particular place, the time, or the species mattered little, except as a stage for good theoretical news. The conclusion to emphasize here is that the answers to the singular queries would have carried little force unless there were reasons to expect that the same pattern was applicable elsewhere – in answering questions about other geographical distributions and other kinds of phenomena.

6. Beyond the Third Dogma of Empiricism

The argument thus far has been that the syntactic-cum-semantic explicates divert our attention from the rationale of explanation. Two findings were instrumental in its recovery, viz., the thin logic of questions (or conversation) and the more bulky structured theory-notion, but both added pragmatic ingredients to the stew. During this search we discovered that unification in its various forms is a fundamental explanatory virtue, and that the rich theory-notion makes it possible for us to explicate that virtue. A number of intriguing questions, both systematic and historical, remain. First, why did it take so long to realize that explanation is a pragmatic affair, and why do we still feel uncomfortable about it? Second, the thin logic of questions, with the annotated constraints on relevance, and the requirement of increase in unification are, in a sense to be made explicit shortly, purely formal. But assuming that they operate in all contexts, are there any pragmatically varying further requirements?

Hempel and Oppenheim's model of scientific explanation grew out of the logical-empiricist research program which aimed at purging all armchair philosophizing from metascience (and science). This was to be achieved by rational reconstructions of notions such as scientific theory, empirical support, and explanation. The program had as its epistemological hard core the emphasis on experience, but equally constitutive were the values of objectivity and intersubjective testability, designed to rule out not just metaphysics but also all manner of psychologism and anthropomorphism. Logical positivism was not the first philosophy to attempt to set philosophy on the secure path of science, but it was the first to emphasize the systematic use of the conceptual tools of modern logic. If there is anything that captured the spirit of the new scientific way of thought in action, it was the idea that all explications made exclusive use of syntactic and semantic concepts. It is therefore appropriate to follow Stegmüller and label this idea the third dogma of empiricism.[34]

It is obvious from Hempel's early works that these concerns were behind his covering-law model. He wrote, in 1942, that there are two notions of understanding, psychological and theoretical.[35] The psychological (and anthropomorphic) sense is exhibited by a "feeling of empathic familiarity," and it is often conveyed by persuasive metaphors and models. Scientific understanding is brought about by knowledge of facts and regularities which make an event expected, or a regularity a matter of course. One way to capture this theoretical mode of understanding, and to capture it within the required level of conceptual hygiene, was to stick with the dogma and bar reference to persons, scientific communities, and other contextual features of inquiry and language use.

It is remarkable that when the three intended applications, the notions of theory, empirical support, and explanation, ran into trouble there were fingers which pointed to the third dogma. But now that we have gone pragmatic in public, have we thereby admitted, as Michael Scriven has, that the notion of explanation derives its meaning entirely from the subjective and psychologist notion of understanding disdained by the positivists?[36] The answer is an emphatic "No!," for we can, and should, deny the equation of the pragmatic with the psychological and subjective. Although the analysandum "S explained to H why E by uttering u in a problem context P" makes reference to an inquirer's successive epistemic states, the acceptability of an answer does not hinge on any idiosyncratic or anthropomorphic sense of intelligibility. The same is true of Gardenfors's refinement of the epistemic construal: although it incorporates subjective belief functions, no subjectivism or psychologism is involved, for *changes* in belief functions are subject to rigorous normative constraints.

How about the pragmatic elements which enter into the interrogative idea *via* its immersion in an explicitly pragmatized theory-notion? Giving and appraising explanations now becomes an explicitly social affair, because the explananda are circumscribed by a disciplinary matrix, subscribed to by a scientific community at given time. But this relativizing need not lead to relativism. Although the selection of questions may be up to the scientific community, picking out the answer need not be, in any damaging sense. There may be contextual requirements of type-appropriacy, along with shared analogies, values, and methodological standards, which make the evaluation of answers contextual and paradigm-dependent, but these restrictions on answers need not be taken at face value. There is always the possibility of denying that a question has a direct answer within the bounds of the matrix. It does not, of course, follow from this that there are unique choices to be made.

But there is an element in the enriched interrogative model which enables a more radical departure from the positivist program. The thin logic of questions and answers, as well as the demand for unification, are analogous to principles of rational decision-making in that they lay down constraints on successive epistemic states but remain purposefully silent on the content of individual states. This leaves open the possibility that there are distinct types of inquiry with distinct demands on modes of understanding and standards of intelligibility to supplement the universal but pale procedural canons. This becomes evident once we leave the area of natural science.

To appreciate this possibility we need to go beyond science, narrowly understood. As an introduction, let us take a brief look at the rise and fall of explanatory monism, starting with Aristotle, for he was not just the father of the interrogative model but also a pluralist with respect to questions.[37] Although he started with a description of the "conditions of natural change" he extended the theory of *aitiai* to mathematics, music, biology, and reasons for human action. Aristotle's theory of why-questions, or his theory of explanation, must be seen against his general theory of substance, and the implicit teleology of ancient Greek thought. Details aside, Aristotle thought that each substance had its natural inclinations, and that why-questions contained four clearly discernible, built-in type requirements which codify these inclinations and (types of) causes of aberration. By the advent of modern natural science during the early Renaissance there had been enough changes in the general *Weltanschauung* to make the implicit teleology lose all remaining appeal. As a result other than efficient causes became suspect and not even worthy of the title.

The new notion of explanation championed by David Hume and John Stuart Mill required that all explanations refer to regularities and antecedent events which together allow us to predict events. To cut a long story very short, Hempel and Oppenheim then canonized this view, and applied the new standards of metatheoretic success, the third dogma. Now the dogma not only requires that analyses be conducted within formal languages, it also requires that the formal languages be interpreted in an empiricist language common to all inquiries. The reason there was no need to refer to persons and scientific communities was simply that all theories and explanations were phrased in a language accessible to all, and once you had a complete description of a state of affairs in that language you had described all there was to an explanandum. But this means that there is essentially but one way to slice the world, and no partisan angles to explananda. Hence the idea that there was a one-one mapping between why-questions and explananda.

The knowledge situations-model of Gardenfors goes beyond this limitation because it individuates explananda, in part at least, in terms of fine-grained belief functions, and explanantia in terms of changes of belief functions. But there are some reasons to think that there is more to an explanandum than is built into knowledge situations in the sense of (11)—there is the possibility that why-questions come in distinct types and that there is no one-one mapping between problem situations and why-questions, either. This would amount to a partial withdrawal of the claim made earlier, viz., that there are no type requirements in why-questions.

If the historical detour above is on the right track, the best candidates for examples are to be found in areas in which the reductive attempt to eliminate all but efficient causes met with greatest resistance, in historical, teleological, and functional explanation, and especially in the explanation of action. I shall take action theory as an example, although parallel stories could be told for other types of explanation.[38] Suppose that you have the question "Why did S go to the market place after lunch?" which, given your knowledge of S's commitments to be in an important meeting at the same time calls for some explaining. A causalist would look for an explanation of the end result of the action, S's going to the market place after lunch, in terms of S's bodily behavior, and beliefs and wants which caused the behavior. S's desire to have fresh fish for dinner and his belief that unless he went to the market place right after lunch he could not have fresh fish for dinner, would do nicely. No two mental cause theories agree on all details, but all agree that if the explanation was correct, it was correct because the want and the belief caused the behavior in S's performing the action, in the ordinary (Humean) sense.

An intentionalist would agree with the causalist that the end result, the explanandum, is explained by wants and beliefs. However, he would disagree over the role of efficient causation in the explanation: on his view actions are understood by construing for them an intelligible background of wants, beliefs, and other determinants such as duties.[39] Two deep disagreements are crucial here. First, by the intentionalist's lights an adequate background story establishes a conceptual, not causal, connection between the explanandum action and its determinants, which means that there is no way to verify their independent existence. Second, the intentionalist insists that he is in the business of understanding intentional action and not bodily behavior. He readily admits that there is the distinct why-question (or perhaps how-question) about the emergence of S's bodily behavior, but when it comes to explaining action we are not interested in it.

The interest of the dispute is not in the outcome but in the initial task of singling out the explanandum. If the intentionalist is right, the causalist makes a category mistake when he identifies the explanandum. There is, in fact, no one-one mapping between the (surface) why-question "Why did S go to the market place after lunch" and explananda, because the explanandum may be either about S's action or the bodily behavior. In the former case the interrogative why-operator ranges over wants, beliefs, and like determinants as "final causes," in the latter they are construed as efficient ones. This gives us one reason to speak of the revival of Aristotle, for to the extent his and our conceptual schemes mesh, the intentionalist proposal aims at rehabilitating one of his four causes or types of explanatory factor.

There is of course no magic in the number four, nor in a romantic return to Aristotle. The strategy I have favored is to impose a minimum of initial structure on explanations and leave the door open for more specific contextual restrictions. This minimalism is not a counsel of despair, although it is motivated by the fate of explanatory models which have made a commitment to historically changing standards of intelligibility or modes of explanation. And the strategy finds further support in the rules for enlightened debates in which disputants agree to withdraw to a common ground.

This explains why I prefer the phrase factual relevance over causal relevance: the function of "because" in explanatory answers is to tie the explanandum with the explanans, and to indicate that the answer is intended to synchronize H's epistemic states. But this "because" may carry a different force in different types of inquiry. It would greatly unify our total world view if it carried the same force in action theory and science, but it would be unwise to judge the issue in the theory of explanation. And there are further types of becauses, in theoretical linguistics and, say, in questions involving legal liability.[40] However, in all these types of inquiry the thin conversational logic says what an anwer is, and the criterion of unification says what a good answer is. And there is a further feature: in action explanation, too, the explanation is no good if the explanandum is not true because the explanans, or answer, is true. Explanations, then, form a family of more specific notions which arise from the common ground of weak procedural rules and the requirement of factual relevance, i.e., that the explanandum E is true *because* the answer is. There is both variety and similarity provided by genetic kinship between members of the family: they all derive from the parent explicandum "S explained to H why E by uttering u in a problem situation P"

Notes

1. For the logic of questions and the conditions of answerhood, see Hintikka (1976), chapter 3. My indebtedness to Hintikka goes deeper than is outwardly visible in the paper, although the interrogative perspective here developed is not his. I also wish to thank Sylvain Bromberger for linguistic and other evidence (in his [1987] Publication and in personal communication) for the peculiarity of those why-questions which seem to require an explanation for an answer. Roughly, why-questions of that type do not have midsentence traces, either in surface structure or in deep structure. Now the

transparent nature of wh-questions, manifest in the mutual deducibility of the questions and what Bromberger calls their attributive presuppositions (presuppositions to the effect that the substitution instances of the matrix of the question are of a certain type), hinges on the existence of such traces. It follows that a Rational Ignoramus who knows that E but does not know why is unable to figure out, from the logical form of the question alone, what the attributive presupposition is, and, hence, what substitution instance counts as answer. This does not mean, of course, that a Rational Ignoramus does not know what counts as an answer, but it does mean that this knowledge is not of a linguistic nature.

2. See Bromberger (1987) for further discussion.

3. There are rival possible pragmatic analysanda, such as Peter Achinstein's (1983) "S explains q by uttering u," Stegmüller's (1983) three-placed "S explains a to H," and Tuomela's (1980) six-placed "S scientifically explains q to H by producing a linguistic token u in situation C, given P" where P stands for a paradigm or a constellation of group commitments. For these and some other alternatives, see Sintonen (1984a).

4. In the jargon of speech-act theory this means that "explaining" is a perlocutionary act. Achinstein (1983) recognizes both the illocutionary and the perlocutionary senses, but adopts the former. On his view the intended effect of understanding why E, which might or might not follow an illocutionary act, is not part of the act but something brought about by the act. There is no denying that there is a sense of "explain" which does not require that the explainee comes to understand why E. I have wished to adopt the perlocutionary sense because it nicely derives both senses from the same analysandum. Thus the achievement reading of "S explained to H why E by uttering u in a problem context P" might be false even when H correctly identifies the intended answer u, u is not known to H, u is of the right type, and u is to the point, if H cannot accept it (for instance because it is contrary to H's other well-entrenched beliefs). Notice that the locution could be false even if u does have the intended effect—if u in fact does not provide a correct anser. See Achinstein (1983), chapter 2, section 1.

5. Hempel and Oppenheim (1948), pp. 277–78.

6. Ackermann (1965), p. 162–63.

7. Eberle, Kaplan, and Montague (1961).

8. See Kim (1963) and Hempel (1965), p. 295.

9. Thorpe has another justification for the MEC-condition: should it happen that, when T is tested, E does not take place, we would rather blame the relatively uncertain T than the initial conditions. I do not see any general justification for this, however. See Thorpe (1974), p. 191. Gardenfors (1976), p. 430, has formulated a condition on explanations which also secures that T is used in explanation: If the pair (T, C) is an explanation of E, then (a) there are no predicates in E which don't occur in T and (b) there are no predicates in C which don't occur in T. Gardenfors leaves the status of this condition (his C5′) open, however.

10. Kyburg (1965). Kitcher (1981), p. 523, has another explanation for the unacceptability of the argument, namely, that accepting it would lead to an undesirable and avoidable proliferation of argument patterns.

11. For discussion, see Omer (1970), Thorpe (1974), Gardenfors (1976), and Sintonen (1984a).

12. Salmon (1984) and Stegmüller (1983). Hempel and Oppenheim (1948, 246) wrote that the question "Why does the phenomenon occur?" is construed as meaning "according to what general laws and what antecedent conditions does the phenomenon occur?" And it is clear from his writings elsewhere that the function of the law is to make things expected. Salmon appears to take the logical form of the covering-law models, deduction or induction, as constitutive of the epistemic notion. The interrogative proposal generalizes this point: a model of explanation in which answers are designed to more expectedness or higher belief value count as epistemic. I also think it is slightly misleading to make, as Salmon does, erotetic or interrogative models a subset of the epistemic ones, because the rival models, the modal and ontic models, can be given a natural erotetic explication. Indeed it does not seem to me that van Fraassen's (1980) is a purely epistemic one.

13. Gardenfors (1980). In presenting knowledge situations I have however, followed Stegmüller's (1983) exposition. I have briefly discussed the model in Sintonen (1984b), and wish to thank the anonymous referee of the paper for pointing out some difficulties in the epistemic construal.

14. Gardenfors attributes this fine-grained principle of individuation to Hansson (1974). However, Gardenfors only regards why-questions as linguistic indicators of the deeper knowledge situa-

tions. As will become evident in section 6 below, my account is closer to Hansson's original proposal in that it leaves room for type requirements which are irreducible to knowledge as it is spelled out in knowledge situations of the form (11). Van Fraassen (1980) makes this reading of why-questions explicit by baking the pragmatic contrast classes and relevance relations into the questions. However, on his view the pragmatic aspects are determined by openly anthropomorphic goals and interests, and this may well trivialize the problem. For further discussion, see section 6 below.

15. Hempel (1965), pp. 430–31.

16. It is arguable that explaining the familiar fact that it is dark at night already goes beyond explaining singular facts, for we can hardly ask "Why was it dark last night?" There is another worry about Hempel's examples: it is not always clear what the explananda are. Hempel has insisted that the relata in the explanation relation are sentences, and the explanandum phenomena characteristically events. But when we explain tides and lightnings, it is unclear if the explananda are sentences which record that there are tides, or particular occurrences of tides and lightnings, or sentences which describe regularities. I shall not, however, dwell on this.

17. This is only partly true of Gardenfors's model, for he gives an elaborate account of the information relevant for a question, when the question is a demand to make the explanandum less surprising. However, the model does not extend to explain regularities and unrelated events.

18. See Balzer and Sneed (1977) and (1978), Sneed (1976), Stegmüller (1979), and, for more recent developments, Balzer, Moulines, and Sneed (1986).

19. For a structuralist exposition of the Newtonian theory and its developemnt, see Moulines (1979).

20. Initially, the structuralists stressed their view as an alternative to the statement view of theories in which theories are construed as sets of sentences axiomatized in a formal language. It is arguable (and has been argued) that this introduces a false contrast, for there can (and must) be a way to translate set-theoretic talk to more traditional model-theoretic talk by assigning the elements in a theory-core a suitable language. Stegmüller makes a concession to this direction by observing that the difference between the two frameworks is one between a very cumbersome formal framework and an intuitively appealing and simple informal way. I do not wish to take a strong metatheoretical stand here — but it does appear to me that any adequate account should be rich enough to represent not only the structure of theories but also their dynamics and the various kinds of intertheory-relations needed in it. For discussion, see Niiniluoto (1980), Pearce and Rantala (1983), Sintonen (1984a), and Stegmüller (1979).

21. Wesley Salmon thinks that one of the most important uses of analogies is to provide prior probabilities needed, e.g., in an objective Bayesian approach. The account given here is but an attempt to show how analogies and models fit in a pragmatically enriched theory-notion. It is a further task to spell out the precise nature of the inductive logic or other elements required, and how they are used in, say, theory-choice. See Salmon (1984), p. 234.

22. The problem has earlier been treated in connection with the problem-solving model of inquiry. Thus for instance Hattiangadi (1979) has argued that all scientific problems have the form of an inconsistency, whereas Laudan (1977), Lugg (1979), Leplin (1980), and Nickles (1981) have tried to find room for a wider notion of a scientific problem. Both Laudan and Nickles have maintained that a mere logical compatibility of a phenomenon with a theory counts as a problem if there is a "premium" for solving it (Laudan) or if the phenomenon falls in the domain of the theory. The account sketched here translates problem-solving talk to the interrogative idiom and provides, I hope, a fuller account of what a "premium" to solving a problem amounts to. See also Sintonen (1985) for further discussion.

23. Why-questions (and how-questions) thus seem to possess both gosh value (the answers provide intellectual pleasure) and golly value (they generate new accessible questions) in Bromberger's sense. See Bromberger (1985), p. 312.

24. Van Fraassen (1980), p. 147. Gardenfors's (1980, p. 412) knowledge-situations model in fact gives an elegant fine-grained account of the knowledge relevant for a (singular) sentence "Qa," for its belief value is determined by the intersection R of all classes R_i such that "R_ia" is true. The belief value B(Qa) is simply P(Q/R) where P is the expected probability function in the knowledge situation. The virture of the more coarse-grained account in terms of the global structure of science is that it accords well with the intuition that the delineation of background knowledge is clearly structured and hierarchically ordered.

25. Friedman (1974) and Hempel (1965), p. 345, where Hempel says that theoretical explanation deepens our understanding by showing that regularities exhibited by a variety of phenomena are "manifestations of a few basic laws."

26. Thagard (1978) shows that it is far from clear that Darwin's theory beat its competitors in deriving more facts, but it was superior in its ability to use a few principles to cover a number of classes of facts, in systematics, embryology, paleontology, morphology, and so forth. Thagard also makes explicit reference to the structuralist idea of intended applications. See also Friedman (1974), pp. 15–16.

27. This gives a rough measure of what I have elsewhere called, following the terminology and intent of William Whewell, the degree of consilience of a theory-net. Unification, it appears, also covers other facets, such as the simplicity of the laws within a particular class of applications. It would go beyond the scope of this paper to go into details, but here we have one reason for the difficulties in attempts to elucidate unification. As I have argued, briefly, in Sintonen (1986), the notion of simplicity covers several distinct intuitions which may be balanced differently in the assessment of the degree of unification or simplicity of a theory.

28. See Moulines (1984) and Balzer, Moulines, and Sneed (1986) for further details.

29. I owe this example and much of the following discussion to Moulines (1984). See also Balzer, Moulines, and Sneed (1986).

30. Kitcher (1981).

31. I owe this pattern of argument and the example to Michael Ruse (1973), chapter 4. Ruse states without reservation that "*population genetics is presupposed by all other evolutionary studies*," and claims that "evolutionary theory is a *unified* theory with population genetics as its presupposed core." Ruse argues that it has a hypothetico-deductive pattern, but it appears to me that the notion of a theory-net which can accommodate a number of distinct kinds of intertheory relationships would be more appropriate.

32. The three types of increase in unification correspond to Stegmüller's three types of evolutionary progress, i.e., confirmational, theoretical, and empirical. See Stegmüller (1979), p. 33.

33. Friedman (1983, section VI.3.) gives a lucid and detailed account of theoretical unification which results when a (characteristically observable) substructure is embedded in a more abstract larger structure. He also notes that unification results in better confirmation: a phenomenological regularity can be well confirmed, but it gains in degree of confirmation if it is derived from a higher-level theory. According to him derivability as such is relatively unimportant; what counts is the interplay between what he calls phenomenological representations and (realistically understood) reduction in time: thus some theoretical assumptions designed to explain (relatively) observable phenomena can be later confirmed by their use in the accommodation of other observable phenomena. This picture is easily translated to queston-theoretic parlance, the way I have sketched. Note, however, that the notions of theory-net and intertheory link allow a more fine-grained classification of the various aspects of unification. For a tentative structuralist classification of intertheory-relations, see Krüger (1980). For unification and confirmation, see also Glymour (1980).

34. See Stegmüller (1983). That logical analysis provided a model to be followed is clearly visible in the Vienna Circle manifesto, *The Scientific Conception of the World: The Vienna Circle*. It quotes (p. 8) Russell's *Our Knowledge of the External world*: the method of logical analysis "has gradually crept into philosophy through the critical scrutiny of mathematics. . . . It represents, I believe, the same kind of advance as was introduced into physics by Galileo: the substitution of piecemeal, detailed and verifiable results for large untested generalities recommended by a certain appeal to imagination."

35. "The Function of General Laws in History." Reprinted in Hempel (1965, pp. 239–40.

36. See Scriven (1975), p.4.

37. According to Julius Moravcsik the popular Aristotelean doctrine of four causes really is "a theory about the structure of explanations." Instead of talking about *aitiai* as causes we should really talk about explanatory factors in a wider sense. See Moravcsik (1974).

38. For a review of the various types of action theory, see Tuomela (1982). Note that there is also a third main type of action theory, the agency theory, which draws on an irreducible type of agent causality.

39. The intentionalist I have in mind is G. H. von Wright, who gives Aristotle credit for the view

that the *aitiai* of human actions are type-distinct from the *aitiai* of things in *rerum natura*. Von Wright has argued that Aristotle had two syllogisms, the theoretical and the practical, with the difference that the theoretical syllogism has as its conclusion a proposition while the practical one has as its conclusion an action (or maybe a proposition which describes an agent's attempt to perform and action). He has also held the very strong view that practical syllogism is a *sui generis* mode of explanation; roughly, it is for human sciences what the covering law is for natural sciences. He has, however, modified his views since. See von Wright (1971).

40. Sylvain Bromberger advances a similar view, with examples from linguistics, in his (1985), p. 321. And it is noteworthy that linguistic arguments against a particular description or explanation standardly refer to its incapability of accounting for this or that generalization. Note also that "because" may carry a different force in different compartments of natural science, as is indicated by the troubles of efficient causation, say, in quantum theory.

References

Achinstein, Peter. 1983. *The Nature of Explanation*. New York and Oxford: The University Press.

Ackermann, R. 1965. Deductive Scientific Explanation, *Philosophy of Science* 32: 155–67.

Balzer, Wolfgang and Sneed, Joseph. 1977 and 1978. Generalized Net Structures of Empirical Theories, *Studia Logica* 36 and 37: 195–211 and 167–194.

——, Ulises Moulines, Carlos, and Sneed, Joseph D. 1986. In The Structure of Empirical Science. *Logic, Methodology and Philosophy of Science VII*, eds. Barcan Marcus et al.. New York: Elsevier, 291–306.

Bromberger, Sylvain. 1965. An Approach to Explanation. In *Analytical Philosophy, 2nd Series*, ed. R. J. Butler. Oxford: Basil Blackwell, 72–103.

——. 1985. On Pragmatic and Scientific Explanation: Comments on Achinstein's and Salmon's Papers. In *PSA 1984*, Volume Two, eds. P. D. Asquith and P. Kitcher. East Lansing, Michigan: The Philosophy of Science Association.

——. 1987. What We Don't Know When We Don't Know Why. In *Scientific Inquiry in Philosophical Perspective*, ed. Nicholas Rescher. Lanham, MD: University Press of America 75–104.

Eberle, R., Kaplan, D., and Montague. R. 1961. Hempel and Oppenheim on Explanation, *Philosophy of Science* 28: 418–28.

Friedman, Michael. 1974. Explanation and Scientific Understanding. *Journal of Philosophy* LVVI, No. 1, 5–19.

——. 1983. *Foundations of Space-Time Theories. Relativistic Physics and Philosophy of Science*. Princeton, NJ: Princeton University Press.

Gardenfors, Peter. 1976. Relevance and Redundancy in Deductive Explanation. *Philosophy of Science* 43: 420–31.

——. 1980. A Pragmatic Approach to Explanation. *Philosophy of Science* 47: 405–23.

Glymour, Clark. 1980, *Theory and Evidence*. Princeton, NJ: Princeton University Press.

Hansson, Bengt. 1974. Explanations–Of What? Stanford University, (Mimeographed).

Hattiangadi, J. N. 1978. The Structure of Problems II, *Philosophy of the Social Sciences* 9: 49–71.

Hempel, Carl. 1965. *Aspects of Scientific Explanation and Other Essays in the Philosophy of Science*. New York: The Free Press.

——, and Oppenheim, Paul. 1948. Studies in the Logic of Explanation. Reprinted in Hempel (1965), 245–95.

Hintikka, Jaakko. 1976. *The Semantics of Questions and the Questions of Semantics*. North Holland: Acta Philosophica Fennica, Vol. 28, No 4.

Kaplan, David. 1961. Explanation Revisited, *Philosophy of Science* 28: 429–36.

Kim, Jaegwon. 1963. On the Logical Conditions of Deductive Explanation, *Philosophy of Science* 30: 286–91.

Kitcher, Philip. 1981. Explanatory Unification. *Philosophy of Science 48*: 507–31.

Krüger, Lorentz. 1980. Intertheoretic Relations as a Tool for the Rational Reconstruction of Scientific Development, *Studies in History and Philosophy of Science* 11, No. 2, 89–101.

Kyburg, Henry E., Jr. 1965. Discussion: Salmon's Paper. *Philosophy of Science*. 32, no. 2, 147–51.

Laudan, Larry. 1977. *Progress and Its Problems. Towards a Theory of Scientific Growth.* London and Henley: Routledge & Kegan Paul.

Leplin, Jarret. 1980. The Role of Models in Theory Construction. In *Scientific Discovery, Logic, and Rationality,* ed. T. Nickles. Dordrecht, Boston, London: D. Reidel, 267-83.

Lugg, Andrew. 1979. Laudan and the Problem-Solving Approach to Scientific Progress and Rationality. *Philosophy of the Social Sciences 9:* 466-74.

Moravcsik, Julius M.E. 1974. Aristotle on Adequate Explanations. *Synthese* 28: 3-17.

Moulines, Carlos Ulises. 1979. Theory-Nets and the Evolution of Theories: The Example of Newtonian Mechanics, *Synthese* 41: 417-39.

——. 1984. Links, Loops, and the Global Structure of Science, *Philosophia Naturalis. Archiv für Naturphilosophie und die philosophischen Grentzgebiete der exakten Wissenschaften und Wissenschaftsgeschichte.* Band 21, Heft 2-4. Meisenheim/Glan: Verlag Anton Hain.

Nickles, Thomas. 1981. What is a Problem that We May Solve It? *Synthese* 47: 85-118.

Niiniluoto, Ilkka. 1980. The Growth of Theories: Comments on the Structuralist Approach. *Proceedings of the Second International Congress for History and Philosophy of Science, Pisa, 1978.* Dordrecht-Holland: D. Reidel. 3-47.

Omer, I. A. 1970. On the D-N Model of Scientific Explanation, *Philosophy of Science* 37: 417-33.

Pearce, David and Rantala, Veikko. 1983. 'New Foundations for Metatheory'. Synthese 56. Pages 1-26.

Ruse, Michael. 1983. *The Philosophy of Biology.* Atlantic Highlands: Humanities Press.

Salmon, Wesley. 1984. *Scientific Explanation and the Causal Structure of the World.* Princeton: Princeton University Press.

The Scientific Conception of the World: The Vienna Circle. In *Otto Neurath: Empiricism and Sociology.* M. Neurath and R. S. Cohen, eds. Dordrecht, Holland: D. Reidel.

Scriven, Michael. 1975. Causation as Explanation, *Noûs* 9, 3-15.

Sintonen, Matti. 1984a. *The Pragmatics of Scientific Explanation,* Acta Philosophica Fennica, Vol. 37. Helsinki: Societas Philosophica Fennica.

——. 1984b. On the Logic of Why-Questions. In *PSA 1984,* Vol. One, eds. P. D. Asquith and P. Kitcher. East Lansing, Michigan: The Philosophy of Science Association, 168-76.

——. 1985. Separating Problems from their Backgrounds: A Question-Theoretic Proposal. *Communication and Cognition, 18, No 1/2,* 25-49.

——. 1986. Selectivity and Theory Choice. *PSA 1986,* Vol. One, eds. A. Fine and P. Machamer. East Lansing, Michigan: The Philosophy of Science Association, 364-73.

Sneed, Joseph. 1976. Philosophical Problems in the Empirical Science of Science: A Formal Approach. *Erkenntnis* 10: 115-46.

Stegmüller, Wolfgang. 1979. *The Structuralist View of Theories. A Possible Analogue of the Bourbaki Programme in Physical Science.* Berlin, Heidelberg and New York: Springer-Verlag.

——. 1983. *Erklärung, Begründung, Kausalität.* Zweite, verbesserte und erweiterte Auflage. Berlin, Heidelberg, New York: Springer-Verlag.

Thagard, Paul. 1978. The Best Explanation: Criteria for Theory Choice, *The Journal of Philosophy* LXXV: 76-92.

Thorpe, D. A. 1974. The Quartercenternary Model of D-N Explanation, *Philosophy of Science* 41: 188-95.

Tuomela, Raimo. 1980. Explaining Explaining. *Erkenntnis* 15: 211-43.

——. 1982. Explanation of Action. In *Contemporary Philosophy. A New Survey,* Vol. 3, ed. G. Floistad. The Hague, Boston and London: Martinus Nijhoff, 15-43.

Van Fraasen, Bas C. 1980. *The Scientific Image.* Oxford: Clarendon Press,

Wright, G. H. von. 1971. *Explanation and Understanding.* Ithaca, NY: Cornell University Press.

Scientific Explanation: The Causes, Some of the Causes, and Nothing But the Causes

1. Introduction

It would be desirable for philosophical accounts of explanation to capture the type of scientific explanation which is exemplified by this case:

In 1981 physicians in Los Angeles and New York began to notice an unusual cluster of cases of formerly rare symptoms—Kaposi's sarcoma, *Pneumocystis carinii* pneumonia, and other opportunistic infections, primarily in young men.[1] Faced with this phenomenon the physicians asked in each case "What is the explanation of this man's illness?"[2] and about the cluster itself "What is the explanation of the multiple incidence of these symptoms?" Because of the absence at that time of any known causes, the response to each question was to begin a systematic search for an explanation. What the investigators were searching for was not an argument or a speech act or a set of sentences, it was a real thing, a cause of the sickness. As we now know, an explanation was found, which was, among other things, a group of retroviruses that cause AIDS. This discovery was made possible only through the systematic use of scientific methodology, involving epidemiology to identify risk groups, theories in molecular biology to identify possible causal factors, and controlled experimentation to isolate specific causal factors which were responsible in each case. Subsequently, and only subsequent to this discovery, were the investigators in a position to answer why-questions, and to gradually fill in the causal story so that groups with different interests—homosexuals, intravenous drug users, public health officials, biomedical researchers, and so on—could be given the parts of the explanatory story in which they were most interested. Most notably, an explanation could be given even though it was incomplete. It was not claimed that there were no other causal factors involved, factors which increased or decreased the risk for an individual, only that a part of the causal mechanism leading to the illness in each case had been found, and a mechanism of transmission found which was causing the cluster.

This activity of searching for and discovering an explanation of a given effect is something which, although not exclusive to scientific research,[3] is a sufficiently important feature of scientific activity that special methods have been developed to isolate such discoveries. These methods may be experimental, they may use statistical surrogates for experimental controls, or they may use theoretical idealizations to mimic such controls. Which method is used will depend upon a complex of conditions, involving the nature of the subject matter, the state of scientific and technological knowledge, ethical constraints, and so on. In each case, however, the emphasis is on isolating causal factors, structures, and mechanisms whose operation may be taken to partially constitute the explanation of the phenomenon at hand. There is also no denying that linguistic explanations are required for conveying this information beyond the point of discovery. So what is the relation between these two quite different uses of the term 'explanation'? It is clear that the first sense, the objective sense in which one can discover explanations, is intimately linked with causation, so closely linked in fact that some writers have wanted to deny that our first sense is in fact a genuine kind of explanation, sometimes because explanations are supposed to generate intensional contexts, whereas causes plausibly do not; sometimes because explanations might have inescapably pragmatic aspects, whereas causes do not; sometimes because the logico-linguistic forms of natural language representations of explanation seem to be different from those of causal claims. Such views tend, I think, to evolve from a philosophical strategy which takes explanatory discourse as a given, a storehouse of factual information about explanations which, after philosophical analysis, will yield the correct form for explanations.[4] Within this approach the logical structure of linguistic explanations is taken as primary, and causal explanations, where applicable, have to conform as a special case to the general logical structure.

I believe that it is worth employing a different kind of strategy. It is significant that most of the counterexamples to Hempel's deductive-nomological and inductive-statistical models of explanation hinge on those models' inability to correctly capture causal relationships.[5] In addition, a significant body of work on the nature of causal relata has exposed difficulties inherent in descriptive representations of events, which suggests that an adequate account of causal relations must involve some direct representation of the causal relata.[6] Increased understanding of probabilistic causality has made it clearer how causal explanations for undetermined effects might be given.[7] And a revival of interest in the role played by experimentation in science tends to make one aware of the limitations imposed by purely logical analyses of causal structure.[8] These considerations lead me to take our project as a synthetic one: it is to see how the analytic methods of science discover the structural form of causal explanations of phenomena, and then to construct an appropriate linguistic representation which faithfully preserves that structure.

Our task is thus a restricted one. It is to provide an account of the nature of singular causal explanations.[9] In various places I shall emphasize the role played by probabilistic causality in explanations, but the framework is designed to apply to non-probabilistic cases as well. Indeed, by examining the similarities between probabilistic causality and causal relations holding among quantitative properties, the probabilistic case can be seen to be rather less idiosyncratic than it first appears.

2. The Multiplicity, Diversity, and Incompleteness of Causal Explanations

We begin by noting that science is called upon to find explanations for phenomena originating in widely differing circumstances. It does best when explaining phenomena produced by science itself in the clean and austere conditions of the laboratory, when there is ordinarily only a single causal factor operating and the law governing the phenomenon is already known. But science is frequently called upon to find the explanation of naturally occurring phenomena such as epidemics, tree diseases, rainfall distributions, migratory patterns, rainbows, the nonexistence of higher forms of life on Mars, and planetary movements. It is also often required to explain the results of applied science, such as rocket explosions, holes in the ozone layer, the properties of artificial elements, the effect of plastics on the environment, and presidential campaigns.

A characteristic feature of these natural and unnatural phenomena is that they are usually the result of multiple separable causal influences. For example, the rate of enzyme-catalyzed reactions is affected by the enzyme concentration, the substrate concentration, the temperature, the pH of the substrate, oxidation of the sulfhydryl groups of an enzyme, and high-energy radiation; the first two increasing the rate of reaction, the last decreasing it, while the actions of the third and fourth have maximal points of inflexion at optimal temperature and pH, respectively.[10] Multiple – because except in specially constructed artificial settings each of these factors will be causally effective in a given reaction. Separable – because the experimental and theoretical devices mentioned earlier enable one to isolate the effects of a single factor. Causal – if they satisfy the invariance requirements of §4 below, where it will be argued that if they do, then we may infer not only that the causes are separable but that they are in fact operating separately.

It then follows that to properly convey the structure of the causal origins of a given phenomenon, a linguistic explanation should preserve the separation between distinct and independent causal influences on that phenomenon. And so we have the first constraint on representations of causal explanations: they must correctly represent the *multiplicity and separateness* of causal influences on a given phenomenon.[11]

Let us now examine a second characteristic of causation. It is by now generally agreed that there are two distinct kinds of probabilistic cause, which I shall call "contributing" and "counteracting."[12] Consider first a simple example.

The bubonic plague bacillus (*Yersinia pestis*) will, if left to develop unchecked in a human, produce death in between 50 and 90 percent of cases. It is treatable with antibiotics such as tetracycline, which reduce the chance of mortality to between 5 and 10 percent.[13] The causal mechanisms involved, the mode of transmission, and the action of the treatment on the infected human are sufficiently well established that there is no doubt that infection with *Yersinia pestis* is a contributing cause of death in a human, and administration of appropriate antibiotic treatment is a counteracting cause of death. It is also true that the contributing cause is not sufficient for death, and that the counteracting cause does not guarantee recovery, as the cited probabilities show. Now suppose that Albert has died, and we ask for an explanation of this event. Once again, it is imperative to separate the different causal influences on the effect to be explained, the reason this time being the diversity of types of cause, rather than mere multiplicity. To do that we shall need to use a new explanatory format. Historically, the standard format for explanations has always been the simple 'Y because X' mode, but one of the striking features of explanations involving contributing and counteracting causes is that this historical format is quite inappropriate. It is absurd to claim that "Albert's death occurred because of infection by the plague bacillus and the administration of tetracyclin." Instead, an appropriate response at the elementary level would be "Albert's death occurred because of his infection with the plague bacillus, despite the administration of tetracycline to him." Thus the second constraint on explanations is: they must correctly represent, where appropriate, the *diversity* of causal influences on a phenomenon.

The third characteristic feature of causal explanations, which this time is epistemic in flavor, is the incompleteness of many of our explanations of causally produced phenomena. Given the multiplicity and diversity of causal influences, it will be rare that we are in a position to provide a complete list of all the influences which affected a given outcome. In both the AIDS example and the enzyme-catalyst example, there is no pretence that a complete explanation has been discovered. Yet we have good reason to suppose that what has been offered as an explanation is true, and if it is, then the incomplete explanation can be partially complemented by successive discoveries which add to our understanding without undermining the accuracy of the previous accounts. Thus the third constraint on linguistic explanations is that they must be able to provide *true yet incomplete* representations of causal explanations.

3. The Canonical Form for Causal Explanations

Here, then, is a linguistic mode for providing explanatory information in the case of specific events which explicitly provides for each of the three features discussed above. If one wishes to request an explanation, the canonical form will be "What is the explanation of Y in S at t?"[14] An appropriate explanation will be "Y in S at t [occurred, was present] because of Φ, despite Ψ" where 'Y', 'S', 't' are terms referring to, respectively, a property or change in property, a system, and a time; 'Φ' is a (nonempty) list of terms referring to contributing causes of Y; and 'Ψ' is a (possibly empty) list of terms referring to counteracting causes of Y.

The explanation itself consists in the causes to which 'Φ' jointly refers. Ψ is not a part of the explanation of Y proper. The role it plays is to give us a clearer notion of how the members of Φ actually brought about Y — whether they did it unopposed, or whether they had to overcome causal opposition in doing so. Thus Ψ may be empty, in which case we have an explanation involving only contributing causes to Y's occurrence, but if Φ is empty (while Ψ is not), then we have no explanation of Y's occurrence, merely a list of factors which lessened the chance of Y's occurring.[15]

We have already seen an elementary application of this format to the plague case. A somewhat more sophisticated example involves the case of enzyme catalyzed reactions mentioned earlier. Thus, if the phenomenon to be explained is an increase in the reaction velocity of a metabolic process, we can assert (omitting references to the system and time) "the increase in reaction velocity occurred because of the increases in enzyme and substrate concentration to optimality, despite the increasing oxidation of the dehydrogenases and irradiation by ultraviolet light." (I note here in anticipation of a later claim that although each of the explanations discussed so far involve phenomena which are plausibly not determined by the cited factors, there is no mention of probability values in their explanations.)

Although I have stressed the way in which probabilistic causality makes us aware of the need for a new explanatory mode, such explanations are also possible for phenomena which we have every reason to suppose are deterministic in character. For instance, in theoretical representations of the value of the angular momentum of the earth, the simplest model treats the sun as fixed. Then, to a good approximation the angular momentum of the earth is constant, and its value is given by the relevant conservation law. But this idealized picture is too simple, and a number of small but important causal influences have to be considered to explain the actual motion of the earth. First, the earth is an oblate spheroid rather than a sphere, and this produces a precession in the orbital plane of the moon, which in turn produces a precession in the earth's angular momentum. Second, tidal friction gradually slows the earth's rotation owing to a couple acting on the equitorial tidal bulges. Third, there are thermodynamical "tides" in the earth's atmosphere owing to periodic heating by the sun, with a consequent gravitational couple from the sun which acts to speed up the earth's rotation. Fourth, the nonuniformity of the sun's gravitational field results in an additional precession of the earth's angular momentum.[16]

Consider how we should respond to a request for an explanation of the increase in angular momentum of the earth over the conserved value. The explanation would be "because of the precession of the moon's orbital plane, the nonuniformity of the sun's gravitational field, and the action of thermodynamical tides, despite the slowing effects of tidal friction." It is important to note that these explanations can be given even though they are incomplete. There is no pretense that all causal factors affecting the angular momentum of the earth have been cited. The omissions are not due to the scientist selecting those factors which interest his listener, or to being constrained by the form of the request for an explanation. It is because there are many influences on the earth's rotation beyond those cited, most of which are as yet unknown. The geophysicist knows that there exist these unknown causal factors, yet the factors cited do provide an explanation, however incomplete, of the explanandum. Nevertheless, the explanation given is true. Every factor cited is a cause of the increase in the earth's angular momentum, and the explanation correctly classifies them into contributing and counteracting causes. (The reason why we can extend this terminology of contributing and counteracting causes beyond the probabilistic realm is given in the next section.)

This example is characteristic of many scientific investigations, both theoretical and experimental. In the theoretical realm, corrections to the ideal gas laws owing to intermolecular forces (by means of virial coefficients), the elaboration of four variable causal models in sociology to five variable models, and time-independent perturbation theory for representing the Coulomb repulsion between electrons in multi-electron atoms all use this cumulative approach to explanation. Sometimes the cumulative filling-in is made via the intermediate device of theoretical models, when influences which have been known about and deliberately omitted for the sake of simplicity are added to refine the model, or to introduce a student to a higher level of sophistication in the explanatory account. The experimental case deserves an essay in its own right: suffice it to say here that one of the principal uses of the experimental method is for causal discovery, confirmation and testing, ordinarily of causal relations which have been singled out from the set of multiple influences by means of controls, randomization, or statistical surrogates such as multiple regression analysis.

4. Ontology

Scientific analysis separates causal influences, and our representations must preserve this separation. So events cannot be identified with spatio-temporal regions[17] because any given spatio-temporal region ordinarily contains many properties and changes, some of them causally relevant to a given effect, many of them not, and those that are relevant may be of different types. For example, the spatio-temporal region containing an increase in temperature from 20°C to 800°C of a sample of magnesium also may contain a change in color and length of the bunsen flame, a change in the sound emitted by the bunsen, a change in the volume of the magnesium sample, the presence of oxygen, the lack of an oxidized layer on the magnesium and so on, each of these factors save the first and last two being causally irrelevant to the effect, which here is the ignition of the magnesium. A similar remark may be made about the spatio-temporal region containing the effect, which also contains numerous irrelevant features including the spatio-temporal location of the effect, the manufacturer's brand name stamped on the sample, the property of being held by Paul Humphreys, and so on.

In our construal of causal explanations I employed this ontology: An event is *the possession of, or change in, a property of a system on a given occasion* (trial). Events are thus taken as concrete, specific entities, actual instantiations of or changes in worldly properties of a system, these properties being possessed by specific structures, themselves a part of the world, with these structures persisting through the change in properties which constitute an event. (For simplicity I restrict myself to monadic properties. Events involving relational properties may be dealt with analogously.)

This approach could be adopted simply on the grounds that it enables us to maintain a separation between causal factors in the desired way. There is, however, a more systematic justification underlying this choice of ontology which I should like to draw out here.[18] In his *System of Logic* Mill argued that the distinguishing feature of a genuine cause was its unconditionalness. Succinctly, for X to be a genuine cause of Y on a given occasion, it must be true that X causes Y whatever other circumstances prevail, for if not, then it is not X *simpliciter* that caused Y on that occasion, but X together with some further factor(s) Z, the presence or absence of which, in combination with X, results in Y appearing on that occasion, but the absence or presence of which, on other occasions, leads to Y's nonoccurrences. This means, of course, that any singular causal claim is also implicitly general. But in what relation does the singular causal sequence stand to the universal causal law? Is the singular sequence causal because it is an instance of the primary universal law, as regularity theorists would maintain, or is the general law nothing more than a collection of singular causal sequences? By considering these two options we can see why permanent structures are required for our ontology, what the correct account of probabilistic causality is, and why it is appropriate to extend the contributing and counteracting causal terminology outside the probabilistic realm.

Regularity theorists require that for an event sequence to count as causal it must be an instance of a universal regularity. But as has often been noted, there are few, if any, observed universal regularities under which singular events can fall. The natural world as observed is simply too chaotic a place for that. However, as Bhaskar (1975) noted, the creation of experimental contexts in the natural world results in just those kinds of regularities which are needed for laws by a regularity theorist. The problem is that these regularities disappear once the experimental controls are lifted. Hence if causal laws are identified with observed universal regularities, and singular event sequences are causal only if instances of causal laws, then singular phenomena occurring outside experimental contexts will rarely, if ever, have causes or causal explanations. (I note in passing that this is a serious problem for any model of explanation which requires deductive subsumption under such regularities.)

There are various ways one might escape this conclusion. To avoid repetition, I refer the reader at this point to my (1988) for detailed arguments on those options, and merely state the conclusion of that paper here: the only plausible account of causation which retains universality and allows causal explanations of singular phenomena in nonexperimental contexts is one which refrains from identifying causal laws with sequences of observed events, but instead allows for the existence of permanent or semi-permanent structures persisting through the creation and destruction of the experimental contents which give rise to whatever regularities are observed. It is important to emphasize that none of the arguments which lead to this conclusion should be unacceptable to an empiricist unless he denies the need for an explanation of the difference between experimental and nonexperimental contexts and rejects *a priori* any entity which does not satisfy a fixed, atemporal criterion of observability. For experimentation is undeniably a central feature of scientific empiricism, and many structures initially discovered in such contexts are later found to persist outside them as well.

I now have to make one unargued assumption to carry the case into the probabilistic realm. It is that there are such things as physical chances grounded in structural features of an indeterministic system. Although I believe that it is possible to extend the above argument to establish the existence of such structural probabilities, one has to make a further explanatory inference from observed relative frequencies to constant physical probabilities as generating conditions of those frequencies, and it is not clear to me under exactly what circumstances this inference is legitimate. Hence I rely on this intuitive picture: physical probabilities are dispositional properties, alterations in the structural basis or in the conditioning variables of which result in an alteration of the associated probability distribution. We can now see how to apply Mill's invariance condition to the probabilistic case. Recall that the characteristic feature of a probabilistic contributing cause was that it raises the chance of the effect, i.e., *it produces an increase in the value of the chance of the effect*. So, assuming the existence of physical chances, the *direct* effect of a contributing cause is an increase in the chance (of some property). But this is no different from any familiar and uncontroversial case of sufficient quantitative causation. Increase the voltage in a wire of a fixed resistance and the effect is an increase in the value of the amperage. Increase the intensity of a classical radiation field on a particle and the effect is an increase in the chance of the induced emission of a quantum of e.m. radiation. This enables us to see why it is possible to naturally apply the contributing cause terminology both to deterministic cases such as the angular momentum of the earth and to probabilistic cases such as the bubonic plague. In both cases a contributing cause increases the value of a quantitative variable – it just happens that in the probabilistic case the variable is the value of the chance.

The application of this approach to the qualitative case is, I believe, quite straightforward, but cases where quantitative variables are specified exactly need some discussion. I shall deal with the deterministic case here, because the examples are clearer. Consider an individual who acquires an extreme fondness for chocolate and who tries to lose weight by taking diuretic pills. The chocolate intake produces an increase of 10 pounds of fat over the level the individual had without the chocolate input. The diuretic pills produce a decrease of 5 pounds of water compared with the weight level the individual had before taking the pills. The net *observed* increase in weight is 5 pounds, but in this case both the contributing and counteracting causes had their full effect. One could actually collect the increased fat and the lost water if one so desired. Thus, not only are the causes operating separately, they produce clearly separable effects which together produce the effect to be explained. Thus it seems appropriate to assert in the qualitative case "the increase in John's weight occurred because of the chocolate's caloric content, despite the diuretic action of the pills." Now consider a case where the contributing and counteracting causes produce their effects through mechanisms that are not so easily separable; for example, when fat is burned off by exercising. Is there any essential difference between this case and the previous one? Suppose that the fat was first put on and then part of it burned off by running. Then there would be no difference between the first situation and this one, and our explanatory form could still be used. How about the case where the chocolate input and the running occur together? Recalling our characterization of an invariant cause, if one reduces the exercise level to zero, the individual will not lose 5 pounds, and if one reduces the chocolate intake to zero, the individual will not gain 10 pounds, in each case compared to the situation with the putative cause present and all other factors as they are. Hence each influence was operating on the system during the trial and each played its role in the way the effect came about. And so "the increase in John's weight occurred because of the chocolate's caloric input, despite the burning off of fat by running" is also correct.

These are all cases where the causes and the effects are taken to have only qualitative properties, and there is little doubt that the approach works well there. The quantitative case, it turns out, is not so transparent, in that ordinary use appears to allow two different representations, the traditional "because" account and the mode suggested here. To decide between them we again need to look at the causal mechanisms underlying the observed phenomenon. Consider a room which is both heated and air-conditioned, and suppose that the temperature rises by 5°C (compared to the situation where neither is operating). Alone the heater would raise the temperature by 10°C; alone the air-conditioning would lower it by 5°C. Again, the qualitative case appears to work: "The temperature of the room rose because of the input from the heater, despite the cooling of the air conditioner." The mixed cases where either the effect is quantified or the causes are quantified, but not both, also seems to fit this pattern. Now consider these two claims:

(1) The increase of 5°C in the temperature of the room occurred because of the input of 10,000 Btu from the heater, despite the extraction of 5,000 Btu by the air conditioning.

(2) The increase of 5°C in the temperature of the room occurred because of the input of 10,000 Btu from the heater and the extraction of 5000 Btu by the air conditioning.

Prima facie, (2) seems more plausible than does (1). Why is this? There are, in fact, two aspects of the explanandum event that need to be explained: the increase in the temperature and the exact value of that increase. Emphasize the former and (1) will have some appeal; emphasize the latter and the superiority of (2) is evident. That (2) was preferable on first inspection is accounted for, I believe, by an entirely justifiable tendency to prefer precise, quantitative, explananda to imprecise, qualitative, explananda within scientific contexts. What does this example show us about our canonical form for causal explanations? Two things, I think. The first point, which I have already discussed and have more to say about in §6, is that it must always be clearly specified which aspect of a spatio-temporal event is the object of an explanation, so that ambiguities can be avoided. The second point is more important. In any deterministic explanation in which the explanandum is the value of a quantitative variable, all causally relevant factors will contribute to the system's having that exact value and hence the traditional 'because' format will be the appropriate one. In contrast, where the causal factors are only probabilistically related to the explanandum, what is crucial are increases and decreases in the chance of that explanandum, even in cases where the explanandum is itself quantitative in form. Consequently, the canonical format for explanations described earlier will frequently need to be used because of the presence of counteracting factors. This difference between the deterministic and the indeterministic cases explains why the inadequacy of the traditional 'because' format is not revealed within a very broad class of deterministic cases. (The reader will have noted that my earlier example involving the earth's angular momentum involved an increase in its value rather than the value itself.) The deficiencies appear only within the domain of qualitative deterministic explanations and qualitative or quantitative indeterministic explanation. This is not to say that the canonical format is the wrong one for quantitative deterministic explanations.

It is simply that the 'despite' clause is not used because of the absence of counteracting causes.

This argument rests on the claim that the precise value of the probability is not something that is involved in explanations of stochastic phenomena. The next section is devoted to establishing this claim.

5. Why Probability Values Are Not Explanatory

We have seen the role played in explanations by the multiplicity, diversity, and epistemic incompleteness of causes. We are now in a position to argue for a fourth thesis — that probabilities have no explanatory role. Let us begin by noting that every other contemporary account of probabilistic or statistical explanation requires that a probability value be cited as an essential part of the explanation of phenomena whose occurrence is, or is taken to be, properly stochastic in nature.[19] The most common reason for this is that they are all versions of a covering law model of explanation, and the covering law is a probability law (i.e., a law which assigns a probability value to the explanandum, either absolutely, or conditionally, or relationally). Because it is standardly required that all the elements of the explanans must be true, the probability law, being part of the explanans, must satisfy this requirement, and hence the true probability must be assigned to the explanandum by virtue of falling under this true covering law.[20]

This consequence, that an essential part of a probabilistic explanation is the attribution of a true probability value, must be rejected. The reasons are twofold. First, the insistence on specifying probability values makes it impossible to separate true explanations from complete explanations, with the dire consequence that it is rare, if ever, that we can in fact provide such an explanation. The situation is clearest if one considers explanations of specific outcomes, with the probability temporarily interpreted as a relative frequency, rather than as a propensity as I prefer. To attribute the correct probability to the explanandum in this situation, the problem of the single case must be solved: that is, the probability attributed to the explanandum must be the appropriate one for that particular case. All such solutions employ essentially the same device — a requirement that all (and only) factors which are probabilistically relevant to the outcome should be used to determine the class or sequence within which the relative frequency is calculated. Omit even one probabilistically relevant factor and a false attribution of probability will be made. To revert to an example I have used elsewhere, if an individual dies from lung cancer, having been a heavy smoker, omitting from the explanation the following probabilistically (and causally) relevant factors will result in a false probability value being given and hence, within the frameworks I am criticizing, a false explanation being given: (i) cosmic radiation from α-Centauri, (ii) a hereditary characteristic inherited from his great-great-grandfather, (iii) particles from a smokestack in Salem, Oregon. Because this completeness condition for probabilistically relevant factors, which surfaces in different ways in maximal specificity conditions, objective homogeneity conditions, and randomness requirements, cannot be separated from the truth conditions for the probability covering law when it is applied to single case explanations, explanations which require the true probability value to be cited cannot omit even absurdly small probabilistically relevant factors and remain true.

In contrast, if one holds that it is causally relevant factors which are explanatory, where a factor is causally relevant if it invariantly changes the propensity for an outcome, i.e., a change in the factor results in a differential change in the propensity irrespective of what other changes or conditions are also present, then specification of one or some of the causally relevant factors will allow a partial yet true explanation even in cases where the other factors are not known and the true probability value cannot be calculated. This distinction between true and complete accounts is similar to the distinction which has been common in English law for centuries between the truth and the whole truth. Of course, there everything is epistemically relativized, whereas the contrast here is between the truth and the complete truth. So we might say that for linguistic explanations we require the truth, nothing but the truth, yet not the whole truth, where for causal explanations this means citing the causes, nothing but the causes, yet not all the causes. Second, and consequently, this approach has the advantage that when a complete explanation is available, i.e., all causally relevant factors have been specified, then a specification of the true propensity or correct reference class is automatically given by the constituents of the Φ and Ψ elements of the explanation (although the probability value may not be calculable from this information, because there is no guarantee that all such values are theoretically computable). This fact that probability values are epiphenomena of complete causal explanations indicates that those values have themselves no explanatory power, because after all the causal factors have been cited, all that is left is a value of sheer chance, and chance alone explains nothing.

This position has a number of immediate consequences. First, it follows that there can be more than one true explanation of a given fact, when different sets of contributing and counteracting causes are cited. This feature of explanations involving multiple factors, while tacitly recognized by many, is equally often ignored in the sometimes acrimonious disputes in social and historical explanations. Very often, a plausible case can be made that a number of supposedly competing explanations of, for example, why the Confederate States lost the Civil War, are all true. The dispute is actually about which of the factors cited was causally most influential, given that all were present, and not about which of them alone is correct.

Second, our account enables us to distinguish between cases where a phenomenon is covered by a probability distribution which is *pure*, i.e., within which no parameters appear which are causally relevant to that distribution (more properly, to the structure to which the distribution applies), and cases where the distribution is affected by such parameters.[21] There is good reason to believe that the traditional resistance to allowing explanations of indeterminate phenomena arose from a naïve belief that all such phenomena were the result of purely spontaneous processes which were covered by pure distributions. While sympathizing with the intent behind this resistance, because as we have argued, pure chance explains nothing, we have also seen an important difference between situations in which the pure chance remains at the end of a comprehensive causal explanation, and situations in which pure chance is all that there is.

Third, the traditional maximal specificity requirements which are imposed on explanations to arrive at a unique probability value must be replaced by the requirement of causal invariance described earlier.[22] This invariance requirement is strictly weaker than maximal specificity because the presence of a second factor can change the propensity for a given factor to produce an effect, without thereby changing that given factor from a contributing cause to a counteracting cause, or vice versa, whereas if the second factor confounds a putative contributing cause and changes it to a counteracting cause, a change in the propensity must accompany this. Of course, epistemically, we can never know for certain that such confounding factors do not exist, but that is an entirely separate matter, although regrettably relative frequentists have often failed to separate epistemic aspects of probabilistic causality from ontic aspects. This rejection of the explanatory value of probabilities is the reason I called my causal account one of "aleatory explanations." This was to avoid any reference to "probabilistic explanations" or "statistical explanations," while still wanting to convey the view that causal explanations are applicable within the realm of chancy, or aleatory, phenomena. It is, perhaps, not ideal terminology, but it serves its intended purpose.

Fourth, aleatory explanations still require laws to ground explanations, but reference to these laws does not appear directly in the explanations themselves, and they are not *covering* laws. The role that the causal laws play here is as part of the truth conditions for the explanatory statement. For something to be a cause, it must invariantly produce its effect, hence there is always a universal law connecting cause and effect. The existence of such a law is therefore required for something to truly be a cause, but the law need only be referred to if it is questioned whether the explanatory material is true. I want to avoid the terminology of "covering laws," however, because the term "covering" carries implications of completeness, which is quite at odds with the approach taken here.

Fifth, there is no symmetry between predictions and explanations. As is well known, the identity of logical form between explanations and predictions within Hempel's inferential account of explanation initially led him to assert that every adequate explanation should be able to serve as a prediction, and vice versa. What we have characterized as causal counterexamples led him to drop the requirement that all predictions must be able to serve as explanations. Arguments due primarily to Wesley Salmon were influential in persuading many philosophers that we can explain without being able to predict. That independence of prediction and explanation is preserved here. We have seen that probability values play no role in the truth of explanations; *a fortiori* neither do high probability values. It is true that we need changes in propensity values to assess degrees of contribution, but even a large contributing cause need not result in a high relative frequency of the effect, for it may often be counteracted by an effective counteracting cause. Thus, as noted earlier, the plague bacillus contributes greatly to an individual's propensity to die, yet the counteracting influence of tetracycline reduces the relative frequency of death to less than 10 percent. It is also worth noting that predictions differ from explanations in that when we have perfect predictive power (a set of sufficient conditions) there is no sense in asking for a better prediction, but perfect sense can be made of giving a better explanation, i.e., a deeper one. The same thing holds for probabilistic predictions. When maximal specificity conditions have been satisfied, there does not exist a better prediction, but again better explanations may exist.

Sixth, aleatory explanations are conjunctive. By imposing the causal invariance condition, we ensure that there are no defeating conditions which turn a contributing cause into a counteracting cause, or vice versa, or which neutralize a cause of either kind. Thus, two partial explanations of E can be conjoined and the joint explanation will be an explanation also, indeed a better explanation by the following criteria: If $\Phi \subset \Phi'$ and $\Psi = \Psi'$, then the explanation of Y by Φ' is superior to that given by Φ. If $\Phi = \Phi'$ and $\Psi \subset \Psi'$ then again Φ gives a superior explanation, in the sense that the account is more complete.[23]

6. Why Ask Why-Questions?

We have seen how to present causal information so that its diversity and multiplicity is properly represented, and if the information is given in response to a request for an explanation, how that request can be formulated. It might seem that there are other, equally appropriate ways of presenting that information and of requesting it. For example, it appears that we might have used instead the form "X because Φ even though Ψ" as in "This individual died because he was exposed to the plague bacillus, even though he was given tetracycline," where X, Φ, and Ψ are sentences describing causes rather than terms referring to them.[24] And, rather than our "What is the explanation of X?," many would prefer "Why is it the case that p?," where again, in the latter, a propositional entity, rather than a term, provides the content of the question.[25] Does anything hinge on our choice of representation, or is it simply a matter of convenience which one we choose?

I believe that it does matter which choice we make. It has become increasingly common to take the why-question format as a standard for formulating explanatory requests. Accompanying this has been an increased emphasis on the need for including pragmatic factors in explanatory contexts. The two are, I think, connected, and because pragmatics have no place in the kind of objective explanations with which I am concerned, linguistic devices which require their introduction ought to be avoided.

Let me begin with three major claims which Hempel made at the very beginning of his [1965] essay, and which go right to the heart of his conception of explanation. The first was that all scientific explanations may be regarded as answers to why-questions (334). The second was that requests for explanations whose explanandum term is a nondescriptive singular referring device such as a noun make sense only when reconstrued in terms of why-questions (334). The third claim was that every adequate response to an explanation-seeking why-question should, in principle, be able to serve as an adequate response to an epistemic why-question (335). All these claims are connected. I begin with the first two.

The first claim is, of course, a fairly weak one, because it suggests only that we can use the why-question format, not that we must. In making this claim, Hempel was clearly influenced by considerations similar to ours, in that a given explanandum event will usually be multifaceted, and one needs to specify which aspect one needs to explain. Here is Hempel's argument:

> Sometimes the subject matter of an explanation, or the explanandum, is indicated by a noun, as when we ask for an explanation of the aurora borealis. It is important to realize that this kind of phrasing has a clear meaning only in so far as it can be restated in terms of why-questions. Thus in the context of an explanation, the aurora borealis must be taken to be characterized by certain distinctive general features, each of them describable by a that-clause, for example: that it is normally found only in fairly high northern latitudes; that it occurs intermittently; that sunspot maxima with their eleven-year cycle are regularly accompanied by maxima in the frequency and brightness of aurora borealis displays; that an aurora shows characteristic spectral lines of rare atmospheric gases, and so on. And to ask for an explanation of the aurora borealis is to request an explanation of why auroral displays occur in the fashion indicated and *why* they have physical characteristics such as those indicated. Indeed, requests for an explanation of the aurora borealis, of the tides, of solar eclipses in general or of some solar eclipse in particular, or of a given influenza epidemic, and the like have a clear meaning only if it is understood what aspects of the phenomenon in question are to be explained; and in that case the explanatory problem can again be expressed in the form 'Why is it the case that p?' where the place of 'p' is occupied by an empirical statement specifying the explanandum. Questions of this type will be called *explanation-seeking why-questions*. (334)

It is evident, however, that one can meaningfully request explanations without resorting to the why-question format.

When John Snow discovered the principal cause of cholera in 1849, he wrote: 'While the presumed contamination of the water of the Broad Street pump with the evacuations of cholera patients affords an exact explanation *of the fearful outbreak of cholera in St. James's parish*, there is no other circumstance which offers any explanation at all, whatever hypothesis of the nature and cause of the malady be adopted." (Snow [1855], 54, my emphasis added)

A more recent example from molecular biology comes from Crick and Watson: "[Wilkins et al.] have shown that the X-ray patterns of both the crystalline and paracrystalline forms is the same for all sources of DNA ranging from viruses to mammals. . . . It seemed to us that the most likely explanation *of these observations* was that the structure was based upon features common to all nucleotides." (Crick and Watson [1954], 83, emphasis added)

Here we have terms "the fearful outbreak of cholera in St. James's parish," "the X-ray patterns of both the crystalline and paracrystalline forms of DNA" which associate properties with systems (in the first case at a particular time, in the second at all times) in the way suggested by our ontology, and the appropriate accompanying question in each case would be "What is the explanation of X in S?" There are many more examples. It is common, and I believe meaningful, to ask for an explanation of such things as the increase in volume of a gas maintained at constant pressure; of the high incidence of recidivism among first-time offenders; of the occurrence of paresis in an individual; of the high inflation rate in an economy; of an eclipse of the sun; and of an influenza epidemic in a population. Indeed, even some requests couched in terms of Hempel's forbidden format appear to be meaningful and legitimate.

It appears to be appropriate for an individual to have asked Galileo or Newton for an explanation of the tides, or a meteorologist for an explanation of the aurora borealis. In the first case, what is being requested is an explanation of the periodic movements of the oceans on earth; in the second an explanation of the appearance of bright displays in the atmosphere in the northern latitudes. In each of these cases, and in each of the previous examples, the explanation requested is usually explicitly, sometimes implicitly, of the occurrence or change of a property associated with a system, and this, rather than the particular linguistic representation, is the important feature. So ordinary usage will not decide between the why-question approach and the one suggested here.[26]

Moreover, a review of some well-known problems accompanying the propositional approach should make us extremely wary of adopting it without being aware of these problems. First, within causal explanations, a propositional representation of the effect (explanandum) will also require a propositional representation of the causes. (A mixed ontology is theoretically possible, but given that most effects are also causes of further phenomena, and vice versa, a symmetric treatment seems advisable.) This then makes the causal relation one holding between propositions, and, as Davidson's well-known adaptation of a

Fregean argument shows (Davidson [1967]), given referential transparency of causal contexts plus substitution *salva veritate* of logical equivalents, propositional causal relations would turn out to be truth functional, which is obviously false. Next, consider what happens when the propositional approach is embedded in a nomological-inferential treatment of causation, as Hempel's was. As Kim (1973) argued, it is extremely difficult to control propositional descriptions of events so that all and only the appropriate inferences are made. Let (x) (Fx → Gx) be any true law which subsumes the cause-proposition Fa and the effect-proposition Ga. Let Hb be any true proposition. Then Hb, H($\rceil$ x[x = b ∧ Fa]) are both propositions describing the same event, i.e., b is H. But then (x)(Fx → Gx) together with H($\rceil$ x[x = b ∧ Fa]) allows us to derive Ga, so Hb causes Ga, according to the subsumption account. This is clearly unacceptable.

Next, recall Kyburg's example of the hexed salt: "All samples of hexed salt stirred in warm water dissolve. This sample of hexed salt was stirred in warm water. Hence this sample of salt dissolved." This satisfies all the criteria of adequacy for a deductive-nomological explanation, yet it is seriously misleading. Again, excess content which is causally irrelevant needs to be excluded from a propositional description to avoid this problem. Similar difficulties lie at the root of some of Salmon's counterexamples to the inductive statistical model of explanation (Salmon [1971], 33–40) although in others all the information given in the explanans is causally irrelevant, as is the case when the spontaneous evolution of a process is accompanied by irrelevant intervention. (Salmon's example of administering Vitamin C to cold sufferers fits this case.)

Each of these cases hinges on the difficulty of keeping out causally irrelevant information from propositions. But the problems are not in fact peculiar to propositional representations of explanation, because similar difficulties infect nominalizations of sentences. Consider Dretske's (1973), (1979) use of emphasis to produce terms referring to event allomorphs. Although his examples involved a combination of relevant and irrelevant causal information, we can construct parallel examples which involve a mixture of different kinds of causes. Consider

(1) The Afghan guerrillas' *surprise attack* after sunrise (caused)(explains) the crumbling of the defenses.

(2) The Afghan guerrillas' surprise attack *after sunrise* (caused)(explains) the crumbling of the defenses.

Given conventional military wisdom, (1) seems clearly to be true, while (2) is false. Compared to an attack with advance warning, a surprise attack increases the chances of victory and is a contributing cause to it. Compared to an attack before sunrise, a postdawn attack lowers the chances of victory and is a counteracting cause of it. What is occurring here, as Dretske noted about his examples involving relevant and irrelevant factors, is that stress markers pick out different aspects of spatio-temporal event descriptions and that far from being a pragmatic feature of ordinary discourse, these aspects are genuine features of the world.[27]

It should by now be clear what the common problem is with all these cases. Mere truth of the explanans and explanandum sentences will not prevent a conflation of relevant and irrelevant factors, or of contributing and counteracting causes. Successful reference to multiaspectival events will also allow such conflations. The most direct way to avoid such problems is to select a linguistic form which directly mirrors the separate structure of causal influences. There may well be other means of doing this than the one I have adopted here, but it will, I think, do the trick.

Appendix
The Causal Failures of the Covering-Law Model

It is now, I think, widely recognized that the original covering-law accounts of explanation were seriously defective in their treatment of causal explanations. Without wishing to retell a story already told, it is worth running quickly through the principal counterexamples to that account to bring out the causal nature of each of the failures.

(a) *The Flagpole Example*. Problem: A flagpole of height h casts a shadow of length l. With knowledge of the length of the shadow, of the angle of elevation of the sun, and of elementary laws of geometry and physics, such as the (almost) rectilinear propagation of light, we can deduce the value of h. But stating l does not explain h, although the deduction is a good D-N explanation.

Solution: The problem is clearly due to the fact that the shadow's length does not cause the flagpole's height, whereas the converse is true (because changing the flagpole's height while keeping other factors such as the sun's elevation constant results in a change in the flagpole's shadow,[28] whereas the converse is false.) This example is a classic case of the failure of the regularity analysis to capture the asymmetry of causal relations.

(b) *The Barometer Example*. Problem: One can deduce (or infer with high probability) from the regularity that falling barometer readings are (almost) always followed by storms, together with the statement that the barometer reading dropped, that a storm occurred.[29] Yet this is no explanation of the storm's occurrence, however good a predictor it might be.

Solution: The problem this time is the inability of the regularity analysis to distinguish a relation between joint effects of a common cause from a genuine causal relation. Here the common cause is a drop in the atmospheric pressure, and the two effects are the drop in the barometer reading and the occurrence of the storm. The explanation has failed to cite the cause of the storm, as is evidenced by the fact that altering the barometer reading, perhaps by heating and cooling the instrument, has no effect on the occurrence or nonoccurrence of the storm.

(c) *The Hexed Salt Example*. Problem: A sample of table salt is dissolved in warm water. The 'explanation' offered is that it was a sample of hexed salt, and samples of hexed salt always dissolve in warm water. Once again, this counts as a legitimate explanation under the D-N account, a clearly wrong conclusion.

Solution: As well as citing the cause of the salt's dissolving, which is its immersion in warm water, a factor which is not a cause (its hexing) has been cited. (The hexing is not a cause because changes in the property 'is hexed', from presence to absence do not result in changes in the property 'dissolves'.) On a causal account of explanation, this makes the explanation false, because something which is not a contributing cause has been (implicitly) claimed as a cause when it is not. Yet everything included in the explanans of the D-N 'explanation' is true, including the regularity that hexed salt always dissolves in water. Thus, the answer to the question raised by Salmon ([1984] p. 92) "Why are irrelevancies harmless to arguments but fatal to explanations?" is "Causal irrelevancies destroy the truth of a causal explanation, whereas arguments, which are concerned only with truth and validity, and not causality, have these features preserved under dilution by noncausal truths in the premises."[30]

(d) *Laws of Coexistence*. It has often been claimed that so-called laws of coexistence are not suitable as the basis of a causal explanation. Indeed, in denying that all explanations were causal Hempel ([1965], 352) cites the oft-repeated example of being able to explain the period of a pendulum by reference to its length, but not conversely, and makes the claim that laws of coexistence cannot provide causal explanations (in terms of antecedent events). Curiously, Hempel briefly discusses and then dismisses the essence of the correct response to this example, which is that the causal influence of the length on the period and the absence of a converse influence for a pendulum with a rigid rod is grounded not in a mathematical law statement, but in a physical procedure. Consider a rigid pendulum, but one whose center of mass can be altered by adjusting a bob. If l is the distance between the fulcrum and the center of mass, then the period is given by $t = 2\pi\sqrt{l/g}$. In this case, by physically changing l by means of raising and lowering the bob, the period t will change. However, because the pendulum is rigid, changing the period by forced oscillations will not result in a change in the distance of the center of mass from the fulcrum. Thus it is correct to say that the length of the pendulum is a cause of its period, whereas the period is not a cause of its length.

Suppose, in contrast, that the pendulum was elastic, and had forced oscillations. Then, in this case, we should correctly be able to say that the period of oscillation explains why the pendulum has the length that it does (strictly, the increase in length over the neutral state of no oscillations), because changing the period changes the length of the pendulum but, in this case where the oscillations are forced from outside the system, the converse does not hold.

(e) *The Causal Potential Overdetermination Problem*. In an example of Scriven's, and cited in Hempel (1965), a bridge is demolished by a charge just before a bomb explodes immediately over the bridge. The bomb is sufficient to destroy the bridge, so that a universal generalization of the form $(x)(Fx \rightarrow Gx)$ is true, where $Fx =_{df} x$ has a bomb explode in the immediate vicinity; $Gx =_{df} x$ is destroyed. Furthermore, both Fa and Ga are true. But the explanation of the bridge's collapse is not given by citing this generalization and the fact Fa. The causal potential overdetermination case is in fact a subcase of the explanatory irrelevancy problem but unlike the case where an extra factor which is *always* causally ineffective is cited, as in the hexed salt case, here it is an accidental feature of the situation that a factor which would ordinarily be explanatory is, because it is causally ineffective in this case, explanatorily ineffective.[31]

In cases of actual overdetermination, the situation is opaque. We have no clear criteria for identifying the cause in those cases unless (and this is in fact more common than admitted) the overdetermined event is different because of the presence of two sufficient factors rather than one. So a death by simultaneous action of cyanide and strychine is a different kind of death ffrom one by either poison alone. In fact, I seriously doubt whether, in the ontic mode, there *are* any genuine cases of overdetermination. A factor which left no trace on the effect would have contributed nothing to that effect, and would violate the principle of causal relevance.

Notes

This essay originated with a suggestion by Philip Kitcher and Wesley Salmon that the exposition of aleatory explanations in my (1981a) and (1983) was too cryptic to fully reveal what its ramifications were and exactly where it differs from other approaches. I should like to thank them for providing the opportunity to write a fuller account, and for stimulating discussions during the NEH seminar. I have also benefited from comments on earlier drafts by Robert Almeder and David Papineau. Earlier versions were read at the Unversity of Minnesota and the London School of Economics.

1. See Gottlieb et al. (1981), Masur et al. (1981), and for an early survey, Gottlieb et al. (1983).

2. Here and occasionally elsewhere I use the definite article for convenience. As we shall see below, the indefinite article is greatly to be preferred, so that we escape the prejudice that there is a unique explanation for any given phenomenon.

3. Etymological note: the original meaning of 'research' is a search for something; it has as source

the French 'recherche.' The modern, much looser meaning would correspond to 'rechercher,' as in re-search.

4. Strawson (1985) p. 115, for example, writes "we also speak of one thing explaining, or being the explanation of, another thing, as if explaining was a relation between the things. And so it is. But it is not a natural relation in the sense in which we perhaps think of causality as a natural relation. . . . It does not hold between things in the natural world, things to which we can assign places and times in nature. It holds between facts or truths. The two levels of relationship are often and easily confused or conflated in philosophical thought." The remarks in section IV of Davidson (1967) also exemplify this position, which is widely held. One aim of the present paper is to argue that this position ought to be reconsidered. The almost exclusive emphasis on linguistic explanations characterizes such otherwise diverse accounts as Aristotle's (*Posterior Analytics*, Book I, ch. 13); Popper (1959), §12; Nagel (1961); Hempel (1965); Bromberger (1966); Friedman (1974); Railton (1978); Fetzer (1981), chs. 5, 6; Kitcher (1981); Niiniluoto (1981); Brody (1973); van Fraassen (1980), ch. 5; Achinstein (1983). The emphasis on linguistic carriers of explanations is particularly striking in the case of those who have noted the central role of causation in explanation, yet have retained a sentential structure for their analyses. For example, Brody (1973), pp. 23, 25, explicitly retains the deductive-nomological explanatory framework, even while supplementing it with causal and essential properties.

5. See Appendix for a list of these failures.

6. See, in particular, Kim (1973) for a convincing set of arguments on this point.

7. Salmon (1984) is the best presentation of a theory along these lines. A formal development of probabilistic causality is given in my (1986a).

8. Probably the most philosophically thorough of these, as well as one of the earliest, is Bhaskhar (1975).

9. I acknowledge here that there are other kinds of explanation than those which cite causes of the explained phenomenon. Achinstein (1983) p. 93 notes three: an explanation of what is occurring; an explanation of the significance of something; an explanation of the purpose of something. There are many more uses of the term "explanation" in English—we ask a miscreant for an explanation of his behavior (give reasons for his actions) and an engineer for an explanation of how a pump works (this is often close to a causal explanation but emphasizes mechanisms). The fact that the English language contains such a variegated set of uses for the term 'explanation' is one reason why it seems preferable to work from causes to causal explanations rather than from a general sense of explanation down to a subcase.

10. See Harper (1975), pp. 139–42.

11. Too great an emphasis on causal sufficiency obscures the role played by multiplicity. Multiple causation within a framework which insists that causes must be sufficient for their effects leads to over-determination, with the consequent difficulty of identifying causes. Of course, sufficient causes may have multiple components, as they do in INUS conditions for example. However, necessary conditions are merely a special case of contributing causes (q.v.), and *sine qua non* accounts in general are unable to represent counteracting causes in the sense intended below. For additional defects in the *sine qua non* approach, see my (1981b), §1.

12. Crudely, contributing causes increase the chance of the effect, counteracting causes lower it. A fuller discussion of these two types of causations may be found in §4 below and in my (1983).

13. The plague is estimated to have killed almost one third of the population of Europe during the Black Death of the fourteenth century. One wonders how much a single vial of tetracycline would have fetched at auction then, had it been available!

14. This is to be compared with Hempel's (1965), p. 334 "Why is it the case that p?" where p is a propositional entity, and Salmon's (1984), p. 34 "Why is it that x, which is an A, is also a B?" where x is an individual, and A and B are properties. I emphasize here, however, that linguistic explanations are not to be necessarily construed as answers to questions. They stand as entities in their own right.

15. This may illustrate one difference between explanation and understanding, for although it is only the contributing causes which explain the outcome, a specification of the counteracting causes is necessary in addition for a full understanding of how the phenomenon came about.

16. For a treatment of some of these factors, see Kibble (1966), pp. 151–54.

17. As, for example, does Quine (1970), p. 30. In Quine (1985) this is reiterated, with some reservations (p. 167).

18. The structure of events that I have used is essentially that which Jaegwon Kim has employed and argued for in a number of subtle and interesting papers (1971), (1973), (1976). I refer the reader to those articles for a detailed exposition of Kim's views. The arguments which led me to adopt this account are rather different from those used by Kim, which emphasize the logical structure of events.

19. These include the accounts given in Hempel (1965) (1968), Salmon (1984), Fetzer (1981), Railton (1978). For Hempel, the probability is a logical probability, which gives the degree of inductive support afforded to the explanandum sentence on the basis of the explanans. In Railton's D-N-P model of explanation, the probability value of the explanandum event's occurrence appears explicitly in both the probabilistic laws occurring in the explanans and in the statement of the probability of the explanandum fact (1978, p. 218). In Fetzer's causal relevance model, the strength of the propensity value occurring in the probabilistic conditional is an essential feature of the explanation. Salmon, I think, has come very close to giving up the probability requirement (1984, p. 266), but not completely because he asserts "we must give serious consideration to the idea that a probabilistic cause need not bear the relation of positive statistical relevance to its effect. . . . The answer . . . lies in the transmission of probabilistic causal influence." (p. 202) "The basic causal mechanism, in my opinion, is a causal process that carries with it probability distributions for various types of interaction." (p. 203) For a more detailed discussion of this point, see Humphreys (1986b).

20. This claim relies on interpreting a standard solution to the problem of the single case in a particular way, viz: A probability attribution $P(A/B) = r$ is *true* if and only if there is no further factor C such that $P(A/BC) \neq r$. Discussions of the single case issue tend not to talk in terms of truth, but of appropriate or correct reference classes or sequences. In fact, when discussing the problem of ambiguity (e.g., Hempel [1968], p. 117), Hempel claims that two incompatible relative frequency statements can both be true. In the sense that the relative frequencies $P(A/B) = r$ and $P(A/B') = r'$ are the frequencies relative to some reference class, this position is correct. However, because we are concerned with the explanandum event itself, and not a representation of it in a class, the appropriate probability is that of the single case, not of the type within a class. One also cannot preserve the ambiguity of probability values by claiming that applications of relative frequencies to single cases are relative to the explanandum sentence which describes the explanandum phenomenon, because then the two different probabilities would not be applied to the same object.

21. I myself doubt whether there are many genuine cases of pure probability distributions. This fact is disguised by the common use of uninterpreted parameters in representations of probability distributions, whereas even in cases such as the binomial distribution for coin tossing, the parameter p is a function of the center of gravity of the coin, and in the exponential distribution for radioactive decay, the parameter λ is a function of the atomic number of the atom. These factors are, it is true, usually structural aspects of the systems, but that does not necessarily rule them out as contributing factors. For more on this matter, see my (1986c).

22. I call a maximal specificity condition any condition which requires that all probabilistically relevant factors must be cited in an objective, nonepistemically relativized explanation.

23. I believe that this kind of causal approach also captures rather better than do traditional accounts how we approach closer to the whole truth. Many accounts of Carnapian verisimilitude use a counting measure on the degree of correspondence between correct state descriptions and proffered state descriptions. That can be replaced by a similar counting measure on $\Phi \cup \Psi$. One can make this more precise, and include a measure of the relative contributions of the causal factors to Y, by using such concepts as explained variance, but I shall not pursue that here.

24. I was myself unconsciously trapped by this in my first paper (1981a) where I unthinkingly used the mixed form of 'A because Φ, despite Ψ', which requires A and Φ to be propositional, yet Ψ to refer to events. I hence slurred over the distinctions by using 'despite the fact that' and the alternative 'even though' for counteracting causes, thus managing to stay within the propositional form. I have not discussed here the variant of the standard form 'Y in contrast to Z because X', which is used by van Fraassen (1980) and other pragmatically oriented writers. Much of what is said in this paper carries over with obvious modifications, although the issue of whether all explanations are comparative in form is not something I can cover here.

25. I use 'sentence' and 'proposition' as stylistic variants of each other here. No ontological distinction is intended.

26. As one might have expected. An earlier debate about linguistic evidence for and against fact-like entities as effects proved inconclusive. See Vendler (1962) and the accompanying papers by Bromberger and Dray. For the arguments that these debates are inconclusive, see Shorter (1965), especially p. 155.

27. The role played by emphasis in posing why-questions, as outlined by van Fraassen (1980), ch. 5 §4, poses different problems, in particular the generation of contrast classes and the related issue of whether questions are implicitly contrastive in form. Although these are important issues, I do not propose to deal with them here. (See note 24 above.)

28. Except for singularities at 0 and $\pi/2$ angles of elevation.

29. It is not evident to me that this regularity is lawlike, for the subjunctive conditional "Were that barometer reading to fall, a storm would occur" requires special constraints on the worlds in which the antecedent is true to preclude, among other things, human intervention from producing an artificial drop in the needle reading. Indeed, because this possibility is often present in these cases of causally spurious associations of common causes, many such regularities will fail the lawlikeness condition.

30. This answer is essentially that given by Salmon. Even though he phrases his answer in terms of statistical-relevance rather than causal relevance (1984, p. 96), it is clear from the causal apparatus employed later in the book that it is causal relevance that is meant.

31. Similar examples have been used by Achinstein (1983, p. 168) to argue that although it is often possible to determine *a priori* whether the formal requirements for a model of explanation have been met (as one can in Scriven's example) one also needs to have *a posteriori* knowledge beyond knowing the truth of the explanans—in Scriven's case we would need to know that no other cause of the bridge's collapse except the explosion of the bomb was present. Indeed, all Achinstein's examples against various models (168–70, 177–78, 180) are of the potential overdetermination type, and the knowledge required to avoid the problem is the knowledge of what the actual causes of the explanandum were. Thus because aleatory explanations require citation of only causal factors, they are not subject to this objection. Nevertheless, Achinstein would rule out aleatory explanations on the grounds that they violate what he calls the "no explanation by singular sentence" (NES) condition. This condition (p. 159) asserts that "no singular sentence in the explanans (no sentence describing *particular* events) and no conjunction of such sentences, can entail the explanandum." The three reasons he cites for imposing this condition are (a) to ban self explanation (b) to require that laws play an essential role in explanations (c) to remove explanatory connectives from the explanans. The first two of these conditions are satisfied by aleatory explanations. No empirical phenomenon is self caused. Our requirement of unconditionalness entails that causal relations ground laws. Regarding the third, no explicit mention of "causes" or "explains" appears in the explanans of an aleatory explanation. However, because Achinstein views the task of an explanatory model to analyze (away) terms such as 'causes' and 'explains', their appearance, implicitly or explicitly, in a model of explanation violates the constraints on such a task. I see no reason to impose the NES condition on aleatory explanations. As previously stated, our task is not to analyze the term 'causation', but to show how causal knowledge can be cumulatively used to provide explanatory knowledge.

References

Achinstein, P. 1983. *The Nature of Explanation*. Oxford: Oxford University Press.

Bhaskhar, R. 1975. *A Realist Theory of Science*. New York: The Free Press.

Brody, B. 1973. Towards an Aristotelean Theory of Scientific Explanation, *Philosophy of Science* 39: 20–31.

Bromberger, S. 1966. Why-Questions. In *Mind and Cosmos*, ed. R. Colodny. Pittsburgh: University of Pittsburgh Press, 86–111.

Crick, F. and Watson, J. 1954. The Complementary Structure of Deoxyribonucleic Acid, *Proceedings of the Royal Society* A, 223: 80–96.

Davidson, D. 1967. Causal Relations, *Journal of Philosophy* 64: 691–703.

Dretske, F. 1973. Contrastive Statements, *Philosophical Review* 82: 411–37.
——. 1979. Referring to Events. In *Contemporary Perspectives in the Philosophy of Language*, P. French, T. Uehling, H. Wettstein, eds. Minneapolis: University of Minnesota Press.
Fetzer, J. 1981. *Scientific Knowledge*. Dordrecht, Holland: D. Reidel.
Friedman, M. 1974. Explanation and Scientific Understanding, *Journal of Philosophy* 71: 5–19.
Gottlieb, M. et al. 1981. *Pneumocystis Carinii* Pneumonia and Mucosal Candidiasis in Previously Healthy Homosexual Men, *New England Journal of Medicine* 305: 1425–31.
——. 1983. The Acquired Immunodeficiency Syndrome, *Annals of Internal Medicine* 99: 208–20.
Harper, H. A. 1975. *Review of Physiological Chemistry* (15th Ed). Los Altos: Lange Medical Publications.
Hempel, C. 1965. Aspects of Scientific Explanation. In *Aspects of Scientific Explanation and Other Essays in the Philosophy of Science*. New York: The Free Press.
——. 1968. Maximal Specificity and Lawlikeness in Probabilistic Explanation, *Philosophy of Science* 35:11 6–33.
Humphreys, P. 1981a. Aleatory Explanations, *Synthèse* 48: 225–32.
——. 1981b. Probabilistic Causality and Multiple Causation. In *PSA 1980*, Vol. 2, eds. P. Asquith and R. Giere. East Lansing, MI: Philosophy of Science Association.
——. 1983. Aleatory Explanations Expanded. In *PSA 1982*, Volume 2, eds. P. Asquith and T. Nickles. East Lansing, MI: Philosophy of Science Association.
——. 1986a. Quantitative Probabilistic Causality and Structural Scientific Realism. In *PSA 1984*, Volume 2, eds. P. Asquith and P. Kitcher. East Lansing, MI: Philosophy of Science Association, 329–42.
——. 1986b. Review of Salmon (1984) in *Foundations of Physics* 16: 1211–16.
——. 1986c. Philosophical Issues in the Scientific Basis of Quantitative Risk Analyses. In *Biomedical Ethics Reviews 1986*, eds. R. Almeder and J. Humber, Clifton: Humana Press, 205–23.
——. 1988. Causal, Structural, and Experimental Realisms. In *Midwest Studies in Philosophy*, Volume XII, eds. P. French, T. Uehling, and H. Wettstein. Minneapolis: University of Minnesota Press.
Kibble, T. 1966. *Classical Mechanics*. New York: McGraw Hill.
Kim, J. 1971. Causes and Events: Mackie on Causation, *Journal of Philosophy* 68: 426–41.
——. 1973. Causation, Nomic Subsumption, and the Concept of Event, *Journal of Philosophy*, 70: 217–36.
——. 1976. Events as Property Exemplifications. In *Action Theory*, eds. M. Brand and D. Walters. Dordrecht, Holland: D. Reidel.
Kitcher, P. 1981. Explanatory Unification, *Philosophy of Science* 48: 507–31.
Masur, H. et al. 1981. An Outbreak of Community-Acquired *Pneumocystis Carinii* Pneumonia, *New England Journal of Medicine* 305: 1431–38.
Mill, J. S. 1856. *A System of Logic*. New York: Harper and Brothers.
Nagel, E. 1961. *The Structure of Science*. New York: Harcourt, Brace and World.
Niiniluoto, I. 1981. Statistical Explanation Reconsidered, *Synthèse* 48: 437–72.
Popper, K. 1959. *The Logic of Scientific Discovery*. London: Hutchinson and Company.
Quine, W. 1970. *Philosophy of Logic*. Englewood Cliffs, NJ: Prentice-Hall.
——. 1985. Events and Reification. In *Actions and Events*, eds. E. LePore and B. McLaughlin. Oxford: Basil Blackwell.
Railton, P. 1978. A Deductive-Nomological Model of Probabilistic Explanation, *Philosophy of Science* 45: 206–26.
Salmon, W. 1971. Statistical Explanation. In *Statistical Explanation and Statistical Relevance*, ed. W. Salmon. Pittsburgh: University of Pittsburgh Press.
——. 1984. *Scientific Explanation and the Causal Structure of the World*. Princeton: Princeton University Press.
Shorter, J. 1965. Causality, and a Method of Analysis. In *Analytic Philosophy* (Second Series), ed. R. J. Butler. Oxford: Basil Blackwell.
Snow, J. 1855. *On the Mode of Communication of Cholera* (2nd Edition). Reprinted in *Snow on Cholera*. New York: The Commonwealth Fund, 1936.

Strawson, P. 1985. Causation and Explanation. In *Essays on Davidson: Actions and Events*, eds. B. Vermazen and M. Hintikka. Oxford: The Clarendon Press, 115–36.

Suppes, P. 1970. *A Probabilistic Theory of Causality*. Amsterdam: North Holland.

van Fraassen, B. 1980. *The Scientific Image*. Oxford: The Clarendon Press.

Vendler, Z. 1962. Effects, Results, and Consequences. In *Analytic Philosophy*, ed. R. J. Butler. Oxford: Basil Blackwell, 1–15.

Pure, Mixed, and Spurious Probabilities and Their Significance for a Reductionist Theory of Causation

1. Introduction

What should we make of such facts as the fact that smoking is probabilistically associated with cancer, or the fact that there is a high correlation between the amount of sun in August in France and the size of the French grape harvest, or the fact that putting a piece of lead under the "1" on a die makes it more likely that the die will land with the "6" face up when thrown? That is, what should we make of facts of the form P(S/C) > P(S/-C)? In particular, I want to ask, what do such correlational facts have to do with causation?

Some philosophers are inclined to say that, give or take a bit, these correlational facts *constitute* causal connections. On this view we have causation as soon as we have positive statistical relevance between two temporally ordered event types, that is, as soon as the probability of the later event given the earlier event is greater than its probability without the earlier event.

I shall call this the "statistical-relevance" view (the S-R view for short). The name most often associated with this view is Wesley Salmon (see in particular Salmon 1970). However, as I shall have occasion to observe later, Salmon's theory of causation has a number of different aspects, and indeed his most recent position is not really a version of the S-R view at all. I shall take Patrick Suppes's (1970) to be the canonical version of the S-R view.

Even the canonical version of the S-R view needs more explanation than has been offered so far. But let me leave this to one side for a moment. For my initial concern is not with the details of the S-R view, but with the whole idea. I want to raise the question of whether we ought to admit indeterministic causation at all.

Contrast the S-R view with the traditional view that causes have to *determine* their effects. On this view, if C doesn't determine E, it doesn't cause it either. Causation isn't just a matter of the cause making the effect more probable, but of its making the effect *certain*.

I want to suggest that this traditional deterministic view is a viable alternative to the S-R view. At first sight this suggestion might seem misguided. Doesn't modern physics show that indeterminism is a basic feature of the physical world? But this isn't the issue. I don't hold any brief for determinism itself. I accept that determinism is false. The view I want to put forward as an alternative to the S-R view is not determinism, but a deterministic view of *causation*. This allows that certain events aren't determined. It simply insists that, insofar as they aren't determined, then they aren't caused either.

Let me spell the point out. Defenders of the deterministic view of causation will claim that circumstances that don't determine some particular E aren't enough to cause E. But this needn't be because they think that there must be more to be said about the case in question, because they think that we haven't yet succeeded in identifying those further features of the situation which made E inevitable. They can accept that there may have been no such further features. Their claim is simply that if there were no such further features, then E wasn't caused: if the prior circumstances didn't make E inevitable, then they shouldn't be counted as causing E. (Of course if *nothing* were ever determined, then a deterministic view of causation would imply that nothing was ever caused, and that would perhaps be indefensible. But, as we shall see, allowing that some things aren't determined doesn't imply that everything is undetermined. In particular, in the paradigm cases where some E is undetermined, it will still be true that E's having a certain *chance* will be determined by the prior circumstances.)

So the initial issue to be discussed here is the choice between the traditional deterministic and the S-R views of causation. In a previous paper I argued in favor of the traditional view over the S-R view (Papineau 1985a). However, I am no longer as persuaded as I was then that the traditional view holds a definite advantage. I now think that, when all the arguments are taken into account, the result is pretty much a stand-off between the two sides. It doesn't make much difference which view we uphold. Both theories can be made to accommodate all the relevant data, albeit in different ways, and in the end the choice is as much terminological as substantial.

The first aim of this paper is to survey the relevant arguments and show that they lead to a stand-off. This survey will simplify some of the arguments of my 1985a paper, and it will also introduce a number of new considerations. In the course of this survey I shall have occasion to turn to a second issue, an issue which arises independently of whether we take causes to make their effects certain or merely probable. This is the question of whether a full *reduction* of causal facts to facts of correlational dependence is possible. In the last few years a number of philosophers (including Salmon) have become persuaded that causation somehow transcends correlational relationships, that there is some extra cement in causal connections that resists reduction to facts of correlational association. I want to argue, against these writers, that causation can be so reduced.

So I intend to show two things. First, that there is nothing to choose between the S-R and traditional deterministic views. And, second, that whichever we do choose, we can reduce causal facts to correlational facts—to facts about some events making others more likely, if we prefer the S-R view of causation, or to facts about some events making others certain, if we prefer the deterministic view of causation.

I now think that this second topic, the possibility of a reduction of causation, is a more substantial philosophical issue than the conflict between the S-R view and the traditional deterministic view of causation. The conflict between the S-R and traditional views eventually degenerates into an unimportant trading of preferences. But, even so, it will be worth exploring at some length. For the issues involved, and the distinctions that need to be made, will prove essential to resolving the question of reduction.

2. Some Initial Intuitions

We have conflicting initial intuitions about causation. On the one hand, there is an initial intuition that causation demands determinism. If, on a given occasion, a full specification of the circumstances left it open that E might *not* have occurred, then how can we say those circumstances *caused* E? On other qualitatively identical occasions E sometimes fails to occur. Doesn't this show that those circumstances aren't enough to cause E on their own?

On the other hand, it also seems intuitively plausible that a probabilistic connection between two properties establishes causation. If smoking makes cancer more likely, doesn't this show that smoking *causes* cancer?

Let us examine the deterministic intuition first. This isn't as straightforward as it looks. For it can be argued that the underlying intuition here is not about causation as such, but about *explanation*: it is the intuition that we haven't fully *explained* E if the circumstances we cite don't make E certain.

To translate this into a conclusion about causation itself we need to assume further that causation is inseparable from full explanation, that E couldn't be caused unless it were fully explainable. But this further assumption is contentious. Thus Philip Kitcher thinks that explanation requires certainty, but that causation only requires increased probability (Kitcher 1985, 638). In a slightly different vein, D. H. Mellor thinks that certainty is required for *full* explanation, but holds that explanation, like causation, comes in degrees; and that we have an explanation, albeit not a full one, whenever the circumstances increase the probability of E (Mellor, forthcoming).

And then, of course, there are also philosophers who simply deny the underlying intuition, and who say that if the prior circumstances increased E's probability, then that's as good an explanation as we ever have. (See the papers by Salmon, Richard Jeffrey, and James Greeno in their 1970 publication. Salmon also argues that if a complete description of the circumstances *reduces* the probability of E, then that amounts to a satisfactory explanation too. But we can leave that to one side for the moment. Whatever other virtues this suggestion might have, it's certainly not supported by any initial intuitions. I shall say something more about such "negative causes" in section 18 below.)

In this paper I want to talk about the objective relationship of causation, not about the anthropocentric idea of explanation. So rather than get bogged down in the connections between the two notions, I shall simply concede the point at issue. In what follows I won't make any further appeal to the intuitive idea that causation requires certainty.

But doesn't this now concede the whole argument to the S-R view? The contrary intuition, that an increase in probability suffices for causation, now seems to have the field to itself. However, I think this intuition is also far less straightforward than it looks. Indeed I think I can show that, despite appearances, this intuition is quite compatible with a deterministic view of causation.

I shall devote the next six sections, 3-8, to explaining this last claim. This will leave the S-R and deterministic views on an equal footing as far as initial intuitions go. I shall then turn, in sections 9-12, to some considerations involving the relation between causation and the rationality of action. These considerations too will leave the issue between the S-R and deterministic views of causation undecided. But they will focus the question of the reducibility of causation. This question will be pursued in sections 13-18. Section 18 will also contain some final comments on the choice between the deterministic and S-R views.

3. Pure and Mixed Probabilities

I shall say that a conditional probability of the form $P(C/S)$ is *mixed* if there exists some Z such that $P(C/SZ) \neq P(C/S)$. If there is no such Z, I shall say the original conditional probability is *pure*.

This distinction is related to the distinction between homogeneous and inhomogeneous partitions of a reference class. If $P(C/S)$ and $P(C/-S)$ are pure, then the partition of the overall reference class into smokers and nonsmokers gives us a homogeneous partition with respect to cancer, in the sense that subdividing the reference class by additional factors Z won't make any further difference to the probability of cancer: given that you are a smoker (or a nonsmoker), nothing else about you makes any difference to the probability of your getting cancer. But if either $P(C/S)$ or $P(C/-S)$ is mixed, then the partition into S and not-S is inhomogeneous, in that further subdivisions, by Z and -Z, will alter the probabilities. The original partition ignored distinctions between S&Z and S&−Z (and −S&Z and −S&−Z) which are in fact relevant to the probability of cancer.

Pure probabilities give the single-case *chances* in particular cases. The chance of C in a particular case will be the probability conditional on all the relevant circumstances: and since "relevant circumstances" here means factors whose presence or absence makes a difference to the conditional probability, such chance-

prob C

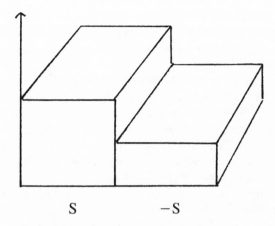

S   −S

Figure 1. S probabilistically relevant to C.

prob C

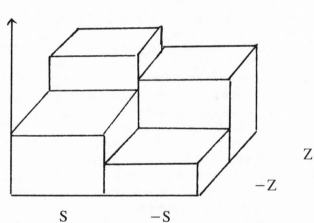

S   −S

Figure 2. Z mixes P(C/S) and P(C/−S).

giving conditional probabilities will always be pure. Conversely, mixed probabilities are in general no good for giving single-case chances, since further conditionalization on circumstances present in the particular case will generally alter a mixed probability.[1]

I take it that it is uncontroversial that the correlations with which I began are mixed. Surely there are other differences between people, apart from whether they smoke or not, which affect the probability of their getting cancer. Surely the amount of sunshine in August isn't the only thing that makes a difference to the size of the grape harvest. Surely, even given a loaded die, there is further room for craps to be "a game of skill," in the Runyonesque sense.

What is more, I take it that this mixedness cannot be eliminated simply by appealing to further things we know about. We do know about some of the further things that make a difference to the probability of cancer, such as working in asbestos factories, or living in dirty air, etc. But even when we have partitioned our reference class by such further known factors, there will still no doubt be some further unknown factors making our probabilities mixed: it is scarcely likely that the medical researchers have identified *all* the factors that are probabilistically relevant to cancer. And similarly in the other cases. Even after we have taken into account all the things that the viniculturists and the dice experts know about, there will still be further unknown factors that make a probabilistic difference, and the probabilities that we will be left with will still be mixed. (Note that I am not making the false deterministic assumption that the only real chances are nought and one. I am simply making the uncontroversial point that we are almost certainly ignorant of some of the factors relevant to the real chances.)

4. Screening Off and Spurious Correlations

What do these last remarks, about the mixedness of our original probabilities, imply about their causal status? If those probabilities are mixed, and if, moreover, our current state of knowledge does not allow us to make them pure by adding in further factors, then it might seem that, on any account of causation, we cannot trust those correlations as indicators of causal conclusions.

But we should not dismiss our original correlations too quickly. At this point we need to make a further distinction. Suppose that both $P(C/S)$ and $P(C/-S)$ are mixed with respect to some Z: Z makes a further difference to the probability of cancer, among both smokers and nonsmokers. We need to distinguish between cases where Z *screens C off from S*, and those where it doesn't.

Z is a "screener-off" in this sense if $P(C/S\&Z) = P(C/Z)$ and $P(C/S\&-Z) = P(C/-Z)$. Once we divide the reference class into Z and -Z, it turns out that smoking makes no real difference to the probability of cancer after all. Intuitively, screening off happens when Z is a common cause of S and C. Suppose there is a gene which, on the one hand, makes people likely to smoke, and, on the other hand, predisposes them to cancer. Among people with the gene, the smokers are no more likely to get cancer than the nonsmokers; and similarly among those without the gene. The original smoking-cancer correlation turns out to be due entirely to the fact that smoking is itself a symptom of the presence of the gene, in that the gene is present more often among smokers than among nonsmokers.

Note that not every Z which renders an S-C correlation mixed is one which

prob C

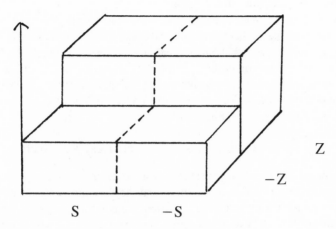

Figure 3. Z screens off C from S.

screens C off from S. A "screener-off" isn't just any old further factor that alters the probability of C. It alters the probability in a quite specific way: in changing it from P(C/S) to P(C/S&Z) it makes it equal to P(C/Z). A "mixer" just alters the probability somehow. A "screener-off" makes it equal to a specific number.

The fact that the existence of mixers does not imply the existence of screeners-off is important. A screener-off shows that S is in fact irrelevant to the probability of C. If Z screens C off from S, then partitioning by Z and -Z shows that the original division into S and -S is no longer necessary. But this irrelevance doesn't necessarily follow whenever Z renders the S-C correlation mixed. That S by itself leaves us with an inhomogeneous partition for C isn't yet any reason for thinking S isn't *part* of what is required for homogeneity. Even if there are unknown physical differences, Z, between people, which mean that some are more likely to get cancer than others, this doesn't mean that smoking doesn't make an extra difference as well. It may still be that P(C/S&Z) is greater than P(C/Z-by-itself). So an S-C correlation can be mixed and S can still be genuinely relevant to C: S can still be needed for the homogeneous partition which gives the pure probabilities of C.

Indeed we can say something stronger. Not only does mixedness allow that S *may* still be needed for the homogeneous partition, but, more strongly, if there aren't in fact any screeners-off, then an original S-C correlation shows that S *must* be part of the homogeneous partition for C. If there are no further factors which yield a partition which has no use for S, then S will still be needed when we get down to the final partition which gives us the pure probabilities of C.

Recall once more the probabilities with which I began. Obviously, as I have already argued, these probabilities are mixed, both by Z's we can identify and, no doubt, by some we can't. But this doesn't mean that there is a Z which screens cancer off from smoking, or the grape harvest off from the August sun, or getting "6" 's off from the die being loaded. And I take it that in fact there aren't any such screeners-off. I accept that our original probabilities are mixed. But I don't think they are *spurious*. I don't think that there are any Z's which will show that the apparent causes are merely symptoms of the apparent effects, which will show that at bottom smoking (or sun, or being loaded) makes no real difference to cancer (or the harvest, or getting a "6").

5. Spuriousness and Statistical Research

How can we have any epistemological confidence that probabilities like the above aren't spurious, once we admit that they are mixed by unknown factors? Doesn't the admission that there are further unknown Z's which alter the probabilities always leave us with the danger that some of those Z's will be screeners-off? It seems as if we will only be able to dispel the epistemological danger of spuriousness if we can identify all the Z's and bring them explicitly into our analysis.

But note that to deal with the epistemological danger of spuriousness we need only take into account those other Z's *that S is itself associated with*, not all the other mixers. Only factors that are themselves associated with S are potential screeners-off. If S makes no difference when you do take Z into account, then it can only seem to make a difference when you don't take Z into account, if S itself happens with Z more often than without.

So to rule out spuriousness, we don't need to take all other Z's into account. We only need consider those that might be associated with S. I take it that this requirement is satisfied in our examples. We haven't identified all the factors that make a difference to the chance of cancer. But this doesn't mean that we haven't looked at all those that might themselves be associated with smoking. Thus, for instance, I presume that tobacco-sponsored researchers have checked to see if anxiety, which is itself associated with smoking, screens off smoking from cancer, and I presume that they have discovered it doesn't. But I expect that even the cigarette companies haven't bothered to take explicit account of, say, pre-natal developmental abnormalities of the lungs, for although these may well increase the chance of cancer, it is hard to see how they could possibly be associated with smoking.

The point I am making here is of fundamental importance for much medical, agricultural, biological, psychological, and sociological research. (Incidentally, it was my interest in the logic of this kind of research that got me interested in the connection between probabilities and causes in the first place.) Statistical data in such areas almost inevitably yield mixed probabilities, not pure ones. If mixing couldn't be separated from screening off without taking all relevant factors into account, then causal conclusions in these areas would be impossible. So it is fortunate that the medical, etc., researchers don't need to take all factors into account, but are entitled to draw causal conclusions once they have explicitly considered all possibly "confounding" factors—that is, once they have explicitly considered all those factors which might themselves be probabilistically associated with the putative cause.

This of course is why "randomized" experiments are so important. If you can split your sample into cases which get the putative cause, and cases which don't, at random, then you can be sure that the other factors relevant to the effect will themselves be probabilistically independent of the putative cause. And this then means that you can be sure that any association between the putative cause and effect won't be spurious. (Of course you might get *sample* associations which were misleading about population probabilities. But that's different. Here and throughout I'm prescinding from evidential issues and assuming we're dealing with real probabilities. I intend spurious probabilities, and, for that matter, mixed probabilities, to be understood as real population probabilities. Spuriousness isn't a matter of misleading samples, but of misleading causal significance.)

We can't always conduct randomized experiments. Often there will be moral or practical reasons why we can't manipulate the data. In such cases we will need to conduct surveys instead, and to try to ensure that all possible confounding factors are taken into explicit account. This demand obviously raises difficulties that are avoided by randomizing. But there is no reason to suppose that this demand can't often be satisfied. Even if we can't experiment on smokers, we can still survey for all the factors that might be confounding the association between smoking and cancer, and, as pointed out above, can achieve some assurance that all potential confounders have been catered for. In principle, no doubt, there always remains a possibility that our causal conclusions are in error. But that is a feature of empirical research in general, not a danger that's peculiar to inferences from surveys.

6. The Importance of the Single Case

I said that my initial intention in this paper was to show how the deterministic view of causation can withstand the intuitive plausibility of the S-R view. So far I have distinguished between spurious and mixed cases of statistical relevance, and pointed out that spurious probabilities are misleading as to causes. But doesn't this simply add weight to the S-R view? After all, no S-R theorist has ever wanted to say that *any* case of probabilistic relevance establishes causation. (Even first-year sociology students don't want to say that correlation always equals causation.) According to Suppes, an initial correlation between S and C shows only that S is a *prima facie* cause of C. For S to be a *genuine* cause of C, there needs in addition to be no Z which screens C off from S.

But isn't that just what I have said? I have urged that we take smoking to cause cancer because, even though we believe the original correlation to be mixed, we don't think it's spurious. This looks just like Suppes. Smoking is a genuine cause, as well as a *prima facie* one, because nothing screens C off from S.

But in fact Suppes's view is untenable. Suppose we have this kind of set-up. A genetic factor, Y, increases the chance of cancer in all those people who possess it. Smoking also, and independently, increases the chance of cancer, both in people with Y and in those without Y. But it only does so in conjunction with some further unknown metabolic condition X (smoking makes no difference at all to your chance of cancer if you lack X). So you have an increased chance of cancer if you either have the gene Y, or smoke and have the metabolic factor X, or both. Suppose nothing else is probabilistically relevant to C. Then, while P(C/S) is mixed, by X and by Y, neither of these further factors screens C off from S (smoking makes an extra difference both among X's and among Y's). So, according to Suppes, S is a genuine cause of C. But now consider this case. Mr. Jones smokes, and gets cancer. Presumably, according to Suppes, his smoking then caused his cancer. But suppose Jones didn't have X, but did have Y. In that case it seems clear that, even though we might misjudge the situation, his smoking wasn't in fact any part of what caused his cancer.

This case ("the case of the misleading cigarettes") shows conclusively that we can't rest with the canonical statistical-relevance view. Even if there is a sense in which unscreened-off correlations are important for causation—as indeed they are—it's not true that A causes B whenever A and B are nonspuriously associated.

But while the case of the misleading cigarettes causes difficulties for Suppes's theory, it is not hard to see what we ought to say about such cases. Jones's example shows that causal relationships in particular cases depend on single case chances, not population probabilities. The chance of *Jones* getting cancer given that he smoked was no greater than the chance he would have had if he hadn't smoked, for he lacked the relevant metabolic factor X. In the population at large nothing screens smoking off from cancer, and so the association between smoking and cancer isn't spurious. But that doesn't matter to the analysis of Jones, given that *his* smoking was irrelevant to *his* chance of cancer.

The moral of the story is that causal facts about particular cases depend on *pure* probabilities. It's not enough to make sure that the probabilities we are dealing with are not spurious. For as long as they remain mixed, we are in danger of being misled about the particular case. We need, so to speak, to get down to a homogeneous partition of the reference class. For unless we know which homogeneous cell Jones himself is in, we are in danger of attributing his chance of cancer to factors which matter in other cells, but not in his.

I might now seem to be taking back the points made in sections 3–5. There I said that we didn't need purity for causal conclusions: mixedness was quite all right, as long as it wasn't spurious. But now I'm saying that we do need purity.

The solution to this conundrum, such as it is, is that there are two kinds of causation at issue. In this section I have been focusing on conclusions about specific cases: did *Jones's* smoking cause *his* cancer? But in earlier sections it was a general, population-relative question: is smoking *a* cause of cancer? It was this latter, population-relative question which I argued earlier can be decided by mixed nonspurious probabilities. For the population-relative question isn't about any particular cell of the homogeneous partition, but about the overall structure of that partition: in effect it is the question of whether smoking is relevant to *any* cell in the homogeneous partition. The point made earlier was that a nonspurious mixed correlation is necessary and sufficient for a positive answer to this question.

So let us distinguish two kinds of causation: "single case causation" and "population causation." The population notion is a perfectly natural notion of causation. It is the notion that we have in mind when we conclude from the statistics that smoking causes cancer. And, as we shall see in section 9, it is, give or take a bit, the notion that is relevant to rational decisions about action. But, even so, single case causation is clearly the more basic notion. In cases like the misleading cigarettes, our intuitions are unequivocally that the smoking didn't *cause* the cancer, even though smoking is a perfectly good population cause of cancer. And indeed population causation is definable in terms of single case causation: to say that smoking is *a* cause of cancer is simply to say that smoking is *sometimes* a single case cause of cancer (namely, when conjoined with the relevant background factors).[2]

7. The Compatibility of Probabilistic Intuitions with a Deterministic View of Causation

In the last section I showed that Suppes's original formulation of the S-R view won't do as an account of the basic single case notion of causation. But this scarcely eliminates the S-R view as such. For the obvious response is to tighten up the S-R view in light of the distinction between single case causation and population causation. Thus, for instance, the S-R view could be put forward specifically as an analysis of the primary notion of single case causation. The S-R view would then be the theory that an increase in probability amounts to a genuine single case cause, as opposed to a prima facie single case cause, as long as that increase in probability is an increase in the pure chance of the result.

On this account, then, the kind of mixed but nonspurious probabilities discussed in sections 3–5 would cease to be of any direct relevance to the basic notion of single case causation. Such mixed nonspurious probabilities would relate only to the derivative notion of population causation. Single case causation would itself be explained entirely in terms of pure probabilities.

From now on I shall understand the S-R view as this thesis about single case causation. The switch to the single case is the obvious way for an S-R theorist to respond to the case of the misleading cigarettes. But now another problem faces the S-R theorist. Once it is allowed that mixed nonspurious probabilities are of no special significance to causation as the S-R theory conceives it, then the initial intuition which seemed to support the S-R view ("Doesn't the smoking-cancer correlation indicate that smoking *causes* cancer?") ceases to count specifically in its favor. For it now turns out that this initial intuition is entirely compatible with a deterministic view of causation.

The crucial point to note here is that the analysis of the last section, about the difference between population and single case causation and about the real significance of mixed nonspurious probabilities, is all *perfectly consistent with the assumption that everything is determined*. To see this, imagine that prior circumstances always leave any individual with either a unitary or a zero chance of cancer. More specifically, imagine that the gene (Y), or smoking in conjunction with the metabolic factor (S&X), or both, make it certain that you will get cancer, and that lacking both Y and S&X makes it certain you won't get cancer. That is, imagine that: S&X or Y ↔ C.

Now in this situation the probability of cancer given smoking will still in general be less than one.[3] Unless Y or X is somehow ensured by smoking, not all smokers will get cancer. What is more, the probability of cancer with smoking will be greater than the probability of cancer without smoking, and both these probabilities will be mixed, and nothing will screen off the difference between them. And all this will show as before that smoking is a population cause of cancer. (Though again we should be careful not to fall into the trap of the misleading cigarettes: that is, we shouldn't take it for granted that any particular smoker's cancer was caused by his or her smoking.)

So everything remains just as it was in the last section. It is easy to see why. To assume determinism is simply to assume that there are only two kinds of cell in the homogeneous partition: those in which cancer's probability is one, and those in which it is zero. So determinism is simply a special case of the situation considered so far, and all the arguments remain valid.

Indeed under the assumption of determinism the nature of the inference from a nonspurious mixed correlation to a population causal conclusion is particularly perspicuous. Let me spell the point out. Among the nonsmokers we'll get cancer if and only if Y. Among the smokers, on the other hand, we'll get cancer if and only if X *or* Y. Which means that cancer will be more likely among smokers than nonsmokers if and only if P(X or Y) is greater than P(Y) — that is, if and only if the background factors which determine cancer together with smoking aren't an empty set.

Actually, that's a bit quick. Even if we leave measure-theoretic and modal niceties to one side (see my 1985a, 73), there's the obvious point that cancer can get to be more likely among smokers than nonsmokers, not because of any nonempty X, but because some elements in Y are disproportionately common among smokers (imagine that Y includes the carcinogenic gene which also induces people to smoke). But this is easily remedied. Consider separately classes of people alike in any such elements of Y, and see whether smoking still makes cancer more likely within them. (Or, to slip back into my original terminology, we need to make sure that the association between smoking and cancer isn't spurious, by taking into explicit account relevant factors themselves associated with smoking.)

The point of all this has been to show that, even if we were dyed-in-the-wool determinists who thought that everything was made certain or impossible by prior circumstances, we would still be very interested in population probabilities that differed from zero and one, and would still be inclined, quite rightly, to infer population causal conclusions from mixed but nonspurious probabilities.

And this now shows how the intuition that increased probabilities establish causation can be accounted for in a way that gives no special support to the theory that single case causation itself is probabilistic. For even if you believed in determinism, and therefore took it for granted that all single case causation was deterministic, you would still be served well by the intuition that an increased probability indicates population causation.

What is more, it seems highly likely that determinism is in fact the basis of that intuition, historically speaking. After all, the view that the world is at bottom indeterministic has only recently been accepted by informed common sense. (Most people probably still believe that macro-events like cancer, or crop yields, or coin tosses, are always determined when they occur.) No doubt writers like Suppes and Salmon were inspired to formulate the S-R view by the indeterminism of modern physics. But the reason their claims struck such an intuitive chord in their readers is surely quite different, namely, our implicit grasp of the fact that in a deterministic universe S can only be nonspuriously correlated with C if S figures among a conjunction of factors that determines C.

This isn't of course an argument against the S-R view. But it does show that the S-R view can't simply appeal to the intuitive connection between probability increases and causes. For this intuitive connection draws most of its strength, not from the S-R view, but from the contradictory view that causation is deterministic.

8. The Deterministic Causation of Chances

Perhaps I can give a genetic account of the power of the intuition that statistical relevance implies causation, by appealing to the long-standing presupposition that macro-events are always determined. But in fact this presupposition is false.

Quantum mechanics tells us that the readings on geiger counters and like devices are undetermined macro-events. And there is no reason not to allow that even such nonlaboratory macro-events, like getting cancer or tossing a coin, might also be undetermined: perhaps they depend on the quantum mechanical breaking of bonds in the DNA molecules in lung tissue cells, or, again, in the breaking of bonds in the air molecules that the dice collide with.

If determinism were indeed true, then nonunitary probabilities would always be due to our ignorance of all relevant factors, and it would be sensible to insist that such probabilities were merely indirect evidence for underlying deterministic single case causes. But this scarcely decides what we should say given that determinism *isn't* true, and that nonunitary probabilities can perfectly well be pure reflections of complete information.

Let me finally face the issue. Suppose cancer is an indeterministic phenomenon. And suppose we are not misled about Jones. Jones does have X. So his smoking does increase the chance of his getting cancer. And then he gets cancer. Surely I have to say that his smoking caused his cancer.

But, as I indicated in the Introduction, there is another possible line here. This allows that something causal is going on. But it denies that Jone's *cancer* is caused. Rather what is caused is his increased *chance* of cancer. And this then is consistent with the view that all causation is deterministic: for, by hypothesis, whenever smoking (and X) occurs, then that increased chance of cancer is determined.

This line might seem unnatural. Doesn't intuition tell us that in such cases Jones gets *cancer* itself because of his smoking? What's the point of insisting that the only real effect is his increased *chance* of cancer, apart from a misplaced hankering for the old deterministic metaphysics?

But I see no reason to accept that we have any real intuitions about situations like this. We are supposing, for the sake of the argument, that we have complete knowledge about Jones, that we aren't being misled by his smoking. But in the real world we are unfamiliar with this kind of situation. Nearly all the actual probabilities different from zero and one that we use in our everyday lives are indubitably mixed, as I stressed in sections 3–5. We recognize intuitively that such mixed probabilities are a sound guide to population causal conclusions (as long as they aren't spurious). But this intuitive recognition, that our incomplete knowledge entitles us to some kind of population causal conclusion, is quite consistent with either analysis of single case causation: either that single case causation is a matter of prior circumstances determining later chances, or that single case causation is a matter of the indeterministic causation of actual results. And so, it seems to me, our intuitive responses in situations of incomplete knowledge leave it quite open which alternative we would adopt if we ever were in the unfamiliar situation of complete knowledge.

9. Rational Action

So far I have argued that our existing intuitions are consistent with the view that single case causation is deterministic. But ought the matter to rest on existing intuitions? Perhaps there are some positive *arguments* for admitting indeterministic single case causation.

In this section I want to consider one such possible argument for indeterministic single case causation. (Let me take the "single case" as read from now on, unless I say otherwise.) This argument derives from the connection between causation and rational action, and goes like this: it is rational to do A in pursuit of B just in case you believe A causes B; but normative decision theory tells us that it is rational to do A in pursuit of B just in case you believe Prob (B/A) > Prob (B/−A); so doesn't it follow that A causes B just in case Prob (B/A) > Prob (B/−A)? (See Mellor forthcoming.)

There is more to this argument than meets the eye. Let us start by looking more closely at the second premise, that it is rational to do A in pursuit of B just in case you believe Prob (B/A) > Prob (B/−A). As is now well known, this premise needs qualification. We know that the probability of avoiding driving accidents (B) if you are a house-owner (A) is greater than the probability of avoiding them if you are not: Prob (B/A) > Prob (B/−A). But this doesn't mean that it is rational to buy a house in order to avoid accidents.

In this case there is no doubt some underlying character type, C, say, which both influences people to buy houses and leads them to drive carefully. Either you have this character type or you don't, and in neither case is buying a house going to make any extra difference to your chance of avoiding an accident. Which is why it isn't sensible to buy a house in order to avoid accidents.

The original probabilities here, Prob (B/A) and Prob (B/−A), are mixed, in that the further factor C is also relevant to the probability of B. Moreover, C is not just a mixer, but is also a screener-off: Prob (B/A&C) = Prob (B/−A&C), Prob (B/A&−C) = Prob (B/−A&−C). What does all this have to do with the rationality of action? How should we qualify decision theory to cope with this case?

One possible response here would be to insist that you should only act on probabilities when you believe them to be pure. The reasoning would be that, as long as your probabilities are mixed, you are in danger of being misled about your actual situation. You can know that more smokers get cancer than nonsmokers. But if this is a mixed probability, you might be a not-X: you might be the kind of person in whom smoking makes no difference to the chance of cancer. In which case there wouldn't be any point in your giving up smoking to avoid cancer.

But the requirement that you believe your probabilities to be pure is surely too strong a condition on rational action. There is a sense in which you might always be misled into wasting your time if your probabilities are mixed. But that doesn't mean that your action would be *irrational*. After all, to repeat the point yet again, nearly all the probabilities we come across in everyday life are indubitably mixed, by factors that we don't know how to identify. The smoking-cancer correlation is just such a probability. So clearly asking for pure probabilities is too strong. If it were irrational to act on probabilities you believed to be mixed, nobody would ever have been right to give up smoking to avoid cancer.

It's not mixed probabilities that are a bad basis for action, but spurious ones. Think of it in terms of homogeneous partitions of reference classes. If your probability is mixed by factors you can't identify, then you don't know which cell of the partition you are in (you don't know whether you have X or not), and so you don't know what difference your action will actually make to the chance of the desired outcome. But, still, you *may* be in a cell where your action makes a difference, and this in itself gives you reason to act. But if your probability is spurious, then your action *can't* make a difference, for *whichever* cell you are in, your action will be rendered irrelevant to the desired outcome by the screener-off (either you have C or not, and either way your house buying won't make any further difference to your accident-proneness).

10. Quantitative Decisions

So the moral is that it is perfectly rational to act on probabilities that you recognize to be mixed, as long as you don't think they are spurious as well. Can we be more specific? So far my comments on decision theory have been entirely qualitative. But normative decision theory deals with numbers. It tells you how *much* probabilistic beliefs should move you to act. You should act so as to *maximize* expected utility. The desirability of an action should be proportional to the *extent* to which it is believed to make desired outcomes likely.

Can't we just say that probabilities will be quantitatively suitable for expected utility calculations as long as you believe they aren't spurious? The thought would be this. As long as you believe your probabilities aren't spurious, the differences between Prob (B/A) and Prob (B/-A) can be thought of as a weighted average of the difference A makes to B across all the different cells of the homogeneous partition. You don't know which cell you are actually in. You might be in a cell where A makes no difference. You might even be in a cell where A makes B less likely. But, even so, the overall difference between Prob (B/A) > Prob (B/ − A) tells you how much difference A makes on weighted average over all the cells you might be in.

But this won't do. So far I have understood spuriousness as an entirely on-off matter. Spuriousness has been a matter of complete screening off, in the sense of a correlation between putative cause A and putative effect B disappearing entirely when we control by some further X. But spuriousness also comes in degrees. A confounding background factor can distort a correlation, without its being the case that the correlation will be completely screened off when we take that factor into account. Rational action needs to be sensitive to the possibility of such partial spuriousness.

Let me illustrate. Suppose once more that there is a gene which conduces, independently, to both smoking and cancer; but now suppose also that smoking makes a slight extra difference to the chance of cancer: both among those with the gene, and among those without, the smokers are slightly more likely to get cancer. In this case the gene won't entirely screen smoking off from cancer. Controlling for the gene won't reduce the correlation between smoking and cancer to zero. Yet the extent to which smoking is associated with cancer in the overall population will be misleading as to its real influence, and therefore a bad basis for decisions as to whether to smoke or not. Smoking will at first sight seem to be much more important than it is, because of its positive association with the more major cause of cancer, possession of the gene.

Technically we can understand the situation as follows. Prob (B/A) and Prob (B/−A) are indeed weighted averages. But they are weighted by the inappropriate quantities for expected utility calculations. Let us simplify by supposing that X is the only other factor apart from A relevant to B. Now,

(1) Prob (B/A) = Prob (X/A) Prob (B/A&X) + Prob (−X/A) Prob (B/A&−X),

and

(2) Prob (B/−A) = Prob (X/−A) Prob (B/X&−A) + Prob (−X/−A) Prob (B/−X&−A)

This is the sense in which Prob (B/A) and Prob (B/−A) are indeed weighted averages of the probability that A (respectively, not−A) gives B in the "X−cell," and the probability that A (not−A) gives B in the "not−X" cell. But the weighting factors here, Prob (X/A) and Prob (−X/A) (respectively, Prob (X/−A) and Prob (−X/−A)), aren't what we want for rational decisions. They depend on the extent to which A is associated with X, and so mean that the difference between Prob (B/A) and Prob (B/-A) reflects not just the influence of A on B, but also the correlation of A with any other influence on B. In the extreme case, of course, this can make for an overall difference between Prob (B/A) and Prob (B/−A) even though A makes no real difference at all: even though Prob (B/A&X) = Prob (B/A&−X), and Prob (B/A&−X) = Prob (B/−A&−X), and X entirely screens off A from B. But the present point is that, even without such complete screening off, any association between A and X will confound the correlation between A and B and make it seem as if A has more influence than it does.

What does this mean in practical contexts? Are quantitative utility calculations only going to be sensible when we have complete knowledge and pure probabilities? Not necessarily. For note that there is nothing wrong with the weighted average argument if we use the right weights, namely P(X) and P(−X), and so really do get the weighted average of the difference A makes in the X-cell and the not-X-cell respectively. That is, the right quantity for utility calculations is

(3) Prob (X) [Prob (B/A&X) − Prob (B/−A&X)] + Prob (−X)
[Prob (B/A&−X) − Prob (B/−A&−X)]

In the special case where A is not associated with X, the weighting factors in the earlier equations (1) and (2) reduce to P(X) and P(−X), and the difference between P(B/A) and P(B/−A) therefore reduces to the requisite sum (3). But if there is an association between A and X, then we have to "correct" for this confounding influence by replacing the conditional weighting factors in (1) and (2) by the correct P(X) and P(−X).

To illustrate with the smoking-cancer-gene example, you don't want to weight the difference that smoking makes within the "gene-cell" by the respective probabilities of *smokers* and *nonsmokers* having the gene, as in (1) and (2), because that will "bump up" the apparent influence of smoking on cancer in line with the positive likelihood of smokers having been led to smoke by the gene. The issue, from the agent's point of view, is precisely whether or not to smoke. And so the appropriate quantity for the agent is the probability of *anybody* having the gene in the first place, whether or not they smoke, not the probabilities displayed by smokers and nonsmokers.

The practical upshot is that anybody interested in quantitative utility calculations needs to take into explicit account any further influences on the result that they believe the cause (action) under consideration is associated with. If you don't believe there are any possible confounding influences, then you can go ahead and act on Prob (B/A) − Prob (B/−A). But if you do think there are associations between other causes X and A, then you will need to turn to the "corrected" figure (3).

11. Causal and Evidential Decision Theory

In the last two sections I have been considering how rational decision theory should respond to the danger of spuriousness. This topic has been the subject of much recent debate. The debate was originally stimulated by Newcomb's paradox (see Nozick 1969), which is in a sense an extreme case of spuriousness. But it has become clear that the underlying problem arises with perfectly straightforward examples, like those I have been discussing in the last two sections.

Philosophers have fallen into two camps in response to such examples: evidential decision theorists and causal decision theorists. Causal decision theorists argue that our decisions need to be informed by beliefs about the causal structure of the world (see Lewis, 1981, for a survey of such theories). Evidential decision theorists, on the other hand, try to show that we can manage with probabilistic beliefs alone: they feel that we ought not to build philosophically dubious metaphysical notions like causation into our theory of rational decisions if we can help it (see Eells 1982; Jeffrey 1983).

At first sight it might seem that I am on the side of evidential decision theory. All of my analysis in the last two sections was in terms of various conditional probabilities, as in equations (1)–(3) of the last section. But this is misleading. For my recommended decisions require an agent to take a view about *spuriousness*, and spuriousness, as I have defined it, depends on an underlying metaphysical picture. (For any sort of effect E, there is a set of factors which yield an *objectively* homogeneous partition of the reference class with respect to E; spuriousness then depends on whether any of *those* factors screen C off from E).

Given the general tenor of evidential decision theory, and in particular given the structure of the "tickle defense" (to be discussed in a moment), it is clear that evidential decision theorists would find my appeal to the notion of objective probability as objectionable as the appeal to the notion of causation. From their point of view my approach would be just as bad as causal decision theory—I'm simply using the notion of objective probability to do the work of the notion of causation.

I am inclined to see things differently. I would say that the possibility of substituting objective probabilities for causes makes causes respectable, not objective probabilities disreputable. And in section 13 I shall begin exploring the possibility of such a reduction of causation to probability at length. But first let me go into a bit more detail about the different kinds of decision theory.

The underlying idea behind evidential decision theory is that we can manage entirely with *subjective probabilities*, that is, with our subjective estimates of how likely one thing makes another, as evidenced in our betting dispositions. This commitment to subjective probabilities is then combined with a kind of principle of total evidence: we should conditionalize on everything we know about ourselves (K), and we should then perform act C in pursuit of E according as $P(E/C.K) > P(E/K)$.

But evidential decision theory then faces the difficulty that the above inequality may hold, and yet an agent may still believe that the correlation between C and E within K is (to speak tendentiously) objectively spurious. And then of course it doesn't seem at all rational to do C in pursuit of E. If I think that some unknown but objectively relevant character trait screens house-buying off from lack of car accidents, then it's obviously irrational for me to buy a house in order to avoid car accidents.

The standard maneuver for evidential decision theorists at this point is some version of the "tickle defense" (see Eells, ch. 7). In effect defenders of evidential decision theory argue that an agent's total knowledge will always provide a reference class in which the agent believes that the C-E correlation is *not* spurious.

The underlying reasoning seems to be this: (a) spurious correlations always come from common causes; (b) any common cause of an action type C and an outcome E will need, on the "C-side," to proceed via the characteristic reasons (R) for which agents do C; (c) agents can always introspectively tell (by the "tickle" of their inclination to act) whether they have R or not; and so (d) they can conditionalize on R or -R), thereby screening C off from E if the correlation is indeed spurious, and so avoid acting irrationally. To illustrate, if the house-buying/car-safety correlation is really due to causation by a common character trait, then I should be able to tell, by introspecting my house-buying inclinations, whether I've got the trait or not. And so the probabilities I ought to be considering are not whether house-buyers as such are more likely to avoid accidents than nonhouse-buyers, but whether among people with the character trait (or among those without) house-buyers are less likely to have accidents (which presumably they aren't).

This is all rather odd. The most common objection to the tickle defense is that we can't always introspect our reasons. But that's a relatively finicky complaint. For surely the whole program is quite ill-motivated. The original rationale for evidential decision theory is to avoid metaphysically dubious notions like causation or objective probability. But, as I hope the above characterization makes clear (note particularly steps (a) and (b)), the tickle defense only looks as if it has a chance of working because of fairly strong assumptions about causation and which partitions give objectively nonspurious correlations. It scarcely makes much sense to show that *agents* can always manage without notions of causation and objective probability, if our *philosophical* argument for this conclusion itself depends on such notions.

Perhaps the defenders of evidential decision theory will say they are only arguing *ad hominem*. *They* don't believe in objective spuriousness, common causes, etc. It's just that their opponents clearly have such notions in mind when constructing putative counter-examples like the house-buying/car-safety story. And so, the defenders of the evidential theory can say, they are merely blocking the counter-examples by showing that even assuming their opponents' (misguided) ways of thinking of such situations, there will still always be an evidentially acceptable way of reaching the right answer.

But this now commits the evidential decision theorist to an absurdly contorted stance. If evidential theorists really don't believe in such notions as causation, objective spuriousness, etc., then they are committed to saying that the mistake you would be making if you bought a house to avoid car accidents would be (a) that you hadn't introspected enough and therefore (b) that you hadn't conditionalized your house-buying/car-safety correlations on characteristics you could have known yourself to have. But that's surely a very odd way of seeing things. You don't need to be introspective to avoid such mistakes. You just need to avoid acting on patently spurious correlations. Pre-theoretically, it's surely their insensitivity to manifest spuriousness that makes us think that such agents would be irrational, not their lack of self-awareness. It seems to me that there must be something wrong with a theory that denies itself the resources to state this simple fact.

One can sympathize with the original motivation for evidential decision theory. The notion of causation is certainly philosophically problematic. And perhaps that does give us some reason for wanting the rationality of action not to depend on beliefs about causal relationships. But, now, given the way I have dealt with rational action, agents don't need causal beliefs, so much as beliefs about whether certain correlations are objectively spurious or not. The fact that evidential decision theorists feel themselves driven to the "tickle defense" shows that they wouldn't be happy with the notion of objective spuriousness either. But putting the alternative in terms of objective probabilities now places evidential decision theory in a far less sympathetic light. For even if the notion of objective probability raises its own philosophical difficulties, modern physics means that we must somehow find space for this notion in our view of the world, and so removes the motivation for wanting to avoid it in an account of rational action. Moreover, if the cost of keeping objective probabilities out of rational decision theory is the contortions of the "tickle defense," then we have a strong positive reason for bringing them in.

I now want to leave the subject of evidential decision theory. The only reason I have spent so long on it is to make it clear that, despite initial appearances, the approach I have adopted is quite different, and indeed has far more affinity with causal decision theory. Let me now consider this latter affinity. On my account rational action requires you to believe that, even if your correlations are mixed, they are not spurious. If you believe your correlations are spurious, to any degree, then you need to correct them, in the way indicated in the previous section: you need to imagine the reference class partitioned into cells within which such spuriousness disappears, and then to average the "within-cells" correlations, weighted by the probability of your being in each cell.

According to causal decision theory, it is rational to act if you believe that your correlations reflect a causal, and not merely an evidential, connection between your action and the desired result. If you believe the correlations are evidential, then you need to consider separately all the different hypotheses about the causal structure of the world you believe possible, and then average the difference that the action makes to the chance of the result under each hypothesis, weighted by the probability that you attach to each hypothesis.

I don't think there is any real difference here. I think that the two approaches simply state the same requirement in different words. This is because I think that facts of causal dependence can be entirely reduced to facts about probabilities in objectively homogeneous partitions of reference classes. But this is itself a contentious thesis. There are various difficulties in the way of this reduction, many of which I have been slurring over so far. Most of the rest of the paper will be devoted to dealing with them.

Note that this issue of reduction is independent of the debate between the S-R and traditional deterministic views of causation. The idea I want to explore (I shall call it the "reductionist thesis" from now on) is that we have causal dependence of E on C if and only if C and all the other probabilistically relevant factors present define a homogeneous cell of the reference class which yields a higher probability for E than is yielded those other relevant factors alone; or, again, if and only if the chance of E given C and all the other relevant factors present is higher than the chance E would have had given those other factors but without C.[4] But now suppose that this reductionist thesis were granted. This would still leave it quite open whether in such cases we should say that C (indeterministically) caused *E*, or whether we should say that C (deterministically) caused the increased *chance* of E.

I'm not going to have much more to say about this latter issue. It seems to me that by now these are pretty much just two different ways of talking (and in discussing the reductionist thesis I shall adopt both indiscriminately). But an earlier argument for the S-R view has been left hanging in the air. Let me briefly deal with this before turning to the general issue of reduction.

12. Action and Causation Again

The argument in question is the one from the beginning of section 9: (a) it is rational to do A in pursuit of B just in case you believe A causes B; (b) it is rational to do A in pursuit of B just in case you believe $P(B/A) > P(B/-A)$; so (c) A causes B just in case $P(B/A) > P(B/-A)$.

I have shown that the second premise (b) won't do as it stands. Not all correlations are a good basis for action. It doesn't matter in itself if a correlation is believed to be mixed. But a correlation is disqualified as a basis for action if it is believed to be spurious. Before we act we need to take into account all the factors that we believe to be confounding the association between A and B, and adjust the correlation accordingly.

It might seem as if this now means that I can respond to the argument at hand as I originally responded (in sections 7 and 8 above) to the initial intuition favoring the S-R view. That is, can't I point out that all the probabilities we actually act on are undoubtedly mixed? We recognize that such probabilities had better not be spurious. But we also recognize that they don't need to be pure. So our intuitions about when it is and isn't rational to act are quite consistent with the supposition that all events are determined and that the only reason we have probabilities other than nought and one is that we are ignorant of various relevant (but non-confounding) factors. And not only are our intuitions so consistent with determinism, they are no doubt inspired by it, since until recently determinism was built into informed common sense. So we can scarcely appeal to such intuitions to decide against a deterministic view of causation.

But this argument won't serve in the present context. For the S-R theorist isn't now appealing to mere intuitions about causation. Rather the appeal is to *facts*, so to speak, about when it's rational to act, over and above any intuitions we may have on the matter. This means that the S-R theorist can now insist that the relevant situation is one where an agent believes that a result is genuinely undetermined. Maybe we don't have any immediate causal intuitions about such indeterministic set-ups, since until recently we didn't believe there were any. But that doesn't stop there being a fact of the matter as to how one ought to act in such situations.

And here the S-R theorist is clearly on strong ground. For there is no question but that knowledge of objective nonunitary chances can be relevant to rational action. If I believed that the effect of not smoking, when all other relevant factors are taken into account, is to increase the chance of avoiding cancer, but without determining it, then obviously this would give me a reason to stop smoking.

So the fact that we were all determinists till recently is irrelevant to the argument from rational action. The issue is not *why* we think that it's rational to act if and only if (nonspuriously) $P(B/A) > P(B/-A)$. Rather the point is that it *is* so rational (and in particular that it is so rational even if $P(B/A)$'s being less than one isn't just due to our ignorance of the relevant determining factors).

But there is still room to resist the S-R view. Even if we concede premise (b), we can still question premise (a). Premise (a) says it is rational to do A in pursuit of B just in case you believe A causes B. But why not say instead that it is rational to do A in pursuit of B just in case you believe A causes an increased chance of B? This will enable us to accommodate all the relevant facts about rational action, while still preserving a deterministic view of causation.

This argument is clearly in danger of degenerating into triviality. But let me just make one observation before preceeding. It might seem *ad hoc* for the traditional theorist to start fiddling with premise (a) when faced by indeterminism. But note that the S-R theorist also has to do some fiddling with (a) in the face of indeterminism. The S-R theorist can't simply say that A causes B whenever it's rational to do A in pursuit of some B. For A can make B more likely, and yet B might not occur. The S-R notion of causation isn't just that A increase the chance of B, but that A increase the chance of B, *and* B occurs. So the S-R theorist has to formulate (a) in some such form as: A causes B just in case it's rational to do A in pursuit of B, *and* B occurs; or, again, it's rational to do A in pursuit of B just in case you're in the kind of situation where A *might* cause B. It's not clear to me that these formulations are any more satisfactory than the deterministic alternative suggested in the last paragraph, according to which it is rational to do A in pursuit of B just in case A invariably causes an increased chance of B.

13. The Metaphysics of Probability

In her (1979) Nancy Cartwright argues against the reducibility of causal relationships to laws of probabilistic association. Her argument depends on the point that probabilistic relationships only indicate causal relationships if they don't get screened off when we conditionalize on relevant background factors – causation demands nonspurious associations, not just any associations. However, the idea of nonspuriousness requires a specification of the class of background factors which need to be taken into account. Cartwright argues that this can only be given as the class of *causally* relevant factors.

So Cartwright allows that a causal relationship is a probabilistic association that doesn't get screened off by any causally relevant factors. This gives us a relationship between causal and probabilistic notions. But the appearance of the notion of "causal relevance" on the right-hand side of this relationship clearly rules it out as a reduction of causation.

Cartwright's argument is often endorsed in the literature (see Eells and Sober 1983, 38; Eells and Sober 1986, 230). But it seems to me that it is easily answered. Why not just say that the factors that need to be taken into account are all those which are *probabilistically*, rather than causally, relevant to the result? This would accommodate the possibility of spuriousness, but without rendering the proposed reduction circular.

What is a "probabilistically relevant" factor for some result E? It's any property which, in conjunction with certain other properties, is relevant to the chance of E. That is, it's any K such that there exists an L such that $P(E/K.L)$ and $P(E/-K.L)$ are pure and unequal.

Putting it like this makes it clear that we don't really need a restriction on the set of factors relevant to spuriousness in the first place. For conditionalizing on probabilistically *ir*relevant factors isn't going to show any probabilistic associations to be spurious, since by definition irrelevant factors don't make any difference to probabilities. So we may as well simply say that a probabilistic association indicates a causal relationship as long as there isn't any background factor which screens it off.

Will we ever have any causal relationships, if a causal relationship is disproved by *any* factor which screens the putative cause C off from the effect E? Surely there will always be *some* way of categorizing things that equalizes the proportions with which E is found with and without C. (See Cartwright 1979, 434.)

At first sight this objection might seem plausible. But it can be countered if we take care to make the distinction between the epistemology and the metaphysics of probability (as Cartwright herself notes, though she has her doubts about the distinction). Certainly if we are dealing with sample frequencies there will always be some way of dividing the sample into two parts that equalizes the relative frequency with E is found with and without C. But that's quite different from the idea that there's always a property that will render C irrelevant to the *chance* of E.

I take it that there are real chances in the world: chances of certain properties being instantiated, in situations defined by certain other properties. Throughout this paper I have intended "probabilities" to be understood either as chances, or as chance mixtures of chances (that is, as the average of chances in different homogeneous cells weighted by the probability of being in each cell).

Probabilities (chances and chance mixtures of chances) manifest themselves in, and are evidenced by, relative frequencies in finite samples. The relationship between probabilities and such frequencies is a deep and difficult issue. But, however that issue is to be resolved, it is clear that not every relative frequency in every sample corresponds to a real probability. And, in particular, it is clear that it doesn't follow, just because sample correlations are always screenable off, that real correlations always are.

It might be objected that by helping myself to an ontology of chances and probabilistically relevant properties, I am begging all the interesting questions. Isn't having chances and the properties they involve tantamount to knowing about the causal structure of the world? What is the difference between probabilistically relevant properties and straightforwardly causally relevant ones?

In a sense I am sympathetic to this complaint. After all, I want to show that the causal facts reduce to probabilistic ones. But this doesn't mean that their relation is trivial, or that there's no point in trying to spell it out. If the reduction is possible, then there is a sense in which causal facts are built into probabilistic facts. But it's certainly not obvious at first sight that this is so.

14. Causal Chains

In this section and the next I want to look at some difficulties to do with the relationship between causation and time.

So far I've been cheating. I've in effect assumed that there are just two times, "earlier" and "later." The effect E happens "later." A number of factors are present "earlier." The reductionist thesis was that an earlier factor C is a single case cause of a later E just in case the chance of E was higher than it would have been if C had been absent and all other relevant earlier factors had been the same.

But of course there aren't just two times, but a whole continuum. And in any case it's not clear why causes should always happen earlier and effects later.

In this section I want to look at a difficulty which arises as soon as we admit that there are more than two times, and even if we continue to assume that causes must precede their effects. In the next section I shall say something about what happens if we admit the possibility of effects preceding their causes.

As soon as we allow that there can be relevant factors temporally intermediate between C and E, there is a difficulty about an earlier C *ever* being a cause of a later E. To be a cause means that you have to make a difference to the chance of E when *all* other relevant factors are taken into account. But now suppose that D is, intuitively speaking, causally intermediate between C and E: C causes E *by* causing D. For example, smoking causes cancer by getting nicotine into the lungs. D is clearly a relevant factor, if C is. But now, according to the reductionist thesis as so far stated, D is going to stop C counting as a cause of E. For C won't make any further difference to the chance of E once we take D into account. Given that you've got nicotine in your lungs, the fact that you smoke doesn't make you *more* likely to get cancer. And similarly if you haven't got nicotine in your lungs (imagine you are a rigorous non-inhaler) smoking won't make you more likely to get cancer either. (I'm assuming here for simplicity that nicotine is the only route by which smoking causes cancer.) The presence or absence of nicotine screens the smoking off from the cancer. And so, according to the reductionist thesis, the nicotine seems to stop smoking from causing cancer. But the argument is quite general. We seem forced to the undesirable consequence that nothing ever causes anything via intermediate causes.

It won't do to say that we shouldn't control for factors temporally intermediate between C and E. For perhaps C isn't in fact a genuine cause of E, but only appears to be so because it is associated with (though not the cause of) some real later cause D. And then it is precisely that C doesn't make a difference when we conditionalize on D that should stop it counting as a genuine cause.

In Cartwright's eyes this provides an additional reason why we can't reduce causes to probabilities (Cartwright 1979, 424). Her original complaint was that we needed to specify the background factors to be taken into account as the set of "causally relevant" factors. I have answered that complaint by arguing that we may as well take *all* background factors into account. But now it seems that we need a further qualification. We shouldn't take all background factors into account after all, but only those which aren't causally intermediate between C and E, lest we end up ruling out all earlier C's as causes of later E's. But now this further qualification threatens to undermine the proposed reduction once more, since as before it seems that we need causal terminology (in particular, the notion of causal intermediacy) to explain which probabilistic relationships indicate causation.

I think there is a way out here. We need to distinguish between direct and indirect causes, and to define the latter in terms of the former.

Let us imagine that the times between C, at t_o, and E, at t_k, consist of a series of discrete instants, $t_1, t_2, \ldots, t_{k-2}, t_{k-1}$. (I shall relax the assumption of discreteness in a moment.)

Then we can say that a factor A at any T_1 is a *direct cause* of some B at the next instant, t_{1+1}, just in case the chance of B given A and all other factors present at t_1, or earlier, is greater than the chance of B given those other factors alone.

Then we can define a *causal chain* as a sequence of events at successive times

t_1 A

t_x D

t_2 B

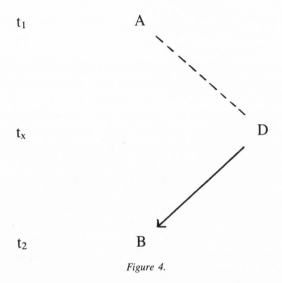

Figure 4.

such that each event is the direct cause of the next. Given any two events on a causal chain we can say that the earlier causes the later *indirectly*, via the intervening stages. In effect this defines causation (direct or indirect) ancestrally, in terms of direct causation: a cause is a direct cause or a cause of a cause.

The obvious objection to all this is that time isn't discrete, but dense. Between any two times there is always another. And this clearly invalidates the proposed approach. For, if we consider only the original discrete sequence of times, a factor A, at t_1, say, might appear to be a direct cause of B at t_2, even though it wasn't really a cause at all. Because even if it's not screened off from B by anything at t_1 or earlier, it might still be screened off from it by some D at t_x, halfway between t_1 and t_2, where D *isn't* in fact causally intermediate between A and B, but merely a confounding factor associated with A (because of some common causal ancestor, say). (See Figure 4: the solid arrow indicates causation, the dotted line probabilistic association.)

Well, we could deal with this case by considering a finer sequence of instants, which included all the times midway between the original times, and so included t_x. Then A would be exposed as not a genuine cause of B, for although D would count as a direct cause of B, A wouldn't be a direct cause of D.

But of course the difficulty would still lurk in the interstices between the half instants. But now the solution should be clear. What we need to consider is the infinite series, $s_1, s_2, \ldots$, of finer and finer sequences of instants between t_0 and t_k. If A really isn't a genuine causal ancestor of B, then at some point in the series we will have a fine enough discrimination of instants for it to be exposed as such, in the way that A was exposed as an imposter by D above. Conversely, if A *is* a genuine causal ancestor of B, then, however far we go down the series, the finer and finer divisions will all present it as a direct cause of a direct cause . . . of a direct cause of B.

Since time is dense there aren't, strictly speaking, any direct causes, and so, given the earlier definition, no indirect causes either. But that doesn't matter. We can regard the idea of direct and indirect causation as defined relative to a given fictional division of time into a discrete sequence of instants. And then we can define genuine causal ancestry as the limit of indirect causation in the infinite series of such fictional divisions, in the way indicated in the last paragraph.

15. Causal Asymmetry

I'm still cheating. I have now given up the earlier simplification that the only relevant times are "earlier" and "later," and explained how to deal with the fact that in between any two different times there are always infinitely many more. But the analysis still depended on a crucial implicit assumption about the relation between causation and time, namely, that the causal direction always lines up with the earlier-later direction.

In effect what I showed in the last section was that genuine causal connections between finitely separated events can be explained in terms of causal connections between, so to speak, infinitesimally separated events. But I simply took it for granted that when we had such an infinitesimal causal connection between some A and B, then it was the earlier A that was the cause of the later B, not vice versa.

I would rather not take this assumption of the temporal directionality of causation for granted. For one thing, there is nothing in the probabilities as such to justify the asymmetry: the relation of having a unscreenable-off probabilistic association is an entirely symmetric one. But that's not the crucial point. If nothing else were at issue, there wouldn't be anything specially wrong with reducing causation to probability and temporal direction, rather than to probability alone. But something else is at issue. There is a good independent reason for being dissatisfied with building temporal direction into the analysis of causation. Namely, that there are obvious attractions to the converse reduction, of temporal direction to causation. After all, what is the past, except those events that can affect the present, including our memories? And what is the future, except those events that the present, including our present actions, can affect? But if we want to expand these thoughts into an analysis, we'd better not build temporal direction into causal direction.

So the problem is to explain causal asymmetry without assuming that causes always precede their effects in time. There isn't any question of treating this problem fully here. But let me try to give some idea of an approach which makes use of some of the notions I have developed in the present paper.

This approach is defended in greater detail in my (1985b). In that paper I begin

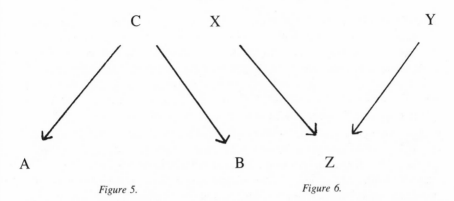

Figure 5.   Figure 6.

with the fact that we often find that probabilistic associations between some A and B are screened off by a third factor C. I then observe that such cases are characteristically those where C is a common cause of A and B (or where C is causally intermediate between A and B). But I point out that we don't find this pattern when some Z, say, is a common *effect* of some X and Y.

The thought I pursue is that the probabilistic difference between Figures 5 and 6 – screened-off associations in Figure 5, but none in Figure 6 – is symptomatic of the differences in causal direction involved. This leads me to look for some independent explanation of the probabilistic differences, which might then serve as an analysis of causal direction.

The explanation I offer is that the screenable-off associations arise because (a) the probabilities involved are mixed, and (b) the mixing factors satisfy certain independence assumptions. Suppose the probability of A given C is a mixture: together with some background conditions C fixes a certain chance for A, together with others it fixes different chances. And suppose that the same is true of the probability of B given C. Then one can argue that, if the background conditions which together with C are relevant to the chance of A are *probabilistically independent* of those which together with C are relevant to the chance of B, then there will be a probabilistic association between A and B, and that association will be screened off by C.[5]

It follows that if, in Figure 6, there are sets of background conditions, together with which Z fixes chances respectively for X and Y, these background conditions can't be probabilistically independent, for if they were then there would be an X-Y association which was screened off by Z. And I confirm the analysis by showing that, in actual cases of joint causes X and Y of a common effect Z, the background factors required to specify laws which run, so to speak, from Z to the respective chances of X and Y, will manifestly not be probabilistically independent.

So I suggest the following account of causal direction. The properties whose causal relationships we are interested in are generally related in "mixed" ways: the chances of one property given another will vary, depending on the presence or absence of various sets of background conditions. The causes can then be differentiated from the effects by the principle that the various sets of background conditions, together with which a cause is relevant to the chances of its various effects, are mutually probabilistically independent, whereas the converse principle got by interchanging "cause" and "effect" is not true.

On this suggestion, the directionality of causation doesn't lie in the structure of the lawlike connections between events themselves, so much as in the further probabilistic relationship between various sets of background conditions involved in such lawlike connections. It may seem odd to attribute causal direction not to the causal links themselves but to the satisfaction of probabilistic independence conditions by (often unknown) background conditions. It is worth noting, however, that quite analogous explanations can be given for two other puzzling physical asymmetries, namely the fact that entropy always increases, and the fact that radiation always expands outward. Although the underlying laws of physics permit the reverse processes to happen, when the laws of physics are combined with certain assumptions about the probabilistic independence of initial, as opposed to "final," micro-conditions, then the asymmetrical behavior can be derived.

It is also worth noting that the analysis of causal direction that I have outlined in this section is *not* committed to the "principle of the common cause": I am not assuming that for every correlation between spatio-temporally separated events there is some common cause that screens off their association. My claim is only that *if* there is such a screener-off, *then* it will be a common cause of its two joint effects, rather than a common effect of joint causes. Note in particular that there is nothing in this to conflict with the existence of unscreenable-off correlations, as in the EPR experiments. (After all, everybody agrees that the intuitive significance of such unscreen-offability is precisely that there couldn't be a common cause of the effects on the two wings.)

Now that the EPR phenomena have been mentioned, it will be worth digressing briefly and saying something more about them. Maybe the EPR phenomena don't causes any difficulties for my analysis of causal direction in terms of screening off. But, even so, they do raise a substantial problem for my overall argument. For they seem to provide direct counterexamples to the reductionist thesis itself.

In the EPR experiments the chance of a given result on one wing is increased by the chance of the corresponding result on the other wing, and this correlation isn't screened off by anything else. Given my overall reductionism, this ought to imply that there is a direct causal connection between the results on the two wings. But we don't want this—apart from anything else, such instantaneous action at a distance would seem to contradict special relativity.

However, I think that the analysis developed so far yields a natural way of ruling out the EPR correlations as causal connections. As a first step, let me add the following requirement to the analysis so far: direct causal connections should be concatenable into causal chains: correlations not so concatenable should be disqualified as causal connections on that account. This might seem trivial: once we accept that A is a cause of B, then won't we automatically conclude, given any D that causes A, that we have a causal chain from D to B through A? But, trivial as it is, this requirement suffices to rule out the EPR correlations as causal connections. For Michael Redhead has shown that part of the weirdness of the EPR correlations is that they are *not* concatenable into causal chains (Redhead, forthcoming).

More precisely, Redhead shows that if A and B are correlated results on the two wings of an EPR experiment, and D is a cause of A, then A doesn't behave probabilistically like a link in a causal chain from D to B: A doesn't screen off the correlation between D and B. This is because, when A comes from D, the A-B correlation is itself altered, in such a way as to undermine the screening-off feature. As Redhead puts it, the A-B correlation is not *robust*, in that it is sensitive to factors which affect the presence or absence of A.

I am assuming here that it is essential to the existence of a causal chain that intermediate links should screen off earlier stages from later stages. I admit that nothing in the earlier discussion of causal chains guarantees this assumption. But it seems to me that it flows naturally from the arguments of the last two sections. Earlier in this section I suggested that it is constitutive of the common-cause-joint-effect relationship that common causes should screen off the correlations among their joint effects. So let me make an analogous suggestion about the causal chains introduced in the last section: namely, that it is constitutive of the idea of one factor being causally intermediate between two others that it should screen off the correlations between them.

The interesting question which remains is whether this last screening-off pattern can be reduced to independence requirements on background conditions, analogous to the suggested reduction of the common-cause-joint-effect pattern. I make some brief comments on this issue in my (1985b). More detailed investigation will have to wait for another occasion.

16. Digression on Independence Requirements

The last section involved certain independence assumptions about background conditions. In this section I would like to make some further points related to such independence assumptions. Most of this is about technical difficulties in my overall argument. Some readers might prefer to skip ahead to the next section.

Let us go back to the idea that mixed probabilities can be reliable guides to population causation. The danger with such probabilities was that they might be spurious, as well as mixed, in which case they would be misleading about population causation. My response was to point out that this threat could be blocked by dividing the overall reference class into cells within which the putative cause C isn't probabilistically associated with any other relevant background factors, and seeing whether C still makes a difference to the probability of E within such cells.

It has been important to a number of my arguments that this doesn't necessarily require dividing the reference class into *homogeneous* cells. It is precisely because not *all* the other conditions relevant to E will in general be associated with C that we are ever able to reach conclusions about population causes from mixed probabilities. Moreover, this fact (or, more accurately, our believing this fact) is also a precondition of our acting rationally on probabilities that we believe to be mixed.

Nancy Cartwright has asked (in conversation) why we should suppose that, once a few confounding factors have been taken into explicit account, the remaining relevant conditions will generally be probabilistically independent of C. That is, why think that once we have made some fairly gross and inhomogeneous division of the reference class, all remaining factors will be independent of C within the resulting cells?

I don't have any basis for this presupposition, beyond a metaphysical conviction, which I may as well now make explicit. This is simply the conviction that in general different properties are probabilistically independent, except in those special cases when they are (as we say) causally connected, either by one being a causal ancestor of the other, or by their having a common causal ancestor.

I can't explain why the world should be like this. But I believe that it is, and, moreover, I believe that if it weren't it would be a very different place. If it didn't satisfy this general principle of probabilistic independence, we wouldn't be able to infer population causes from mixed probabilities, nor therefore would we be able to act on such probabilities. And indeed, if there is anything to the arguments of the last section, there wouldn't be any causal direction in such a world either.

There is, however, a difficulty which arises in connection with this independence principle, and which I rather slurred over in my (1985a). Consider the old chestnut of the falling barometer (B) and the rain (R). Suppose, for the sake of the argument, that rain is always determined when it occurs, either by a fall in atmospheric pressure (A) and high humidity (H), or by one of a disjunction of other factors, which I'll write as Y:

(1) A.H or Y $\leftrightarrow$ R.

Suppose also that A and X (the barometer is working) determine B, and so does Z (the kind of barometer malfunction which makes the barometer fall even though the atmospheric pressure hasn't):

(2) A.X or Z $\leftrightarrow$ B.

(Throughout this section I shall assume that all events have determining causes. Most of the arguments I give will be generalizable to indeterministic causes, or, equivalently, to the deterministic causation of chances.)

Now, if (1) and (2) are true, there is a sense in which falling barometers are, in Mackie's terminology, "inus conditions" of rain.

For it immediately follows from (1) and (2) that B. $-$Z.H $\rightarrow$ R. This is because B. $-$Z ensures A, by (2): if the barometer is not malfunctioning ($-$Z) and it falls (B), then the atmospheric pressure must have fallen. And so if we have high humidity (H) as well, it'll rain, by (1).

Moreover, we can no doubt cook up a Q which covers all the other causes of rain apart from drops in atmospheric pressure (Y), and also covers those cases where the barometer doesn't fall when the pressure falls in high humidity. Which will give us:

(3) B. $-$Z.H or Q $\leftrightarrow$ R.

Which is what I meant by the barometer being an "inus condition" of rain.

Equivalence (3) means that there is a sense in which $-$Z, H and Q are background conditions which, along with B, fix the chance of rain. Which means, given everything I've said so far, that once we've divided our reference class up enough to ensure that B is no longer associated with $-$Z, H and Q, we can draw conclusions about whether or not B is a population cause of R by seeing whether it is still probabilistically relevant to it.

The trouble, in this particular case, is that we won't ever be able to divide up our reference class in such a way as to get rid of confounding associations. For $-$Z was specified as the absence of the kind of malfunction which makes the barometer fall, and clearly that's going to remain (negatively) associated with B, the barometer's falling, however much dividing up of the reference class we do.

But this is now somewhat paradoxical. For surely it is intuitively clear, quite apart from all these messy equivalences, that we *can* find out, from appropriate mixed probabilities, whether or not barometers cause rain. There is an initial probabilistic association between falling barometers and rain: falling barometers mean that rain is likely. But, by looking more closely at different kinds of cases, though without necessarily identifying all factors relevant to rain, it is in practice perfectly possible to show that this initial association is spurious.

Let me spell out the paradox. We know that we can expose the barometer-rain correlation as spurious without getting down to pure probabilities. But (3) and my general argument seem to imply that we oughtn't be able to do this, since we can't get rid of the confounding association with -Z without dividing the reference class *ad infinitum*. True, -Z is rather different from most of the confounding factors we have met so far, in that -Z is *negatively* associated with B, and so threatens to produce a spurious null or negative correlation between B and R, rather than the spurious positive correlation threatened by the usual kind of confounding factor. But the point remains. How can we be confident that the statistics show that B and R are *genuinely* null correlated, rather than only spuriously so, even though the negative confounding factor -Z hasn't been controlled for?

We need to take a few steps back to disentangle all this. The first thing to note is that, even if there's a sense in which barometers are "inus conditions" of rain, we certainly don't want to count them as causes of rain on that account. Barometers don't cause rain. The moral of the equivalences (1)-(4) isn't that barometers cause rain, but simply that inus conditionship isn't enough for causation.

It is obvious enough how we might strengthen the idea of inus conditionship to rule out cases like the barometer. The barometer gets to be an inus condition only by proxy, so to speak: as the derivation of (3) makes clear, it only suffices for rain by virtue of the fact that the atmospheric pressure has already always fallen on the relevant occasions. What is more, the atmospheric pressure sometimes suffices for rain on occasions when the barometer doesn't fall. In my (1978a) and (1978b) I said that in such situations the atmospheric pressure *eclipses* the barometer as an inus condition of rain. And I hypothesized that in general causation required *uneclipsed* inus conditionship, rather than inus conditionship alone.[6]

The importance of the suggestion that causation is equivalent specifically to *uneclipsed* inus conditionship is that it enables us to make a rather stronger claim about the connection between probabilities and population causation than anything we've had so far. In a number of places I've assumed that for any putative cause C and effect E there will be background conditions X which together with C ensure E, and also that there will be other sets of factors, Y, which don't include C, which also ensure E: C.X or Y ↔ E. And then I've claimed that, once we've divided up our reference class into cells within which C is no longer associated with X or Y, then we'll get remaining correlations between C and E if and only if C is a population cause of Y.

But if causes are uneclipsed inus conditions we can say something stronger. Namely, that as soon as we've divided our reference class into cells in which C is no longer *associated with Y*, we'll get remaining correlations if and only if C is a population cause of E. The point here is that with uneclipsed inus conditionship we don't any longer have to control for X. As I show in my (1978a) and (1978b), the condition required for remaining correlations to be necessary and sufficient for uneclipsed inus conditionship is only that C not be associated with the other, independent factors ensuring E — it doesn't matter if C is associated with the factors that it acts in concert with.

Let us return to the barometer. The puzzle I raised was that, while in practice we can clearly use mixed probabilities to find out whether or not barometers cause rain, the condition required for the drawing of such inferences seems not to be satisfiable. For barometer falls will continue to be (negatively) associated with -Z, however much dividing up of the reference class we do.

But we now see that the requirement that any association with −Z be eliminated is too strong. Given that not all inus conditions are causes, but only uneclipsed ones, we can relax the requirement on eliminating confounding factors.

We are in a position to draw conclusions about the barometer as long as we have assured ourselves that it is not associated with any of the other, independent sources of rain. And that we can easily do, by looking separately at classes of cases where the barometer falls, but the atmospheric pressure doesn't, and vice versa. Within these classes we will find that there are no remaining correlations between barometers and rain. And this then entitles us to conclude that barometers don't cause rain.

If what we wanted were conclusions about inus conditionship simpliciter, then we *would* need to control for $-Z$. And so we wouldn't be able to decide that the barometer wasn't an inus condition from the null partial correlations appealed to in the last paragraph, because those correlations are in classes where F is still negatively associated with $-Z$. But that's just as well, since the barometer *is* an inus condition of rain, albeit an eclipsed one. Indeed this shows, so to speak, why the association with $-Z$ has to be ineliminable, given that F has null partial correlations with R: for if the association with $-Z$ were ever eliminated, then we would be entitled to infer, from the null correlations, the false conclusion that F isn't an inus condition of R.

17. Causal Processes and Pseudo-Processes

Some recent writers working in the probabilistic tradition have turned against the reductionist idea that causal relationships can be defined in terms of probabilistic relationships between properties. Instead they suggest that causation is primarily a matter of *causal processes*, and that generic causal relationships between properties need to be defined in terms of causal processes, rather than conversely. Thus Wesley Salmon says, "The basic causal mechanism, in my opinion, is a causal process that carries with it probability distributions for various kinds of interactions" (1984, 203). And Elliott Sober says, "Connecting processes exist independently of the events they connect. Such processes are like channels in which information flows: the existence of the channel does not imply what information (if any) actually flows over it" (1986, 111).

In this section I want to show that, while the notion of a causal process is indeed an important notion, it can be defined straightforwardly enough in terms of the concepts already developed in this paper, and does nothing to suggest those concepts are inadequate for a reductionist theory of causation.

Let me begin with Salmon's distinction between causal processes and pseudo-processes (1984, 141–42). A causal process can transmit marks. Pseudo-processes cannot. So a moving shadow on a wall is a pseudo-process, for you can't alter the later characteristics of a shadow by operating on its earlier stages. The expansion of radiation, or the persistence of a normal physical object, on the other hand, are causal processes, since the later stages of such processes can generally be altered by acting on their earlier stages.

This is indeed a significant distinction. But it would be a mistake to think that it is somehow primitive to the theory of causation. Suppose we call any sequence of space-time points a space-time *worm*. Some space-time worms will be distinguished by the fact that some of the properties possessed by earlier stages will be correlated in various ways with some of the properties possessed by later stages. Let us call these worms *processes*. Processes are those worms which carry sequences of correlated properties.

So far this goes for causal and pseudo-processes alike. The difference is that the sequences of properties carried by *causal* processes comprise causal chains, in the sense outlined in Section 14 above: insofar as the probabilistic relevance of earlier properties in a given sequence to later ones is screened off, it will only be by properties intermediate in that sequence. The sequences of properties carried by pseudo-processes, on the other hand, are not causal chains: for the probabilistic relevance in such sequences will often be screened off by facts which are not part of the pseudo-process at all.

To illustrate, think of people as causal processes. Many of the earlier properties of people are relevant to their later properties. Childhood weight is relevant to adult weight, childhood hair color to adult hair color, etc. And in these cases nothing fully screens off adult weight (hair color, etc.) from childhood weight (hair color, etc.) except the intermediate weights (etc.). But in the case of a moving shadow, say, the intermediate shape of the object casting the shadow will screen off the shadow's later shape from its earlier shape: which stops the sequence of shapes of the shadow being a causal chain, since the intermediate shape of the object casting the shadow isn't itself part of that sequence.

So we can define causal *processes* in terms of causal *chains*. Causal processes are space-time worms that carry ensembles of causal chains. If we like, we can think of causal processes as bundles of causal chains.

There is nothing in anything said so far to motivate any revision of my overall reductionist thesis. However, both Salmon and Sober seem to think that once we have a notion of causal process, then this allows for a new kind of causation. They hold the later features of a causal process should be deemed to be caused by the relevant earlier features, even in cases where our reductionist notions would give the opposite answer. Thus, to pursue the above example, they would say that if a fat child grows into a thin adult, the adult thinness is caused by the childhood fatness.

The idea here seems to be that, since weight in humans is a property carried by a causal process, the later weights of humans are always caused by their earlier weights. But there is nothing in the notion of causal process, as I have explicated it, to warrant this. Causal processes are picked out as such because they carry causal chains. Causal chains are a matter of certain earlier properties making certain later ones more likely than they would otherwise be. Nothing in this requires us to say that when, in a particular case, the relevant later property doesn't occur, that its noninstantiation was caused by the earlier property. We can still insist that an earlier property only affects another if it makes its chance higher than it would otherwise have been.

18. Negative Causes

In this final section I want to look more closely at the possibility of "negative causes," causes that make their effects less likely. In the last section I argued that the notion of a causal process as such yields no argument for such negative causes. But in a sense this puts the cart before the horse. For perhaps there are independent arguments for admitting negative causes. And if there are, then this will itself provide a reason for wanting causal processes as a primitive component in the theory of causation, over and above any arguments considered in the last section.

The reason would be this. As long as we assume that all causes are positive, then we can plausibly define causation in terms of one event fixing an increased chance for another, and we can define causal processes in terms of causal chains as in the last section. But if causes can either increase or decrease the chances of their effects, then it seems highly unlikely that probabilistic relationships alone will distinguish cause-effect pairs from others—for all kinds of events either increase *or* decrease the chances of other events. (See Sober 1986, 99.) So if we allow negative causes, it seems that we will need some independent notion, such as that of a causal *process*, to provide the cement in the causal relationship, that is, to tell us which pairs of (positively or negatively) probabilistically related events are in fact related as cause and effect.

This then is an argument *from* negative causes *to* causal processes. But, still, why should anybody accept negative causes in the first place? If somebody grows up thin after being a fat child, it seems quite counterintuitive to say that they are now thin *because* they were once fat. The natural thing to say, surely, is that they are now thin *despite* once being fat.

Even so, a surprising number of philosophers working on probabilistic causation have been prepared to override this intuition and admit negative causes. To understand this tendency we need to recognize an implicit theoretical rationale, which goes back to the original reasons for switching from the traditional deterministic theory of causation to the S-R view. One initial attraction of the S-R view of causation was that it allowed us to go on having causation in the face of indeterminism. However, this initial attraction loses some of its force once we attend to the fact that in general causes produce their effects only via chains of intermediate events. For it always seems possible that one or more such intermediate events might reduce the chance of the final result, not increase it. If we allow such intermediate negative causes to break the overall chain, then it seems that we won't have much causation in an indeterministic world after all. So to preserve the attractions of the S-R over the traditional deterministic view, we had better allow negative causes in a causal chain to count as intermediate causes. (See Salmon 1984, ch. 7, where this line of thought is clearly implicit.)

The kind of example at issue was first discussed in Suppes's original (1970) formulation of the S-R view. A man hits a golf ball. In its flight it strikes a tree. It lands in the hole. Golf balls that hit trees are less likely to go in holes than those that don't. But, still, wasn't the ball caused to go in the hole by hitting the tree?

A natural initial reaction here is to say that given a full enough description of the golf ball's initial path and the precise way it hit the tree, the ball's hitting the tree would no doubt have made it more likely to go in the hole, not less. But the example can be tightened up. The initial shot is a good one. But a squirrel runs out onto the fairway and kicks the ball as it goes past. The kick, when described in full detail, yields an objective probability distribution over subsequent trajectories for the ball which make it far less likely than before that the ball will go in the hole. But the ball does go in the hole. Didn't the squirrel's kick cause it to go in the hole? (See Sober 1986, 99.)

But now that the case has been tightened up to make it clear that the kick objectively *reduced* the chance of the result, then it seems to me as before that it is highly counterintuitive to say that the kick caused the ball to go in the hole. True, the ball ended up in the hole. But that was just a matter of luck, given the kick. It wasn't because of it.

If the price of defending the S-R view over the traditional view is admitting negative causes, then so much the worse for the S-R view. Surely we would be better off simply resting with the traditional view, which didn't offer to transmit causation across indeterministic gaps in the first place. At least that would remove the temptation to adopt the curious view that results can be caused by events that lower their chances.

What is more, defending the S-R view by allowing in negative causes reduces the power of the S-R view to explicate causation in the first place. For, as explained at the beginning of this section, once we admit negative causes, then we will also need to introduce some independent notion of a causal process to distinguish causal connections from others. So in order to defend the S-R view via negative causes it will be necessary to relegate it to a relatively minor role in the theory of causation.

I don't want to suggest that all this yields a conclusive argument against the S-R view. For it seems to me that the original rationale for admitting negative causes is weak, even from the perspective of the S-R view. That is, I think that S-R theorists can perfectly well accommodate the squirrel-type cases without resorting to negative causes.

There are a number of alternative moves open to S-R theorists here. For instance, they could hold that, whenever an earlier event makes a later one more likely, and the later event happens, then the earlier event causes the later event, even if an intermediate event along the way later shifted the probability of the eventual result downward to some degree (and so isn't itself to be deemed a cause). But perhaps this raises difficulties of its own. If the probability-lowering intermediate event isn't a cause, then we will be committed to a kind of causal action at a temporal distance, for the initial event will cause the final event without the causation being continuously transmitted in between. Such causal action at a distance seems unattractive.

A better move for the S-R theorist seems simply to give up the squirrel-type cases as causal chains, and allow that, if there are probability-lowering intermediaries, then the causal chain is broken. This will mean that S-R theorists will have to accept that there is rather less causation in the world than we used to suppose before quantum mechanics. But there is no reason why the S-R view should be committed to preserving universal causation.

Exactly how many *prima facie* causal chains would in fact be broken by probability-lowering stages is an interesting question, which seems to me to deserve rather more consideration than it has received in the literature. For it by no means immediately follows from quantum mechanics that every sequence of events contains some probability-lowering stage.

Even if it were concluded that *most* apparent causal chains get broken by probability-lowering stages, the S-R theory would scarcely be worse off than the traditional deterministic view on this score. For the traditional view will regard causal chains as broken by *any* indeterministic stages, whether or not they lower the probability of the eventual result. The traditional view may well still have *some* causal chains. (Quantum mechanics doesn't imply that every sequence of events contains indeterministic stages, any more than it implies that every sequence contains probability-lowering stages. It yields indeterminism only where wave functions "collapse" in "measurements," and it's an open question exactly which events should be so conceived.) But the traditional view certainly won't have *more* causal chains than the S-R view.

Doesn't this now yield a kind of quantitative argument for the S-R view over the deterministic view? The S-R view preserves more *prima facie* causal chains than the deterministic view. But why suppose that the maximal preservation of *prima facie* causal chains as real causal chains is a desideratum in the theory of causation? Some S-R theorists seem to think that it is, and indeed it is this preservationist concern that moves them to countenance negative causes. But I have argued that negative causes are a bad idea, and that the best way to do without them is to recognize that certain sequences of events that we took to be causal before quantum mechanics might well not be causal. Given this, it is scarcely decisive against the traditional view of causation to say that it denies continuous causation to all sequences involving indeterministic stages. For the traditional theorist can counter by pointing out that the S-R view also denies continuous causation to certainly apparently causal sequences — namely, sequences like the squirrel case where an intermediate event lowers the chance of the eventual result. At which point we can discuss exactly which class of *prima facie* cases really ought to be deemed to lack continuous causation. All the old arguments will come back into play, and once more we will be led to a stand-off.

In developing the ideas in this paper I have benefited from a long history of discussions with Nancy Cartwright and Hugh Mellor. An early version of the paper was delivered in 1985 at a seminar at the Minnesota Center for the Philosophy of Science in the program run by Wesley Salmon and Philip Kitcher; I would like to thank both of them for inviting me, and for much helpful discussion. Some of the later parts of the paper emerged from a symposium between Elliott Sober and myself, chaired by Hugh Mellor, at the joint session of the Aristotelian Society and Mind Association in 1986, and were improved by what they said there. I would also like to thank Paul Humphreys for discussing a draft of the paper with me.

Notes

1. To avoid confusion I should say that I regard *chance* as the fundamental notion, rather than long-run relative frequency in a reference class, or anything along such lines. Probabilities are always chances, or chance combinations of chances. A reference class should be thought of as a construction out of its homogeneous subdivisions: the probability in an inhomogeneous cell C is then the weighted average of the chances in the homogeneous subdivisions of that cell, with the weighting factors as the probabilities, given C, of being in each such homogeneous subdivision.

2. Perhaps I should make it clear that in talking about "single cases" and "single case causation" I am not endorsing the view that causation needs to be analyzed in terms of some irreducible relation between particular events. I still intend to show that causation reduces to generic (and not explicitly causal) relations of association between properties (together, obviously, with particular, but nonrelational, facts about which properties are instantiated on which occasions). The point of talking about single cases is simply to make it clear that the generic associations relevant in any particular case are those involving *all* probabilistically relevant properties present on that occasion, and not any mixed-probability-yielding subset thereof.

3. If determinism is true, how can there be any probabilities other than nought and one? But note that the probabilities in question need only the *initial conditions*, X or Y, to have probabilities different from nought and one, given S. And that's certainly consistent with C always being determined, in the sense of there always being complete determinism, so to speak, *between* S, X and Y, on the one hand, and C, on the other. But still, the objection might be pressed, how can X and Y have real probabilities different from nought and one, if everything's always determined? I used to think that this problem could be pushed back to infinity, with probabilities always deriving from probability distributions over prior initial conditions, which themselves were determined by prior initial conditions, etc. But now I'm not so sure. Unless there were some real chance events in the background somewhere, I don't think the probabilities over the initial conditions would display the kind of randomness and stable long-run frequencies characteristic of real probabilities. But we can by-pass this worry: for the purposes of this paper let us understand determinism not as the cosmic claim that there are no chances anywhere, but as the context-relative thesis that the events under consideration are always determined by their recent local histories.

4. In fact the two definitions aren't quite equivalent. Imagine a case where the chance of a given result is overdetermined, in the sense that an alternative chain of events, not present in the actual circumstances, would have come into play and ensured the same chance for E if C had been absent. In this case it is true that C and the other factors actually present fix a probability-increasing reference

class for E (and correspondingly E does intuitively depend causally on C). Yet it's not true that E would have had a lower chance if C had been absent. This case shows that to preserve the counterfactual formulation of causal dependence, we need to qualify it so as to make it sensitive to the result's actual causal ancestry in the circumstances. Though I shall not explicitly discuss this issue any further, the discussion of "causal chains" in section 14 provides the materials for the necessary qualification. See David Lewis's treatment of this issue in his (1973).

   5. This conclusion now seems to me somewhat less compelling than I supposed in my (1985b). In that article I concentrated on the deterministic case where C plus background factors (X, say) determines A, and C plus Y similarly determines B. And in that case it is certainly true that an appropriate probabilistic independence condition on X and Y will force A and B to have a probabilistic association that is screened off by C. But if C.X only determines that A should have some chance p, and similarly C.Y only determines that B should have chance q, where p and q are less than one, then there is still room, so to speak, for A and B to coordinate themselves in such a way as to prevent screening off by A. I am thinking here of the kind of phenomenon observed in EPR situations. But, even so, the basic idea that screening off comes from mixtures still seems to me sound. For even in the indeterministic case the independence of X and Y will guarantee that A's *having chance* p will be probabilistically independent within C of B's *having chance* q, and to that extent at least we will tend to get screening off of A-B correlations by C.

   6. The claim that causation is specifically uneclipsed inus conditionship faces its own difficulties. Suppose an intermediate cause, D, of some result, E, always comes from a given prior cause, C. And suppose also that there's an alternative route from C to E, that is, that C sometimes causes E via a chain that doesn't include D. Then C eclipses D as a cause of E. But, by hypothesis, D does sometimes cause E. Here I am inclined to appeal to our intuition in such cases that if D *were* ever to be present without C, then E would still occur. (Compare the contrasting intuition, that even if a properly working barometer were ever to fall without a drop in atmospheric pressure, that wouldn't produce rain.) In terms of chances rather than counterfactuals, I am suggesting that we should look to the cell of the reference class defined by D & not-C to see that D rather than C is on occasion the direct cause of E (and thus to see that D isn't "really" eclipsed by C). The remaining problem, of course, is that there aren't any actual cases in the relevant "D & not-C" cell: I am invoking a probabilistic law without any this-worldly instances. But I'm not too worried. Quite apart from the present problem, there is clearly a general need for some philosophical account of uninstantiated laws. I don't want to belittle the philosophical difficulties involved in giving such an account. But since they will have to be dealt with anyway, there's no reason not to appeal to uninstantiated laws to get out of the present problem.

References

Cartwright, N. 1979. Causal Laws and Effective Strategies, *Noûs*, 13:419–37.
Eells, E. 1982. *Rational Decision and Causality*. Cambridge: the University Press.
——, and Sober, E. 1983. Probabilistic Causality and the Question of Transitivity, *Philosophy of Science*, 50:35–57.
——. 1986. Common Causes and Decision Theory, *Philosophy of Science*, 53:223–45.
Jeffrey, R. 1983. *The Logic of Decision*, 2nd ed. Chicago: University of Chicago Press.
Kitcher, P. 1985. Two Approaches to Explanation, *Journal of Philosophy*, 82:632–39.
Lewis, D. 1973. Causation, *Journal of Philosophy*, 70:556–67.
——. 1981. Causal Decision Theory, *Australasian Journal of Philosophy*, 59:5–30.
Mellor, D. forthcoming. Indeterministic Causation.
Nozick, R. 1969. Newcomb's Problem and Two Principles of Choice. In *Essays in Honour of Carl G. Hempel*, ed. N. Rescher. Dordrecht: D. Reidel, 114–46.
Papineau, D. 1978a. *For Science in the Social Sciences*. New York: Macmillan.
——. 1978b. Salmon, Statistics and Backwards Causation. In *PSA 1978*, eds. P. Asquith and I. Hacking vol. 1, 302–13.
——. 1985a. Probabilities and Causes, *Journal of Philosophy*, 82:57–74.
——. 1985b. Causal Asymmetry, *British Journal for the Philosophy of Science*, 36:273–89.
Redhead, M. Forthcoming. Relativity and Quantum Mechanics – Conflict or Peaceful Coexistence, *Annals of the New York Academy of Sciences*.

Salmon, W. 1970. Statistical Explanation. In Salmon, Jeffrey, and Greeno.

———. 1984. *Scientific Explanation and the Causal Structure of the World.* Princeton: Princeton University Press.

———. Jeffrey, R., and Greeno, J. 1970. *Statistical Explanation and Statistical Relevance.* Pittsburgh: University of Pittsburgh Press.

Sober, E. 1986. Causal Factors, Causal Inference, Causal Explanation, *Aristotelian Society Supplementary LX*:97–113.

Suppes, P. 1970. *A Probabilistic Theory of Causality.* Amsterdam: North Holland.

Capacities and Abstractions

1. The Primacy of Singular Causes

Recently I have been arguing that singular causes are primary and that general causal laws — such as 'Aspirins relieve headaches' — are best rendered as attributions of capacities for single case causation. I have been defending this position by painfully and methodically chipping away at the dominant opposition view, that causal laws are statements of some kind of regular association.[1] In its most sophisticated, and I think defensible, form, the association is taken to be probabilistic, something expressed in terms of partial conditional probabilities, or a probabilistic analogue of path coefficients.

On one front the attack has proceeded in two stages: First, to show that these probabilistic measures already require reference to singular causes if they are to work at all; second, to argue that even then they are not reliable, and the kind of unreliability is one that can't be removed by looking at the probabilities in a randomized experiment, as one might hope.

On the second front, I have argued that with the same kind of background causal information — both singular and general — that is necessary to establish causal laws via probabilities, it is perfectly possible, contrary to Hume, to establish singular causal claims without first establishing regularities, and to do so reliably. And, even more strikingly, in the so-called one-shot experiments of physics, there is generally no distinction made between establishing the singular claim and establishing the corresponding causal law. For example, in the Einstein-de Haas experiment described in "An Empiricist Defense of Singular Causes" (note 1) the move is automatic from the singular claim "orbiting electrons caused the magnetism in *this* iron bar" to the general law "orbiting electrons cause magnetism." This fits well with my reading of the causal law: the electrons in this iron bar show us what orbiting electrons are capable of.

Third, from (1) above, probabilistic measures, even sophisticated ones like the path coefficient, are not entirely reliable in their implications about causal laws. So it would be very wrong to take the probabilistic association to constitute the causal facts. Probabilities at best provide evidence about causes. But what exactly are they evidence for? On the third front I maintain that the simplest account aligns the probabilistic methods with those of the one-shot experiment: in both cases the methods are used to ensure that we have seen at least one good single case of the kind of causal connection in question, a case where a capacity was elicited to produce the appropriate effect. If we take it that this is what we want to find out, then all the pieces fall into place to see why various probabilistic measures provide the evidence that they do for our causal claims.

2. The Failure of the Defeasibility Account

The considerations described in the introduction point to the primacy of singular causal claims in thinking about causal laws. But exactly what is the relationship between a law statement and a singular claim? I propose that the causal law attributes a capacity to the featured characteristic — say, being an aspirin — a capacity to produce the appropriate effect in individual cases. How then are we to think about capacities? It is this question that I want to concentrate on here.

One promising account, at first appearance, is modeled on ideas about defeasibility: the cause will produce the effect; if it doesn't there is a reason. But it seems that when one tries to give a more precise statement of this idea, it always leads to formulations that are either trivial or false. In fact, given the conclusions sketched above, this should not be surprising. For attempts to explicate what capacity ascriptions are true in terms of what things regularly do, or even regularly do in the unlikely circumstances where they get into the right conditions — is still an attempt to explicate causal laws in terms of regularities. This is exactly what I have been arguing can't be done in the painstaking sequence described at the beginning. There is no probabilistic theory of causality. There is no deterministic theory either. Associations are at best evidence for causal laws, and which particular associations are evidence and under what conditions depends finely on the concrete ways in which the causal feature is realized.

Consider an example. To make it clear that the problems of causality are not confined to the fuzzy realm of the social sciences, the example is taken from a branch of applied physics, the study of lasers. Laser theorists teach that the fundamental principle by which a laser operates is this: an inversion in a population of atoms causes amplification in an applied signal. The amplification is narrow band and coherent.[2] Inversion means there are more atoms in the upper state than in the lower, contrary to the way atoms come naturally. This already suggests that accounts in terms of 'normal conditions' have the wrong starting idea, since the conditions must be very abnormal before the causal antecedent is ever instantiated.

The example serves a number of purposes. First, it concerns a capacity which we learned about at a particular time. It is new with quantum mechanics. In classical mechanics inversions could not produce signal amplifications. Second, the language appropriate to the manifestation of capacities—that of harnessing them—fits perfectly here. Quantum physicists have known since Einstein's paper on the *A* and *B* coefficients that inversions *could* amplify. But not until the microwave research of World War II did anyone have an idea how to produce an inversion and harness its capacity.

Third is a point that will turn out to be my central thesis—a point about abstraction and defeasibility. We do know ways to produce inversions now. So suppose you want to amplify some signal. It won't do just to ring up the shop and order an inverted population. Suppose it is delivered to you in a brown paper bag! Now you might be more cautious. You may know some general defeating conditions for amplification. For instance, the inverted atoms must not have a strong absorption line of the right sort from the lower state of the inversion. In fact, physicists take this term absorption, which has a particular concrete meaning here, and use it more abstractly, as a catch-all, to describe anything that might take up the energy that the inversion should be putting out—scattering, thermal agitations, or whatever. So imagine even that you use this catch-all phrase, and order an inverted population (at a given frequency) in a nonabsorptive medium (for that frequency). Won't you want to say more?

The point is yes, but what you will want to say depends intimately on what medium the inversion is in and how it was brought about, as well as on specific features of the signal you want to amplify. For example, depending on what the inverted medium is, the temperature of the amplifier may have to be very finely tuned to the temperature of the source.

One of the early reports by P. P. Kisliuk and W. S. Boyle[3] on successful amplification in a ruby laser provides a good illustration. Kisliuk and Boyle report a gain factor of 2, using a second ruby laser as a source for the signal. The use of a laser source gave them a signal strong enough not to get lost in the noise generated by the amplifier. To get significant amplification, the frequency of the signal should be matched to the transition frequency in the amplifying ruby. But the transition frequency in ruby depends on its temperature, so the temperature of the source and the amplifier must be about the same. Not exactly, however, for the ruby line width is also temperature dependent, and the gain is a function of the line width; so the optimum condition for a given temperature of the source is reached not when the amplifier is at exactly that temperature, but rather when it is somewhat colder. The authors report: "The expected difference is of the order of a few degrees near room temperature. . . . As the temperature is lowered, its control becomes more critical because of the narrowing of the line."[4]

There is nothing special about the laser case. Consider a second example. Donald Glaser built the first bubble chambers, using diethyl ether. He operated on the principle that a passing charged particle has the capacity to cause bubbling in a liquid in a superheated state. (The liquid is ready to boil and just needs a catalyst.) He also was successful with hydrogen and most of the heavy liquid hydrocarbons, like propane or freon. But surprisingly, the bubble chamber didn't work with xenon. The reason depends on the peculiar nature of the xenon—here the passing charged particles were exciting optical transitions and the energy was being sent off as light rather than producing the heat necessary to catalyze the boiling.[5]

The point again is that once general defeasibility conditions are laid out, there remains a very large number of enabling and stopping conditions that are dependent on the particular realization. The least that follows from this is that there is no hope to convert the abstract "T's have the capacity to A" into a specific law "if I and C_1 and . . . C_n and not-R_1 and . . . not-R_m then A" even where A itself might already be a causal verb. I do not deny that there are regularities of this form, in fact a vast array of them. Otherwise two lasers identically machined would behave in different ways. But, as with my original remarks, these regularities do not constitute the truth of the general causal claim. They are merely evidence for it; and evidence that is not entirely reliable, for just the kinds of reasons I mentioned at the start.

Equally important, one must notice that the kinds of regularities pointed to by the defeasibility account—even if we use causal language in their expression—are not all, or even the bulk of, the regularities associated with the capacity claim. We have been looking at the kind of behavior that best provides evidence for a capacity. But there is also all the behavior that the capacity explains—what happens when there are preventatives depends on the capacity, just as much as what happens when there are no preventatives. So there is a very large stock of concrete regularities associated with the capacity claim, both regularities explained by the capacity and regularities that are evidence for it.

3. Abstractions and Idealizations

I want to turn now to something that on the face of it looks like a very different topic—the relation of abstract models in physics to the more concrete laws that they are supposed to explain. Consider the two diagrams, Figures 1 and 2, and their labels.[6] The first is a schematic diagram of a laser constructed from specified materials in a specified way—it is a diagram of the first helium-neon gas laser. The diagram is schematic; it leaves out a large number of features, possibly even a large number that are relevant to its operation. The concrete object pictured here operates, I believe, under some law. I call the law a phenomenological law because, if we could write it down, it would literally describe the features of this concrete phenomenon and the nomological links among them. It would be a highly complex law and would include a specific description of the physical structure and surrounds of the concrete device. Possibly it could not be written down; perhaps the features that are relevant to its operation make an open-ended list that

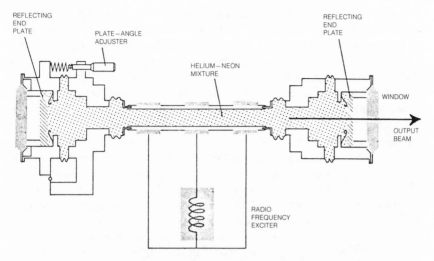

Figure 1. Sketch of He-Ne gas laser, Reprinted from William H. Louisell, Quantum Statistical Properties of Radiation. *New York: Wiley & Sons (Figure 8.1, p. 445).*

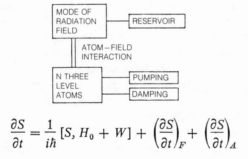

$$\frac{\partial S}{\partial t} = \frac{1}{i\hbar} [S, H_0 + W] + \left(\frac{\partial S}{\partial t}\right)_F + \left(\frac{\partial S}{\partial t}\right)_A$$

Figure 2. Block diagram of laser model. Reprinted from William H. Louisell, Quantum Statistical Properties of Radiation. *New York: Wiley & Sons (Figure 9.1, p. 470).*

cannot be completed. But because we are talking about a concrete object, there is some such law true of its operation.

Now look at the second drawing, and its accompanying equation. Here we do not have a schematic diagram for a particular laser, but rather a block diagram for *any* laser. In the block diagram we consider no particular medium, no specific way of pumping, no determinate arrangement of the parts. We abstract from the concrete details to focus, as we say, "on the underlying principles"; the equation we see here records one of these principles.

I want to distinguish this kind of abstraction from what I will call "idealization."[7] In (ur-) idealization, we start with a concrete object and we mentally discount some of its inconvenient features—some of its specific properties—before we try to write down a law for it. Our paradigm is the frictionless plane. We start with a particular plane. Since we are using it to study the inertial properties of matter, we ignore the small perturbations produced by friction. Often we do not just delete factors. Instead we replace them by others which are easier to think about, usually easier to calculate with. That is what I call idealization, or "uridealization," since I think most of what are called idealizations in physics are a combination of this ur-idealization with what I am calling "abstraction."

By contrast, in the kind of abstraction we see in the second example, we do not subtract or change some particular features of properties; rather mentally we take away any matter that it is made of, and all that follows from that. This means that the law we get by material abstraction functions very differently from idealized laws. For example, it is typical in talking about idealizations to say (here I quote specifically from Ernan McMullin): the "departure from truth" is often "imperceptibly small" or "if appreciably large" then often "its effect on the associated model can be estimated and allowed for."[8] But where it is the matter that has been mentally subtracted, it makes no sense to talk about the departure of the remaining law—a law like the one we see in Figure 2—from truth, about whether this departure is small or not, or about how to calculate it. After a material abstraction, we are left with a law that is meant not literally to describe the behavior of objects in its domain, but, rather, as I said, to reveal the underlying principles by which they operate—though saying this is just to label a problem.

What I want to focus on is the relationship between the abstract law and the vast network of laws that fall under it, which do literally describe the behavior of the concrete objects in its domain. For short, I call this network of complex laws, laws with highly detailed concrete antecedents, the "phenomenal content" of the abstract law. As with the causal example of inversions and amplifications with which we began, the phenomenal content of an abstract law is highly realization-dependent, and there is no recipe for going from the abstract law to its phenomenal content. The exact features which must be put back vary from one realization to another, and there is no further legitimate concept that covers them all. When I say legitimate I mean a concept that has independent criteria for its application, and does not just mean "whatever in the specific case is relevant."

I do not have a theory about what this relationship is or how it works. It seems to me that it is a crucial one; but there is no near-satisfactory philosophical account. In *Understanding Physics Today*, W. H. Watson gives a kind of Wittgensteinian account in terms of the unformalized and unformalizable training of physicists. I mention him because he points to the same problem that worries me: "In this way," he says, "we pass easily from one level of theory to another, usually a more abstract one. . . . Thus mathematical abstraction can be offered without its having to face the question, 'How is this relevant to the real world?' "[9]

Return now to capacities. Causal laws are best rendered, I claimed, as capacity ascriptions. Now what I want to say is that when capacity ascriptions play this role, they are functioning as material abstractions. Like our block diagram and its associated equation, they are abstract claims that cover and and explain a vast network of complex, detailed laws that do describe the behavior of the concrete objects in their domain; and like these other abstract claims, their phenomenal content is highly realization-dependent. The difference is that the realization-dependent laws that fall under a capacity ascription are laws about causings: "in such and such particular circumstances, x's cause y's," although, as with the non-causal case, these laws may not be finitely expressible.

4. Conclusion

I begin with an assumption that I do not defend here: contrary to Hume, we need some concept of singular causing already if we are to talk about causal laws. Then I present a structure with three tiers. At the bottom we have individual singular causings. At the top we have general causal claims, which I render as statements associating capacities with properties—"aspirins have the capacity to relieve headaches"; "inversions have the capacity to amplify." In between stands the phenomenal content of the capacity claim—a vast matrix of detailed complicated causal laws. Once the concept of singular causing has been let in, this middle level is unproblematic. These are just laws, universal or probabilistic, about what causings occur in what circumstances what percentage of the time.

The third level is more troubling. It sounds doubly non-Humean. First we have causal connections in the single case; then we add mysterious, or occult powers. But this is not what is happening. The capacity claim has two facets: first it is a material abstraction of a kind familiar in physics. It resembles the block diagram we looked at, and its accompanying equation; and it has the same peculiar and ill-understood relationship to the phenomenological laws that it covers. This relationship is indeed troubling, but it has nothing special to do with causation. Second, unlike the equation for the block diagram, the laws which make up the phenomenal content of the general capacity claim are laws about causings, and not about co-association. The use of capacity language marks this special feature of the phenomenal content. We see that the picture is not doubly non-Humean. Once we let in singular causation—as I think we must for other reasons—we have no more special problems with capacity ascriptions than with other abstract descriptions in physics. The laws of physics are indeed mysterious; but causal laws are not more mysterious than others.

Notes

1. Nancy Cartwright, "An Empiricist Defense of Singular Causes," forthcoming in a volume edited by Hide Ishiguro in honor of G. E. M. Anscombe. Also "Regular Associations and Singular

Causes," forthcoming in a volume edited by William Harper and Brian Skyrms. The fullest development of my views on these topics can be found in Nancy Cartwright, 1989. *Nature's Capacities and Their Measurement.* Oxford: the University Press.

2. Anthony E. Siegman, 1986. *Lasers.* Mill Valley, CA: University Science Books, section 1.1, "What is a Laser?"

3. P. P. Kisliuk and W. S. Boyle, 1968. "The Pulsed Ruby Maser as a Light Amplifier." In *Lasers,* ed. J. Weber. New York: Gordon and Breach, pp. 84–88.

4. Kisliuk and Boyle, note 3, pp. 84–85.

5. Peter Galison, 1985. "Bubble Chambers and the Experimental Workplace." In *Observation Experiment and Hypothesis in Modern Physical Science*, eds. P. Achinstein and Owen Hannaway. Cambridge, Mass.: MIT, Bradford, pp. 309–74.

6. From William H. Louisell, 1973. *Quantum Statistical Properties of Radiation.* New York: Wiley and Sons, p. 445; p. 470 respectively.

7. These views about abstraction versus idealization are heavily influenced by Henry Mendell. See Henry Mendell, 1985. *Aristotle and the Mathematicians.* Stanford University Ph.D. dissertation, especially 11.7, "The Abstract Objects."

8. Ernan McMullin, 1985. "Galilean Idealization," *Studies in the History and Philosophy of Science* 16:247-73.

9. W. H. Watson, 1963. *Understanding Physics Today.* Cambridge: the University Press.

The Causal Mechanical Model of Explanation

Wesley Salmon's *Scientific Explanation and the Causal Structure of the World* (*SE*) presents a sustained and detailed argument for the causal/mechanical conception of scientific explanation which Salmon has developed in a series of papers over the past decade. *SE* ranges over a wide variety of topics – in addition to the material discussed below, it contains discussions of probabilistic theories of causality, scientific realism, of the notion of objective reference class homogeneity, and of much of the recent literature on causality and explanation. In my judgment, *SE* is the most interesting general treatment of scientific explanation since Hempel's *Aspects of Scientific Explanation*. Even those who are not persuaded by many of Salmon's conclusions will find this book eminently worth careful study.

According to Salmon, causality plays a central and fundamental role in scientific explanation. To explain a particular occurrence is to show how it "fits into" the causal network of the world (276). Causality has three fundamental "aspects" (179). The most basic causal notion is that of a *causal process*. A causal process is characterized by the ability to transmit a mark or the ability to transmit its own structure, in a spatio-temporally continuous way.[1] Examples include the movement of a free particle or an electromagnetic wave through empty space. Causal processes are to be distinguished from pseudo-processes (e.g., the successive positions occupied by a shadow cast by a moving object) which lack the above abilities. *Causal interactions* occur when one causal process (spatio-temporally) intersects another and produces a modification in its structure. A typical example is a collision of two particles. Interactions commonly involve correlated changes in structure governed by conservation laws. *Conjunctive forks* involve correlations among spatio-temporally separated effects which are explained in terms of separate causal processes deriving from a common cause, as when food poisoning is invoked to explain the indigestion experienced by a number of people who attended a picnic.

I would like to thank Paul Humphreys for helpful comments on an earlier draft of this paper.

A scientific explanation of some particular outcome will consist of citing (some portion of) the causal processes and interactions leading up to that outcome. The explanation of a type of outcome or a generalization is a derivative notion; such an explanation will describe what the particular causal processes and interactions responsible for instances of that type or generalization have in common. Purported explanations which fail to cite genuine causal processes and interactions — Salmon suggests thermodynamic explanations which appeal to the ideal gas laws as cases in point — are spurious or at least seriously defective precisely because of their noncausal character. Salmon claims that it is a virtue of his approach that it enables us to see why such noncausal explanations are defective, while the traditional DN model does not.

While Salmon thus no longer subscribes to the idea, expressed in his well-known monograph "Statistical Explanation and Statistical Relevance," that information about statistical relevance relations is by itself explanatory, his discussion in *SE* makes it clear that he by no means wishes to entirely abandon the SR model. Instead Salmon suggests that scientific explanation is typically a "two-stage affair" (22). In the first stage, one assembles information about statistical relations of the sort demanded by the SR model; in the second stage, one provides a causal account in terms of processes and interactions of why those relations obtain. Information from the first stage is said to play the role of providing evidence for claims about causal connections introduced in the second stage, but it is only claims of the latter sort which have explanatory import. Providing a correct characterization of the SR basis, as Salmon now prefers to call it, is accordingly still an important part of the project of providing an adequate model of scientific explanation, and Salmon devotes a substantial part of *SE* to developing such a characterization and defending it against objections.

While this new formulation improves in interesting respects on Salmon's earlier formulation, the fundamental idea of the earlier account is retained: one begins with a reference class A and a set of explanandum properties $\{B_i\}$ and then, by introducing a set of statistically relevant factors $C_1 \ldots C_s$, partitions the reference class A into a set of mutually exclusive and jointly exhaustive cells $A.C_i$; each of which is required to be "objectively homogeneous" with respect to the set $\{B_i\}$ — i.e., none of the cells "can be further subdivided in any manner relevant to the occurrence of any B_i."[2] (37) Moreover, Salmon still wishes to retain a number of characteristic features of the SR model — for example, he still thinks of statistical theories like elementary quantum mechanics (QM) as providing explanations of individual outcomes in circumstances in which they merely permit a calculation of the probability with which those outcomes will occur. Thus, according to Salmon one can explain why some individual carbon-14 atom decayed during a certain short time interval by citing, among other things, the probability of decay during this interval, even though this probability will be quite low. Various alternative theories of explanation (such as Hempel's IS model) are still judged as inadequate in part because of their inability to accommodate supposed explanations of this kind.

Salmon's discussion takes place against the background of a provocative contrast between different general conceptions of scientific explanation. *Epistemic conceptions* represent attempts to characterize explanation in epistemic terms — the best known example is the DN model, in which the key epistemic notion is nomic expectability. While the epistemic conception, at least in the DN version, leads one to think of an explanation as an argument or an inference, and accordingly to look for a "logic" of explanation valid for all possible worlds, Salmon holds that this is a "futile venture, and . . . little of significance can be said about scientific explanation in purely syntactic or semantic terms" (240). By contrast, the *ontic conception* which Salmon favors represents an attempt to characterize explanation in terms of the fundamental causal mechanisms which, as a matter of contingent fact, operate in our world, with no suggestion that similar explanatory principles would be useful in all logically possible worlds.

In what follows I shall explore four general questions raised by Salmon's discussion: (1) Is it reasonable to suppose that all scientific explanations will meet the requirements of the causal/mechanical model? (2) Does Salmon's model capture all of the features which are relevant to the assessment of scientific explanations? (3) Should we take statistical theories like quantum mechanics as providing explanations of individual outcomes; and, relatedly, does Salmon's SR basis capture the way in which statistical evidence is relevant to the construction of explanations in QM and elsewhere in science? (4) Should we abandon epistemic conceptions of explanation in favor of Salmon's ontic conception? Is there a single logic of explanation valid in all possible worlds or, for that matter, everywhere in our world?

I

Salmon's conception of explanation seems to fit most neatly simple physical systems, whose behavior is governed by the principles of classical mechanics and electromagnetism (including such common-sense paradigms of causal interactions as the collision of a golf ball with a tree limb). I think it is clear that Salmon has captured a number of the central features of the notion of causation, as it is applied to such systems. In connection with such systems one typically thinks of causation as involving the transfer of energy and momentum in accordance with a conservation law, and it is a consequence of Special Relativity that if, when transferred, such quantities are conserved, they must be conserved locally — that there is no causal action at a distance and that causal processes will be spatio-temporally continuous, just as Salmon claims.[3]

However, even in physical contexts, the idea that explanation must involve the tracing of causal processes and interactions in Salmon's sense becomes increasingly problematic, as we move away from the above paradigms. Consider, for example, explanations which make reference to the geometrical structure of spacetime as in Special and General Relativity. Even if one thinks, as Salmon does, of a particle moving along a time-like geodesic as a causal process, it seems fairly clear that the *explanation* one gives in General Relativity for why the particle moves as it does is not a causal explanation. The explanation one gives will make reference to facts about the affine and metrical structure of spacetime and the variational equation of motion $\delta \int d_s = 0$, and these facts about the geometrical structure of spacetime will in turn be explained in terms of the distribution of mass and energy (as expressed in the stress-energy tensor). Neither the fact that the particle moves along a geodesical path nor facts about the geometrical structure of spacetime are themselves explained in terms of continuous causal processes or interactions between such processes, or in terms of such characteristically causal notions as forces, or transfers of energy and momentum. A similar point holds with respect to the explanations of length contraction and time dilation in a moving inertial frame provided by Special Relativity. A rapidly moving clock is a continuous causal process, but the time dilation and length contraction it will exhibit have their explanation in the structure of Minkowski spacetime and are geometrical effects rather than causal effects whose origin is to be sought in causal processes and their interactions. Indeed the contrast between the explanation of such phenomena in Special Relativity and their explanation in a theory like Lorentz's, in which these effects are regarded as a consequence of the operation of electromagnetic forces, seems to be precisely the contrast between a noncausal, geometrical explanation and a causal one.[4]

Explanation in elementary quantum mechanics represents a second class of explanations which do not fit comfortably within Salmon's framework. Salmon is aware of this fact and includes in *SE* an interesting discussion of explanation in QM. This focuses largely on the difficulties EPR type correlations create for the common-cause explanatory principle. Salmon seems to suggest that it is an explanatory deficiency in QM that it fails to explain such correlations in terms of spatio-temporally continuous processes connected to a common cause and that it is an advantage of his account of explanation that it allows us to recognize this, while the standard DN model would not.[5]

This is hardly the place to embark on a detailed treatment of the notoriously difficult and intractable issues surrounding the interpretation of quantum mechanics. I confine myself to two general observations concerning Salmon's discussion. The first is that it is at least arguable that any plausible theory of scientific explanation must make room for a possibility which Salmon seems to reject: that we can learn, as a result of empirical inquiry, not just that various particular candidates for an explanation of some phenomena are defective, but also that the demand for any explanation (or at least any explanation of some very general type) is inappropriate or misguided. In many cases the question of whether the demand for a certain kind of explanation of some explanandum is a reasonable one is not something we should expect to settle just via appeal to an abstract model of explanation (whether of the deductive-nomological or causal/mechanical variety) but will depend rather on the results of empirical inquiry. There is thus a difference between a theory that fails to successfully explain some potential explanandum, where it is agreed that the demand for such an explanation is perfectly appropriate, and a theory that provides us with good reasons for rejecting the demand for such explanations — only in the former case is talk of an explanatory lacuna appropriate.[6]

It seems to me that one of the central issues raised by Salmon's discussion is whether the former or the latter characterization is the more appropriate way to think of the failure of quantum mechanics to explain phenomena like EPR correlations in terms of a common-cause mechanism. This issue is in turn bound up with more general issues concerning the interpretation of quantum mechanics. Although Salmon does not explicitly commit himself to a general view about the interpretation of QM, much of his discussion seems to suggest or presuppose a rejection of the standard, Copenhagen interpretation of the theory in favor of a "realistic" interpretation according to which all measurable quantities have precise values prior to or independently of measurement.[7] Given the expectations generated by this sort of realism, it is particularly natural, if not irresistible to inquire, as Salmon does, after the details of the mechanism by which nature arranges things so that the EPR correlations hold and to suppose that it is a limitation on the explanatory power of quantum mechanics that it fails to specify such a mechanism.

However, as is well known, various no hidden variable theorems seem to show in principle that no common-cause mechanism and indeed no realistic theory of the sort that Salmon envisions could exist which reproduces the experimentally observed correlations.[8] It is tempting to think — I take this to be the attitude of many physicists — of such impossibility results as one of a number of considerations which show that there is something misguided or inappropriate about the demand for a common-cause mechanism of the sort described above. Rather than concluding that there is an explanatory gap in quantum mechanics, those who adopt this view will conclude instead that it is our theory of explanation and our prior expectations that requests for explanation of a certain general kind are always in order which need revising. More needs to be said by Salmon to support his alternative characterization of the failure of QM to explain EPR type correlations and to show that this characterization can be motivated without the assumption of a kind of realism which seems empirically false and which would be rejected by the majority of physicists.

The second and more general observation is that because Salmon's discussion focuses largely on the EPR problem, the implications of his views for the assessment of the explanatory power of QM generally are not drawn as sharply as they might be. Consider an elementary textbook example, referred to briefly by Salmon: the derivation of the probability that a charged particle will penetrate a potential barrier by solving the time-independent Schrödinger equation for the system. This strikes me as a paradigm of quantum mechanical explanation – it represents the (highly idealized) basis for the usual elementary treatment of Salmon's often repeated example of the radioactive decay of a carbon-14 atom. However, as Salmon acknowledges (256–57), this explanation certainly does not seem to involve the explicit tracing of continuous causal processes leading up to the penetration of the barrier. If we adopt Salmon's model of explanation, it is not clear that we can avoid the conclusion that QM is nonexplanatory in this sort of typical application (and indeed in virtually all of its applications), and not just that it fails to explain EPR type correlations. Various scattered remarks suggest that Salmon is perhaps willing to accept this conclusion, in part because he thinks (or at least holds that it is not obviously false that) more recent theoretical developments in quantum electrodynamics (QED) and quantum chronodynamics (QCD) more closely approximate the demands of his model.

This strikes me as unpersuasive for several reasons. First, I think that most physicists and chemists think of elementary, nonrelativistic QM as a satisfactory explanatory theory as it stands in the domains to which it is taken to apply and that it does not require supplementation by some more causal theory (in Salmon's sense) before it becomes explanatory. Quite apart from this, it is unclear how a theory like QED, which employs such notions as virtual particles and vacuum fluctuations or in which one does calculations by summing over all possible past histories of a particle or over all possible ways in which a certain interaction can occur, is one in which, as Salmon claims, a "broadly causal picture seems to emerge" (255). Salmon's brief remarks in support of this claim (which he seems to intend just as tentative suggestions) are, I think, simply not detailed enough to enable one to see very clearly what he has in mind. Moreover, although the issues involved here are murky, the no hidden variable results alluded to above seem not merely to undercut the possibility of a realistic interpretation of quantum mechanics as presently formulated, but to strongly suggest that, however our understanding of microphysics may change, we should not expect a future theory in which phenomena like EPR type correlations are explained in terms of spatiotemporally continuous processes. Salmon seems (254–55) to envision a theory which provides causal explanations (in his sense) of such phenomena but which is at the same time not a local hidden variable theory, but it is not obvious how this is possible.

More also needs to be said about how Salmon's model applies to complex physical systems which involve large numbers of interactions among many distinct fundamental causal processes. In such cases it is often hopeless to try to understand the behavior of the whole system by tracing each individual process. Instead one needs to find a way of representing what the system does on the whole or on the average, which abstracts from such specific causal detail. It is not clear how Salmon's model which is formulated in terms of individual causal processes and their binary interactions, applies to such cases.

Consider again Salmon's example of the explanation of the behavior of a gas in terms of the statistical mechanics of its component molecules and the contrast Salmon wishes to draw between this and a (noncausal and hence defective) explanation which appeals to the ideal gas laws. Plainly it is impossible to try to trace the trajectories and interactions of each individual molecule (the individual causal processes involved here) and to exhibit the behavior of the gas as in any literal sense the sum of these. Moreover, although some of Salmon's remarks perhaps suggest that he takes a contrary view, it seems equally clear that it is a trivial, nonserious explanation of the behavior of the gas to say simply that it is composed of molecules, that these collide with one another in accordance with the laws of Newtonian mechanics and that (somehow) the behavior of the gas results from this. The usual sort of treatment found in textbooks on statistical mechanics does neither of these things, but instead proceeds by making certain very general assumptions about the distribution of molecular velocities, the nature of the forces involved in molecular collisions and so forth, and then deriving and solving the Maxwell-Boltzman transport equation for the system in question. One then shows how various facts about the behavior of the gas (such as that it obeys the ideal gas laws) follow from the solution of this equation.

This sort of treatment does not literally describe in detail the collisions of any individual molecule, let alone the entire collection of molecules comprising the gas. Instead, in constructing an explanation one abstracts radically from details of such individual causal processes and focuses on finding a way of representing the aggregate behavior of the molecules. In this treatment, such characteristically "epistemic" or "inferential" concerns as finding techniques for actually solving the relevant equation governing this aggregate behavior and for avoiding computational intractabilities are of quite central importance. Rather than merely mirroring facts about causal interactions and processes the relevance of which for inclusion in the explanation is determined on other grounds, such epistemic considerations seem to have an independent role in determining why this sort of explanation takes the form it does. One omits causally relevant detail about individual interactions for the sake of such epistemic considerations as derivability and generality.

Examples such as this raise a number of natural questions in connection with Salmon's discussion. Just what does the causal/mechanical model require in the case of complex systems in which we cannot trace in detail individual causal processes? How, in terms of Salmon's fundamental notions of causal processes and interactions, does the statistical mechanical explanation sketched above amount to successfully specifying a mechanism? How much of the above explanation (how much detail and of what sort) does one have to provide before one has satisfied the constraints of the causal/mechanical mode? To what extent can one capture or account for the salient features of the above explanation without appealing to epistemic or inferential notions? On what grounds can we say — as I think that we must — that the trivial explanation described above fails to provide a mechanism or to explain the behavior of the gas? A fully worked out version of the causal mechanical model needs to provide a principled answer to such questions.

II

Even if we agree that some or all good explanations involve the postulation of continuous causal processes, it is doubtful that all of the features of an explanation which are relevant to its assessment have to do with the extent to which it successfully postulates such processes. There are important explanatory desiderata which are not captured by an account which just focuses on the idea that to give an explanation is to cite a cause. One such desideratum is that a good explanation should diminish one's sense of arbitrariness or contingency regarding the explanandum. For example, an important feature of many explanations in statistical mechanics (and indeed of explanations elsewhere that successfully invoke equilibrium-type considerations, as in evolutionary biology)[9] is that they proceed by showing that for a great many possible sets of initial conditions, an outcome like the actual outcome would have ensued. In this way, one's sense that the actual outcome was fortuitous or arbitrary is at least partly removed. This feature is missed in an account like Salmon's which just focuses on describing the actual causal history leading up to the explanandum-outcome and leaves out questions about whether other possible but not actual histories would have led to a similar outcome.

The idea that a successful explanation ought to diminish arbitrariness is also reflected in the common distaste of scientists for theories that contain many free parameters, the values of which are not determined by the theory itself but rather must, as it is commonly expressed, "be put in by hand" — introduced with no other rationale than that they are required by the data. The idea that a theory having this sort of feature is defective as an explanation because it is left arbitrary why the parameters in question should have just these values, rather than some other set of values, is very common, especially among physicists. Various theories of the weak and strong interactions (including QCD) are frequently criticized on just this score and one of the central attractions of recent "superstring" theories is that they seem to go a long way toward eliminating this sort of arbitrariness.[10] Here again, it is at least not obvious how this sort of dimension of explanatory assessment is captured by Salmon's model, with its focus on individual causal processes and its de-emphasis on the role of inferential considerations in explanation.

Another consideration which is quite central to the assessment of explanations but is insufficiently emphasized in Salmon's model has to do with the idea that a good explanation will provide a unified account of phenomena previously thought to be unconnected.[11] In my view, the idea that such unification is an important goal in scientific explanation provides a much more natural account of the salient features of, say, contemporary high energy physics than the idea that theory construction in this area is driven by the demand that microphysical phenomena be explainable in terms of continuous causal processes. It is certainly the former notion and not the latter which is generally emphasized by physicists in their own accounts of their activities.[12] The demand for such unification is evident in such recent achievements as the Weinberg-Salam electroweak theory, which unifies the electromagnetic and weak forces, and in more recent proposals for a unified treatment of the strong and electroweak forces. One can plausibly point to many central mathematical features of such theories, such as the imposition of various gauge symmetry requirements, as providing a concrete implementation of this demand for unification.

Salmon, after critizing epistemic versions of accounts of explanation which assign a central role to theoretical unification, suggests that his own treatment can also take unification to be an important aspect of explanation since on his view, "unity lies in the pervasiveness of the underlying mechanisms upon which we depend for explanation" (276). This makes it sound as though unification is an incidental (but welcome) byproduct of the search for causal mechanisms. I believe that this gets matters backwards—in many areas of science, it is the demand for unification which is primary, and one determines what the relevant mechanisms are in the light of this demand. Spontaneous symmetry breaking or the Higgs mechanism are regarded as important mechanisms in high energy physics because of the role they play in current unification programs and not because they are independently required by the constraints of the causal/mechanical model.

The significance of this last point becomes even clearer when we consider the implications of the causal/mechanical model for explanatory theorizing in biology, psychology, and the social sciences. A great deal of theorizing in these disciplines proceeds on the assumption that systems which may differ significantly from the perspective of some fine-grained, microreductive causal theory may nonetheless exhibit interesting common patterns or regularities at a more macroscopic level of analysis and that one can construct explanatory theories by focusing on such patterns and regularities. Put crudely, the basic idea is that complex systems can exhibit different levels of organization and that, corresponding to these, different levels of explanation are appropriate. Thus systems that differ in underlying physical or chemical respects can nonetheless be treated as similar for the purposes of biology, psychology, or economics.[13] In the absence of such an assumption, it is hard to see how serious explanatory theories in these disciplines are possible.

Thus, explanations in evolutionary biology of why quite different organisms possess similar traits or behaviors will often make reference to the fact that these organisms face quite similar adaptive situations or selection pressures, despite the fact that the proximate mechanisms underlying these traits may be quite different in the case of different organisms. For example, one finds, in evolutionary theory quite general game-theoretical explanations of various behavioral strategies (regarding defense of territory, parental investment in offspring, and so forth) although the immediate causal antecedents of such behavior in different organisms will be quite different. Similarly, consider such general results in evolutionary theory as Fisher's fundamental theorem on natural selection (that the rate of increase in fitness in a population at a time equals the additive genetic variance in fitness at that time) or the standard explanation of how heterzygote superiority can lead to a stable polymorphic equilibrium. Here too we have quite general results which apply to a wide range of different organisms acted upon by natural selection despite important differences of causal detail. Or consider the use of quite general formal models in ecology — say, the Lotka-Volterra equations for prey/predator interactions. Here the attempt is to model and explain quite general features of such interactions, despite the fact that the causal details will differ greatly from population to population. Finally consider explanations of people's cognitive capabilities (e.g., chunking and recency effects in memory or tendencies to make certain inferential errors) in terms of the way they process and store information. It is a central claim of much theory in cognitive psychology that such accounts can provide genuine explanations even though they do not describe in detail the operation of neurophysiological or biochemical mechanisms and even though similar information-processing strategies may have interestingly different neurophysiological realizations in different subjects.[14]

None of these explanations seems to explain by tracing in detail continuous causal processes or underlying physical mechanisms. Rather, as the statistical mechanical case, they explain by abstracting from such detail and finding general patterns. An account of explanation which attaches a central role to theoretical unification and nonarbitrariness (and which recognizes a connection between explanation and inference and is sensitive to the ways in which computational intractabilities can interfere with attempts to explain) can make sense of and legitimate the above explanations. The demand for explanations that unify (and diminish arbitrariness, and so forth) and the demand for explanations that specify the details of fundamental physical mechanisms or that trace continuous causal processes can conflict, and when this happens, it is often the former demand that ought to prevail. By contrast, I think it is at least not obvious (even if one leans *very* heavily on the idea that the causal/mechanical model is an "ideal" which is only partially realizable in practice or to which successful actual theories represent an approximation — see *SE*, 263ff.) how Salmon can avoid the conclusion that many of the above theories are pretty dubious as explanations, in virtue of their apparent failure to specify continuous causal processes.

It would have been very useful to have had an explicit discussion of such examples in *SE*. Does Salmon think (appearances to the contrary) that the above examples satisfy the requirements of the causal mechanical model?[15] If so, what sort of biological and psychological explanations does the model rule out? Is there some natural way, within the context of the causal/mechanical model (even in its ideal text version) of avoiding the (seemingly wrong-headed) conclusion that the best way to improve the above explanations would be to add more specific detail about proximate mechanisms and continuous causal processes (even if generality and other explanatory desiderata are lost) and that the ability of the above treatments to unify disparate phenomena has little to do with their explanatory power at least if this unity is not a result of a sameness of underlying mechanisms? Or is this perhaps a conclusion which Salmon would wish to endorse and defend?

III

Like Hempel (and indeed most other writers who have discussed the matter) Salmon retains the idea, associated with the original SR model, that statistical theories like QM, which in many circumstances merely specify probabilities strictly between 0 and 1 that certain outcomes will occur, nonetheless can be thought to explain those outcomes.[16] Salmon is quite effective in showing that *if* this idea is accepted it will have very important implications for how one ought to think about explanation. For example, Salmon argues convincingly that given this conception of statistical explanation, it is arbitrary to suppose (as Hempel did in his original IS model) that an explanation which assigns a high probability to some outcome is for that reason better than an explanation which assigns a low probability to some outcome. He also shows that given the above conception it will be hard to avoid the conclusion that the same explanans can explain both the occurrence of some event E and the nonoccurrence of E (113) — a conclusion that is certainly inconsistent with a great many philosophical accounts of explanation.

However, like other writers on the subject, Salmon devotes comparatively little attention to arguing for the antecedent of the above conditional — i.e., that statistical theories like QM *do* provide us with explanations of individual outcomes in the sense intended. From his brief discussion, it appears that he is in part influenced by the idea that the only plausible alternative to his own view is that statistical theories explain facts about approximate relative frequencies in large numbers of outcomes (e.g., that roughly 1/10 of the atoms in some large collection decay in a certain interval or that approximately 375 out of 500 plants have red blossoms (216)). I agree with Salmon that this alternative is inadequate. First, one wants such statistical theories to be applicable to or able to explain facts about small populations or individual outcomes. Second, given a theory which predicts (P) that an outcome E will occur with a certain probability k, the claim that the approximate relative frequency of E in even a large population will be "close" to k does not of course follow deductively from (P), but is, according to the law of large numbers, merely probable given P. Thus to claim that facts about relative frequencies are explained in such a case is not to avoid the notion of statistical explanation, but to embrace a particular version of it—in this case, something like the IS model in which an explanandum is explained by finding an explanans which confers a high probability on it. If this were the only alternative to Salmon's approach, his treatment would be vindicated.

It seems to me, however, that Salmon fails to emphasize a rather natural third possibility: one can distinguish between claims about probabilities and claims about relative frequencies (probabilities cannot literally be relative frequencies since, among other things, the latter have the wrong formal properties—e.g., they are not countably additive). Facts about relative frequencies of outcomes instead have the status of evidence for claims about the probabilities of those outcomes and it is such facts about probabilities of outcomes (e.g., the probability that a particle with a certain kinetic energy will penetrate a potential barrier of a certain kind) that are explained by statistical theories like QM. The evidential connection between information about relative frequencies and claims about probabilities is established in QM, as elsewhere, by the use of standard statistical tests such as tests of significance—tests whose role would be quite opaque if we did not make something like the above distinction. On this conception one thinks of the explanations provided by theories like QM as having something like an ordinary DN structure or as what Hempel calls deductive-statistical (DS) explanations. What is explained by such theories is (just) what can be deduced from them—claims about probabilities of individual outcomes. So construed, such theories do explain facts about (do apply to) individual outcomes and not just large collections—although what is explained is not why such particular outcomes occur, but rather why they occur with a certain probability.

I suspect that one reason Salmon does not take this possibility as seriously as he might is that the paradigms of statistical theorizing with which he works involve very low-level phenomenological generalizations about the behavior of particular kinds of systems, e.g., generalizations about the half-life of particular radioactive isotopes. If one thinks that generalizations of this sort are typical of the generalizations which figure in the explanans of microphysical explanations—that they (perhaps when supplemented in the appropriate way with information about continuous causal processes) are what do the explaining—it will be hard to avoid the conclusion that if anything is explained by microphysical explanations, it is the occurrence of particular outcomes. If the information that a carbon-14 atom has a probability of 1/2 of emitting an electron within a period of 5730 years and that such atom is a continuous causal process explains anything at all (as Salmon supposes (46–47, 202–4)), what could it explain but why such an emission will occur on a particular occasion?

However, although paradigms of this sort have dominated philosophical discussions of statistical explanation, it seems to me doubtful that they are good examples of the sorts of explanations provided by a statistical theory like QM. It is much more natural and in accord with scientific practice to think that explanation of the behavior of some system in QM characteristically involves the solution of the Schrödinger equation for that system given facts about the Hamiltonian governing the system and other facts about initial and boundary conditions. On this sort of conception, the sorts of low-level probabilistic generalizations about the behavior of particular kinds of systems described above are among the explananda and not part of the explanans in typical quantum-mechanical explanations, for it is such generalizations that one derives in solving the Schrödinger equation for a particular kind of system. Explanation in QM does not come in when, e.g., one subsumes some particular episode of barrier penetration under a generalization that tells us that whenever a particle of such and such kinetic energy encounters a potential barrier of such and such a shape, it has such and such a determinate probability of tunneling through. (Nor does it come from seeing such an episode as the result of a continuous causal process.) Rather, one possesses an explanation when facts like those expressed in the above generalization are derived in an appropriate way from a generalization like Schrödinger's equation of much wider scope and when one comes to see how this derivation is an instance of a much wider pattern of derivation, in which Schrödinger's equation is solved with respect to a variety of different microphysical systems. On this alternative conception QM would not explain anything—it would not be a serious candidate for a physical theory at all—if it just consisted of a vast collection of generalizations about such matters as the half lives of various kinds of atoms, the behavior of electrons in particular kinds of potential wells, and so forth. Understanding these particular kinds of probabilistic behavior as part of a much more general pattern is essential to the sort of understanding QM provides.[17]

Clearly, drawing attention to this alternative picture of what a statistical theory like QM explains does not by itself show that Salmon's own views are misguided. However, I think it does suggest that it is unlikely that Salmon's conception can be supported just via appeal to what philosophers find it reasonable or intuitive to say about specific examples of the use of statistical theories (after all, these "intuitions" have been formed largely by reading the philosophical literature on the subject). Such examples—including those Salmon appeals to in *SE*—can always be reconstrued as ones in which what is explained is a probability or else denied to be cases of statistical explanation at all.[18] Instead, it seems to me that the best way to approach the question of what statistical theories explain is to ask whether there is any real work that is done by Salmon's model which could not be done by the alternative conception elaborated above.

Consider, for example, the problem of assessing the explanatory power of rival quantum mechanical theories—say Bohr's early theory of 1913 vs. elementary nonrelativistic QM in its modern formulation, or the latter theory vs. quantum electrodynamics. Are there any considerations that are relevant to assessing the explanatory power of these theories which would be left out if one were to reject Salmon's conception of statistical explanation in favor of the alternative DS conception considered above? I think that much of the appeal of the DS conception derives from the suspicion that the answer to this question is "no" and that the sorts of considerations which are relevant to the assessment of the explanatory power of statistical theories are just the sorts of considerations that are relevant to assessing the explanatory power of deterministic theories — familiar considerations having to do with what can be derived from the two theories, with how unified, non-ad hoc each theory is, and so forth. Thus, for example, one might say, within the framework of a purely DS account of statistical explanation, that Bohr's quantum theory provided a better explanation of spectral emissions of hydrogen (where it at least made accurate predictions) than it did of nonclassical barrier penetration (which it did not predict at all). Again, one might say that Bohr's theory does not explain very well, if at all, why an electron in a potential well will occupy only discrete energy levels (Bohr's "quantum conditions" are imposed ad hoc, without any real justification besides the fact that they yield experimentally correct results — another nice example, incidentally, of arbitrariness in explanation), while modern quantum mechanics provides a much better explanation of this phenomenon (the quantization of allowable energy levels arises in a natural way — as the solution to an eigenvalue problem — out of the imposition of certain boundary conditions on Schrödinger's equation).

In making comparisons of this sort, we do not seem to need to appeal to the notion of statistical explanation of individual outcomes. Adherents of the DS conception will think that this is true generally and that there is nothing about our practices of methodological assessment of how well real-life statistical theories explain which requires the introduction of such a notion. If this is so, and if, as Salmon's discussion leads one to suppose, any plausible conception of statistical explanation of particular outcomes will have counterintuitive features (e.g., that E_1 will explain both E_2 and not E_2), this seems to me to be a good reason to try to get along without such a conception. Conversely, to defend his claim that statistical theories like QM explain individual outcomes, it seems to me that Salmon must show that we need his model for purposes of methodological assessment in connection with serious, realistic examples — that, say, we need to appeal to a model of the statistical explanation of particular outcomes if we are to make plausible comparisons of the explanatory power of the different quantum-mechanical theories described above.

IV

Finally, I want to conclude by commenting on Salmon's remarks on logic-oriented or epistemic conceptions of explanation. While much of what Salmon has to say on these topics is interesting and convincing, I thought that his discussion suffered from (a) a tendency to run together a number of distinct questions and (b) especially in connection with his discussion of the SR basis, a failure to fully come to terms with the apparent implications of these remarks for his own model of explanation. With regard to (a), one might well want to distinguish the following questions: (i) Can one determine on a priori grounds, and quite independently of contingent facts about the world (from logic alone), what sort of criteria a good scientific explanation must satisfy? (ii) Even if one cannot do this, can one discover, as a result of an a posteriori investigation of features possessed by paradigms of good explanation in our world, necessary and/or sufficient conditions statable in purely syntactic or semantic terms for any purported explanation (regardless of subject matter) to be acceptable? (iii) Even if the answers to (i) and (ii) are negative, do good explanations often possess a fair amount of explicit deductive structure and should one think of them as explaining in part in virtue of providing such deductive structure? Does whether or not we have an explanation of some phenomenon (e.g., the behavior of some gas or of barrier penetration) have something to do with whether one knows how to write down and solve an equation associated with that phenomenon rather than just providing looser specification of an underlying mechanism without such an explicit derivation? (iv) Should one in trying to characterize the notion of explanation assign a central role to such characteristically "epistemic" or "inferential" notions as diminishing arbitrariness or achieving unification?

It seems to me that Salmon is surely right in answering (i) and (ii) in the negative, and that it is an important insight on his part that many previous accounts of explanation are defective because they imply an affirmative answer to (i) and (ii). However, it does not follow from the rejection of (i) and (ii), (and is arguably false that) that the correct answer to (iii) and (iv) is "no." One can share Salmon's skepticism regarding content-free, subject matter-neutral, purely "logical" characterizations of explanation and agree with his suggestion that whether a certain explanatory strategy or criterion for explanatory goodness is likely to be reasonable or fruitful is heavily dependent on various empirical facts about the domain to which it is to be applied and yet continue to believe that explanation is in important respects an epistemic notion and that in many but not all areas of investigation (physics, but not history) whether one has a good explanation has a lot to do with whether (and what sort of) derivation one has.[19]

With regard to (b), Salmon's general claims about the contingent character of explanation raise an obvious question concerning the range of applicability of the causal/mechanical model. Although Salmon's discussion is nondogmatic in tone and is tempered by various qualifications, it also exhibits a clear tendency to suppose that the causal/mechanical model will apply to most domains of scientific investigation, at least outside of quantum mechanics.[20] But if one gives up the expectation that a model of explanation should be universal in the sense that it must apply to all logically possible worlds, why should one continue to expect that such a model must be universal in the sense that it applies to all or even most domains of investigation in our world? Given the general point that which explanatory principles we will find "useful and appealing" will depend on general facts about the causal structure of our world (240), why not take the further step of admitting that different explanatory principles may be useful and appealing in, say, physics, psychology, and sociology, depending on general, contingent facts about the characteristic subject matters of these disciplines? Taking this step would allow one to recognize that the causal/mechanical model captures important features of explanation in classical physics and yet to deny that the model is likely to be illuminating in connection with, say, information-processing explanations in psychology. It would also allow one to resist the temptation to say that physical theories like elementary quantum mechanics are unexplanatory to the extent that they fail to specify continuous causal processes.

Salmon's discussion of the SR basis seems to me to represent a striking illustration of this general point. Although, as we have noted, the SR basis is plainly intended to be applicable to examples in QM such as radioactive decay, it is also extensively illustrated by reference to examples from other sciences such as sociology and epidemiology. Indeed Salmon's first and only fully worked out illustration of the SR basis involves a sociological explanation of juvenile delinquency in terms of such factors as class and religious background and parents' marital status (37–43). Here too, it seems to me that—quite apart from my critical remarks about the appropriateness of thinking of quantum mechanics as providing explanations of individual outcomes in section III above—Salmon is much too willing to assume that there is some single, unitary notion of statistical explanation or some single way in which statistical information is relevant as a first stage in the construction of explanations (as in "the" SR basis), which all of his examples illustrate.

In sociological cases (and often in epidemiological cases as well) researchers typically employ one or another so-called causal modeling technique—e.g., regression analysis or path analysis. Here it seems to me one really is interested in something resembling (what Salmon calls) "statistical relevance relations," although in scientific practice such relations are expressed in the form of data regarding variances and covariances and the correlation and regression coefficients calculable from such information and not in the form prescribed in Salmon's SR basis.[21] When one uses such techniques one is, in effect, making inferences about causal connections on the basis of information about statistical relevance relations in conjunction with certain other nonstatistical information just as Salmon's discussion of the SR basis suggests.[22] Like the SR stage itself, causal modeling techniques are, in important respects, data-driven or inductivist procedures for theory construction. They are used when one has a great deal of statistical information about the incidence of various factors of interest in specific populations, but lacks a precise, predictively powerful theory of the sort found in many areas of physics and chemistry. From statistical information, including information about statistical relevance relations as expressed in claims about covariances and various other assumptions, one infers, say, values for the coefficients in a regression equation, which are taken to reflect facts about structural causal connections.

Furthermore, causal modeling techniques possess a number of other features at least roughly resembling those Salmon assigns to the SR basis. For example, as the SR basis requires, it is natural to think, in connection with such techniques, that one begins with a reference class which is specified by the population from which one is sampling (American teen-agers). One will accordingly have statistical data specifying a determinate "prior probability" (or at least prior frequency) of occurrence of the explanandum variable (e.g., juvenile delinquency) in that class. It is then natural to think, as Salmon's discussion suggests, of explanation in terms of the introduction of further variables (class, religious background, and so forth) which "partition" this original reference class and which are relevant to the likelihood of incidence of juvenile delinquency. The conclusion one reaches, just as in the case of the SR basis, will be relative to the specific population with which one begins – it will be, at best, a conclusion about the causes of delinquency among contemporary American, but not contemporary Chinese, teen-agers.

Now contrast the above features with the sorts of explanations provided by a statistical theory like QM. The statistical information which is used to test QM is not what one would naturally think of as information about statistical relevance relations between the dependent and independent variables in the theory (that is, information about covariances or correlations) at all, but is rather in effect information about the frequency distribution of measured variables (position, momentum, etc.) which can be used to test the probabilistic predictions of the theory. QM is a precise, integrated, predictively powerful theory of a kind one typically does not have available in the sort of context in which one uses regression analysis. One would not expect (or need) to infer the values of key parameters in such a theory, reflecting claims about causal connections, from the data as in the case of regression analysis; rather the theory itself prescribes values for such parameters. It would be mad to try to construct, arrive at, or confirm QM by running regression analyses on, say, statistical information about the incidence of radioactive decay in various populations of atoms (or on spectroscopic data or on any other body of information which constitutes the evidential basis for QM). Moreover, the claims of QM are not specific to particular populations of microphysical systems, as is characteristic of the SR basis, but are intended to apply universally to all systems (anywhere in the universe) possessing certain very general features. It is *not* natural to think of explanation in QM in terms of beginning with information about the incidence of an explanandum variable in some specific reference class and then partitioning this reference class into sub-classes on the basis of information about statistical relevance relations.[23]

If this were the end of the matter, it would be reasonable to conclude that the SR basis captures important aspects of the use of causal modeling techniques, but not (even as a "first stage") of explanation in quantum mechanics. But other features that Salmon assigns to the SR basis seem not to fit at all with the use of causal modeling techniques, although they do in some cases correspond to features of explanation in quantum mechanics. Consider, for example, the requirement that the explanans variables in the SR basis effect an objectively homogeneous partition—roughly that no further statistically relevant subdivision (meeting certain other natural conditions) of the partition established by the explanans variables be possible. The various no-hidden variable results perhaps ensure that (something like) this requirement is satisfied in quantum mechanics. On the other hand, it seems to be generally accepted that such a requirement will virtually never be satisfied in the sorts of contexts in which causal modeling techniques are typically used. Given a regression equation containing certain variables, it will always be possible to find other variables the inclusion of which in the model will change previous claims about statistical relevance relations (in this case, the values of the original coefficients in the regression equation).[24]

One can see this clearly in the case of Salmon's own example concerning juvenile delinquency. Given a partition created by explanans variables having to do with class, religious background, and so forth, it looks as though there will be an indefinite number of further variables (finer gradations in personal income, child-rearing practices, geographical location, schooling, characteristics of the criminal justice system, and so forth) which will be statistically relevant. This fact certainly raises interesting methodological problems regarding the use of causal modeling techniques and points to the need for restrictions on the class of candidates for explanans variables when such techniques are employed. But to impose the requirement that any explanation of a phenomenon like juvenile delinquency must employ a partition such that no further statistically relevant subdivsions are possible is in all likelihood to ensure that no explanation of this phenomenon will ever be forthcoming. Instead, what one wants is an account of explanation which makes it clear how the citing of general explanans variables like class can be explanatory, even though further relevant partition is plainly possible.

Finally, we should note another important disanalogy between Salmon's SR basis and techniques like regression analysis. As noted above, Salmon thinks that statistical information, whether in quantum mechanics or sociology, is used to explain individual outcomes. Thus, in connection with his juvenile delinquency example, he suggests that statistical information about the incidence of juvenile delinquency can be used to explain why some particular boy, Albert, commits a delinquent act (37). But even if one wants to think of a statistical theory like QM as explaining individual outcomes, it seems doubtful that this is the correct way to think about the juvenile delinquency example. When one uses a technique like regression analysis (or more generally assembles information about statistical relations for the purposes of constructing an explanation), it seems more plausible to suppose that one is trying to explain facts about population level parameters — facts about changes in the mean incidence of juvenile delinquency in the American population or perhaps facts about the variance of this variable. One is interested in trying to explain such facts as the great increase in juvenile crime in the 1960s and 1970s, or why, on the average, youths from urban areas commit more crimes than youths from rural areas, and not in explaining why any particular boy became a juvenile delinquent.

I thus suggest that the kind of explanations provided by QM and by the use of causal modeling techniques (which represent the way in which information about statistical relevance relations is used to make causal inferences in actual scientific practice) differ in a number of quite fundamental respects, which in turn reflect differences in subject matter, in available information, and so forth. In my view, Salmon's SR basis represents an attempt to join together in one unified account conditions on explanation drawn from several quite different ways in which statistical consideration can figure in the construction of explanations. Many of these conditions correspond to quite real features of either explanations in QM or of explanations based on statistical relevance relations, but there is no single, unitary notion of statistical explanation (or of the evidential basis for such explanations) which combines all of these disparate conditions. Once one abandons the demand for a single, universal logic of explanation it seems to me to be equally natural to abandon the idea that there must be some single model of statistical explanation possessing all of the features Salmon assigns to the SR basis.

Although I have been critical of a number of Salmon's claims, I hope it is clear from my discussion that *SE* is a valuable and provocative book. It provides a clear and perhaps definitive statement of a distinctive conception of scientific explanation, a conception that unquestionably captures central aspects of causal explanation in many physical contexts, even if it is less generally applicable than Salmon supposes. And it develops a number of general themes — the role of individual causal processes, the limitations of purely formal models, the importance of contingent facts about how nature actually works in the characterization of explanation — that one suspects will be at the center of philosophical discussion for some time to come.

Notes

1. Salmon apparently regards these various features of causal processes as co-extensive but this is not obviously correct. An election, for example, when represented quantum mechanically as a wave packet can be marked and transmits a determinate structure and a determinate probability of being found in various states on measurement, but there are certainly limits on the extent to which it can be appropriately regarded as a spatio-temporally continuous process, with a definite location. Similarly consider a macroscopic sample of gas. This can transmit a mark and, when in a certain state

of temperature and volume, transmits a determinate probability of exerting a certain pressure. If our standards are not too fine-grained, it can even be regarded as a spatio-temporally continuous process. Yet, as I note immediately below, Salmon denies that an explanation of the pressure of the gas in terms of the ideal gas law (or, one presumes, an explanation which appeals to the fact that the macroscopic sample is a causal process with a determinate probability of exerting a certain pressure) satisfies the constraints of the causal/mechanical model. Here the idea seems to be that a satisfactory explanation must cite more fundamental microphysical mechanisms underlying the behavior of the gas, or perhaps explicitly cite processes which involve the local transfer of energy and momentum, as a statistical mechanical analysis would do. The various criteria on which Salmon seems to rely in these cases for determining whether an explanation appropriately cites causal processes do not necessarily coincide. An explanation may cite processes which transmit structure (which are not pseudo-processes) without detailing fundamental mechanisms or explicitly detailing mechanisms which are mechanical in the sense that they involve the transfer of energy and momentum. Indeed, on pp. 202-3 of *SE* several examples (including one in which a subject in a psychological experiment is said to have a determinate probability to respond in a certain way) are given of processes which transmit structure but which do not cite, at least in any detail, any fundamental mechanisms and these are apparently regarded by Salmon as acceptable explanations (or at least as examples of causal processes or the transmission of causal influence).

I think that more needs to be said by Salmon about the interrelations among these various features of causal processes and about which features should be taken as criterial. Why, for example, do the ideal gas laws fail to cite causal processes, while the above description of the psychological experiment apparently does? Why not take the ability to transmit a mark as the fundamental feature of a causal process (whether or not such transmission satisfies a spatio-temporal continuity constraint) and thus take the evolution of the wave function in QM to represent a legitimate causal process?

2. A factor C is statistically relevant to the occurrence of B in reference class A if and only if $P(B/A \cdot C) \neq P(B/A)$. I should add that this is just the bare bones of the SR basis; for the full characterization, which is much more complex, see *SE*, pp. 36–37.

3. It is a further question, however, whether it is appropriate to think of the notion of a spatio-temporally continuous causal process as a primitive notion for purposes of explanation, even in contexts in which the notion applies. Salmon seems to hold, or at least to suggest, especially in his discussion of explanation in quantum mechanics (see below) that there is just something intrinsically more intelligible about a process that involves local rather than remote conservation or action at a distance. Processes of the former sort are, as it were, natural stopping places in explanation. My own quite different view (and also, I think, the view of many physicists) is that the existence of spatio-temporally continuous causal processes is itself something that stands in need of explanation – one can perfectly intelligibly ask *why* the world should have this feature.

One natural suggestion is that the explanation for this fact is, as already indicated, bound up with Special Relativity and with the fact that the structure of spacetime is locally Minkowskian. This alternative conception fits naturally with the idea, urged briefly below, that a more fundamental element in our conception of a good physical explanation is that such explanations should satisfy certain symmetry and invariance (e.g., a requirement of Lorentz invariance). It is because we think that theories satisfying such requirements provide especially good explanations that we are led, derivatively, to value explanations containing continuous causal chains and local transfers of energy and momentum. A somewhat similar view is defended by Brian Skyrms in his *Causal Necessity*, (1980) pp. 110–27, and by Hans Ohanian in his *Gravitation and Spacetime* (1976). Ohanian writes, a passage quoted by Skyrms, that

Nowadays, all the fundamental interactions are regarded as due to local fields. . . . Why do we prefer fields to action-at-a-distance? The answer is simple: we need fields in order to uphold the law of conservation of energy and momentum [given the requirement of Lorentz invariance]. (36).

4. For a defense of the idea that the ascription of geometrical structure (particularly affine and metrical structure) to spacetime is explanatory, see Graham Nerlich's *The Shape of Space* (1976) and Michael Friedman's *Foundations of Space-Time Theories* (1983), and for an explicit defense of the idea that such explanations are noncausal, see Nerlich. That explanation in General Relativity does not seem to fit the causal/mechanical model very well is also noted by Clark Glymour in his "Causal Inference and Causal Explanation" (1982).

I might also note that the question of whether the ascription of structure to spacetime is explanatory is presumably closely bound up with the adequacy of various "causal" theories of the structure of space time and with whether various "conventionalist" or anti-realist theses regarding spacetime are correct. Someone who thinks that claims about spacetime structure, if taken literally or realistically, must be reducible to claims about actual and perhaps possible causal processes and interactions and who thinks that the ascription of space-time structure which is not so reducible is a matter of convention, will presumably deny that such structure can figure in explanations. If the structure in question is conventional, it can hardly figure in explanations, and if it is not conventional, and is reducible to talk about causal processes, then on a view like Salmon's, one should presumably just appeal to facts about such processes in constructing explanations. The idea that the ascription of spacetime structure can be explanatory in its own right seems to require realism about such structure and the rejection of the claim that such structure is reducible to causal relations.

Since Salmon has defended conventionalist views elsewhere and has explicitly claimed that the attribution of affine and metric structure in General Relativity is a matter of convention (1977), I strongly suspect that that he would just deny that General Relativity explains, via the postulation of geometrical structure, in the sense I have claimed. My own view is that in view of the widespread rejection of conventionalism and of purely causal theories of spacetime (at least in conjunction with General Relativity) in recent philosophical work, the apparent philosophical connection between the causal/mechanical conception of explanation and causal theories of space-time structure represents a liability of the former conception. At any event, the connection deserves further examination. Both are expressions of the idea that the fundamental furniture of the universe is (just) causal processes and their interactions. I suspect that both stand or fall together.

5. At least I take this to be the tendency of Salmon's discussion, which is tentative in character and perhaps does not reach an unequivocal conclusion. (Salmon tells us, on the final page of *SE*, that he has "not offered any account of quantum mechanical explanation and I do not believe that anybody else has done so either." (279)). I should also acknowledge that while Salmon seems to regard quantum mechanics, at least on its presently available interpretations, as not fully satisfactory from the point of view of furnishing explanation or understanding, he certainly recognizes the possibility adumbrated below: that instead of revising microphysics to bring it more into accord with the requirements of the causal/mechanical model, perhaps we ought to revise the requirements on explanation embodied in the causal/mechanical model and change our views about the fundamental mechanisms in the world on the basis of the apparent microphysical facts. Thus, for example, he writes at one point:

> Does this kind of conservation [that is the kind of "remote" conservation apparently involved in the EPR correlations] require explanation by means of some special mechanism? Or is this one of the fundamental mechanisms by which nature operates? These questions strike me as profound and I make no pretense of having an answer to them. (258)

6. I have no general account of when it is reasonable to conclude that the demand for explanation of some general kind is misguided or inappropriate. This will depend upon a variety of complex empirical and conceptual considerations and must be determined on a case-by-case basis. But plausible illustrations are not hard to find: for example, demands for explanations which appeal to essences or essential properties or to final causes are certainly inappropriate in many, if not all, domains. It is not a limitation on the explanatory power of current physical or chemical theories that they do not provide answers to all the kinds of explanatory questions Aristotle would have regarded as appropriate. More controversially, it is arguably no limitation on the explanatory power of theories in neurobiology or psychology that they provide no explanation of what it is like to be a bat, in the sense intended by Thomas Nagel in his well-known paper (1974) of that title.

7. Salmon rather closely follows Bernard d'Espagnat's recent discussion (1979) which seems to adopt this assumption. Thus Salmon says that the measurement of a component of spin of one particle in the EPR experiment "alters" (251) the spin component of the correlated particle and appears to endorse d'Espagnat's suggestion that "nature seems to exhibit action at a distance in the quantum domain" (250). But the standard view, in any event, is that such talk of "altering" or of action at a distance in the sense of the transmission of causal influence or information makes sense only if the correlated particle has a definite spin component prior to measurement. It is precisely by rejecting this assumption that the standard view attempts to avoid interpreting quantum phenomena as involving action at

a distance or violations of "local" causality. Moreover, it is well known that if one adopts the contrary, realistic assumption, one is led to the conclusion that various quantum phenomena like barrier penetration actually violate conservation laws, a conclusion Salmon would presumably wish to avoid.

8. See, for example, Bell (1964), Kochen and Specker (1967), Gleason (1957), Aspect and Roger (1982). The connection between those results and Salmon's discussion is perhaps not as clear as one would like—in part because there is disagreement about the interpretation of the results themselves, and about just what they rule out, but also because Salmon himself is not very explicit about what he thinks a common-cause explanation of the EPR correlations would involve. For example, there are a number of results like Bell's theorem which rely on various versions of a locality condition to derive inequalities in contradition with the statistics predicted by quartum mechanics and actually obtained in experiment. In versions of Bell's theorem which apply to deterministic local hidden variable theories, the locality condition is often interpreted as (or motivated by reference to) the special relativistic requirement that faster than light signals be impossible (but see Earman [1986] for the contrary view that the locality condition required for Bell's theorem has nothing to do with Einstein locality and instead represents semantic locality). Moreover, if the correlations in EPR type experiments are regarded as obtaining strictly and universally, one can show that the local hidden variable theory producing this result must be deterministic. On the other hand, if one is willing to regard the strict obtaining of the correlations as an idealization, which holds only approximately in the real world, then one can consider stochastic hidden variable theories. To derive a Bell-like inequality for such theories, one must assume an (apparently) stronger locality condition, the motivation and physical significance of which is a source of some controversy (see, e.g., Hellman [1982], Jarrett [1984]). As Jarrett shows, this condition has a natural decomposition into relativistic locality and an additional condition that he calls completeness and that is also sometimes called factorizability of conditional stochastic independence—roughly that measurement results on the two correlated particles be independent, conditional on the values of the hidden variables.

Now Salmon gives no indication that he supposes that conservation laws hold only approximately in EPR type correlations and seems to accept that such correlations do not involve superluminal action (SE, 250). More generally, his conception of a common cause explanation seems to be such that it satisfies some fairly strong version of a locality condition: the idea is that a common cause operates via a local interaction and the subsequent propagation of the results of the interaction via distinct continuous causal processes to the correlated measurement results, each process containing the "instructions" relevant to the result to which it is connected. One would think that this sort of conception rules out, say, a purported explanation of the outcome of performing a measurement on particle L in which the characterization of the common cause requires reference to the measurement outcome obtained for the correlated particle R or reference to the experimental arrangement employed at R. (Compare van Fraassen (1982) for arguments showing how one can derive a Bell type inequality given reasonable restrictions of this kind on what a common-cause explanation for the EPR correlations would have to look like.) Given all of this, it is very hard to see how there would be any sort of common-cause explanation of the EPR correlations meeting other conditions Salmon would accept.

On the other hand, if sufficiently weak restrictions are placed on what may count as a common cause—if a stochastic, contextual hidden variable theory which doesn't satisfy a strong locality constraint will qualify, then of course it will be possible to reproduce the observed correlations. It would have been very useful to have had a more explicit discussion of all this in SE. Just what is the relationship between the demand for a common-cause mechanism and results like Bell's? In which, if any, of the several different senses of locality must a common-cause explanation satisfy a locality requirement? Which theories in the space of possible hidden variable theories (deterministic vs. stochastic, local vs. nonlocal, contexual vs. noncontexual) would, if empirically adequate, count for Salmon as providing common-cause explanations?

9. For relevant see Elliot Sober's "Equilibrium Explanation" (1983).

10. For example, one prominent high-energy physicist comments that it is a "strength" of gauge theories such as a GCD in comparison with earlier theories that "they require comparatively few free . . . parameters." Nonetheless

> even if the free parameters have been reduced to a managable number, they remain an essential part of the theory. No explanation can be offered of why they assume the values they do. (t'Hooft 1980, 136)

By contrast an attraction of recent superstring theories is that they lack such arbitrariness—such theories apparently can only be consistently formulated in 10 dimensions and with one of two gauge symmetry groups. They thus yield the hope of explaining why spacetime has the number of dimensions it does and why nature prefers the symmetry groups it does. As a recent discussion in *Physics Today* puts it:

> Unlike the (nongravitational) grand-unified point-field theories (GUTS) of recent years, where the gauge symmetry groups and coupling parameters could be chosen with considerable freedom and fit the data, the superstring theories offer almost no free choices. (Schwarzschild 1985)

The correspondent in *Science* expresses a similar idea

> [in superstring theory, the] symmetry groups were defined by the underlying model, instead of being adjusted by hand to fit the data. For the first time there seemed to be a mechanism [although of course not a mechanism in Salmon's sense—JW] for nature to choose her symmetry group. (10)

11. The idea that theoretical unification plays a central role in scientific explanation is defended by, for example, Michael Friedman, (1974), Philip Kitcher (1981), and James Woodward (1979).

12. See, for example, Stephen Weinberg "The Search for Unity: Notes Toward a History of Quantum Field Theory" (1977).

13. This general point is made very clearly and elegantly by Philip Kitcher in the context of a discussion of the relation between Mendelian genetics and molecular biology in his "1953 and All That: A Tale of Two Sciences" (1984). One of Kitcher's illustrations is particularly apt: Mendelian genetics accounts for various facts about gene transmission in part in terms of meiosis and the independent assortment of genes on nonhomologous chromosomes. However, the molecular processes which underlie meiosis in different organisms are quite heterogeneous. An account which just traced such detail in the case of a particular species would lose the more general pattern embodied in Mendel's laws.

14. The claim that complex systems exhibit levels of organization and that explanation often proceeds by abstracting from certain kinds of lower-level causal detail and finding general patterns does not imply that one can simply ignore constraints owing to lower-level causal facts in attempting to discover such general patterns. How much two systems can offer in causally relevant detail at some lower level of analysis and still fruitfully be regarded as relevantly similar at a higher level of analysis and how much lower-level features constrain possible upper-level patterns is not something that can be stipulated a priori but requires detailed empirical inquiry on a case-by-case basis. Philosophical defenses of functionalist doctrines in psychology notoriously ignore this point.

15. Of course it is perfectly true that a particular gene, organism, or psychological subject is a continuous causal process, but presumably it does not follow *just* from this observation that the above explanations successfully explain by citing (or just by citing) continuous causal processes and interactions. Or, if this conclusion does follow, it is not clear what sorts of explanations the causal/mechanical model rules out.

16. Another reason (regarding which I comment below) has to do with Salmon's failure to distinguish between statistical theories like QM and the use of information about statistical relevance relations to make causal inferences, and his tendency to think that techniques of the latter sort are used to establish claims about individual causal connections.

17. Salmon's intuition here is, as I understand it, a very different one: it is that (if I may put it this way) all that there fundamentally is in the world is particular atoms with particular determinate propensities to decay, particular electrons, particular electromagnetic fields, and so forth. On an "ontic" conception of explanation it must be such facts about particular causal processes and interactions which constitute the raw material out of which our explanations are constructed—it is these that are really most central or significant for purposes of explanation. The significance of a high-level generalization like Schrödinger's equation is, if I have understood Salmon's view correctly, derivative from or parasitic on these facts about particular causal processes—the equation is, at best, an abstract description of what lots of individual processes have in common.

18. Salmon does (117–18) describe a number of concrete scientific examples (including the Davisson-Germer electron diffraction experiment, Rutherford's scattering experiments, Compton scattering, and an example from genetics) which he claims must be understood as cases in which a statistical theory is used to explain an individual outcome. While I lack the space to discuss these in detail, I found this claim unpersuasive, both for the general reasons given in the main body of this paper and because all the examples are quite underdescribed and, in a number of cases, do not clearly

involve subsumption under a statistical generalization at all. For example, Salmon does not spell out what he has in mind when he speaks of providing a theoretical explanation of the Rutherford scattering or Davisson-Germer diffraction results, but the usual elementary textbook treatments (and the treatments originally given by Rutherford and Davisson-Germer themselves) are essentially classical and appeal to fundamental laws which are deterministic rather than statistical. In the Davisson-Germer experiment the analysis makes use of the relation λ = h/p but otherwise proceeds along the same lines as ordinary optical diffraction and results in the derivation of an expression (the Bragg relation: $n\lambda$ = 2d cos θ) which gives the intensity maxima in the reflected beam. It is the existence of these maxima which seem to be explained via the derivation, and the derivation seems to have a straightforward DN structure. In the case of Rutherford scattering, in the usual treatment one assumes that the scattering is due to a repulsive coulomb force and derives, from classical considerations, the equation of motion for an incident particle and from this an expression for the differential scattering cross-section (or for the fraction of particles scattered at a given angle). Here again the derivation looks like a straightforward DN derivation which involves no subsumption under irreducibly statistical laws. A similar point holds for Compton scattering: here one uses the conservation of energy and momentum and the relation p = h/λ to derive an expression (λ' $-$ λ = $\frac{h}{m_0c}$ (1 $-$ cos θ)) for the shift in wavelength of the reflected photon as a function of the scattering angle θ. (See, e.g., Anderson [1971] for relevant discussion.) It is perfectly true that in a full quantum mechanical treatment of scattering that one represents both the incident and reflected flux of particles as wave functions and assumes that they must obey the Schrödinger equation, so that the treatment is, in this sense, irreducibly statistical. But here again what seems to be explained is just what can be derived—e.g., facts about the scattering cross-section or about the probability of finding a particle (or the particle flux) at a given scattering angle. At the very least, I think that Salmon needs to spell out his examples in much more detail and make it clear just exactly what in his view is being explained, what is doing the explaining, what sort of theoretical treatment is envisioned, and so forth. Just because the processes involved in scattering experiments are "irreducibly statistical" it does not follow that the explanations we give for features of those processes must be statistical explanations, in Salmon's sense.

19. While it is plausible that there are no necessary and perfectly general or sufficient conditions statable in purely syntactic or semantic terms for explanatory goodness, it is quite compatible with this that various features of good explanations in specific domains can be given interesting formal (but domain-specific) characterizations which fall short of specifying necessary and sufficient conditions. For example, we have already noted that theoretical unification represents an important goal of physical explanation. While there is arguably no perfectly general formal characterization of explanatory unification, a natural way of partially characterizing such unification in the context of fundamental physical theories is via the imposition of symmetry and invariance requirements. I am inclined to think that there may be other domain-specific formal criteria for explanatory goodness in other areas of investigation—see, for example, Clark Glymour's "Causal Inference and Causal Explanation" (1982) and his "Explanation and Realism," (1984) for some interesting suggestions along this line appropriate to causal modeling techniques.

20. At several points in his discussion, Salmon acknowledges the possibility that the causal/mechanical model may not apply everywhere in our world. Thus he writes: "I make no claim for universal applicability of my characterization of scientific explanation in all domains of our world, let alone for universality across all possible worlds" (240). At a number of other points, however, he reveals expectations that are considerably more universalist regarding explanation in our world: "I hope that the causal theory of scientific explanation outlined above in this book is reasonably adequate for the characterization of explanation in most scientific contexts—in the physical, biological and social sciences—as long as we do not become involved in quantum mechanics." (278)

21. For helpful discussion of connections between Salmon's notions of statistical relevance and screening off and various aspects of the use of causal modeling techniques see Suppes (1970) and especially Irzik (1986). The analogy seems closest in connection with causal models with dichotomous (two-valued) variables and in connection with procedures like those developed by Herbert Simon and Herbert Blalock for choosing among causal models on the basis of information about correlation coefficients. As one moves to more general linear models with continuous variables, the analogy becomes much more attenuated if not positively misleading. In these models the relevant notion of cau-

sation seems to be the notion of counterfactual supporting functional dependence, and to have little to do with positive statistical relevance or the tracing of causal processes in Salmon's sense. For additional discussion, see Woodward (1988).

22. There is, however, an important disanalogy between Salmon's treatment and causal modeling techniques at this point, which deserves a more detailed exploration than I can give it here. Both Salmon's treatment and causal modeling techniques reject the idea that one can infer claims about causal connections just from facts about patterns of statistical association. Both agree that additional information is required to support such an inference. However, they disagree in an important way about the character of the additional information which is required. For Salmon, the additional information has to do centrally with facts about individual causal processes and their interactions and thus with facts about temporal order and spatio-temporal continuity. By contrast, while such considerations certainly sometimes play a role in the sorts of contexts in which causal modeling techniques are used, they are often not of central or decisive importance. Thus, although one's conclusions about whether two variables of interest are causally related can be greatly affected by which additional variables one includes or excludes from a regression equation (see note 24), considerations of spatio-temporal continuity are often not very useful in considering which variables to exclude and often are not appealed to by users of regression equations. For example, whether or not a variable measuring the execution rate seems to causally affect a variable measuring the murder rate may depend on whether one includes variables reflecting the employment rate and poverty level in the regression equation. But considerations regarding spatio-temporal continuity or demands for the tracing of continuous causal processes are often simply not helpful in deciding whether it is reasonable to include these variables — instead users of causal modeling techniques typically appeal to general theoretical considerations bearing on whether the poverty level is a "possible cause" of changes in the murder rate.

Similarly, given the sort of data one often has available when one uses causal modeling techniques (e.g., cross-sectional data at a relatively high level of aggregation) and the nature of the models themselves which are not dynamical in character or precise about temporal relations between variables, it is often impossible to use information about temporal order to disambiguate the direction of causality. Given two correlated variables X and Y, arguments about which is the cause and which is the effect instead will turn on general theoretical or common-sensical claims about whether X is the sort of thing which could cause Y or vice versa. Thus in studies of voting behavior it is commonly (and I think falsely) assumed that a voter's judgement that a political candidate has views similar to his own can cause a decision to vote for that candidate but not vice versa.

Extra-statistical assumptions of these sorts about possible causes or about causal order are plainly already causal in character, and so causal modeling techniques are emphatically not techniques for deriving causal claims from purely noncausal premises; they are rather techniques which show us how we can test causal claims on the basis of statistical information if we are willing to make other causal assumptions. To the extent that Salmon's treatment involves the idea that purely statistical information (the SR basis) when supplemented just with considerations of spatio-temporal order and continuity will allow us to sort out genuine causal connections from noncausal sorts of association, it is more "Humean" and reductionist in spirit than techniques like regression analysis. It is also at odds with the actual practice of those who use such techniques.

23. Consider, for instance, Salmon's example of explaining the decay of a carbon-14 atom. What should we understand the reference class A to be here? If we make what might seem to be the natural choice of taking A to be the class of all carbon-14 atoms, then it appears that no further statistically relevant partition of this reference class is possible — there are no further factors that are statistically relevant to the decay and so the basic requirement of constructing the SR basis by introducing a further partition is not met. To even get started constructing the SR basis, we need to construe the demand for explanation as relative to some reference class we can partition — say, the class of all carbon atoms in a certain sample. If we have information about the prior frequency of decay in this reference class, we can then partition the reference class into, e.g., those atoms that are radioactive carbon atoms and those that are stable isotopes — this will be a further statically relevant factor and will thus form part of the SR basis and part of the basis for any explanation we construct.

But even as a "first-stage" account of explanation in QM (or of the role of statistical evidence in constructing such explanations) this seems misdirected. The prior probability of decay in the above reference class will depend upon quite idiosyncratic and contingent facts about the frequency with which carbon-14 and stable carbon atoms occur in the sample, and as we choose different reference classes or different samples, this probability will vary greatly. (I take it that something like this point underlies Nancy Cartwright's claim that the S-R model involves an undesirable kind of relativization to a reference class in her [1979]). The sorts of explanations of decay and related phenomena found in quantum mechanics textbooks do not seem to invoke the above information about prior probabilities (indeed in many cases this probability will be entirely unknown) and give no indication of involving this sort of relativization to a reference class. If our aim is the characterization of how explanation works in QM, it is not plausible that if someone asks why this particular carbon-14 atom decayed, we can give no explanation (or assemble no statistical information bearing on the construction of an explanation), but if someone asks why this carbon atom from a mixed sample decayed (and we happen to know the prior probability of decay and this happens to be different from the posteriori probability), we do have the basis for constructing an explanation and the explanation we give is that it is a carbon-14 atom with such and such a probability of decay. The explanation we give for the behavior of the atom in question should be the familiar quantum mechanical one involving the penetration of a charged particle by a potential barrier and it should be the same in both cases. While some causal explanations genuinely do involve relativization to a reference class and proceed in effect by partitioning it—this is characteristic of many explanations produced by causal modeling techniques—this sort of conception just does not seem to fit QM.

24. Suppose we are trying to estimate the coefficients in the general linear regression equation

$$Y = B_1X_1 + B_2X_2 + \cdots B_nY_n.$$

The basic point is that the partial regression coefficient B_i, which (supposedly) tells us about the causal or functional connection between the dependent variable Y and the independent variable X_i, depends not just on the covariance between X_i and Y but also on the covariance between X_i and the other independent variables in the equation. If, say, we originally do the regression just using variable X_1 and then add X_2 to the regression equation, the coefficient for X_1 will change as long as X_1 and X_2 exhibit a non-zero correlation in the data. Given any finite body of data—and it is worth emphasizing that of course this is the only kind of data we ever actually have access to—it is very likely that we are always going to be able to find such additional, correlated variables. Regression analysis thus always requires additional extra statistical assumptions about which variables to include in the regression equation—assumptions which will draw on prior theoretical ideas about causal and explanatory relevance. The idea, perhaps suggested by Salmon's treatment of the role of the SR basis, that one should proceed by first assembling "all" information about statistical relevance relations, without causal or theoretical presuppositions, and only then trying to construct a causal explanation on the basis of this information, is not something it would be sensible (or even possible) to carry out in practice.

References

Anderson, Elmer. 1971. *Modern Physics and Quantum Mechanics*. Philadelphia: W. B. Saunders.

Aspect, A., Dalibard, J., and Roger, D. 1982. Experimental Test of Bell's Inequalities Using Variable Analyzers. *Physical Review Letters* 49:1804–7.

Bell, J. S. 1964. On the Einstein-Podolsky-Rosen Paradox. *Physics* I: 195–200.

——. 1966. On the Problem of Hidden Variables in Quantum Mechanics. *Reviews of Modern Physics* 38:447–75.

Cartwright, Nancy. 1979. Causal Laws and Effective Strategies. *Noûs* 13:419–37.

D'Espagnat, Bernard. 1979. The Quantum Theory and Reality. *Scientific American* 241, no. 5 (November):158–81.

Earman, John. 1986. *A Primer on Determinism*. Dordrecht, Holland: D. Reidel.

Friedman, Michael. 1974. Explanation and Scientific Understanding. *Journal of Philosophy* 71:5–19.

——. 1983. *Foundations of Space-Time Theories*. Princeton, NJ: Princeton University Press.

Gleason, A. M. 1957. Measures on the Closed Subspaces of a Hilbert Space. *Journal of Mathematics and Mechanics* 6:885–93.

Glymour, Clark. 1982. Causal Inference and Causal Explanation. In *What? Where? When? Why?* ed. Robert McLauglin. Dordrecht, Holland: D. Reidel.

——. 1984. Explanation and Realism. In *Scientific Realism*, ed. Jarret Leplin. Berkeley: University of California Press.

Hellman, Geoffrey. 1982. Stochastic Einstein-Locality and the Bell Theorems. *Synthèse* 53:461–504.

Hempel, Carl G. 1962. Deductive-Nonological vs Statistical Explanation. In *Minnesota Studies in the Philosophy of Science*, Vol. 3, eds. Herbert Feigl and Grover Maxwell. Minneapolis: University of Minnesota Press.

——. 1968. Maximal Specificity and Law Likeness in Probabalistic Explanation. *Philosophy of Science* 35:116–33.

Irzik, Gurol. 1986. Causal Modeling and Statistical Analysis of Causation. *PSA*, Vol. I. East Lansing, MI: Philosophy of Science Association.

Jarrett, Jon. 1984. On the Physical Significance of the Locality Conditions in the Bell Arguments. *Nous* 18:569–89.

Kitcher, Philip. 1981. Explanatory Unification. *Philosophy of Science* 48:507–31.

——. 1984. 1953 and All That: A Tale of Two Sciences. *Philosophical Review* 93, no. 3, 335–73.

Kochen, S., and Specker E. P. 1967. The Problem of Hidden Variables in Quantum Mechanics. *Journal of Mathematics and Mechanics* 17:59–87.

Nagel, Thomas. 1974. What Is It Like to be a Bat? *Philosophical Review* 83, No. 4, 435–50.

Nerlich, Graham. 1976. *The Shape of Space*. Cambridge: the University Press.

Ohanian, Hans. 1976. *Gravitation and Space Time*. New York: W. W. Norton.

Salmon, Wesley. 1971. Statistical Explanation and Statistical Relevance. In *Statistical Explanation and Statistical Relevance*, ed. Wesley Salmon. Pittsburgh: University of Pittsburgh Press.

——. 1977. The Curvature of Physical Space. In *Minnesota Studies in the Philosophy of Science*, Vol. VIII, *Foundation of Space-Time Theories*, eds. John Earman, Clark Glymour, and John Stachel. Minnesota: University of Minnesota Press.

Schwarzschild, Bertram. 1985. Anomaly Cancellation Launches Superstring Bandwagon. *Physics Today* 38, No. 7, 17–20.

Skyrms, Brian. 1980. *Causal Necessity*. New Haven, Ct: Yale University Press.

Sober, Elliot. 1983. Equilibrium Explanation. *Philosophical Studies* 43:201–10.

Suppes, Patrick. 1970. *A Probabilistic Theory of Causality*. Amsterdam: North-Holland.

t'Hooft, Gerard. 1980. Gauge Theories of the Forces between Elementary Particles. *Scientific American* 242, no. 6 (June):104–38.

van Fraassen, Bas. 1982. The Charybdis of Realism: Epistemological Implications of Bell's Inequality. *Synthese* 52:25–38.

Weinberg, Stephen. 1977. The Search for Unity: Notes Toward a History of Quantum Field Theory. *Daedalus* 106, no. 4, 17–35.

Woodward, James. 1979. Scientific Explanation. *The British Journal for the Philosophy of Science* 30:41–67.

——. 1988. Understanding Regression in *PSA 1988*. East Lansing, MI: Philosophy of Science Association.

Explanation in the Social Sciences

1. Introduction

Disagreements about explanation in the social sciences are closely bound up with views about whether or not the so-called social sciences really are sciences. The dispute has a long history. J. S. Mill, following Hume and the philosophers of the French Enlightenment, maintains that a science of human nature is possible (1874, 586). He believes that the thoughts and feelings of humans are the causes of their actions. On this basis, he argues that we can investigate the causal connections between thoughts and actions by employing the same canons of inference (Mill's Methods) that we use to discover and justify causal regularities in the physical world.

Mill recognizes that the complexity of human behavior impedes the development of causal explanations. Nevertheless, he believes that at least an *inexact* science of human behavior is possible. Whether a science is "exact" or "inexact" depends on how accurate the predictions of the science are. Mill doubts that the science of human behavior will ever become as exact as the physical science of astronomy, for example, because human actions are subject to so many unknown, and possibly unknowable, circumstances. In addition, even when the circumstances surrounding behavior are known, we are sometimes unable to describe or measure them accurately.

Mill points out, however, that the accumulation and interaction of many minor causal forces simlarly hinder accurate prediction in some physical sciences. Thus, the main laws that govern the movement of tides are known, but, because of irregularities in shorelines and ocean floors, as well as changes in direction of winds, precise predictions of tidal movements are not possible. Still, Mill says, no one doubts that tidology is a science, and hence the failure of prediction is no barrier in principle to the development of an *inexact* human science. With this

The author is grateful to the editors of this volume, and to colleagues Fritz Ringer, Jeremy Sabloff, and, particularly, Mikael Karlsson for their comments on an earlier version of this paper.

limitation in mind, Mill urges scientists to investigate human behavior with the aim of uncovering general laws.

Mill says that to discover patterns of connection between human thought and human action we must first study history to discern some regular connections. In Mill's version of what later came to be called a *covering-law model of explanation*, he suggests that the regularities revealed by historical studies ("the lowest kind of empirical laws") are themselves to be explained by showing that they are derived from laws of character development. Laws of character development, in turn, are to be explained by showing how they result from the operation of general laws of the mind (1874, 589).

Mill believes that fundamental causes of human behavior are probably mental rather than physical, but he does not think that this makes explanation of human behavior significantly different from explanation in the physical sciences. For Mill, subsumption under causal generalizations is at the heart of explanation. When we explain an event (either a physical occurrence or a human action) we must subsume it under an appropriate causal generalization; when we explain a generalization we must subsume it under a more general law or set of laws. *Explanation* thus is possible in inexact as well as exact sciences, though in the latter more precise *predictions* are possible.

C. G. Hempel's work on explanation in the social sciences lies squarely in the tradition of Mill. Hempel argues that insofar as explanations in history and other social sciences are complete, rather than elliptical or partial, these explanations require relevant universal or statistical generalizations. Hempel recognizes that many of the generalizations invoked in explanations in the social sciences are vague, common-sensical claims, unlike the more precise and well-confirmed generalizations in the physical sciences. He points out, though, that when the generalizations are not well founded, the explanations that they underlie are accordingly weakened (1962, 15–18). Both Hempel and Mill insist that explanatory laws have empirical content, but Hempel, unlike Mill, countenances *noncausal* explanatory laws.

Hempel agrees with Mill that human action can be explained by reference to mental causes, such as motives, beliefs, desires, and reasons. Similarly, he agrees with Mill that future investigations might show that mental concepts are "reducible" in accord with some materialist program. However, Both Hempel and Mill recognize that even if succesful materialist reductions were forthcoming, explanations in terms of mental causation would not thereby be rendered obsolete. Just as in the physical sciences, an explanation at one level (such as, for example, explanation of the behavior of gases in terms of pressure and temperature) can be correct and informative even when the phenomena can also be explained at a "deeper" level (as when the behavior of gases is explained in terms of molecular motion).

Although both Mill and Hempel embrace nonmaterialist causal explanations of behavior, their shared belief that explanations of human behavior are fundamentally similar to explanations of physical phenomena is challenged by critics who argue that the ability of humans to exercise free choice sets them apart from the rest of nature. Since humans are able to make decisions and carry out plans in accord with their own reasons rather than some external constraint, the critics say, voluntary human behavior, unlike physical phenomena, cannot be subsumed under laws. Thus, the critics conclude that covering-law explanations cannot account for human behavior.

Mill was aware of this objection, but thought that it was based on a misunderstanding about the possibility of *accurate prediction* in science. He believed he had solved the problem by showing that even in the physical sciences precise predictions are not always possible. He argued that since the absence of precise predictions is not an impediment to the construction of inexact physical sciences, it cannot prevent the construction of inexact human sciences.

Hempel responds differently to the criticism that some distinctive form of explanation is required to account for human actions. He argues that explanations in terms of mental causes, such as motivating reasons, have the same *logical* structure as covering-law explanations in the physical sciences. A human action, like a physical event, he says, is explained when it is shown to follow from explanatory facts that include at least one law statement. Hempel presents the following model for explaining behavior that is a result of rational deliberation:

Agent A was in a situation of type C. [Initial Condition]
A was disposed to act rationally. [Initial Condition]
Any person who is disposed to act rationally will, when in situations of type C, invariably (with high probability) do X. [General law]

A does X. [Event to be explained] (1962, 27)

In Hempel's model, agents are considered rational if they are disposed to take appropriate means to achieve their chosen ends. In contrast to Dray's (1957) account, Hempel does not assume that agents always act rationally, but instead he regards the attribution of rationality to an agent as an explicit initial condition in the explanation. Hempel adds that covering-law explanations with the same logical form as explanations of rational behavior can be framed for behavior that is not a result of "rationality and more or less explicit deliberation, but . . . other dispositional features, such as character and emotional make-up" (1962, 27).

Hempel's models display clearly what he means when he says that explanations in physical and social sciences are similar to one another: Each has the logical structure of an *argument* (inductive or deductive) in which the event to be explained is the conclusion and some initial conditions and law or laws constitute the explanatory premises. Hempel's models so aptly clarify and refine respectable popular intuitions about the nature of explanations in the physical sciences that they have become a focus for discussion of alternate views about explanation in social sciences. In what follows, I will look at some objections to his approach and some alternative accounts of explanation in the social sciences.

The objections to Hempel's account that I will discuss fall roughly into three categories. The first position, *interpretativism*, regards explanations of human purposive action as having an entirely different structure from causal explanations or any other explanations that appeal to *laws*. Interpretativists deny that there are empirical laws connecting reasons with actions; they say that to suppose there could be such laws involves committing a *logical* error.

The second type of objection can be called "nomological skepticism." It is dominated by the worry that there are no *laws* in social science available for constructing covering-law explanations. Some nomological skeptics admit the eventual possibility of discovering laws, while others are less sanguine. The skeptics' doubts about the possibility of finding appropriate laws are pragmatic in contrast to the logical concerns of the interpretativists. Skeptics believe that the great variability and complexity of human behavior pose practical barriers to framing generalizations that are at once informative and true. Although both Mill and Hempel have addressed these practical concerns, the skeptics remain unconvinced.

The third type of objection, proposed by *critical theorists*, sees lawful explanation as a threat to human autonomy. Critical theorists worry about the ethical implications of trying to explain human behavior in the same manner that we explain the actions (or movements) of nonconscious physical objects.

Aside from the three types of objections just described, other fundamentally important criticisms of Hempel's models raise questions about whether laws are *parts* of explanations or whether instead laws justify or underlie explanations (Scriven 1959; Humphreys 1989), whether explanations are arguments (Jeffrey 1969), whether a statistical explanation must show that the event to be explained is highly probable (Salmon 1965), and whether the pragmatic features of explanation have been adequately addressed (van Fraassen 1980). As the present volume attests, the models originally proposed by Hempel more than twenty years ago cannot survive these criticisms intact. Not surprisingly, an adequate account of scientific explanation must go beyond that early work.

Nevertheless, we can look at criticism of explanation in the social sciences without exploring these refinements. For insofar as newer models are causal models of explanation or covering-law models of explanation (see, for example, Glymour et al. 1987) the objections raised to them by interpretativists, nomological skeptics, and critical theorists will not be very different from their objections to Hempel's original models. Since this paper focuses on lawful explanation in the social sciences, I will limit the scope of this discussion to the criticisms raised by interpretativists, nomological skeptics, and critical theorists.

2. Interpretativism

Interpretativists reject as logically confused the claim that explanations in the social sciences are fundamentally similar to those in the physical sciences. Their position goes back at least to Dilthey, whose main work begins just after Mill's death. A forceful exposition of interpretativism is found in R. G. Collingwood (1946).

Collingwood uses the terms "history" and "historical thought" to refer to all studies of human affairs, including not only history, but also (parts of) anthropology, sociology, political science, psychology, and economics. The contrast class, called here "physical science," includes physics, chemistry, and even the so-called historical sciences, such as evolutionary biology and geology.

The events studied by history have, according to Collingwood, an "inside" as well as an "outside." The outsides of events consist of everything belonging to them that can be described in terms of bodies or their movements. The insides of events can be described only in terms of the thoughts of the agent (or agents) who are responsible for the events. Human behavior requires for its complete description not only the account of bodily motions involved but also an account of the beliefs and desires of the agent. For example, the physical description of one person cutting another with a knife does not distinguish an act of surgery from an act of assault, a ritual act, or an accidental cutting. A more detailed physical description of the "outside" may provide clues that will help to ascertain the intention of the cutter, but until that intention is uncovered, Collingwood would say, we simply do not know what action took place.

Collingwood argues on the basis of the difference between events with only an outside and those with both an inside and an outside that there is a fundamental difference between the search for so-called causes of human behavior (the *reasons* for the behavior) and causes of physical events. In the physical sciences, the event is perceived, and its cause (a separate event) is sought. This investigation takes the scientist beyond the original event in order to relate it to other separate events, thus bringing it under a general law of nature.

In contrast, when studying human actions, the historian must look inside the event to discover the thought that is expressed *in* the event. The event studied is not really separate from the thought, but is the mere expression of the thought. The thought is the event's "inside," that which makes the event what it is. From his claim that the reasons for actions are not external to the actions, Collingwood concludes the relation between the two cannot be governed by laws of cause and effect. Thus, he says, causal laws play no essential role in the explanation of human behavior.

When an action is described as the sort of action it is, the description itself includes the reason for the action. In the case of purposive human behavior, the acts of describing and explaining are therefore one and the same. The reason for the act is not viewed as some cause that is separate from it, but rather is *logically* or *meaningfully* related to the act. The reason gives meaning to the act, and makes it the sort of act it is. Accordingly, Collingwood would say that the model of searching for regular connections between actions and reasons, as proposed by Mill and, later, Hempel, simply makes no sense.

Among contemporary philosophers, Peter Winch (1958) is a leading advocate of the interpretativist position. Like Collingwood, he denies the role of anything analogous to a law of nature in explaining human behavior, and he regards the term "social science" as misleading for this reason. Winch's view is that "social relations really exist only in and through the ideas that are current in a society; [and] . . . that social relations fall into the same logical category as relations between ideas" (1958, 133). Causality is thus no more an appropriate category for understanding social relationships than it is for understanding mathematical relationships. Whereas Collingwood pays special attention to individual beliefs and desires, Winch, following Wittgenstein, emphasizes the social character of human action and thus he focuses on the importance of *rules* or norms of behavior.

Winch's broad notion of a rule covers not only formal regulations, such as traffic rules and tax deadlines, but also unstated cultural norms or conventions, such as those governing the appropriate distance between speakers engaged in a face-to-face conversation. In addition, "rule" embraces practices and institutions such as religion, democratic government, and money. Some rules are regulative, such as the rule to stop for red lights, while others are constitutive of the practices they embrace.

Consider, for example, the act of offering a sacrifice. This act is possible only within a certain type of institution in which particular kinds of behavior, such as killing animals in a prescribed way and under certain conditions, count as offering sacrifices. Without such an institution or social practice, it makes no sense to call the killing of the animal in that way a sacrifice. The point of these constitutive rules is that whether an action is performed and what kind of action it is depends not only on individual intentions, but also on the social set-up. Social relations, however, according to Winch, do not *cause* the act; they rather constitute it by giving it the meaning it has.

Obviously, we cannot understand what is going on in a sacrifice if we do not have the concept of a sacrifice. To say that the concept of sacrifice depends on the social set-up, however, goes far beyond the claim that, as individuals, we can acquire the concept only through socialization. For, not only how we come to learn the concept, but also the very meaning of the concept depends on the possibility of social relations of a certain type.

To understand human behavior, Winch says, we require more than just abstract knowledge of the rules of a society. We also need to understand what counts as following a rule in a particular case. To grasp this, he says, we must somehow come to share the viewpoints, attitudes, and feelings of the actors. For example, consider a society which has a norm that requires showing respect to elders. Knowing this rule and having a physical description of some bit of behavior are not enough to figure out whether or not this behavior in the presence of an elder counts as showing respect. We also require some knowledge of the beliefs and attitudes of the person engaging in the behavior. Was the person aware an elder was present? Did the person know that form of address was considered disrespectful? Was the person merely careless in his manner? Was the form of address considered disrespectful by the speaker or the elder?

Even in one's own society, working out such matters can require complicated negotiations. Understanding human behavior in exotic societies is doubly problematic, and those who attempt it must be on guard against ethnocentrism. Winch's sensitivity to the importance of rules and how they are applied, as Papineau notes (1978, 96–97), accords well with the concerns of contemporary cognitive anthropologists. Ward Goodenough (1957), for example, says that culture consists not of "things, people, behavior or emotions but the forms or organizations of these things in the minds of people" (quoted in Frake 1969, 38). Cognitive anthropologists like Goodenough see their task as uncovering these forms (i.e., rules) primarily on the basis of what people say about how they categorize and organize the furniture of their worlds in applying the rules.

Winch does not deny that we can predict human behavior. He admits that with a knowledge of the rules of a society and an understanding of how the rules are applied, reliable prediction is often possible. He insists, however, that successful prediction of behavior is unlike predictive success in the physical sciences. The difference is not just that physical science yields more accurate predictions, for in some cases, as in tidology, it does not. The point is that in the physical sciences, *causal regularities* are the basis of the predictions, whereas insofar as behavior is rule-governed, it is not subject to causal laws. Rules, according to interpretativists, function as standards or norms of behavior; they give meaning to behavior, but do not cause it. Although we can predict behavior from a knowledge both of the rules of a society and of how those rules are translated into actual behavior, the prediction is not based on a causal relation between the rule and the behavior. The concepts of cause and effect, Winch insists, simply do not apply to the relation between a reason for action and the action, any more than the concepts of cause and effect apply to the relation between being a Euclidean triangle and having internal angles with a sum of 180 degrees.

Interpretativists acknowledge that discerning the rules that underlie human behavior is an important goal. They insist, however, that these rules have a logically different character from causal laws—or other empirical laws—discovered in the physical sciences. Because interpretativists see the relation between reasons and behavior as a logical relation rather than a causal relation, they reject the covering-law models of explanation.

Drawing support from interpretative philosophers, some social scientists have also rejected covering-law models of explanation. Clifford Geertz, for example, claims to do so in *The Interpretation of Cultures* (1975).

Anthropology, for Geertz (in contrast to cognitive anthropologists like Goodenough), is ethnography. Ethnography, he says, consists in interpreting the flow of social discourse. This interpretation itself consists of inscribing or recording the flow, "fixing" it so that it can be shared and reexamined long after the actual events take place. A major problem for the ethnographer is that of finding the appropriate general concepts for describing or classifying the observed behavior. For example, an ethnographer might observe that meetings are taking place and that political issues are being discussed. Further features noted might include low voices and concern with arrangements to secure secrecy. Is the appropriate *thick description* (Geertz borrows Ryle's expression) or interpretation of such activities that of "fomenting a rebellion" or something less inflammatory?

Geertz pointedly calls the ethnographer's activity of fixing the flow "interpretation" rather than explanation. Following Dray's (1957) advice to historians, Geertz believes that anthropologists should primarily be concerned with assigning observed behavior to appropriate concepts. Like Dray, he denies that covering laws play any role in this activity. He adds that any true statements that are general enough to serve in covering-law explanations in anthropology are either hopelessly vague or trivial. In such passages, Geertz's rejection of the possibility of lawful explanation in anthropology echoes Scriven's criticism of Hempel's models in "Truisms as the Grounds for Historical Explanation" (1959), and also reflects the position I have characterized as nomological skepticism.

In this same work, however, Geertz also says that anthropologists are interested not only in particular ethnographic studies (which he calls microscopic work) but also in going beyond ethnographies to compare, to contrast, and to generalize. Indeed, he acknowledges that some generalization occurs even within microscopic studies. As Geertz says, "We begin with our own interpretations of what our informants are up to, or what they think they are up to, and then systematize these." In another place, where he is talking about doing ethnography, he says:

> Looking at the ordinary in places where it takes unaccustomed forms brings out not, as has so often been claimed, the arbitrariness of human behavior . . . but the degree to which its meaning varies according to the pattern of life by which it is informed. Understanding a people's culture exposes their normalness without reducing their particularity. . . . It renders them accessible; setting them in the frame of their own banalities, it dissolves their opacity. (1975, 14)

However, Geertz's remarks are ambiguous, for they lend themselves either to an interpretativist or to a nomothetic (covering-law) construction. The nomothetically inclined reader can say: "What can it mean to expose the normality of a people if we do not know what normality is and have no concern with the normal? Some account of what is normal provides the framework for our recognition and understanding of traits in another culture. But to have an account of what is normal is to be in possession of at least some statistical laws of human nature, to know that humans usually believe, say, or do thus and so under such and such circumstances. Moreover, at least part of what it is to interpret or to understand something is to place it within an intelligible framework or to see how it fits into a pattern. But intelligible frameworks and patterns are the sorts of regularities that, according to Hempel, are expressed in statistical or universal laws.

Winch could object to such a nomothetic reading of Geertz's position on the grounds that the reading involves a serious misunderstanding of the type of regularity or normality present. The appropriate regularities, according to the interpretativist, are not causal regularities, but rules or norms of behavior. A grasp of these is crucial for understanding behavior, but norms and rules are not explanatory laws, and so cannot play that role in explanation.

Geertz, however, in contrast to Winch, does not want to deny that anthropology is a science, albeit a different and "softer" kind of science than physics. In discussing how anthropological theories are constructed, Geertz says that whereas in covering-law models individual cases are subsumed under *laws*, the generalizations that constitute theory in anthropology are generalizations *within* cases rather than generalizations *across* cases. Amplifying this point, Geertz goes on to say that the type of generalization he refers to is called, in medical science, "clinical inference." He says that anthropologists should look for "intelligible patterns" into which otherwise unrelated bits of information can be fitted rather than "universal laws" that can be combined with initial conditions to *predict* cultural patterns.

Geertz's rejection of prediction as a goal for anthropology provides an important clue to his dismissal of the covering-law models. He apparently believes that covering-law explanation commits one to an objectionable view about the close relationship between explanation and prediction.

Hempel, it is true, argues for a kind of symmetry between scientific explanation and scientific prediction. He regards scientific explanations of individual events as arguments to the effect that the event to be explained was to be expected on the grounds of laws and initial conditions, and scientific predictions as arguments to the effect that the event predicted is to be expected on the grounds of laws and initial conditions. Given Hempel's analysis, explanation and prediction do not differ in their logical structure, but only in the temporal order of the events to be explained or predicted and in the starting point of our knowledge of the situation. In explanation, we are aware that the event has occurred, we search for the relevant laws, and we try to reconstruct the initial conditions from which the event can be derived. In prediction, we are aware of the initial conditions, and with knowledge of the relevant laws, we derive the occurrence of the predicted future event.

It is important, however, not to misread Hempel's thesis of the symmetry of explanation and prediction as asserting that whenever an event can be explained it could have been predicted. Hempel, in response to criticisms of his symmetry thesis by Scriven and others, defends "the conditional thesis that an adequate explanatory argument must be such that it could have served to predict the explanandum event *if* the information included in the explanans had been known and taken into account before the occurrence of that event" (1965, 371). At the same time, he insists that his symmetry thesis does *not* require that we always can know independently of the occurrence of the explanandum event that all the conditions required for an explanation are realized. There are clearly situations in which it is only "after the fact" that we are aware that some of the initial conditions were present. For example, before a murder occurs there may have been no way to recognize that the rage of the murderer was sufficient to motivate him to commit the act, or to know that he was capable of murder.

Hempel's version of the symmetry thesis does *not* claim that because the act can be explained in terms of laws and initial conditions, that these laws and initial conditions could have been discovered before the act occurred. To hold that any act which can be explained after the fact could have been predicted before the fact (without further qualification) requires a much stronger symmetry thesis. This stronger version of symmetry is indefensible, and should not be attributed to Hempel. (Hempel's symmetry thesis is closely linked with his understanding of explanations as *arguments*. For a weaker version of symmetry, compatible with the view that explanations are not arguments, see Salmon 1970.)

Geertz apparently regards the indefensible strong symmetry thesis as a feature of "the covering-law model of explanation." He specifically rejects attempts to establish covering laws in anthropology to *predict* behavior because he sees the goal of social science as understanding rather than prediction. He then goes on to reject explanation as a goal because, guided by the strong symmetry thesis, he erroneously identifies lawful explanation with the ability to make predictions.

Ironically, when Geertz says that clinical inference provides a model for understanding anthropology, he seriously undercuts his interpretativist stance. Clinical inference involves diagnosis of symptoms, such as the diagnosis of measles on the basis of its telltale signs. Diagnosis, as Geertz correctly notes, is not concerned with predicting. It is directed toward analyzing, interpreting, or explicating the complex under consideration. Geertz sees this activity as the essential concern of anthropologists and therefore argues that anthropology is a diagnostic science rather than a nomothetic one.

Hempel shows, however, that cases of diagnosis or clinical inference fall easily within the scope of covering-law explanation (1965, 454–55). Moreover, in several places where Hempel discusses "explanation by concept" and other forms of explanation in the social sciences, he offers an account of explanation that deemphasizes the relation between explanation and prediction, and focuses instead on an explanation's power to increase our understanding. Near the end of "Aspects of Scientific Explanation," for example, he says that explanation "seeks to provide a systematic understanding of empirical phenomena by showing how they fit into a nomic nexus" (1965, 488). The same emphasis on understanding rather than prediction is apparent in "Explanation in Science and in History" (1962, 9).

Hempel's arguments strongly support the position that scientific "understanding" is just as dependent on laws as scientific explanation. If Hempel is correct, Geertz's form of interpretativism is very different from Winch's, for it is just as dependent on laws as nomological explanation. Geertz, however, as noted earlier, has also questioned whether there can be any nontrivial laws in anthropology. This nomothetic skepticism is a sticking point for many social scientists. In the next sections, we will consider some attempts to find and characterize appropriate laws for explaining (and understanding) human behavior.

3. Rationality and Explanations of Behavior

When Hempel wrote "The Function of General Laws in History" (1965, 251–52), he believed that it would be possible to discover causal laws of human behavior though he did not specify any particular form that these laws would take. At that time there was widespread hope that behaviorism would soon deliver what it promised. Hempel reflects this optimism when he suggests that we may someday find behavioristic stimulus-response laws or laws of learned behavior that connect circumstances and actions. If statements describing such regularities— which I take to be causal—were available, they would be suitable covering laws for constructing explanations of human actions. Similarly, though Hempel does not say so here, if causal laws connecting brain states and human *actions* (not mere bodily movements) were available, these could also be used in covering-law explanations of behavior. However, neither of these materialist programs has so far succeeded in supplying the sorts of laws which the programs' early adherents hoped would be forthcoming.

Hempel clearly regards instances of his proposed law schema—"Any person who is disposed rationally will, when in circumstances of type C, invariably (with high probability) do X" (1962, 28–29)—as having empirical content, whether or not the law is causal. In keeping with his understanding of rationality and other character traits as *dispositional* properties, Hempel does not regard them as completely definable in terms of manifest behavior. He says instead that they may be partially definable by means of reduction sentences. Following Carnap's account of reduction sentences, Hempel says that the connections between dispositional properties (e.g., rationality, fearlessness) and the actions that are symptomatic of these dispositions are stated in claims that express either necessary or sufficient conditions for the presence of the given disposition in terms of the manifest behavior. Reduction statements seem to be analytic, for they offer at least partial definitions of dispositional properties. Yet not all these statements can qualify as analytic, Hempel says, for in conjunction they imply nonanalytic statements of connection between various manifest characteristics. For example, if a specific form of behavior *A* is a sufficient condition for the presence of rationality, and another specific form of behavior *B* is a necessary condition of rationality, then the claim "Whenever *A* is present *B* will be also," which is a consequence of the conjunction of the two reduction sentences, and which asserts that two types of behavior are always found together, "will normally turn out to be synthetic" (1962, 28–29). On this basis, Hempel claims that the generalizations which state that in situations of a certain sort, rational agents will act in specified ways are empirical.

The success of Hempel's defense of the empirical character of the laws that connect dispositions with behavior is obviously tied to the possibility of finding suitable reduction sentences for characterizing dispositional properties such as rationality, fearlessness, and the like. This is, I believe, so closely linked with behaviorism that it inherits the problems associated with that program. Moreover, Hempel admits in "A Logical Appraisal of Operationism" (1965, 133) that the ability to derive synthetic statements from pairs of reduction sentences casts some doubt on the "advisability or even the possibility" of preserving the distinction between analytic and synthetic sentences in a logical reconstruction of science. Despite his concession to the blurring of this traditional empiricist distinction, he maintains his original standard requiring empirical content in explanatory laws.

It is fair to say that behaviorism no longer enjoys the sway it once held. In a more contemporary attempt to understand the nature of rationality, many philosophers have rejected the behaviorist approach and turned to a decision-theoretical analysis of rationality. However, if we try to understand "rationality" in decision-theoretic terms rather than behavioristic terms, the empirical content in the law (or law schema) – "Any person who is disposed rationally will, when in circumstances of type C, invariably (with high probability) do X" – remains elusive.

The decision-theoretic approach to rationality is intended to be applicable under the assumption that an agent is acting independently and with only probabilistic knowledge of the outcomes of various actions. Under such circumstances, a person is said to behave rationally just in case the person acts so as to maximize expected utility. If we adopt this criterion, then to say that if an agent is disposed to act rationally in circumstances C, the agent will do X, is equivalent to saying that in circumstances C, action X maximizes expected utility (or, in Papineau's (1978) phrase "expected desirability").

The expected desirability of an action is the sum of the products of the probabilities and the values for each of the possible outcomes of the action. Maximizing expected desirability just means choosing the action with a sum that is not lower than the sum of any other action. The statement that an agent is in circumstances C (i.e., the agent holds various beliefs to which probabilities are assigned and also has desires with values attached to them) is clearly empirical. However, whether or not actions maximize expected desirability under such specified circumstances is not an empirical matter, but rather a judgment based on a calculation which is determined by the criterion of rationality.

The decision-theoretic account of rationality does not actually require agents to perform calculations. However, to be rational, actions must accord with the results of such calculations, had the calculations been performed. This means that *rational* agents must choose actions that maximize expected desirability, for they could not do otherwise and still be rational agents.

Thus, if we use the decision theorist's analysis of rationality, instances of Hempel's schema for a law ("Any person who is disposed to act rationally will when in circumstances C invariably (with high probability) do X") fail to meet his criterion of empirical content. But if the laws have no empirical content, they cannot ground genuine scientific explanation, even though they may be the basis for another form of explanation, similar to explanation in mathematics, where the laws are not empirical.

The criterion of rationality offered by decision theory is minimal in the sense that agents need only take some available means to achieve whatever ends they desire, or at least to avoid frustrating those ends. The decision-theoretical account of rationality does not assume that agents make good use of available evidence when forming beliefs. Even if agents' beliefs are based on prejudice or ignorance, or if their desires are peculiar or hard to comprehend, their behavior can be rational. Moreover, as mentioned before, to be rational in this sense, agents need not assign explicit probabilities to beliefs or quantify values, or even make rough or precise calculations of expected desirability. It is enough for agents to act *as if* they were maximizing desirability, given their beliefs and desires.

If we strengthen the decision theorist's standard of rationality to require that rational agents use an objective physical basis instead of a purely personal consideration to assign probabilities to various possible outcomes of an action, we invoke some probabilistic laws. However, these are not covering laws; the probabilistic laws are used only to assess the truth of the initial condition which states that the agent is disposed to act rationally. As such they are not even part of the covering-law explanation itself.

David Papineau adopts the decision-theoretical approach to rationality in his (1978) attempt to reconcile the interpretativist position with nomothetic explanation of human action. Papineau presents a covering-law model of explanation of human action that differs from Hempel's in several respects.

According to Hempel, an agent's disposition to act rationally is an initial condition that must be established empirically. For Papineau, agents always act rationally. According to him, the lawlike generalization that grounds explanations of behavior is: "Agents always perform those actions with greatest expected desirability" (1978, 81). This law, he believes, is implicit in ordinary explanations of individual human behavior. Accordingly, for Papineau, acting out of a character trait, such as fearlessness, would not contrast with his acting rationally, as it might for Hempel. (For Hempel, humans may be rational at some times, and not at others; for Papineau, it is a contingent, but universal truth that humans behave rationally. We must remember though that the two differ about the meaning of rationality.)

Papineau agrees with Winch that rules do not *cause* behavior, for otherwise, he says, humans would be mere puppets of their cultural milieu, unable to violate social norms. Obviously people can disregard norms, whereas causal laws cannot be violated. Papineau acknowledges that the existence of rules plays an important role in forming agents' beliefs and desires, and that the interpretativist has something important to say about this (1978, ch. 4; also see Braybrooke, 1987, 112–16). However, Papineau says that the information which guides our attributions of beliefs and desires to agents is not actually part of the explanation of human behavior. All that is required for explanation, in addition to the generalization about maximizing expected desirability, is an account of what beliefs and desires an individual has (these are the initial conditions), not an account of how the agent came to have them. Papineau does not deny the importance and relevance of the interpretativists' concerns, but he sees them as supplementing, rather than conflicting with, lawful explanation.

Papineau is aware of the apparent lack of empirical content in his proposed law: "Agents always perform those actions with greatest expected desirability," and he tries to defend his model of explanation against this criticism. Although, he says, it is true that we use this general principle to infer the nature of agents' beliefs and desires, it is legitimate to do so as long as we do not include the action we are trying to account for in establishing those attitudes. We infer beliefs and desires from agents' past actions, aided by knowledge of the norms of their society, and then – attributing to them those attitudes – explain their present actions in the light of that knowledge.

Nevertheless, Papineau recognizes that sometimes people do not act in accord with the beliefs and desires that have been correctly attributed to them in the past. And when this happens, he admits that we do use the anomalous action to infer the beliefs and desires that underlie it. His argument to legitimize this move appeals to Lakatos's understanding of a theoretical "core statement" that is maintained in the face of presumptive counterexamples by revising various auxiliary assumptions (see, for example, Lakatos 1970). It is a commonplace that people's beliefs and desires do change over time, so we must be prepared to take account of such revisions. However, he says, "this preparedness does not condemn the overall theory as unscientific – none of the most revered theories in the history of science would ever have survived if their proponents had not been similarly prepared to defend their central tenets from the phenomena by revising auxiliary hypotheses" (1978, 88).

Papineau does not mean to countenance ad hoc revisions to save his generalization that agents act always so as to maximize expected desirability. He says that if we attribute to an agent extraordinary desires or beliefs that we would not expect to be available to the agent, we should be prepared to give some account of the circumstances that will lead to new and independently testable propositions. We are bound to do this, just as any physical scientist is required to do so in a "progressive" research program (1978, 88).

Ultimately, however, Papineau urges us to accept this theory of human action because we do not have a better theory available. Behaviorism is impoverished, he says; physiological accounts are woefully undeveloped, and the rule-governed account of behavior must be integrated into the decision-theoretical model if humans are not to be understood as cultural puppets who rigidly conform to norms, rules, and conventions.

Even if interpretativists were willing to accept Lakatos's account of the nature of theories in physical science, I doubt that Papineau's arguments would undermine their insistence that human behavior is not subject to causal laws, for he does not address the claim that reasons for behavior are *logically* different from causes. At the same time, those who urge nomological explanations of human behavior will be disappointed in Papineau's failure to put forth an explanatory generalization with empirical content.

One can accept a decision-theoretical account of the nature of rationality, while nevertheless recognizing, as Donald Davidson (1980) does, that no *criterion* of rationality, however satisfactory it is, can function as a law in covering-law explanations of human behavior.

In a series of papers (1980, especially essays 1, 7, 12, 14), Davidson refutes the interpretativist claim that it is a logical error to suppose that reasons can cause actions. Interpretativists have typically argued that if things are logically related (or connected by relations of meaning), then they cannot be causally related. Davidson shows that whether or not two events are related causally depends on what the world is like, whereas various linguistic accounts of the relationship may be classified as "analytic" or "synthetic." For example, suppose that a rusting understructure causes a bridge to collapse. The sentence "The rusting of the iron understructure caused the bridge to collapse" would normally be regarded as synthetic, whereas "The cause of the bridge's collapse caused the bridge to collapse," would normally be regarded as analytic. Yet the causal relationship between rusting and collapsing is a feature of the world, not of either sentence.

However, Davidson does not believe that reason explanations can be understood as covering-law explanations that invoke an implicit law connecting *types of reasons* with *types of behavior*. Thus, suppose that Jason's running in the marathon can be (truly) explained by his desire to prove his self-worth. Davidson does not deny that there is a causal law connecting Jason's reason (his desire to prove his self-worth) for running the marathon with his running, for he agrees with Hempel that "if *A* causes *B*, there must be descriptions of *A* and *B* which show that *A* and *B* fall under a law" (1980, 262), and he also agrees that explanations in terms of reasons are—if correct—causal explanations. However, he does say that we hardly ever, if ever, know what that empirical law is, and also argues that the (unknown) law does not have the form of a regular connection between a psychological cause and an action, such as "Whenever anyone wants to prove his worth, he runs a marathon." He argues further that no matter how carefully the circumstances surrounding, for example, someone's wanting to prove his worth and the opportunities to do so may be qualified, the resulting expression is not a psychophysical law.

Obviously it is not possible here to develop or to criticize Davidson's arguments with the care they deserve. It is worth pointing out, though, that whereas Davidson refutes a major premise of the interpretativists (i.e., that events cannot be causally related if descriptions of them are logically related), he seems to agree with Winch in saying that Mill was barking up the wrong tree in his search for explanatory *laws* that have the form of connecting mental causes with behavioral effects.

Davidson argues that explanations that cite reasons are informative because they tell us a lot about the individuals whose acts they are invoked to explain, rather than a lot about general connections between reasons and actions. For example, if the explanation of why Jason ran the marathon is correct, then we have achieved some understanding of what motivates him and how he expresses this. Furthermore, with this knowledge, we can make reliable conditional predictions of how he would behave in other circumstances.

Although Davidson says that his reflections reinforce Hempel's view "that reason explanations do not differ in their general logical character from explanation in physics or elsewhere" (1980, 274), his arguments against the possibility of covering laws that connect descriptions of psychological states and behavior in such explanations represent a significant departure from the tradition of Mill and Hempel. Although Davidson delineates the special character of reason explanations differently from the interpretativists, they can perhaps take some comfort in his recognition of the special or anomalous nature of explanations in terms of reasons.

4. The Existence of Appropriate Laws

Those who deny that there is a strong similarity between *reason* explanations of human action and explanations in physical science should remember that not all explanations in the social sciences appeal to laws connecting reasons (or dispositional properties such as fear) with actions. Consider, for example, the tentative laws concerning social structure proposed by G. P. Murdock (1949). These statements do not refer to beliefs or desires of any individuals. The generalizations relate various systems of kinship in different societies to differing forms of marriage, to patterns of postmarital residence, to rules of descent, and to forms of family. These generalizations are not causal, for they do not attempt to assign temporal priority, nor do they cite any mechanisms for the regularities they describe. In many cases, Murdock's generalizations state the coexistence of some rules with other rules, but they do not seem themselves to express norms or rules of any society. By analogy with some physical *structural* laws, such as "All copper conducts electricity," Murdock's proposed laws can also be classified as "structural laws." Other candidates for structural laws in the social sciences are the "law of evolutionary potential" (the more specialized the system, the less likely it is that evolution to the next stage will occur) proposed by Sahlins and Service (1960) and the "law of cultural diffusion" (the greater the distance between two groups in time and space, the more unlikely it is that diffusion will take place between them) (Sanders and Price 1968).

Murdock's proposed laws are generally regarded by anthropologists as problematic. He tried to verify his generalizations by using information that had been recorded by scores of anthropologists who differed widely in methods and theoretical presuppositions. As a result, serious questions can be raised about whether terms used to characterize the data on which Murdock based his generalizations were employed consistently. While anthropologists have justifiably criticized the design of Murdock's studies, there seems to be no reason in principle why structural generalizations in the social sciences cannot be framed and tested.

Not all social regularities that are well established are good candidates for *explanatory laws*. Some regularities raise more questions than they answer; they require rather than provide explanations. This may be true of Murdock's generalizations. Current examples of nonexplanatory generalizations are those puzzling but well-supported statements that describe patterned connections between birth order, sibling intelligence, and achievement. However, when correlations like these are strongly supported across many cultures, social scientists are stimulated to search for deeper regularities to explain them (Converse 1986). The deeper regularities, if discovered, might be causal laws or other structural laws. It is not clear in advance of their discovery whether these deeper generalizations will refer to any individual human reasons for acting.

In opposition to Geertz's pessimism about the ability of the generalizations framed by social scientists to "travel well," that is, to apply to situations other than those which gave rise to their formulation, Converse is optimistic. He believes that social scientists will eventually discover useful high-level generalizations. He says, on the basis of his own (admittedly limited) experience, that "we shall discover patterns of strong regularity, for which we are theoretically quite unprepared, yet which reproduce themselves in surprising degree from world to world and hence urgently demand explanation" (1986, 58).

Anthropologist Melford Spiro also says that it is too early to abandon hope of finding any panhuman generalizations that are not either "false—because ethnocentric—or trivial and vacuous." He says:

> That any or all of the generalizations and theories of the social sciences (including anthropology) may be culture-bound is the rock upon which anthropology, conceived as a theoretical discipline, was founded. But the proper scholarly response to this healthy skepticism is not, surely, their a priori rejection, but rather the development of a research program for their empirical assessment (1986, 269).

Despite the optimism of Spiro and Converse about the future of the social sciences, neither they nor anyone else can deny the superior status of the laws now available to contemporary physical sciences. Two standard responses to the paucity of interesting, well-supported generalizations in the social sciences are that the data are far more intractible than in the physical sciences, and that we have not been trying long enough. Converse, for example, says that given the complexity of the data, "it would not surprise me if social science took five-hundred years to match the accomplishment of the first fifty years of physics" (1986, 48). However, Alasdair MacIntyre (1984), analyzes the situation rather differently.

First of all, MacIntyre claims that the salient fact about the social sciences is "the absence of the discovery of any law-like generalizations whatsoever" (1984, 88). MacIntyre thinks that this reflects a systematic misrepresentation of the aim and character of generalizations in the social sciences rather than a failure of social scientists. He says that although some "highly interesting" generalizations have been offered that are well supported by confirming instances, they all share features that distinguish them from *law-like* generalizations:

(1) They coexist in their disciplines with recognized counterexamples;
(2) They lack both universal quantifiers and scope modifiers (i.e., they contain unspecified *ceteris paribus* clauses);
(3) They do not entail any well-defined set of counterfactual conditionals (1984, 90–91).

MacIntyre presents four "typical" examples of generalization in the social sciences including Oscar Newman's generalization that "the crime rate rises in high-rise buildings with the height of a building up to a height of thirteen floors, but at more than thirteen floors levels off."

In response to a rather obvious objection to (1)—that most laws of social science are probabilistic rather than universal—MacIntyre replies that this misses the point. He says that probabilistic generalizations of natural science express universal quantification over sets, not over individuals, so they are subject to refutation in "precisely the same way and to the same degree" as nonprobabilistic laws. He concludes from this that

> [W]e throw no light on the status of the characteristic generalizations of the social sciences by calling them probabilistic; for they are as different from the generalizations of statistical mechanics as they are from the generalizations of Newtonian mechanics or of the gas law equations. (1984, 91)

MacIntyre's position represents a serious misunderstanding of the nature of statistical laws. These laws do not state some universal generalization about sets, they state the *probability* (where this is greater than 0% and less than 100%) that different types of events will occur together. Of course, the probabilistic connections in the physical sciences are usually stated numerically, and in this they differ from many generalizations in the social sciences. But statistical generalizations need not be stated numerically; the probabilistic connection can be conveyed with such expressions as "usually" and "for the most part."

Moreover, statistical generalizations in physics, or in any other field, are not confirmed or disconfirmed in exactly the same way as universal generalizations. A universal generalization can be overthrown by a single genuine counterexample that cannot be accommodated by a suitable revision of auxiliary hypotheses. In contrast, *any* distribution in a given sample is compatible with a statistical generalization.

MacIntyre is correct in saying that statistical generalizations in social science are different from those in statistical mechanics. But this is because the probabilistic laws in statistical mechanics are based on a theory that is well tested and supported by far more elaborate and conclusive evidence than is presently available for any statistical generalizations in the social sciences. The retarded development of theories in the social sciences may be at least partly attributed to a scarcity of resources for investigation along with greater complexity of data.

MacIntyre's point about "scope modifiers" is one that has received much attention from those concerned to point out the differences between the social and physical sciences (e.g., Scriven 1959). In the physical sciences, the exact conditions under which the law is supposed to apply are presumably explicit, whereas in the social sciences, vague clauses specifying "under normal conditions" or some such equivalent, are substituted. This difference can be interpreted in several ways. MacIntyre regards the *ceteris paribus* clauses as required because of the ineliminability of *Fortuna*, or basic unpredictability, in human life. Hempel points out, however, the widespread use in physical science of *provisoes*, which is his term for assumptions "which are essential, but generally unstated presuppositions of theoretical inferences" (1988). Hempel supports his point with an example from the theory of magnetism:

The theory clearly allows for the possibility that two bar magnets, suspended by fine threads close to each other at the same level, will not arrange themselves in a straight line; for example if a strong magnetic field of suitable direction should be present in addition, then the bars would orient themselves so as to be parallel to each other; similarly a strong air current would foil the prediction, and so forth.

Hempel says that the laws of magnetism neither state precisely how such conditions would interfere with the results, nor do they guarantee that such conditions will not occur. Yet such *ceteris paribus* clauses are surely implicit.

Ceteris paribus clauses sometimes result from inadequate information (the complexity of the data again) about the precise boundary conditions under which a given lawlike statement is applicable. *Fortuna* plays no role here. With *ceteris paribus* clauses, proposed laws can be stated tentatively, while research proceeds to attempt to sharpen and refine the spheres of application. As Converse points out (1986, 50), such tidying up occurs in the physical sciences as well, as shown by work that had to be done by astrophysicists as a result of information brought back from recent space explorations of distant planets. Whether or not *ceteris paribus* clauses are *stated*, then, rather than whether they are required, seems to distinguish the social sciences from the physical sciences.

MacIntyre's third point—the failure of generalizations in the social sciences to support counterfactuals—raises a complicated issue. One common way of attempting to distinguish genuine laws from "coincidental" generalizations (a thorny, and as yet unresolved problem) is by appealing to the former's ability to support counterfactuals. So, on this view, saying that a generalization cannot support counterfactuals is just another way of saying it is not a law. However, since we do not have any widely accepted account of what it is to support counterfactuals that is independent of our understanding of causal laws, it is not clear what MacIntyre's remarks about the inability of generalizations in the social sciences to support counterfactuals adds to his claim that these generalizations are not laws.

In any case, MacIntyre admits that the probabilistic laws of quantum mechanics do support counterfactuals. It is reasonable to claim that these laws also contain elements of "essential unpredictability," for the laws cannot predict the behavior of individual atoms or even sets of atoms.

Furthermore, if we do not equate the possibility of explanation with that of accurate prediction (and MacIntyre agrees that the strong symmetry thesis is indefensible) then essential unpredictability poses no barrier to lawful explanation in the social sciences.

Nothing that is said here supports the view that Oscar Newman's generalization about crime rates in high-rise dwellings, quoted above, is a genuine law or that it could play a role in a covering-law explanation. (However, it may have

considerable practical predictive value for those contemplating designs of hous-
ing projects.) As it stands, Newman's generalization states an interesting correlation. We want to know how well it stands up in new situations. More than that,
even if it does not apply beyond the observed instances — if the generalization is
no more than a summary — we want to know *why* the correlation exists for those
instances. This is the kind of generalization that can lead us to form interesting
and testable causal hypotheses about connections between criminal behavior and
features of living situations. These can stimulate the acquisition of new data and
further refinements of the hypotheses, or the formulation of additional hypotheses. Ultimately, this process could lead to the establishment of laws that are very
different in form (not merely refined in terms of scope) from the generalization
that initiated the inquiry.

5. Ethical Issues

Central to MacIntyre's discussion of the character of generalizations in social
science is an attack on systems of bureaucratic managerial expertise. He says that
those who aspire to this expertise misrepresent the character of generalizations
in social science by presenting them as "laws" similar to laws of physical science.
Social scientists do this, MacIntyre claims, to acquire and hang on to the power
that goes along with knowledge of reliable predictive generalizations.

The ethical problems that trouble MacIntyre are of paramount importance to
the critique of social science put forth by a group of philosophers, known as critical theorists, who are associated with the Frankfurt School. While the views of
this group — which includes Horkheimer, Adorno, Marcuse, Habermas, Apel,
and others — are not monolithic, certain themes are pervasive. These writers regard any attempt to model social science on the pattern of the physical sciences
as both erroneous and immoral. Like the interpretativists, they believe that
covering-law explanations in social science involve a fundamental confusion between natural (causal) laws and normative rules. In addition, they complain that
explanations in the physical sciences are divorced from historical concerns, and
that this cannot be so in the social sciences. Most critical theorists would agree,
for example, with Gadamer's characterization of physical science: "It is the aim
of science to so objectify experience that it no longer contains any historical element. The scientific experiment does this by its methodological procedure"
(1975, 311, quoted in Grünbaum 1984, 16).

Grünbaum refutes this characterization, using examples from classical electrodynamics and other fields (1984, 17–19) to show that laws of physical sciences
do embody historical and contextual features. Grünbaum also criticizes the attempts of critical theorists to argue that the historical elements in physical theories are not historical in the relevant sense (1984, 19–20), and points out that

Habermas bases this so-called lack of symmetry between physics and psychoanalysis on the platitude that Freudian narratives are *psychological* (1984, 21).

Karl-Otto Apel's defense of the asymmetry between history and physical science departs somewhat from the statements of Habermas that Grünbaum criticizes. It brings out, perhaps, more clearly the worry about loss of human autonomy that is the real concern of critical theorists:

> It is true, I think, that physics has to deal with irreversibility in the sense of the second principle of thermodynamics. . . . But, in this very sense of irreversibility, physics may suppose nature's being definitely determined concerning its future and thus having no history in a sense that would resist nomological objectification.
>
> Contrary to this, social science . . . must not only suppose irreversibility—in the sense of a statistically determined process—*but irreversibility, in the sense of the advance of human knowledge influencing the process of history in an irreversible manner.* (Apel 1979, 20, emphasis mine)

Apel then goes on to talk about the problem presented to social science (but not to physical science) by Merton's theorem concerning self-fulfilling and self-destroying prophecy.

Apel's remarks suggest that critical theorists' talk about the special "historic" character of social science, here and elsewhere, really amounts to the recognition that humans are often able to use their knowledge of what has happened to redirect the course of events, and to change what would have been otherwise had they not been aware of what was going on and had they not formed goals of their own. Since humans are agents with purposes, they enter into the molding of their own histories in a way not possible by any nonthinking part of nature.

Grünbaum does include examples of "feedback" systems in his account of how past states count in the determination of present behavior (1984, 19), but it is at least arguable that the concept of "purposive behavior" is not entirely captured in the descriptions of mechanical feedback systems (Taylor 1966). Apparently, the critical theorists use the term "history" in a special way to refer to accounts of autonomous human behavior. In this, they are similar to Collingwood, who does not apply the term to any processes of nature—even geological and evolutionary processes—that do not involve human intentions:

> The processes of nature can therefore be properly described as sequences of mere events, but those of history cannot. They are not processes of mere events but processes of actions, which have an inner side, consisting of processes of thought; and what the historian is looking for is these processes of thought. All history is the history of thought. (Collingwood 1946, 215)

In addition to using "history" in this special way, critical theorists are concerned that the "regularities" observed in our (corrupt) social system—that are the result of unfortunate historical circumstances and that can be changed—are in danger of being presented by a nomothetic social science as exactly analogous to unchangeable laws of nature. The reaction of critical theorists to the attempt to discover laws and to construct nomological explanations in the *physical* sciences ranges from Horkheimer's acceptance of the goal to Marcuse's outright condemnation (see Lesnoff 1979, 98). Critical theorists, however, agree in rejecting nomological explanation in the social sciences.

Deductive-nomological explanations (covering-law explanations in which the laws are universal generalizations) are supposed to show that the event to be explained *resulted from* the particular circumstances, in accord with the relevant laws cited in an explanation of it (Hempel 1962, 10). Since the description of the event in a successful deductive-nomological explanation follows logically from the explanatory statements, it is plausible, using a *modal* conception of explanation, to say that, *given the circumstances and the operative laws, the event had to occur*. Leaving aside the point that in the social sciences explanations are much more likely to be probabilistic than deductive-nomological, the critical theorists, I believe, mistakenly read this feature of "necessity" in deductive-nomological explanations as an attempt to take what is the case (i.e., the event to be explained), and show that it *must be* the case, in the sense that the event was inevitable and could not have been otherwise, even if circumstances had been different. The mistake here is similar to the incorrect belief that any conclusion of a correct deductive argument is *necessary* just because the conclusion follows *necessarily* from the premises. If deductive-nomological explanation is misinterpreted in this way, it seems to present a challenge to humans' abilities to intervene and change circumstances.

However, such an understanding of scientific explanation is mistaken. A social science that is committed to providing scientific explanations is not thereby committed to serve the ends of regimes that want to maintain their dominance by making any existing social arrangements seem *necessary*.

In the same vein, critical theorists protest that scientific explanations are merely descriptions of the status quo, since scientific explanations fail to present a range of possible alternatives to what is in fact the case. However, it is not at all clear that explanations should tell us what could be or might have been; the goal seems rather to say *why* things are as they are. Understanding why things are as they are is, after all, often a prerequisite for changing the way things are.

In part, the complaint that nomological science is oriented only toward description rather than understanding or explanation is based on incorrectly identifying science with technology, and mistaking the goals of technology—prediction and control of the environment—for the goals of science. In the grips of this mistake some version of the following argument is adopted by critical theorists:

Physical science aims only at prediction and control of the physical environment. Therefore, a social science that is similar to physical science in its methods and aims has as its goal the prediction and control of the behavior of other humans.

Such a science would be an inherently manipulative—and thus ethically unsavory—enterprise (Habermas 1984, 389).

The picture of science as mere technique—prediction and control—is obviously inadequate as well as somewhat at odds with the critical theorists' own view that science is committed to the status quo and insensitive to what might be. Manipulation is, after all, often directed toward other ends than the maintenance of the status quo.

We have already discussed the differences between explanation and prediction, and have rejected the strong symmetry thesis. We can sometimes reliably predict outcomes on the basis of regularities that give no understanding of the situation. We can also have significant understanding and be unable to use this knowledge for reliable prediction. It is difficult to make the case that all of science is directed toward prediction and control of the environment. Certainly those scientists who are engaged in pure research cannot always spell out immediate practical applications of that research when asked to do so.

If, as I believe, the critical theorists' assessment of the nature of physical science is grossly inaccurate, their ethical worries may nevertheless be well founded. For certainly predictive knowledge and control are highly valued byproducts of scientific knowledge. Furthermore, if scientists pretend to have the power of prediction and control when they do not, or if they capitalize on the laymen's respect for science to claim that scientific expertise grants them moral expertise as well, then they behave unethically.

It would be naïve to suppose that an increase in understanding is the only aim of scientists or even the chief aim of most scientists. Fame and money motivate scientists as they do all humans. It has been argued (Bourdieu 1975; Horton 1982) that the struggle for status rather than a pure concern for truth is dramatically more pronounced in the social sciences than in the physical sciences. Concern for status is often shown in attacks on credentials and other forms of name-calling. Horton recognizes these features in the conduct of social scientists and blames this behavior on the comparative lack of agreement about what constitutes normal social science, and consequently what counts as outstanding achievement. However, this cannot be the whole story, for among physical scientists as well, attempts to increase one's status by denigrating the credentials of others is all too common. *The Double Helix* (Watson 1968) shocked many nonscientists with this revelation, but it came as no surprise to those working in the field. The heated dispute about whether or not the impact of a comet caused the extinction of the dinosaurs provides a current public example of name-calling among physical scientists that can hardly be overlooked by anyone who reads newspapers.

Social scientists are painfully aware that it would be a serious deception to put forth the present findings of their disciplines in the same light as well-founded physical theories. Unfortunately, overconfidence in and misuse of the predictive power of science is a feature of bureaucracy, as MacIntyre notes. However, bureaucratic overconfidence is not confined to the pronouncements of social scientists, as the investigation into the tragic failure of the space shuttle in 1986 attests. MacIntyre and the critical theorists do raise our awareness that such abuses occur, and that is helpful. But the occasional occurrence of abuses does not prove that a search for scientific laws and scientific explanations in physical science or in social science is unethical.

6. Conclusion

Hempel's account of covering-law explanation in the social sciences, which is similar to his account of explanation in the physical sciences, was chosen because of its clarity and importance as a point of departure for discussion of contemporary views of explanation in the social sciences. Responses to Hempel's models of explanation by interpretativists, nomological skeptics, and critical theorists were presented and criticized.

From the array of accounts of scientific explanation presented in this volume, it should be apparent that no consensus about these matters exists or is likely to be reached any time soon. In the absence of a completely acceptable account of scientific explanation, we have only approximations. Yet, despite protests of the critics of causal and nomological explanation in the social sciences, the best approximations to a satisfactory philosophical theory of explanation seem to embrace successful explanations in the social sciences as well as successful explanations in the physical sciences. None of the critics, I believe, has demonstrated that the admitted differences between our social environment and our physical environment compel us to seek entirely different methods of understanding each.

References

Apel, K.-O. 1979. Types of Social Science in the Light of Human Cognitive Interests. In Brown, pp. 3–50.
Braybrooke, D. 1987. *Philosophy of Social Science*. Englewood Cliffs, NJ: Prentice-Hall.
Bourdieu, P. 1975. The Specificity of the Scientific Field and the Social Conditions of the Progress of Reason. *Social Science Information* 14:19–47.
Brown, S. C., ed. 1979. *Philosophical Disputes in the Social Sciences*. Sussex; Harvester Press; Atlantic Highlands, NJ: Humanities Press.
Canfield, J., ed. 1966. *Purpose in Nature*. Englewood Cliffs, NJ: Prentice-Hall.
Collingwood, R. G. 1946. *The Idea of History*. Oxford: the University Press.
Colodny, R., ed. 1962. *Frontiers of Science and Philosophy*. Pittsburgh: University of Pittsburgh Press.

Converse, P. 1986. Generalization and the Social Psychology of "Other Worlds," In Fiske and Shweder, pp. 42–60.

Davidson, D. 1980. *Essays on Actions and Events*. Oxford: Clarendon Press.

Dray, W. 1957. *Laws and Explanation in History*. Oxford: the University Press.

Fiske, D. W., and Shweder, R. A. 1986. *Metatheory in Social Science*. Chicago and London: University of Chicago Press.

Frake, C. O. 1969. The Ethnographic Study of Cognitive Systems. In Tyler, 28–41.

Gadamer, H. G. 1975. *Truth and Method*. New York: Seabury Press.

Gardiner, P., ed. 1959. *Theories of History*. New York: The Free Press.

Geertz, C. 1975. *The Interpretation of Cultures*. London: Hutchinson.

Glymour, C., et al. 1987. *Discovering Causal Structure*. San Diego: Academic Press.

Goodenough, W. 1975. Cultural Anthropology and Linguistics. In *Georgetown University Monograph Series on Language and Linguistics* No. 9, 167–73.

Grünbaum, A. 1984. *The Foundations of Psychoanalysis*. Berkeley and Los Angeles: University of California Press.

Habermas, J. 1981. *The Theory of Communicative Action*, Vol. 1. Trans. by T. McCarthy. Boston: Beacon Press.

Hempel, C. G. 1962. Explanation in Science and in History. In Colodny, 1–33.

——. 1965. *Aspects of Scientific Explanation*. New York: The Free Press.

——. 1988. Provisoes: A Problem Concerning the Inferential Function of Scientific Theories. *Erkenntnis* 28:147–64. (Also to appear in *The Limitations of Deductivism*. A. Grünbaum and W. Salmon, eds., Berkeley and Los Angeles: University of California Press.)

Hollis, M., and Lukes, S. eds. 1982. *Rationality and Relativism*. Cambridge, MA: The MIT Press.

Horton, R. 1982. Tradition and Modernity Revisited. In Hollis and Lukes, 201–60.

Humphreys, P. 1989. Scientific Explanation: The Causes, Some of the Causes, and Nothing but the Causes. This volume, chap. 4.

Jeffrey, R. 1969. Statistical Explanation vs. Statistical Inference. In Rescher, ed., 104–13.

Lakatos, I. 1970. Falsification and the Methodology of Scientific Research Programmes. In Lakatos and Musgrave, 91–195.

——, and Musgrave, A., eds. 1970. *Criticism and the Growth of Knowledge*. Cambridge: the University Press.

Lesnoff, M. 1979. Technique, Critique, and Social Science. In Brown, 89–116.

MacIntyre, A. 1984. *After Virtue*. Notre Dame: University of Notre Dame Press.

Mill, J. S. 1874. *Logic*, 8th Ed. New York: Harper Bros.

Murdock, G. P. 1949. *Social Structure*. New York: Macmillan

Papineau, D. 1978. *For Science in the Social Sciences*. New York: St.Martin's Press.

Rescher, N., ed. 1969. *Essays in Honor of Carl G. Hempel*. Dordrecht: D. Reidel.

Sahlins, M., and Service, E. 1960. *Evolution and Culture*. Ann Arbor: University of Michigan Press.

Salmon, W. C. 1965. The Status of Prior Probabilities in Statistical Explanation. *Philosophy of Science* 32:137–46.

——. 1970. Statistical Explanation. In Colodny, ed., 173–231.

Sanders, W. T., and Price, B. 1968. *Mesoamerica: The Evolution of a Civilization*. New York: Random House.

Scriven, M. 1959. Truisms as the Grounds for Historical Explanation. In Gardiner, 443–71.

Spiro, M. 1986. Cultural Relativism and the Future of Anthropology. *Cultural Anthropology* 1, 3:259–86.

Taylor, R. 1966. Comments on a Mechanistic Conception of Purposefulness. In Canfield, 17–26.

Tyler, S., ed. 1969. *Cognitive Anthropology*. New York: Holt, Rinehart, Winston.

van Fraassen, B. 1980. *The Scientific Image*. Oxford: the University Press.

Watson, J. 1968. *The Double Helix*. Boston: Atheneum Press.

Winch, P. 1958. *The Idea of a Social Science and its Relation to Philosophy*. London: Routledge & Kegan Paul.

Explanatory Unification and the Causal Structure of the World

1. Introduction

The modern study of scientific explanation dates from 1948, the year of the publication of the pioneering article by C. G. Hempel and Paul Oppenheim. Nearly forty years later, philosophers rightly continue to appreciate the accomplishments of the covering-law models of explanation and the classic sequence of papers in which Hempel articulated his view. Even though it has become clear that the Hempelian approach to explanation faces difficulties of a number of types, the main contemporary approaches to explanation attempt to incorporate what they see as Hempelian insights (with distinct facets of the covering-law models being preserved in different cases), and they usually portray themselves as designed to accommodate one or more of the main problems that doomed the older view. My aim in this essay is to compare what I see as the chief contemporary rivals in the theory of explanation, to understand their affiliations to the covering-law models and their efforts to address the troubles of those models, and to evaluate their success in doing so. Ecumenical as this may sound, the reader should be forewarned that I shall also be interested in developing further, and defending, an approach to explanation that I have championed in previous essays (1981, 1985c).

1.1 Hempel's Accounts

Let us start with Hempel. The principal features of Hempel's account of explanation are (i) that explanations are arguments, (ii) that the conclusion of an expla-

I owe a long-standing debt to Peter Hempel, who first inspired my interest in the study of scientific explanation and whose writings on the topic seem to me paradigms of what is best in twentieth-century philosophy. My own thinking about explanation was redirected by Michael Friedman's seminal essay on explanation and scientific understanding, and I have also learned much from the comments, encouragement, and advice of Paul Churchland, Paul Humphreys, David Papineau, Kenneth Schaffner, and Stephen Stich. Above all I am deeply grateful to Wesley Salmon, for the depth and lucidity of his ideas and the kindness and patience of his conversation. The present essay continues a long dialogue, and, because that dialogue has been so pleasant and so instructive, I trust that it is not yet over.

nation is a sentence describing the phenomenon to be explained, and (iii) that among the premises of an explanation there must be at least one law of nature. Although the original treatment (1948) focused on cases in which the argument is deductive and the conclusion a singular sentence (a sentence in which no quantifiers occur), it was clear from the beginning that the account could be developed along two different dimensions. Thus there can be covering-law explanations in which the argument is nondeductive or in which the conclusion is general. *D-N* explanations are those explanations in which the argument is deductive and the conclusion is either a singular sentence or a nonstatistical generalization. Hempel assigned deductive explanations whose conclusion is a statistical generalization a special category — *D-S* explanations — but their kinship with the official cases of D-N explanation suggests that we should broaden the D-N category to include them (see Salmon, 1984 and this volume). Finally, *I-S* explanations are those explanations in which the argument is inductive and the conclusion a singular sentence to which the premises assign high probability.

The motivation for approaching explanation in this way stems from the character of the explanations given in scientific works, particularly in those texts that are intended to introduce students to the main ideas of various fields. Expository work in physics, chemistry, and genetics (and, to a less obvious extent, in other branches of science) often proceeds by deriving descriptions of particular events — or, more usually, descriptions of empirical regularities — from sets of premises in which statements identified as laws figure prominently. Among the paradigms, we may include: the demonstration that projectiles obtain maximum range on a flat plain when the angle of projection is 45°, the Newtonian derivation of Galileo's law of free fall, Bohr's argument to show that the frequencies of the lines in the hydrogen spectrum satisfy the formulas previously obtained by Balmer and others, the kinetic-theoretic deduction of the Boyle-Charles law, computations that reveal the energy required for particular chemical reactions, and the derivation of expected distributions of traits among peas from specifications of the crosses and Mendel's laws. In all these cases, we can find scientific texts that contain arguments that come very close indeed to the ideal form of explanation that Hempel describes.

1.2 Hempel's Problems

There are four major types of objection to the Hempelian approach. The first is the obverse of the motivational point just canvassed. Although we can identify some instances in which full-dress covering-law explanations are developed, there seem to be many occasions on which we accept certain statements as explanatory without any ability to transform them into a cogent derivation of a sentence describing the phenomenon to be explained. This objection, made forcefully in a sequence of papers by Michael Scriven (1959, 1962, 1963), includes several different kinds of case, of which two are especially important for our purposes here. One source of trouble lies in our propensity to accept certain kinds of historical narrative — both in the major branches of human history and in evolutionary studies — as explaining why certain phenomena obtain, even though we are unable to construct any argument that subsumes the phenomena under general laws. Another results from the existence of examples in which we explain events that are very unlikely. Here the paradigm is Scriven's case (later elaborated by van Fraassen) of the mayor who contracts paresis. Allegedly, we hold that the question "Why did the mayor get paresis?" can be answered by pointing out that he had previously had untreated syphilis, despite the fact that the frequency of paresis among untreated syphilitics is low.

A second line of objection to the covering-law models is based on the difficulty in providing a satisfactory analysis of the notion of a scientific law. Hempel is especially forthright in acknowledging the problem (1965, 338). The challenge is to distinguish laws from mere accidental generalizations, not only by showing how to characterize the notion of a projectible predicate (and thus answer the questions raised by Goodman's seminal 1956) but also by diagnosing the feature that renders pathological some statements containing only predicates that are intuitively projectible (for example, "No emerald has a mass greater than 1000 kg.").

The first objection questions the necessity of Hempel's conditions on explanation. The third is concerned with their sufficiency. As Sylvain Bromberger made plain in the early 1960s (see especially his 1966), there are numerous cases in which arguments fitting one of Hempel's preferred forms fail to explain their conclusions. One example will suffice for the present. We can explain the length of the shadow cast by a high object (a flagpole or a building, say) by deriving a statement identifying the length of the shadow from premises that include the height of the object, the elevation of the sun, and the laws of the propagation of light. That derivation fits Hempel's D-N model and appears to explain its conclusion. But, equally, we can derive the height of the object from the length of the shadow, the elevation of the sun and the laws of the propagation of light, and the latter derivation intuitively *fails* to explain its conclusion. Bromberger's challenge is to account for the asymmetry.

A close cousin of the asymmetry problem is the difficulty of debarring Hempelian arguments that appeal to irrelevant factors. If a magician casts a spell over a sample of table salt, thereby "hexing" it, we can derive the statement that the salt dissolved on being placed in water from premises that include the (apparently lawlike) assertion that all hexed salt dissolves on being placed in water. (The example is from Wesley Salmon's seminal 1970; it originally comes from Henry Kyburg [1965]). But, it is suggested, the derivation does not explain why the salt dissolved.

Finally, Hempel's account of statistical explanation was also subject to special problems. One trouble, already glimpsed in the paresis example, concerns the requirement of high probability. Among the guiding ideas of Hempel's account of explanation is the proposal that explanation works by showing that the phenomenon to be explained was to be expected. In the context of the statistical explanation of individual events, it was natural to formulate the idea by demanding that explanatory arguments confer high probability on their conclusions. But, as was urged by both Richard Jeffrey (1969) and Wesley Salmon (1970), this entails a whole class of counterintuitive consequences, generated by apparently good explanations of improbable occurrences. Moreover, the high-probability requirement itself turns out to be extremely hard to formulate (see Hempel 1965 for the surmounting of preliminary difficulties, and Coffa 1974 for documentation of residual troubles). Indeed, critics of Hempel's I-S model have charged that the high-probability requirement can only be sustained by supposing that all explanation is fundamentally deductive (Coffa 1974, Salmon 1984, 52–53).

Even a whirlwind tour of that region of the philosophical landscape occupied by theories of explanation (a region thick with syphilitic mayors, flagpoles, barometers, and magicians) can help to fix our ideas about the problems that an adequate account of scientific explanation must overcome. Contemporary approaches to the subject rightly begin by emphasizing the virtues of Hempel's work, its clarity, its connection with parts of scientific practice, its attention to the subtleties of a broad range of cases. When we have assembled the familiar difficulties, it is appropriate to ask "What went wrong?" The main extant rivals can be viewed as searching for the missing ingredient in the Hempelian approach, that crucial factor whose absence allowed the well-known troubles that I have rehearsed. I shall try to use the four main problem-types to chart the relations among Hempel's successors, and to evaluate the relative merits of the main contemporary rivals.

2. The Pragmatics of Explanation

Not all of the problem-types need be viewed as equally fundamental. Perhaps there was a basic mistake in Hempel's account, a defect that gave rise directly to one kind of difficulty. Solve that difficulty, and we may discover that the remaining troubles vanish. The suggestion is tantalizing, and it has encouraged some important proposals.

One approach is to regard the first type of problem as fundamental. Hempel clearly needed an account of the pragmatics of explanation. As his own detailed responses to the difficulties raised by Scriven (Hempel 1965, 359–64, 427) make entirely clear, he hoped to accommodate the plausible suggestion that narratives can serve an explanatory function even when we have no idea as to how to develop the narrative into an argument that would accord with one of the models. The strategy is to distinguish between what is said on an occasion in which explanatory information is given and the ideal underlying explanation.[1] Although the underlying explanation is to be an argument including laws among its premises, what is said need not be. Indeed, we can provide some information about the underlying argument without knowing all the details, and this accounts for the intuitions of those (like Scriven) who insist that we can sometimes say explanatory things without producing a fully approved Hempelian argument (or without knowing much about what the fully approved argument for the case at hand would be).

Instead of backing into the question of how to relate explanations to what is uttered in acts of explaining, we can take the characterization of explanatory acts as our fundamental problem. This strategy has been pursued in different ways by Peter Achinstein and Bas van Fraassen, both of whom believe that the main difficulties of the theory of explanation will be resolved by gaining a clear view of the pragmatics of explanation. Because van Fraassen's account introduces concepts that I take to be valuable to any theory of explanation, I shall consider his version.[2]

2.1 Van Fraassen's Pragmatics

Van Fraassen starts with the claim that explanations are answers to why-questions. He proposes that why-questions are essentially contrastive: the question "Why P?" is elliptical for "Why P rather than P^*, P^{**}, . . . ?" In this way he can account for the fact (first noted in Dretske 1973 and further elaborated in Garfinkel 1981) that the same form of words can pose different contrastive why-questions. When Willie Sutton told the priest that he robbed banks because that is where the money is, he was addressing one version of the question "Why do you rob banks?," although not the one that the priest intended.

With this in mind, van Fraassen identifies a why-question as an ordered triple $<P_k,X,R>$. P_k is the topic of the question, and an ordinary (elliptical) formulation of the question would be "Why P_k?" X is the contrast class, a set of propositions including the topic P_k. Finally R is the relevance relation. Why-questions arise in contexts, where a context is defined by a body of background knowledge K. The questions have presuppositions: each why-question presupposes that its topic is the only true member of the contrast class (intuitively, the question "Why P_k in contrast to the rest of X?" is inappropriate if P_k is false or if some other member of the contrast class is true), and also that there is at least one true proposition A that stands in the relation R to $<P_k,X>$. A why-question arises in a context K provided that K entails that the topic is the only true member of the contrast class and does not entail that there is no answer to the question (more exactly, that there is no true A bearing R to $<P_k,X>$).

Van Fraassen recognizes that the theory of explanation ought to tell us when we should reject questions rather than attempting to answer. His pragmatic machinery provides a convincing account. We reject the why-question Q in context K if the question does not arise in this context, and, instead of trying to answer the question, we offer corrections. If Q does arise in a context, then a direct answer to it takes the form "Because A," where A is a true proposition that bears R to $<P_k,X>$. The proposition A is the core of the direct answer.

2.2 Why Pragmatics Is Not Enough

Because he hopes to avoid the tangles surrounding traditional approaches to explanation, van Fraassen places no constraints on the relations that can serve as relevance relations in why-questions. In consequence, his account of explanation is vulnerable to trivialization. The trouble can easily be appreciated by noting that it is *prima facie* possible for any true proposition to explain any other true proposition. Let A, B both be true. Then, given van Fraassen's thesis that explanations are answers to why-questions, A will explain B in context K provided that there is a question "Why B?" that arises in K for which A is the core of a direct answer. We construct an appropriate question as follows: let $X = \{B, -B\}$, $R = \{<A, <B,X> >\}$. Provided that K entails the truth of B and does not contain any false proposition entailing the nonexistence of any truth bearing R to $<B,X>$, then the question $<B,X,R>$ arises in K, its topic is B, and its only direct answer is A.

Wesley Salmon and I have argued (Kitcher and Salmon 1987) that van Fraassen's account cannot avoid this type of trivialization. We diagnose the absence of constraints on the relevance relation as the source of the trouble. Intuitively, genuine why-questions are triples $<P_k,X,R>$ where R is a genuine relevance relation, and a large part of the task of a theory of explanation is to characterize the notion of a genuine relevance relation.

In fact, van Fraassen's account of the pragmatics of explanation can be used to articulate the Hempelian approach, and the articulation enables us to see how many problems remain to be resolved. Following Railton's development of Hempel's own embryonic pragmatics, let us suppose that acts of explanation provide information about underlying ideal explanatory texts. We say that $<P_k,X,R^*>$ is an *ideal* Hempelian why-question just in case R^* is the set of pairs $<A, <B,C> >$ such that (i) C is a finite set of propositions, one of which is B, (ii) A entails $B\&D$ where D is the conjunction of the negations of the remaining members of C, (iii) A contains at least one general law, and (iv) the general law is essential for the derivation of $B\&D$. An answer to an ideal Hempelian why-question is a D-N explanans for an explanandum of form $B\&D$ where B is the topic of the question and D the conjunction of the negations of the remaining members of the contrast class.[3] Now suppose that *actual* episodes of explanation involve why-questions that are *actually* answered by providing something that falls far short of an answer to an ideal why-question. Following Railton, we view the actual answers as providing information about the answer to the ideal question. So, let us say that $<P_k,X,R>$ is a genuine why-question in context K just in case there is an ideal why-question with topic P_k and contrast class X that arises in context K such that there is an answer to $<P_k,X,R>$ that would provide those who accept K with information about an answer to the ideal why-question. Hempelian explanation-sketches are answers to genuine why-questions that do not answer the associated ideal questions.

Van Fraassen contends that his pragmatic approach to explanation solves the problem of asymmetry that arises for the Hempelian account. His solution consists in showing that there is a context in which the question "Why is the height of the tower h?" is answered by the proposition that the length of the shadow cast by the tower at a certain time of day is s. That proposition answers the question by providing information about the intentions of the builder of the tower. Thus it has seemed that van Fraassen does not touch the Hempelian problem of distinguishing the explanatory merits of two derivations (both of which satisfy the conditions of the D-N model), and that the claim to have solved the problem of asymmetry is incorrect (see Salmon 1984, 95, and Kitcher and Salmon 1987, for arguments to this effect).

The previous discussion enables us to say more precisely why van Fraassen bypasses the real problem of asymmetry. An ideal answer to the ideal why-question "Why was the height of the tower h?" (with some appropriate contrast class) would derive a description of the height of the tower from premises about the plans of the builder, the effectiveness of those plans, and the stability of the resulting structure. If indeed considerations of shadow length figured in the builder's plans, then it is no surprise to learn that there is a genuine why-question with the same topic and contrast class that can be answered by citing the length of the shadow. Given the right context, in which we know that shadow length was important to the builder in planning the tower, we can use the proposition about shadow length to gain information about the ideal answer to the ideal question. None of this touches the issue of why there is not an *ideal* answer to the ideal question that consists of the conjunction of the ascription of shadow length, the specification of the elevation of the sun, and the laws of the propagation of light. Translating Hempel's approach into van Fraassen's idiom, we have construed ideal Hempelian why-questions in terms of the relation R^*, and R^* allows answers that are intuitively unsatisfactory. Thus the problem of asymmetry re-emerges in van Fraassen's framework because (a) there is a serious problem of characterizing the genuine why-questions (and, correspondingly, the genuine relevance relations) and (b) the Hempelian constraints are just as inadequate in coping with this problem as they were in solving the asymmetry problem in its original guise. (*A fortiori*, an approach like van Fraassen's, that imposes no constraints on genuine relevance relations, inherits all the failures of the Hempelian approach with respect to sufficiency and more besides.)

I suggest that van Fraassen's illuminating discussion of why-questions is best seen not as a solution to all the problems of the theory of explanation, but as a means of tackling the problems of the first type (see section 1). Given solutions to the difficulties with law, asymmetry, irrelevance, and statistical explanation, we could embed these solutions in van Fraassen's framework, and thus handle the general topic of how to relate idealized accounts of explanation to the everyday practice of answering why-questions. This is no small contribution to a theory of explanation, but it is important to see that it cannot be the whole story.

2.3 Possible Goals for a Theory of Explanation

Van Fraassen's work also enables us to see how to concentrate the three residual problems that arise for Hempel's account into one fundamental issue. The central task of a theory of explanation must be to characterize the genuine relevance relations, and so delimit the class of genuine why-questions. To complete the task it will be necessary to tackle the problems of asymmetry and irrelevance, to understand the structure of statistical explanations, and, if we suppose that genuine relevance involves lawlike dependence, to clarify the concept of law.[4] However, the formulation of the task is ambiguous in significant respects. Should we suppose that there is a single set of genuine relevance relations that holds for all sciences and for all times? If not, if the set of genuine relevance relations is different from science to science and from epoch to epoch, should we try to find some underlying characterization that determines how the different sets are generated, or should we rest content with studying a particular science at a particular time and isolating the genuine relevance relations within this more restricted temporal and disciplinary area?[5]

It appears initially that Hempel sought a specification of the genuine relevance relations that was time-independent and independent also of the branch of science. However, in the light of our integration of Hempel's approach with van Fraassen's treatment of why-questions, I think we can achieve a more defensible view of the Hempelian task. Plainly, the set of ideal relevance relations (or of ideal why-questions) may be invariant across times and sciences, even though different actual questions become genuine in the light of changing beliefs. Thus one conception of the central problem of explanation — I shall call it the *Hempelian conception* — is the question of defining the class of genuine relevance relations that occur in the ideal why-questions of each and every science at each and every time. We can then suppose that variation in the why-questions arises partly from differing beliefs about which topics are appropriate, partly from differing views about the character of answers to underlying ideal why-questions, and partly from differing ideas about what would yield information about those answers.

An illustration is provided by the changing attitudes toward functional/teleological questions in biology. Consider the form of question "Why do O's have P?"

Posed in a pre-Darwinian context, we might explicate the question in terms of a contrast class including propositions that ascribe different properties to the organisms and a relevance relation that relates propositions of the form "P enables O's to do X and *doing* X promotes the welfare of O's by bringing benefit B" to the ordered pair of topic and contrast class. We can understand the legitimacy of the question in terms of an associated why-question whose ideal answer presents the historical causes of O's being as they are. In a context in which it is assumed that the history traces back to the planning of an omniscient, omnipotent, and benevolent creator, the relevance relation that figures in the functional/teleological why-question will appear appropriate because the proposition about the benefits brought by having P will yield information about the nature of the creator's intentions. Even though this defense is undermined after Darwin, the functional/teleological questions can still remain as legitimate. Once again, we assume that the ideal answer to the associated ideal why-question will trace the history of the origins of O's. Now, however, we may assume that the having of P either originated or is maintained by natural selection.[6] The functional/teleological question again seems appropriate because the specification of benefits to the organism provides information about the selection pressures originating or maintaining the trait. However, it is easy to understand why functional/teleological questions disappeared from some areas of science in which they had previously seemed appropriate. Without the underlying idea that the features to be explained were directly designed or formed/maintained through a selection process, there is no obvious way to regard the specification of their beneficial effects as providing information about the ideal causal history. Hence we can see how functional/teleological why-questions lapsed in some areas of science, while, in others, they survived with an altered conception of the relationship to issues of causation.[7]

Because philosophical attention to the history of science has exposed numerous important shifts in methodological ideals, the Hempelian conception of the theory of explanation may seem far too ambitious and optimistic. However, one way to respond to claims about shifting standards is to argue that there are overarching principles of *global* methodology that apply to all sciences at all times. As particular scientific fields evolve, the principles of global methodology are filled out in different ways, so that there are genuine modifications of *local* methodology.[8] The version of the Hempelian conception that I have just sketched assigns to global methodology a characterization of ideal why-questions. Shifts in admissible why-questions, corresponding to changes in local methodology, can occur against the background of constancy in the underlying ideals—witness my brief discussion of functional/teleological questions.

Perhaps this picture makes the Hempelian conception somewhat less at odds with current thinking about the modification of methodology in the history of science. But can anything positive be said in favor of that conception? I believe it can. The search for understanding is, on many accounts of science, a fundamental goal of the enterprise. That quest may take different forms in different historical and disciplinary contexts, but it is tempting to think that there is something that underlies the various local endeavors, something that makes each of them properly be seen as a striving after the same goal. The Hempelian conception proposes that there is an abstract conception of human understanding, that it is important to the development of science, and that it is common to the variety of ways in which understanding is sought and gained. Scientific explanations are intended to provide objective understanding of nature. The task of characterizing the ideal notions of explanation, why-question, and relevance is thus one of bringing into focus one of the basic aims of science.

I do not suppose that these remarks provide any strong reasons for thinking that the Hempelian conception is correct. It might turn out that there is nothing but ritual lip movements in the avowal of explanation as an aim of the sciences. Nonetheless, there is an obvious motivation for pursuing the Hempelian conception, for, if it is correct, then we can hope to obtain some insight into the rationality and progressiveness of science. Since I know of no conclusive reasons for abandoning my preferred version of the conception, I propose to consider theories of explanation that undertake the ambitious task of characterizing the ideal relevance relations. More modest projects can come once ambition has failed.

3. Explanation as Delineation of Causes

There are two main approaches to explanation that can be seen as undertaking the project just outlined. One of these can be motivated by considering the problems of asymmetry. Intuitively, the length of the shadow cast by a flagpole is causally dependent on the height of the flagpole, but the height is not causally dependent on the shadow-length. Thus we arrive at the straightforward proposal that Hempel's failure to solve problems of asymmetry (and irrelevance) stems from the fact that causal notions are avoided in his analyses. Diagnosis leads quickly to treatment: genuine relevance relations are causal relations, explanations identify causes.

Of course, the invocation of causal notions has its costs. Hempel's account of explanation was to be part of an empiricist philosophy of science, and it could therefore only draw on those concepts that are acceptable to empiricists. If causal concepts are not permissible as primitives in empiricist analyses, then either they must be given reductions to empiricist concepts or they must be avoided by empiricists. Hempel's work appears to stand in a distinguished tradition of thinking about explanation and causation, according to which causal notions are to be understood either in terms of the concept of explanation or in terms of concepts that are themselves sufficient for analyzing explanation. Empiricist concerns about the evidence that is available for certain kinds of propositions are frequently translated into claims about conceptual priority. Thus, the thesis that we can only gain evidence for causal judgments by identifying lawlike regularities generates the claim that the concept of law is prior to that of cause, with consequent dismissal of analyses that seek to ground the notion of law in that of cause.

One of Hume's legacies is that causal judgments are epistemologically problematic. For those who inherit Hume's theses about causation (either his positive or his negative views) there are obvious attractions in seeking an account of explanation that does not take any causal concept for granted. A successful analysis of explanation might be used directly to offer an analysis of causation — most simply, by proposing that one event is causally dependent on another just in case there is an explanation of the former that includes a description of the latter. Alternatively, it might be suggested that the primitive concepts employed in providing an analysis of explanation are just those that should figure in an adequate account of causation.

Because the invocation of causal dependency is so obvious a response to the problems of asymmetry and irrelevance, it is useful to make explicit the kinds of considerations that made that response appear unavailable. One central theme of the present essay is that there is a tension between two attractive options. Either we can have a straightforward resolution of asymmetry problems, at the cost of coming to terms with epistemological problems that are central to the empiricist tradition, or we can honor the constraints that arise from empiricist worries about causation and struggle to find some alternative solution to the asymmetries. The two major approaches to explanation respond to this tension in diametrically opposite ways. As we may anticipate, the central issues that arise concern the adequacy of proposed epistemological accounts of causation and of suggestions for overcoming problems of asymmetry and irrelevance without appealing to causal concepts.

Before we pursue these questions, it will help to have a more detailed view of both approaches. The remainder of this section will be devoted to the causal approach. I shall examine its rival in section 4.

3.1 Causal Why-Questions and Causal Explanations

Let us start with the explanation of particular facts and events. We restrict our attention to singular propositions that describe particular facts and events — paradigmatically such things as the mayor's having paresis or an electron's tunneling through a potential barrier. Call any why-question that has a singular proposition as its topic a *singular* why-question. An admissible contrast-class for a singular why-question is a set of propositions, among which is the topic of the why-question, such that the propositions exhibit *homogeneous variation* with respect to some property or object described in the topic. A set of propositions exhibits homogeneous variation just in case each pair of propositions in the set has some common constituent (property or object) and the common constituents of any two propositions in the set are the same. This restriction of contrast-classes is intended to permit such classes as {"Sutton robbed the bank," "Sutton robbed the grocery store," "Sutton robbed the church"} and {"Sutton robbed the bank," "Sutton worked as a mechanic," "Sutton worked as a neurosurgeon"} but to debar such bizarre collections as {"Sutton robbed the bank," "Shakespeare died in 1916," "Babe Ruth hit more home runs than Hank Aaron"}.

An ideal singular why-question is a triple $<P,X,R>$ where P is a singular proposition, X is an admissible contrast-class (including P), and R is the relation of *complete causal relevance*. This relation obtains between A and $<P,X>$ just in case A is the proposition expressing all and only the causal information relevant to the truth of P and the falsity of the other members of X. Intuitively, A tells the complete causal story about why P is true and the other members of the contrast-class are false. Although there may be room for skeptical worries about the coherence of the notion of a complete causal story, I shall suppose that there is indeed some proposition that relates that part of the history of the universe culminating in the obtaining of the state decribed by P and in the nonobtaining of the states described by the other members of the contrast-class.[9]

Actual why-questions are rarely aimed at eliciting the entire causal history underlying a particular event or state. Those who ask why typically want to elicit a certain type of information about the causal antecedents of the event/state described in the topic. So I shall take an actual singular why-question to be a triple $<P, X, R>$ where P is a singular proposition, X an admissible contrast-class, and R a relation of *particular* causal relevance. Each particular causal relevance relation R is associated with some condition C such that A bears R to $<P,X>$ just in case (i) A is a logical consequence of the proposition A^* that bears the relation of complete causal relevance to $<P,X>$, (ii) A satisfies C, and (iii) A is not a logical truth. I shall assume that no restrictions need to be placed on the appropriate associated conditions C. In effect, this allows for genuine actual why-questions that are directed at eliciting information about any aspect of the causal history behind an event or state. Typically, we focus on some temporal period that interests us and on some state that obtained (or some event that occurred) during this period. There is an actual why-question, whose topic describes the height of the tower, that is answered by citing the length of the shadow because it is pertinent to inquire about the intentions of the builder at the time the tower was designed. More formally, we ask for a consequence of the complete causal history that describes the desired effect that a tower of the given height would have.

As I have emphasized, an obvious virtue of the causal approach is that it handles problems of asymmetry and irrelevance. How does it fare with respect to the other difficulties that beset the Hempelian account? Some of the objections raised by Scriven evaporate immediately once we have an adequate pragmatics of explanation (as indeed Hempel pointed out in his own seminal discussions of them). Others require us to make explicit a point about causal determination.

Consider the case of the mayor's paresis. If we inquire why the mayor, rather than some one of the other townspeople, contracted paresis, then an adequate answer may be to point to the mayor's prior history of untreated syphilis. Here, the topic of our question is the proposition that the mayor contracted paresis, the contrast-class contains propositions ascribing paresis to the other (more fortunate) citizens, and the relevance relation is a particular relation of causal relevance based on the associated condition of describing some antecedent feature of the mayor's medical state that distinguishes him from the other townsfolk. *It is not required that identification of that prior state should enable us to deduce that the mayor later contracted paresis, or to infer with high probability that he would later contract paresis, or even that it should raise the prior probability of his contracting paresis.* There is no demand that answers should make the topics of the question more expectable than they were previously.[10]

It is not hard to see that the causal approach bypasses the problem of providing an analysis of scientific laws. While proponents of the approach may believe that, in general, complete causal histories will mention laws of nature — or even that the structure of these histories may sometimes accord with the requirements of the covering-law models — it is not incumbent on them to provide an analysis of the notion of law, at least not for the purposes of giving an account of scientific explanation. Moreover, they may even remain agnostic with respect to the question whether all ideal explanations (ideal answers to why-questions) involve general laws.

The causal approach also has little difficulty in overcoming the problems that confronted Hempel's account of statistical explanation. As already indicated in the discussion of the paresis case, the approach has no commitment to the idea that there is any statistical relationship between the information given in an answer to a why-question and the topic (or ordered pair of topic and contrast-class) of the why-question. What is important is the provision of causal information, and once this has been achieved the effects on the probability of the topic (or on other members of the contrast class) is of no concern.[11]

These successes for the causal approach are impressive, and they provide some motivation for considering whether the traditional empiricist arguments that dissuaded Hempel (and others) from using causal concepts in the analysis of explanation are cogent. We shall pursue this question in some detail below. For the moment, however, it is necessary to examine whether or not the causal approach is vulnerable to different kinds of objection.

3.2 Are There Noncausal Explanations of Singular Propositions?

One important worry about the causal approach is that, as I have so far characterized it, it is restricted to giving an account of the explanation of singular propositions. Before we try to remedy this deficiency, it is useful to consider whether there are some cases in which singular propositions are explained in noncausal ways. More exactly, we should ask if it is always true that the ideal answer to an ideal question with a singular topic is the associated complete causal history. The doubt divides into two parts: (i) are there singular propositions that give rise to why-questions and that can be explained by answering those why-questions, but which describe phenomena that do not have causal histories?; (ii) are there singular propositions that describe phenomena which have causal histories and which give rise to why-questions, but which are ideally explained without relating the complete causal histories?

There are areas of inquiry, formal sciences, in which investigators appear to put forward explanations despite the fact that the phenomena to be explained do not have causes. Two obvious examples are formal syntax and pure mathematics. Explanations of the grammaticality or ungrammaticality of particular strings in particular natural languages are given by identifying the constraints set by the underlying rules of syntax.[12] Note that it will not do to suggest that the formal explanation is a placeholder for a description of causal processes that occur in the brains of speakers—for part of the point of the enterprise is to distinguish between explanations of competence and explanations of performance.

I shall not pursue the examples from formal linguistics, because it seems to me that, in the case of mathematics there are several instances in which we can distinguish between the explanatory worth of arguments that yield a particular theorem, even though we cannot claim that one of the arguments provides insight into the *causes* of the fact reported in the theorem. Hence, if this is indeed correct, the problems that provoke the introduction of causal concepts in the theory of explanation for singular propositions about physical states of affairs (paradigmatically problems of asymmetry and irrelevance) have counterparts in domains to which causal notions are inapplicable.

Before examining some examples, I shall try to forestall an objection, forcefully presented to me by Paul Humphreys. Can we legitimately talk about explanation in mathematics, and, even if we can, should we suppose that attention to mathematical cases will indicate anything about *scientific* explanation? I reply to the skepticism that provokes the question in two ways. First, given my own views of the nature of mathematics (see Kitcher 1983, 1987b, 1988) mathematical knowledge is similar to other parts of scientific knowledge, and there is no basis for a methodological division between mathematics and the natural sciences. Second, the importance of mathematical explanation in the growth of mathematical knowledge is appreciated not only by philosophers who do not share my heterodoxies (see, for example, Steiner 1978) but also by mathematicians when they are engaged in critical discussion of specific mathematical reasoning. The examples that follow are intended to show that proofs and axiomatizations are, at least sometimes, assessed for their explanatory merits, and it seems to me that the burden of argumentation is on skeptics who wish to campaign for some sort of methodological dualism.

A. The theorem that a function with derivatives of all orders that takes values of opposite sign at the endpoints of an interval has a zero within the interval (the *intermediate zero theorem*) can be justified as follows. Any function meeting the conditions given can be represented as a smooth curve, in the intuitive sense of a curve that can be drawn without lifting the pencil from the paper. Since the curve must lie below the axis at one end of the interval and above the axis at the other end of the interval, there must be a point within the interval at which it crosses the axis. This point corresponds to a zero of the function. Bolzano complained that this argument inverts the true order of the sciences (see Kitcher 1975 for exegesis), and he set out to find a more satisfactory derivation. While Bolzano recognized that the appeal to the properties of smooth curves produces conviction *that* the theorem is true, he contended that it fails to identify "the reason for the fact" (in Aristotle's famous terminology), and he suggested that the failure stems from the use of considerations that are extraneous to claims about numbers and functions. Instead, he sought, and Weierstrass later completed, a proof that would identify the properties of the real line on which the intermediate zero theorem depends. I suggest that the problem posed by Bolzano is a mathematical analogue of the irrelevance problem for explanations in the natural sciences, and that we cannot diagnose the flaw in the intuitive geometrical argument by suggesting that geometrical considerations are *causally* irrelevant to truths about numbers and functions. Indeed, Bolzano's insight is that there is a broader notion of objective dependency to which correct explanations must conform—an insight that derives ultimately from Aristotle.

B. It is possible to axiomatize the theory of finite groups in a number of different ways. The standard approach takes a group to be a finite set which is closed under an associative operation, *multiplication*. There is an idempotent element, *1*, such that, for any element a of the group $a1 = 1a = a$; and, for any element a there is an inverse, a^{-1} such that $a^{-1}a = aa^{-1} = 1$. On this basis one can prove that division is unique wherever it is possible. Alternatively, a finite group can be identified as a finite set, closed under an associative operation, *multiplication*, such that division is unique wherever it is possible. On this basis, one can show that finite groups have all the properties attributed in the usual axioms. Mathematicians distinguish the two axiomatizations—if only in their practice of choosing the standard axioms—and they sometimes express the preference by saying that the usual axioms are more "natural" than the nonstandard ones, or that the division property is "less fundamental." I suggest that what they are recognizing is a case of the asymmetry problem: we can explain why finite groups satisfy the division property by using the axioms about the existence of inverses and idempotent elements to demonstrate that division is unique wherever it is possible. But the derivation of the existence of an idempotent element and of inverses from the division property is nonexplanatory, and, I think, nonexplanatory in just the same way as the derivation of heights from shadowlengths. If this is correct, then the example reveals that the asymmetry problem can arise in cases where causal considerations are quite beside the point. Moreover, in this case and in that discussed in A, it is not hard to see a reason for the distinguishing of the derivations: the preferred derivation can be generalized to achieve more wide-ranging results. Thus, by using the ideas developed by Bolzano, Weierstrass, and Dedekind, we can show that the intermediate zero theorem holds in the much weaker case in which the function is merely continuous. Similarly, if we drop the restriction to *finite* groups, we can show that any group satisfying the standard axioms has the division property, but it is not the case that any set closed under an associative operation with the division property has an idempotent element and inverses (that claim need not hold when the set is infinite). In both instances, the explanatory derivation is similar to derivations we could provide for a more general result; the nonexplanatory derivation cannot be generalized, it applies only to the local case. I shall try to show later how this diagnosis can be made more precise, and how we can use it to offer a picture of explanation that differs from the causal approach.

C. There are numerous classes of equations in one variable for which we can specify solutions as rational functions of the coefficients. This is trivial in the case of the class of linear equations ($ax + b = 0$), and familiar in the case of quadratic equations ($ax^2 + bx + c = 0$). It also holds for cubic equations and for quartic equations. To solve the general cubic, $ax^3 + bx^2 + cx + d = 0$, it suffices to note that setting $x = z - b/3a$, generates an equation of the form

(*) $pz^3 + qz + r = 0$

where p, q, and r are all rational functions of a, b, and c. If we now set $z = y - q/3py$, and substitute in (*), we obtain an equation of the sixth degree whose coefficients are rational functions of p, q, r, (and therefore of a, b, and c), and which is a quadratic in y^3. Here we can apply the ordinary formula for quadratic equations to obtain an expression for y, and the expression of z, and x, follows in two easy steps.

Since a similar trick works for the quartic equation, we can show, for each class of polynomial equations up to and including degree 4 that the roots can be written as rational functions of the coefficients. So much was appreciated by the end of the eighteenth century (in fact, even earlier), but mathematicians concerned with the theory of equations — most notably Lagrange — believed that the mere ability to provide the derivations alluded to in the last paragraph does not show us why equations in these classes permit expression of the roots as rational functions of the coefficients. Some insight into the structure behind the phenomenon is needed, and this was provided partially by Lagrange's investigations of the effects of permutations of the roots on various functions in the roots, ultimately in Galois's development of the theory that bears his name. After Galois, we have a criterion for the expressibility of roots as rational functions of the coefficients, to wit the solubility of the Galois group of the equation, and we can see just why this applies in the four special cases. Nonexplanatory special derivations give way to an explanatory proof drawn from a general theory about the properties of classes of equations.

I turn now to the second type of concern about the causal approach to the explanation of singular propositions. As we shall see, this worry has affinities with the complaint made in C above, namely that why-questions are frequently posed in a search for theoretical understanding. In the case described in C, there was no issue of finding causal histories of the phenomena to be explained. However, even when causal histories are available, they may not be what the explanation requires. Two examples will illustrate the moral.

D. There is a party trick in which someone "knots" a telephone cord around a pair of scissors. In fact, no genuine knot is produced, and the scissors can easily be removed (and the cord returned to its standard configuration) if the victim makes a somewhat unobvious twist at the start. Those who do not make the right initial twist can struggle for hours without getting anywhere. What explains their failure? In any such case, we could, of course, provide the causal details, showing how the actions actually performed lead to ever more tangled configurations. But this appears to omit what should be central to the explanation, namely the fact that the topological features of the situation allow for disentanglings that satisfy a specifiable condition, so that sequences of actions which do not satisfy that condition are doomed to failure. We need to know the topological structure that lies behind the vicissitudes of the particular attempt and the particular failure.[13]

E. We discover that, for a particular city, over a period of a century, the sex-ratio at birth, combined over all the hospitals, is always very close to 1.04 to 1, with males being more common. There is a complete causal history underlying this fact: it involves vast numbers of details about the production of sperm and eggs, circumstances of mating, intra-uterine events, and so forth. However, in explaining the sex-ratio, we do not want any of this information. Instead, it suffices to point out that there are selection pressures on individuals of *Homo sapiens* that result in the approximate attainment of a 1–1 sex ratio at reproductive age, and that higher male mortality between birth and reproduction requires a natal sex-ratio of 1.04 to 1. Having a 1:1 sex-ratio at reproductive age is an evolutionary equilibrium for a species like ours, and we explain demographic data from a large local population by showing how they approximate the evolutionary equilibrium.[14]

In both instances, the causal approach seems to err by overlooking the fact that the particular phenomenon to be explained is one example of a class, all of whose members instantiate a general regularity. Post-Hempelian philosophers of science have sometimes delighted in attacking the covering-law models by noting that the mere citation of a covering law is often a very poor explanation: if we hope to explain why a particular sample of copper expanded when heated, then we gain very little from the statement that all samples of copper expand when heated. But, in examples D and E, the identification of the regularity is an ingredient in the explanation. We explain a victim's frustration with the telephone cord by identifying the topological features of the "knot," and noting that only certain kinds of actions will produce the desired result. We explain the birth sex-ratio by offering a general claim about sex-ratios in large populations of *Homo sapiens*, and then going on to explain why that claim holds. To accommodate both the features of these examples and the legitimacy of the complaint about the triviality of providing covering laws, I suggest that singular why-questions are often concerned to relate the phenomenon described by the topic to other similar phenomena, rather than to fathom the causal details of a particular situation. In some cases, the delineation of a class of instances in which similar things happen is only the first stage in explanation, for the original question was concerned with why the phenomenon is to be found throughout the class. In such cases, the intent of the question is implicitly general and we could say that, while the *apparent* topic of the question is singular, the *real* topic concerns a regularity. Hence, merely stating the regularity is explanatorily worthless and the poverty of the simplest forms of covering-law derivation can easily be understood. However, there are other occasions on which the identification of a phenomenon as belonging to a class—typically defined in terms of some language that does not occur in the initial posing of the question—suffices to explain it. The failure to untangle the telephone cord is explained by using topological notions to characterize the initial configuration of the wire. Or the delineation of the class can be a preliminary step in explanation, as when we formulate the general regularity about population sex-ratios in *Homo sapiens* and then proceed to the evolutionary explanation.

The negative point of these examples is that the account of singular explanation needs to be amended to allow that explanations need not, and sometimes should not, deliver information about the causal history of a particular occurrence. The positive point concerns the penetration of singular explanation by theoretical explanation. It might be tempting to believe that the explanation of singular propositions can be studied autonomously without worrying about the character of theoretical explanation. Examples D and E are intended to show that this is an illusion. Even when we are interested in explaining a particular event or state, the explanation we desire may well be one that would also explain something quite general, and any attention to the local details may be misguided and explanatorily inadequate. The same point is made in the mathematical context by example C.

Proponents of the causal approach can respond by revising their account of genuine relevance relations and genuine why-questions. For example, the conditions on genuine relevance relations can now be viewed as disjunctive allowing either for answers that provide information about the complete causal history or for answers that provide information about constraints on causal processes of a particular kind, to wit processes that generate phenomena similar (in some specified way) to the phenomenon reported in the topic. The task of working this out in full detail is not trivial, but, even were it to be done successfully, there would still be concern about the liberality of the proposal. Although the explanations I have claimed to be preferable in cases like D and E would now be accommodated, the recitations of causal detail would not have been debarred. Whether they should be seems to me to be an interesting open question.

However, the issue of how to integrate an account of causal singular explanation and an account of theoretical explanation (the issue bruited in the last paragraph) appears less fundamental than the problem of providing a characterization of theoretical explanation itself. The plausibility of the causal approach derives chiefly from its handling of singular propositions—as we turn our attention to the explanation of general propositions, the talk of causation comes to seem forced or even inapplicable.

3.3 Causal Explanation and Theoretical Explanation

Theoretical explanation provides some support for the Hempelian idea that explanation is derivation. For, when we consider the paradigms of theoretical explanation, the Newtonian derivation of Kepler's laws, quantum chemical accounts of the propensities of elements for forming compounds, molecular biology's explanation of the copying of genetic material, plate tectonic accounts of the presence of earthquake zones, the derivations found in standard sources seem to provide ideal explanations. Nonetheless, Hempel's account of theoretical explanation is underdeveloped, and with good reason. As Hempel and Oppenheim clearly recognized (1948, note 33), if theoretical explanation is conceived in terms of the derivation of laws using laws as premises, then it is unclear why we cannot explain any law by deriving it from the conjunction of itself with any other law. The obvious response to the difficulty is to say that we explain laws by deriving them from more fundamental laws. If the causal approach to explanation is to be fully developed, it must provide some way of saying what is meant by the intuitive (but murky) thesis that some laws are more fundamental than others.

Consider the quantum mechanical account of the characteristics of the periodic table. At the first (and most informal) step one introduces the idea of discrete energy levels in the atom and applies the Pauli exclusion principle (see, for a classic source, Pauling 1960, 47ff.). This provides an attribution to each element of the pattern of shell-filling around an atomic nucleus, and the sequence reveals the periodicity originally established by Mendeleev. Thus, to cite one example, we discover that the noble gases (helium, neon, argon, and so forth) have in common the property that their outermost electron shells (i.e., the highest energy levels in which electrons occur around the nucleus) are completely filled. Their stability is then interpreted in terms of the lack of opportunity for the formation of ionic bonds (involving transfer of electrons) or covalent bonds (sharing of electrons). The characterizations thus achieved are subsequently obtained by applying the formalism of quantum mechanics to provide a rigorous account of shell-filling, of stability, and of bond-formation.

The generalization that noble gases do not form compounds with other elements is explained by beginning with the laws governing bond formation, deriving the conclusion that bond formation requires complementation with respect to the filling of the outermost electron shells, concluding, as a special case, that when the outermost shells are already filled there are no opportunities for complementation in terms of electron transfer or electron sharing, and finally using the law that noble gases have their outermost shells filled to infer that noble gases do not form compounds with other elements. The derivation can be enriched by treating one (or more) of the generalizations employed in it from a formal quantum-mechanical perspective (this amounts to what I shall call *explanation extension*—see section 4.5 below). How is the explanatory power of the derivation to be understood from the perspective of the causal approach?

For those, like Salmon, who think of the explanation of singular propositions as primary, the obvious response is to suggest that the explanation of regularities involves the identification of mechanisms that are at work in all the cases covered by the regularity. The problem is to make this formulation more precise. What is it for a "mechanism to be at work in an event, state, or process"? What is it for an event, state, or process to be "covered by the regularity"? Let us look at the difficulties involved in clarifying these notions by considering in more detail the example discussed in the last paragraph.

The states and events covered by the law that noble gases are chemically inert are, it would appear, episodes in which samples of noble gas occur in the presence of samples of other elements (or of compounds). On each occasion we can recognize a time at which the noble gas molecules first enter the scene, and we can imagine a complete causal history that traces the interrelations between these molecules and molecules of the other substances that are present. Now we ask what makes the shell-filling properties of the noble gas molecules the fundamental mechanism that accounts for the fact that, in all these episodes, we find no chemical combination between the noble gas sample and the other substances. Those properties are, to be sure, one aspect of all the causal histories—as is the fact that no bonds form—but it is hard to see what distinguishes it as crucial to the explanation.

The obvious response is to claim that we have a general account of what occurs when elements (or compounds) *do* combine, an account that makes reference to electron transfer, electron sharing, and shell-filling. This general account reveals to us that the shell-filling properties of the noble gases operate as a constraint on the causal processes that can occur when they are brought into the presence of other substances. What counts as a fundamental mechanism is not dictated by the local details of the individual causal processes that occur in these episodes. Rather, the fundamental mechanisms are disclosed by our most general theoretical accounts of a range of regularities. The task of understanding the explanatory adequacy of these general accounts thus remains as a presupposition of the causal approach, and I do not see that it can be completed in terms of the local studies of individual cases of causation that the causal approach takes as primary.

In an earlier essay (Kitcher 1985c), I contrasted "top down" and "bottom up" approaches to explanation. That contrast resurfaces in the present context. Top down approaches will attempt to provide an account of what theoretical explanation is, use this as a basis for underwriting talk about "fundamental mechanisms," and so proceed toward the identification of causes in particular cases. Bottom up approaches view us as having the ability to discern causal relations in specific episodes, and see theoretical explanation as stitching together results about the causation of individual states and events. The considerations I have just advanced are attempts to show that the project of proceeding from singular explanation to theoretical explanation is more problematic than one might have thought. Those considerations obtain greater force in light of the earlier point that theoretical explanation penetrates singular explanation. Finally, waiting in the wings is the complaint that explanation can go forward in areas of discourse in which causal notions are inapplicable and that these areas sometimes order the phenomena in ways that have similar features to the explanatory orderings of the natural sciences.

Nonetheless, as I emphasized earlier in this section, the causal approach has obvious merits. Does it have any serious rival? Let us see.

4. Explanation as Unification

On both the Hempelian and the causal approaches to explanation, the explanatory worth of candidates—whether derivations, narratives, or whatever—can be assessed individually. By contrast, the heart of the view that I shall develop in this section (and which I shall ultimately try to defend) is that successful explanations earn that title because they belong to a set of explanations, the *explanatory store*, and that the fundamental task of a theory of explanation is to specify the conditions on the explanatory store. Intuitively, the explanatory store associated with science at a particular time contains those derivations which collectively provide the best systematization of our beliefs. Science supplies us with explanations whose worth cannot be appreciated by considering them one-by-one but only by seeing how they form part of a systematic picture of the order of nature.

4.1 The Ideal of Unification

All this is abstract and somewhat metaphorical. To make it more precise, let us begin with the proposal that *ideal* explanations are derivations. Here there is both agreement and disagreement with Hempel. An argument can be thought of as an ordered pair whose first member is a set of statements (the premises) and whose second member is a single statement (the conclusion). Hempel's proposal that explanations are arguments appears to embody this conception of arguments as *premise-conclusion* pairs. But, on the systematization account, an argument is considered as a derivation, as a sequence of statements whose status (as a premise or as following from previous members in accordance with some specified rule) is clearly specified. An ideal explanation does not simply list the premises but shows how the premises yield the conclusion.

However, the systematization approach retains the Hempelian idea that to explain a phenomenon is to produce an argument whose conclusion describes the phenomenon, and this would appear to founder on the difficulties adduced by Jeffrey (1969) and Salmon (1984) concerning the explanation of objectively improbable events. I shall postpone discussion of this point to the next section.

For a derivation to count as an *acceptable* ideal explanation of its conclusion in a context where the set of statements endorsed by the scientific community is K, that derivation must belong to the explanatory store over $K, E(K)$. At present, I shall assume that K is both consistent and deductively closed, and that the explanatory store over a set of beliefs is unique. $E(K)$ is to be the set of derivations that best systematizes K, and I shall suppose that the criterion for systematization is unification.[15] $E(K)$, then, is the set of derivations that best unifies K. The challenge is to say as precisely as possible what this means.

We should be clear about just what is to be defined. The set of derivations we are to characterize is the set of explanations that would be acceptable to those whose beliefs comprised the members of K. At this stage, the project does not provide an account of *correct* explanation, and it will be important to remedy that deficiency later. To avoid metaphysical complications, my attempt will be postponed to the final section.

The idea that explanation is connected with unification has had some important advocates in the history of the philosophy of science. It appears to underlie Kant's claims about scientific method[16] and it surfaces in classic works in the logical empiricist tradition (see Hempel [1965] 345, 444; Feigl [1970] 12). Michael Friedman (1974) has provided the most important defense of the connection between explanation and unification. Friedman argues that a theory of explanation should show how explanation yields understanding, and he suggests that we achieve understanding of the world by reducing the number of facts we have to take as brute.[17] Friedman's motivational argument suggests a way of working out the notion of unification: characterize $E(K)$ as the set of arguments that achieves the best tradeoff between minimizing the number of premises used and maximizing the number of conclusions obtained.

Something like this is, I think, correct. Friedman's own approach did not set up the problem in quite this way, and it proved vulnerable to technical difficulties (see Kitcher 1976 and Salmon, this volume). I propose to amend the account of unification by starting from a slight modification of the motivational idea that Friedman shares with T. H. Huxley (see note 17). Understanding the phenomena is not simply a matter of reducing the "fundamental incomprehensibilities" but of seeing connections, common patterns, in what initially appeared to be different situations. Here the switch in conception from premise-conclusion pairs to derivations proves vital. *Science advances our understanding of nature by showing us how to derive descriptions of many phenomena, using the same patterns of derivation again and again, and, in demonstrating this, it teaches us how to reduce the number of types of facts we have to accept as ultimate (or brute).*[18] So the criterion of unification I shall try to articulate will be based on the idea that $E(K)$ is a set of derivations that makes the best tradeoff between minimizing the number of patterns of derivation employed and maximizing the number of conclusions generated.

4.2 Argument Patterns

First we need the notion of pattern. A *schematic sentence* is an expression obtained by replacing some, but not necessarily all, the nonlogical expressions occurring in a sentence with dummy letters. Thus, starting with the sentence "Organisms homozygous for the sickling allele develop sickle-cell anemia," we can generate a number of schematic sentences: for example, "Organisms homozygous for A develop P" and "For all x, if x is O and A then x is P" (the last being the kind of pattern of interest to logicians, in which *all* the nonlogical vocabulary gives way to dummy letters). A set of *filling instructions* for a schematic sentence is a set of directions for replacing the dummy letters of the schematic sentence, such that, for each dummy letter, there is a direction that tells us how it should be replaced. For the schematic sentence "Organisms homozygous for A develop P," the filling instructions might specify that A be replaced by the name of an allele and P by the name of a phenotypic trait. A *schematic argument* is a sequence of schematic sentences. A *classification* for a schematic argument is a set of statements describing the inferential characteristics of the schematic argument: it tells us which terms of the sequence are to be regarded as premises, which are inferred from which, what rules of inference are used, and so forth. Finally, a *general argument pattern* is a triple consisting of a schematic argument, a set of sets of filling instructions, one for each term of the schematic argument, and a classification for the schematic argument.

A particular derivation, the sequence of sentences and formulas found in a scientific work for example, instantiates a general argument pattern just in case: (i) the derivation has the same number of terms as the schematic argument of the general argument pattern, (ii) each sentence or formula in the derivation can be obtained from the corresponding schematic sentence in accordance with the filling instructions for that schematic sentence, (iii) the terms of the derivation have the properties assigned by the classification to corresponding members of the schematic argument. Later in this section I shall offer some examples that are intended to show how this conception of argument patterns functions in actual cases to capture the structure that underlies the derivations put forward within particular scientific fields.

Derivations may be similar either in terms of their logical structure or in terms of the nonlogical vocabulary they employ at corresponding places. The notion of a general argument pattern allows us to express the idea that derivations similar in either of these ways have a common pattern. However, similarity is a matter of degree. At one extreme, a derivation is maximally similar to itself and to itself alone; at the other, any pair of arguments can be viewed as having a common pattern. To capture the notion that one pair of arguments is more similar than another pair, we need to recognize the fact that general argument patterns can demand more or less of their instantiations. If a pattern sets conditions on instantiations that are more difficult to satisfy than those set by another pattern, then I shall say that the former pattern is more *stringent* than the latter.

The stringency of an argument pattern is determined in part by the classification, which identifies a logical structure that instantiations must exhibit, and in part by the nature of the schematic sentences and the filling instructions, which jointly demand that instantiations should have common nonlogical vocabulary at certain places. If both requirements are relaxed completely then the notion of pattern degenerates so as to admit of *any* argument. If both conditions are simultaneously made as strict as possible then we obtain another degenerate case, a "pattern" which is its own unique instantiation. Relaxing the demands on nonlogical vocabulary (the conditions set by the schematic sentences and the filling instructions) while also requiring that the classification determine the precise inferential status of each term in the schematic argument yields the logician's notion of pattern.

Plainly, we make the global problem more tractable if we compare the relative stringency of argument patterns that differ only with respect to their classifications and assess relative stringency where the classifications are the same and the differences are confined to the demands made by the schematic sentences and the filling instructions. The general problem would require us to make comparisons in cases where the classification of P_1 sets more exacting conditions than the classification of P_2 but the asymmetry is reversed with respect to the conditions on substituting dummy letters. In this essay, I shall not try to offer a general account of relative stringency. As we shall discover in section 7, the relevant comparisons often involve only the straightforward cases.

4.3 Systematization of Belief

We want *E(K)* to be the set of arguments that best unifies *K*. Typically, there are many ways of deriving some statements of *K* from others. Call any set of arguments that derives some members of *K* from other members of *K* a *systematization* of *K*. *E(K)* will be the best systematization of *K*.

The initial requirement that must be met by *E(K)* is that all the arguments it contains should be acceptable relative to *K*. Say that a set of derivations is acceptable relative to *K* just in case each step in each derivation is deductively valid and each premise of each derivation belongs to *K*. In considering ways of systematizing *K* we restrict our attention to those sets of derivations that are acceptable relative to *K*. This is an idealization, since we often admit as explanatory arguments that do not accord with our present beliefs: for example, in using Newtonian dynamics to explain trajectories or Mendelian rules to account for a distribution of traits. I shall show how the idealization can be relaxed in the next section.

Continuing with the central ideas of the account, let us define a *generating set* for a set of derivations to be a set of argument patterns such that each derivation in the set instantiates some pattern in the generating set. If we have a set of derivations *D* and a generating set for it *G*, then *G* will be said to be complete with respect to *K* just in case every derivation that is acceptable relative to *K* and which instantiates a pattern in *G* belongs to *D*. In determining the explanatory store *E(K)* we first narrow our choice to those sets of arguments that are acceptable relative to *K*, the acceptable systematizations of *K*. Then we consider, for each such set of derivations, the various generating sets that are complete with respect to *K*. (The importance of the requirement of completeness is to debar explanatory deviants who use patterns selectively. If someone claims that an argument instantiating a particular pattern explains why Mars follows the trajectory it does, admits that there is an acceptable derivation instantiating the same pattern that will yield as its conclusion a description of the trajectory of Venus, but refuses to allow the latter derivation as explanatory, then, I suggest, that person has incoherent views about explanation.) Now, having associated with each acceptable systematization a collection of complete generating sets that generates it, we pick out for each acceptable systematization that member of the associated collection of complete generating sets that does best according to criteria for unification (to be indicated shortly). Call this the *basis* of the systematization in question. Finally, we rank the bases of the acceptable systematizations in terms of their unifying power. *E(K)* is that acceptable systematization whose basis ranks highest.

The intuitive idea behind unification is the generation of as many conclusions as possible using as few patterns. It is also important that the instantiations of the patterns should genuinely be similar, that is, that the patterns in question should be stringent. With this in mind, define the *conclusion* set of a set of derivations *D*, *C(D)*, to be the set of statements that occur as conclusions of some member of *D*. The unifying power of a complete generating set for *D* varies directly with the size of *C(D)*, directly with the stringency of the patterns in the set, and inversely with the number of patterns in the set. As in the case of our discussion of stringency, I shall not explore the ways in which tradeoffs among these factors might be made. I am prepared to allow for the possibility that, with respect to some possible corpora *K*, there might be genuine indeterminacy in deciding how to weigh relative stringency, paucity of patterns and range of conclusions against one another, with consequent indeterminacy about *E(K)*. But I shall argue below that with respect to actual (present or past) sets of beliefs it is possible to use the incomplete criteria I have given to judge the merits of rival systematizations and so resolve the difficulties that problems of asymmetry and irrelevance pose for theories of explanation.

4.4 Why-Questions Revisited

The account that I have offered needs to be integrated with the approach to pragmatic issues given in section 2.1. We can proceed in a fashion parallel to that adopted in articulating the causal approach (3.1), with the difference that there is no need to restrict ourselves to singular why-questions. Using the notion of homogeneous variation sketched in section 3.1, let us say that an admissible contrast-class must be a set of propositions satisfying van Fraassen's conditions and subject to the further demand that all the propositions exhibit homogeneous variation. An ideal why-question acceptable relative to *K* is a triple $<P,X,R>$ where *P* is expressed by some member of *K*, *X* is an admissible contrast-class, and *R* obtains between a sequence of propositions *A* and $<P,X>$ just in case *A* is expressed by a derivation in *E(K)* whose conclusion expresses the conjunction of *P* and the negations of the remaining members of *X*. An actual why-question acceptable relative to *K* is a triple $<P,X,R>$ where *P*, *X* must satisfy the same conditions as before and *R* holds between *A* and $<P,X>$ just in case *A* is a subsequence of a sequence of propositions expressed by a derivation in *E(K)* whose conclusion expresses the conjunction of *P* and the negations of the remaining members of *X*.

To say that a why-question is acceptable relative to *K* is not to imply that other why-questions should be rejected in a context where the background beliefs are the members of *K*. For we ought to allow for the possibility that a why-question might be answered by producing a derivation among whose premises is some proposition (or propositions) that is not expressed by any statement in *K* but which would be rationally accepted by those who believe the members of *K*. It is even possible that why-questions should be answered by derivations instantiating patterns that are not in the basis of *E(K)* but that would be included in the basis of *E(K*)* where *K** would be rationally accepted by those who accept *K* and who recognize the validity of the derivations in question. We want to allow that science should make progress by appreciating new possibilities for explanation. So I shall say that why-questions that are included under these possibilities are acceptable relative to *K in the extended sense.*

4.5 Explanatory Unification and Causal Dependence

So far, I have developed the view of explanation as unification in a rather abstract way, and it may seem rather ethereal by contrast with the causal approach. One obvious attraction of the latter is that it seems to give a compelling diagnosis of the asymmetry and irrelevance problems. As we shall see, the resolution of these problems is a major obstacle that the unification view must try to overcome, but, before we proceed any further, it is worth relating the claims of the unification approach to the idea that the explanatory asymmetries signal causal asymmetries. *For this is not something that a proponent of the unification view ought to deny.* What is distinctive about the unification view is that it proposes to ground causal claims in claims about explanatory dependency rather than vice versa. So we account for the intuition that appeals to shadows do not explain the heights of towers because shadow lengths are causally dependent on tower heights, by suggesting that our view of causal dependency, in this and kindred cases, stems from an appreciation of the explanatory ordering of our beliefs.

How can this be? Surely few people have any explicit knowledge of the explanatory patterns immanent in the practice of scientists, and fewer still could articulate the factors that contribute to unifying power. So the idea that any one individual justifies the causal judgments that he/she makes by recognizing the patterns of argument that best unify his/her beliefs is clearly absurd. However, in claiming that causal dependence is grounded in explanatory dependence, the champions of the unification approach need commit themselves to no such implausible story. Our everyday causal knowledge is gained by absorbing the lore of our community. The scientific tradition has articulated some general patterns of derivation — sometimes explicitly considering how the phenomena within a domain could be unified, sometimes only under the tacit guidance of the methodological directive to use the minimum of patterns in generating the maximum of conclusions. Derivations that accord with these patterns become accepted as explanatory, and the phenomena described in their conclusions are viewed as objectively dependent on the phenomena described in their premises. So there passes into our common ways of thinking, and our common ways of talking, a view of the ordering of phenomena, and this picture of how phenomena are ordered is expressed, often though not invariably, in our recognition of causal dependencies. Thus the picture advanced by the unification approach shows the concept of causal dependence as derivative from that of explanatory dependence, but it does not promote the dubious idea that each of us gains explicit knowledge of causal dependencies through recognition of the structure of the explanatory store.

My claim that the structure of the explanatory store gives rise to views of objective dependence which are *often though not invariably* expressed in our causal discourse signals the ability of the unification approach to accommodate some of the difficulties that we noted at the end of section 3.2. For even in areas of investigation where causal concepts do not apply—such as mathematics—we can make sense of the view that there are patterns of derivation that can be applied again and again to generate a variety of conclusions. Moreover, the unification criterion seems to fit very well with the examples in which explanatory asymmetries occur in mathematics. Derivations of theorems in real analysis that start from premises about the properties of the real numbers instantiate patterns of derivation that can be used to yield theorems that are unobtainable if we employ patterns that appeal to geometrical properties. Similarly, the standard set of axioms for group theory covers both the finite and the infinite groups, so that we can provide derivations of the major theorems that have a common pattern, while the alternative set of axioms for the theory of finite groups would give rise to a less unified treatment in which different patterns would be employed in the finite and in the infinite cases. Lastly, what Lagrange seems to have aimed for is the incorporation of the scattered methods for solving equations within a general pattern, and this was achieved first in his pioneering memoir and later, with greater generality, in the work of Galois.

The fact that the unification approach provides an account of explanation, and explanatory asymmetries, in mathematics stands to its credit, but this may be viewed as too small a benefit to count very heavily in its favor. By contrast, the problem of understanding theoretical explanation is surely of the highest importance. In the rest of this section I shall try to show that the unification approach gives us insight into the ways in which scientific theories yield explanations. Not only will the discussion of particular examples provide concrete illustrations of the notions of pattern and unification. It will also reveal how the unification approach leads to an improved understanding of important metascientific concepts, revising our ideas about theories, laws, and reduction.

4.6 Unification and Theoretical Explanation

I shall begin by recalling an important point made by Thomas Kuhn (1970, 23–51, 181–191) and, in a somewhat different way, by Sylvain Bromberger (1963). When we conceive of scientific theories as sets of statements (preferably finitely axiomatized) then we naturally think of knowing a scientific theory as knowing the statements—typically knowing the axioms and, perhaps, some important theorems. But, as Kuhn points out, even in those instances where there are prominent statements that can be identified as the core of the theory, statements that are displayed in the texts and accompanied with names—as, for example, Maxwell's equations, Newton's laws, or Schrödinger's equation—it is all too common for students to know the statements and yet to fail to understand the theory, a failure signaled by their inability to do the exercises at the end of the chapter. Scientific knowledge involves more than knowing the statements. A good account of scientific theories should be able to say what the extra cognitive ingredient is.

I claim that to know a theory involves the internalization of the argument patterns associated with it, and that, in consequence, an adequate philosophical reconstruction of a scientific theory requires us to identify a set of argument patterns as one component of the theory. This is especially obvious when the theory under reconstruction is not associated with any "grand equations" and when reconstructions of it along traditional lines produce a trivialization that is remote from the practice of scientists. I shall consider three examples: two (classical genetics and neo-Darwinian evolutionary theory) in which traditional philosophical approaches seem to me to have provided very little insight and one (the theory of the chemical bond) that is rarely discussed but that would seem to offer a more ready application of standard methods of analysis.

4.6.1 Classical Genetics

When we read the major papers of the great classical geneticists or when we read the textbooks in which their work is summarized, we find it hard to pick out any laws about the transmission of *all* genes. These documents contain information about the chromosomal arrangement of particular genes in particular organisms, about the effect on the phenotype of particular mutations, about frequencies of recombination, and so forth. In works from the pre-Morgan era we do find two general statements about gene transmission—Mendel's Laws (or "Rules" as they are sometimes called)—but, in the heyday of classical genetics (1910–1953), the writings of classical geneticists are predominantly concerned either with cases to which Mendel's simple laws do not apply or with instances in which they are false. The heterogeneous collection of particular claims advanced about linkage, position effect, epistasis, nondisjunction, and so forth, in *Drosophila, Zea mays, E. coli, Neurospora,* etc. seem more like illustrations of the theory than core principles of it. That, I suggest, is precisely what they are.

Classical genetics is centrally focused on (though by no means confined to) a family of problems about the transmission of traits. I shall call them *pedigree problems,* for they are problems of identifying the expected distributions of traits in cases where there are several generations of organisms related by specified connections of descent. The questions that arise can take any of a number of forms: What is the expected distribution of phenotypes in a particular generation? Why should we expect to get that distribution? What is the probability that a particular phenotype will result from a particular mating?, and so forth. Classical genetics answers such questions by making hypotheses about the relevant genes, their phenotypic effects, and their distribution among the individuals in the pedigree. Each version of classical genetic theory contains a problem-solving pattern exemplifying this general idea, but the detailed character of the pattern is refined in later versions so that previously recalcitrant cases of the problem can be accommodated.

I shall consider four examples of explanatory schemata that have been employed in genetics in our century. By exhibiting them in detail, I hope to illustrate concretely how explanatory unification of a field works and how changes within a field can show cumulative modification of an underlying pattern. Obviously, picking out four points in a long historical development gives the illusion that the history proceeds in large jumps. Those familiar with the history of genetics in the twentieth century should be able to see how to interpolate and to extrapolate.[19]

[1] **Mendel** *(1900)*

(1) There are two alleles A, a. A is dominant, a recessive.

(2) AA (and Aa) individuals have trait P, aa individuals have trait P'.

(3) The genotypes of the individuals in the pedigree are as follows: i_1 is G_1, i_2 is G_2, . . . , i_N is G_N. {(3) is accompanied by a demonstration that (2) and (3) are consistent with the phenotypic ascriptions in the pedigree.}

(4) For any individual x and any alleles yz if x has yz then the probability that x will transmit y to any one of its offspring is ½.

(5) The expected distribution of progeny genotypes in a cross between i_j and i_k is D; the expected distribution of progeny genotypes in a cross . . . {continued for all pairs for which crosses occur}.

(6) The expected distribution of progeny phenotypes in a cross between i_j and i_k is E; the expected distribution of progeny phenotypes in a cross . . . {continued for all pairs in which crosses occur}.

Filling Instructions: A, a are to be replaced with names of alleles, P, P' are to be replaced with names of phenotypic traits, i_1, i_2, . . . , i_N are to be replaced with names of individuals in the pedigree, G_1, G_2, . . . , G_N are to be replaced with names of allelic combinations (e.g. AA, Aa, or aa), D is replaced with an explicit characterization of a function that assigns relative frequencies to genotypes (allelic combinations), and E is to be replaced with an explicit characterization of a function that assigns relative frequencies to phenotypes.

Classification: (1), (2), and (3) are premises; the demonstration appended to (3) proceeds by showing that, for each individual i in the pedigree, the phenotype assigned to i by the conjunction of (2) and (3) is that assigned in the pedigree; (4) is a premise; (5) is obtained from (3) and (4) using the principles of probability; (6) is derived from (5) and (2).

Comments: **Mendel** is limited to one locus, two allele cases with complete dominance. We can express this limitation by pointing out that the pattern above does not have a correct instantiation for examples which do not conform to these conditions. By refining **Mendel**, we produce a more complete schema, one that has correct instantiations in a broader class of cases.

[2] **Refined Mendel** *(1902?-1910?)*

(1) There are n pertinent loci $L_1, \ldots, L_n$. At locus L_i there are m_i alleles $a_{i1}, \ldots, a_{imi}$.

(2) Individuals who are $a_{11}a_{11}a_{21}a_{21} \ldots a_{n1}a_{n1}$ have trait P_1; individuals who are $a_{11}a_{12}a_{21}a_{21} \ldots a_{n1}a_{n1}$ have trait P_2; . . . {Continue through all possible combinations.}

(3) The genotypes of the individuals in the pedigree are as follows: i_1 is G_1, i_2 is $G_2, \ldots, i_N$ is G_N. {Appended to (3) is a demonstration that (2) and (3) are consistent with the phenotypic ascriptions given in the pedigree.}

(4) For any individual x and for any alleles y, z, if x has yz then the probability that a particular one of x's offspring will have y is $\frac{1}{2}$.

(5) The transmission of genes at different loci is probabilistically independent.

(6) The expected distribution of progeny genotypes in a cross between i_j and i_k is D; the expected distribution of progeny genotypes in a cross . . . {continued for all pairs in the pedigree for which crosses occur}.

(7) The expected distribution of progeny phenotypes in a cross between i_j and i_k is E; the expected distribution of progeny phenotypes in a cross . . . {continued for all pairs in the pedigree for which crosses occur}.

Filling Instructions: very similar to those for **Mendel** (details are left to the reader).

Classification: (1), . . . , (5) are premises; (6) is derived from (3), (4), and (5) using principles of probability; (7) is derived from (2) and (6).

Comments: **Refined Mendel** can cope with examples in which a phenotypic trait has a complex genetic basis; it is freed from the limitation to complete dominance and recessiveness and it can allow for epistasis. But **Refined Mendel** does not take account of linkage and recombination. The next step is to build these in.

[3] **Morgan** *(1910-1920)*

(1)-(4) As for **Refined Mendel**.

(5) The linkage relations among the loci are given by the equations $Prob(L_i, L_j) = p_{ij}$. $Prob(L_i, L_j)$ is the probability that the alleles at L_i, L_j on the same chromosome will be transmitted together (if L_i, L_j are loci on the same chromosome pair) and is the probability that arbitrarily selected alleles at L_i, L_j will be transmitted

together (otherwise). If L_i, L_j are loci on the same chromosome pair, then $0.5 \leq p_{ij} \leq 1$. If L_i, L_j are on different chromosome pairs, then p_{ij} is 0.5.

(6) and (7). As for **Refined Mendel**.

Filling Instructions and *Classification*: As for **Refined Mendel**.

Comments: Whereas in **Refined Mendel**, we had two general laws about the transmission of genes—namely (4) and (5)—one of these has given way to a schematic sentence which can be differently instantiated in different situations (depending on underlying cytogological details). Within Classical Genetics, **Morgan** is further refined after 1920 to allow for nondisjunction, duplication, unequal crossing over, segregation distortion, cytoplasmic inheritance, and meiotic drive. I shall leave to the reader the task of showing how all but one of these phenomena are accommodated by modifying **Morgan**. The exception—no more important than the others for the history of classical genetics, but especially interesting for the present study—is meiotic drive. Consideration of meiotic drive requires the abandonment of the *other* general law about the transmission of genes that has so far figured in our schemata, (4), and its replacement by a schematic sentence that can be instantiated differently in different cases. I shall introduce this modification simultaneously with developing another, namely the embedding of classical genetics within molecular biology.

[4] Watson-Crick

(1) There are n loci $L_1, \ldots, L_n$. At locus L_i there are m_i alleles $a_{i1}, \ldots, a_{imi}$.

(2) (a) The DNA sequence of a_{11} is $XYUV \ldots$, the DNA sequence of . . . {continue through all alleles}.

(b) Details of transcription, post-transcriptional modification, and translation for the alleles in question.

(c) The polypeptides produced by $a_{11}a_{11}a_{21}a_{21} \ldots a_{n1}a_{n1}$ individuals are $M_1, \ldots, M_k$, the polypeptides produced by . . . {continue for all allelic combinations}.

(d) Details of cell biology and embryology for the organisms in question.

(e) Individuals who are $a_{11}a_{11}a_{21}a_{21} \ldots a_{n1}a_{n1}$ have phenotype P_1, individuals who are . . . {continue through all possible combinations}.

(3) The genotypes of the individuals in the pedigree are as follows: i_1 is $G_1, \ldots, i_N$ is G_N. {Appended to (3) is a demonstration that (2e) and (3) are consistent with the phenotypic ascriptions given in the pedigree.}

(4) If an individual x has $a_{11}a_{12}$ at locus L_1 then the probability that a particular offspring of x will receive a_{11} is q_{112}, if an individual x has . . . {continue through all heterozygous combinations}.

(5)–(7). As for **Morgan**.

Filling Instructions: As for **Morgan**, with the further condition that $X,Y,U,V, . . .$ in (2a) are to be replaced with names of bases (Adenine, Cytosine, Guanine, Thymine) and that the M_i in (2c) are to be replaced with names of polypeptides.

Classification: (2c) follows from (2a) and (2b); (2e) is derived from (2c) and (2d). Otherwise, as for **Morgan**.

Comments: The replacement of (4) with a schematic sentence allows us to accommodate cases of meiotic drive, for these will be represented as situations in which one of the q_{ijk} is different from 0.5. The contribution of molecular biology consists in the *extension* of **Morgan** through the derivation of what was previously a premise (2), which now appears as a conclusion (2e). There are several instances in which the molecular derivation can proceed as far as (2c), but the information about cell physiology and embryology (2d) is always too sparse to permit us to make a *complete* derivation of (2e). One of the closest approximations is furnished by studies of human sickle-cell anemia (and of the molecular structure of the genes for globin chains), and such instances show how the derivation would *ideally* be carried out for phenomena that are developmentally more complex.[20]

Summary. These examples of four main explanatory patterns are intended to show concretely how explanation-seeking questions about the transmission of traits through pedigrees – specifically questions about why we should expect to find particular distributions in specified generations – are addressed within genetic theory. As I have already noted, the successive articulations of the initial pattern, **Mendel**, appears to be cumulative, and, as we proceed, it becomes hard to identify any general laws about the transmission of genes. I shall return to the significance of this case for our understanding of various important metascientific concepts after we have looked more briefly at two other examples.

4.6.2 Darwinian Evolutionary Theory

Darwin's theory of evolution by natural selection addresses a number of general questions about the characteristics of life. These questions include problems of biogeography, of the relationships among organisms (past and present), and of the prevalence of characteristics in species or in higher taxa. As I have argued elsewhere (Kitcher 1985a), Darwin's principal achievement consisted in his bringing these questions within the scope of biology, by showing, in outline, how they might be answered in a unified way.

In its most general form, Darwin's proposal is to make history central to the understanding of biological phenomena. We explain the distribution of organisms in a particular group – the Galapagos finches or the ring-tailed lemurs, for example – by tracing a history of descent with modification that charts the movements of the organisms in the lineage that terminates with the group in which we are interested. In similar fashion, history can be made relevant to the explanation of relationships among organisms or of the presence of prevalent traits. Historical explanations of this general type divide into two subsidiary classes: in some cases there is an attempt to give a causal account of the modifications in the lineage; in other instances we simply record the modifications without any attempt to identify their causes. In providing (or sketching) explanations in the latter class, Darwin draws only on the less controversial part of his theory, contending that organisms have evolutionary histories but not committing himself to claims about the agents of evolutionary change. When he and his descendants attempt to give the more ambitious historical explanations that specify causes of particular evolutionary changes — most prominently in their efforts to understand why particular traits are prevalent in particular groups — then they venture into areas that were highly controversial long after the idea of evolution was broadly accepted and that still excite (different) debates today.[21]

I shall illustrate the explanatory structure of Darwinian evolutionary theory by displaying two patterns, one from the less ambitious class and one that uses the key Darwinian idea of natural selection. Consider first the explanation of homologous characteristics in related groups. Here the question that confronts us takes the form "Why do G and G^* share the common property P?" In general, questions of this type may be answered by instantiating any of several different patterns — for common traits may occur as a result of parallelism or convergence, instead of being homologies. However, Darwin proposes that many questions of the form can be answered by discovering true premises that instantiate schemata in the following pattern.

Homology

(1) G and G^* descend from a common ancestral species S.

(2) Almost all organisms in S had property P.

(3) P was stable in the lineage leading from S to G: that is, if S was ancestral to S_n and S_n immediately ancestral to S_{n+1} and S_{n+1} ancestral to G, then if P was prevalent in S_n almost all members of S_{n+1} were the offspring of parents, both of whom had P.

(4) P was stable in the lineage leading from S to G^*.

(5) P is heritable: that is, almost all offspring of parents both of whom have P will have P.

(6) Almost all members of G have P and almost all members of G^* have P.

Filling Instructions: P is to be replaced by the name of a trait, G and G^* by the names of groups of organisms (populations, species, genera, higher taxa), S by the name of a species.

Classification: (1)–(5) are premises; (6) is derived from (1)–(5) using mathematical induction on the lineages.[22]

Let us now turn to the primary Darwinian pattern, the pattern underlying *many* explanations of the presence of traits in groups of organisms. Here, the explanation-seeking question with which we are concerned is "Why do almost all the organisms in *G* have *P?*"

Simple Selection

(1) The organisms in *G* are descendants of the members of an ancestral population *G** who inhabited an environment *E*.

(2) Among the members of *G** there was variation with respect to *T*: some members of *G** had *P*, others had *P#*, *P##*, . . .

(3) Having *P* enables an organism in *E* to obtain a complex of benefits and disadvantages *C*, making an expected contribution to its reproductive success *w(C)*; having *P#* enables an organism to obtain a complex of benefits and disadvantages *C#*, making an expected contribution to its reproductive success *w(C#)*; . . . {continued for *P##* and all other variant forms of *T* present in *G**}. *w(C) > w(C#)*, *w(C) > w(C##)*, etc.

(4) For any properties P_1, P_2, if $w(P_1) > w(P_2)$ then the average number of offspring of organisms with P_1 that survive to maturity is greater than the average number of offspring of organisms with P_2 that survive to maturity.

(5) All the properties *P*, *P#*, *P##*, . . . are heritable.

(6) No new variants of *T* arise in the lineage leading from *G** to *G* (i.e., the only variation with respect to *T* comprises the properties *P*, *P#*, *P##*, . . . already present in *G**). All the organisms in this lineage live in *E*.

(7) In each generation of the lineage leading from *G** to *G* the relative frequency of organisms with *P* increases.

(8) The number of generations in the lineage leading from *G** to *G* is sufficiently large for the increases in the relative frequency of *P* to accumulate to a total relative frequency of 1.

(9) All members of *G* have *P*.

Filling Instructions: *T* is to be replaced by the name of a determinable trait (a "character-type"), *P*, *P#*, *P##*, . . . are to be replaced with names of determinate forms of the trait, *G** with the name of an ancestral species, *E* with a characterization of the environment in which members of *G** lived, *C*, *C#*, and so forth are to be replaced with specifications of sets of traits, and *w(C)*, *w(C#)* are replaced with non-negative numbers.

Classification: (1)–(6), (8) are premises; (7) is derived from (1)–(6); (9) is derived from (7) and (8).

Comments: **Simple Selection** is a pattern that can be attributed to Darwin and his early followers — in the sense that although the formalism that I have offered is absent from their writings, it is nonetheless implicit in the explanations of the prevalence of traits that are found in the *Origin* and other Darwinian texts. However, **Simple Selection** does not cover all of the examples studied by early Darwinians. In some instances (as with Darwin's explanation of the role of selection in the evolution of eyes) the pattern employed seems to involve a number of iterations of simple selection. Moreover, there are Darwinian explanations – notably explanations of quantitative characteristics – that relax (6), allowing that as an advantageous variant is increasing in frequency, a yet more successful trait may arise in the population. A pattern capturing this idea can be presented relatively easily (call it **Directional Selection**). Both iterations of **Simple Selection** and **Directional Selection** can be subjected to a further, gradualistic, constraint, in that we can demand that the properties that increase in frequency at successive stages in the derivation should form a "continuous" sequence (in the obvious, but imprecise, sense that the differences among adjacent members of the sequence are small).

There is no reason to insist that all Darwinian answers to questions of the form "Why is P prevalent in G?" should be instantiations of **Simple Selection**, iterations of **Simple Selection**, or **Directional Selection**. Darwin's insightful appeals to "Correlation and Balance" can be accommodated by recognizing a pattern of **Correlated Selection**, in which the increase in frequency of a characteristic P is explained by using one of the selectionist patterns to show that another trait Q will increase in frequency, and using a premise asserting the correlation of P and Q to derive the conclusion that P increases in frequency.

When we turn from Darwin to contemporary evolutionary theory, we observe the same kind of explanatory extension that we have already seen in the case of classical genetics. In some ways, the extension is more impressive in the present case, for there are several places in **Simple Selection** (and in the other selectionist patterns) in which schematic premises can be derived as the conclusions of detailed schematic arguments. So, for example, instances of (5) can be derived from specifications of the genetic basis of the trait under study, and the use of population genetics can yield a more precise version of (4) and the derivation of a conclusion about rates of increase in the relative frequency of P that will imply (7). Using population genetics and a premise specifying the number of generations in the lineage between G^* and G, it is also possible to obtain (8). Finally, the use of ecological models (such as optimality models or game-theoretic analyses) enables us to derive precise claims that imply (3).[23] Thus **Simple Selection** becomes embedded in a much larger argument pattern that can, in principle and occasionally in practice, be instantiated to give neo-Darwinian explanations of the prevalence of traits.

4.6.3 The Theory of the Chemical Bond

The examples that I have been considering involve theories that have proved difficult to reconstruct using the traditional kinds of philosophical formalism. The last case to be discussed focuses on a theory that one might initially regard as far more amenable to those formalisms. I want to suggest a different way of looking at the structure of a part of chemistry as it has developed from the early nineteenth century to the mid-twentieth century.

One leading question of post-Daltonian chemistry concerned the ratios of the weights of substances that form compounds together. Within nineteenth-century chemistry, we can discern the following simple pattern for answering questions of the form "Why does one of the compounds between X and Y always contain X and Y in the weight ratio *m:n?*":

Dalton

(1) There is a compound Z between X and Y that has the atomic formula X_pY_q.

(2) The atomic weight of X is x; the atomic weight of Y is y.

(3) The weight ratio of X to Y in Z is $px:qy$ ($= m:n$).

Filling Instructions: X,Y,Z are replaced by names of chemical substances; p,q are replaced by natural numerals; x,y are replaced by names of real numbers.

Classification: (1) and (2) are premises, (3) is derived from (1) and (2).

Dalton is elementary — although it was, of course, instantiated in many different ways during the early years of the nineteenth century by chemists who had very different ideas about the formulas of common compounds! What makes it important for our purposes is the way in which **Dalton** was extended by subsequent work.

The first step in the extension was the introduction of the concept of valence and rules for assigning valences that enabled chemists to derive conclusions about which formulas characterized possible compounds between substances. Thus instances of (1) could be derived from premises ascribing valences to the substances under study (i.e., to X and Y), from premises stating the constraints on formulas for possible compounds in terms of valence relationships, and from the principle that all compounds corresponding to formulas that meet all the constraints can be formed. At this first stage, the attributions of valence are unexplained and there is no understanding of why the constraints hold. However, the original explanations of weight relationships in compounds are deepened, by showing regularities in the formulas underlying compounds.

The second stage consists in the introduction of a shell model of the atom to explain the hitherto mysterious results about valences. From premises attributing shell structure to atoms, together with principles about ionic and covalent bonding, it is now possible to provide derivations of instances of (1). These derivations provide a deeper understanding of the conclusions than was given by the simple invocation of the concept of valence because they show us *in a unified way* how the apparently arbitrary valence rules are generated. Moreover, the appeal to the model of the atom enables us to derive instances of (2) from premises that characterize the composition of atoms in terms of protons, neutrons, and electrons.

Finally, the derivations given at the second stage can be embedded within quantum mechanical descriptions of atoms and the shell structures and possibilities of bond-formation revealed as consequences of the stability of quantum-mechanical systems. Although this is only mathematically tractable in the simplest examples, it does reveal the ideal possibility of a further extension of our explanatory derivations.

4.6.4 Conclusions from the Examples

My three examples are intended to illustrate the notion of pattern that is employed in my account of explanation, and to show that it is not so remote from the practice of science as the abstract description of the earlier parts of this section might suggest. But I hope they do more. Specifically, I would like to suggest that they shed light on some important metascientific concepts.

Laws. As we have seen, Hempel's account of scientific explanation faced serious difficulties in giving a characterization of scientific laws. One way to try to understand the notion of a scientific law is to turn the Hempelian account on its head. So we can suggest that the statements accepted as laws at a given stage in the development of science (recall that our approach has been focusing on *acceptable* explanations not on *correct* explanations) are the universal premises that occur in explanatory derivations. Many of these will be "mini-laws": such as the statement that specific genotypes regularly give rise in particular environments to certain phenotypes or the statement that sodium and chlorine combine in a one-one ratio. My earlier claim that some sciences may not be identifiable by concentrating on a few grand equations may now be made more precise by proposing that a maxi-law is a nonschematic universal premise that occurs in an explanatory pattern. As the example of genetics shows, there were stages in classical genetics at which it seemed that there were maxi-laws in genetics—general principles about the transmission of genes—namely Mendel's laws. But all our examples point to the possibility that sciences may have no maxi-laws, and that their generality may consist in the patterns of derivation that they bring to the explanation of the phenomena.

Reduction. The classical way of thinking about reduction is to consider the derivability of laws of the reduced theory from laws of the reducing theory. If theories are viewed as constituted by the patterns of derivation they put forward (by the ordering of the phenomena that they propose) then the important notion becomes that of *explanation extension*. As the examples reveal, explanation extension can go forward even when some of the concepts of the extended theory cannot be formulated in terms of the concepts of the extending theory. Hence, even when the conditions demanded by reductionists cannot be met, we can still make clear the relations among successive theories, and so capture the idea of an accumulation of knowledge which often seems to make reductionism attractive.

So I propose that the outmoded concept of reduction, which is tied to an inadequate account of scientific theories, should be replaced with the notion of explanation extension, and disputes about the virtues of reductionism reformulated accordingly.

Unification. When the view that explanation is unification is intially presented, I think that it strikes many people as invoking a rather ethereal ideal. However, in the examples I have discussed, we do find that a single pattern of derivation (or several closely related patterns of derivation) is (are) used again and again to derive a variety of conclusions. Thus I take the examples to provide *prima facie* support for the view that unification is important to explanation and that unification works in the way that I have suggested.

With two approaches to explanation now before us, we are in a position to see how they fare with respect to the problems adduced in section 1. In the next section I shall consider whether the unification approach can handle statistical explanation.

5. A Defense of Deductive Chauvinism

One obvious deficiency of the systematization account, as I have so far presented it, is that nothing has been done to accommodate the two kinds of difficulty that surround the issue of probabilistic explanation. First, while there is no bar in principle to the use of nondeductive arguments in the systematization of our beliefs, so that we can make sense of the notion of a general argument pattern even in cases where the instantiations are not deductive, the task of comparing the unifying power of different systematizations looks even more formidable if nondeductive arguments are considered. Second, and perhaps more obvious, is the objection that the systematization account is fundamentally wedded to the old Hempelian idea that to explain is to derive, so that it will be forced to adopt some version of the high-probability requirement, and thereby prove vulnerable to the criticisms leveled by Jeffrey, Salmon, and Coffa against Hempel's account of inductive-statistical explanation. I believe that both problems can be sidestepped by a simple and radical step. The explanatory store contains only deductive arguments. *In a certain sense*, all explanation is deductive.

Following Salmon (who took the label from Coffa), I shall call the thesis that all explanation is deductive *deductive chauvinism*. As I have just noted, deductive chauvinism is logically independent of the systematization account, but a successful elaboration of deductive chauvinism would enable the systematization account to circumvent what appear to be serious problems about probabilistic and statistical explanation. My goal in the present section is to elaborate deductive chauvinism, and show that it is not so absurd a view as it might initially appear.

Now there are some kinds of explanation involving probabilities that pose no threat to the deductivist ideal of explanation. When a geneticist explains why the probability of obtaining a particular phenotype from a specified cross takes on a certain value, what is given is a deductive argument in which the identification of the probability is derived from a set of premises including some claims about the organisms under study and the mathematical principles of probability (see 4.6.1). It is no accident that such arguments are solicited in student exercises in applied probability theory. To revert to Hempelian terminology, if all explanation were *D-S* explanation, there would have been no need to confront the problems that Hempel faced in developing his *I-S* account.

Trouble arises because there are areas of science — most obviously in applied science — where particular occurrences or states of affairs are explained by appeal to probabilities. The classic examples involve the recovery of patients who have been administered drugs that have known frequencies of success, the committing of crimes by people with a specified age, sex, and socioeconomic background, the occurrence of particular phenotypic traits in organisms whose parents have particular phenotypes, and so forth. Now in all of these cases the deductivist has a relatively straightforward gambit: we treat the probabilistic account as a place-holder for an underlying, unknown, deductive explanation. Thus, if we explain why an individual a has property P by pointing out that a has properties Q, R, S, T and showing that $Pr(P/Q\&R\&S\&T)$ takes some (high) value p then we are treating the high value of p as an indication that there are further, as yet unspecified properties of a, X, Y, Z such that *all* entities that have Q, R, S, T, X, Y, Z have P. The probabilistic argument is pragmatically successful because we have grounds for thinking that it exhibits part of a *deductive basis* for the phenomenon to be explained, where a deductive basis consists of those properties that would be attributed to the object(s) mentioned in the *explanandum* in the premises of a complete deductive explanation. Those grounds consist in the fact that the probability value p is high. Hempel's high-probability requirement returns as a claim about what would provide evidence for thinking that we have identified part of the deductive story.[24]

While I think that this straightforward gambit expresses part of what a deductive chauvinist should say about statistical explanation of individual events, it cannot be a completely satisfactory solution to the problem. Even if there are legitimate hopes that our uses of probability in medical and social contexts are only expressions of ignorance, there is at least one area of theoretical science in which a case can be made for the necessity of appealing to probabilities in explaining individual events and states. Quantum mechanics (QM) seems to be indeterministic. Moreover, an opponent of deductive chauvinism may contend that the indeterminacies of QM ultimately affect the macroscopic phenomena of chemistry, biology, medicine, and social science, so that the dream of deductive explanations in these areas may prove to be quite unrealistic. The point can be further reinforced by suggesting that there may be additional sources of indeterminacy beyond those already recognized in QM. I shall try to meet these challenges in turn.

5.1 The Objection from Quantum Mechanics

One familiar response to QM is the belief (hope, wish) that it will some day be replaced by a physical theory that will enable us to derive descriptions of individual events from underlying theoretical principles and initial and boundary conditions. The version of deductive chauvinism that I want to defend does not adopt any such attitude. It may well turn out that successor theories to QM retain its indeterministic character. Deductive chauvinism should distinguish between two senses in which an explanatory account might be ideal. In one sense, an ideal explanatory account is a deductive derivation (more exactly, a deductive derivation that instantiates a pattern in the explanatory store). In another sense, an ideal explanatory account is the best that the phenomena will allow. Deductive chauvinists should concede that QM (or some other essentially indeterministic theory) might be the best there is, that it might provide ideal explanatory accounts (of individual events) in the second sense, but they should deny that QM provides ideal explanatory accounts (of individual events) in the first sense.

The issues here can be treated more precisely by returning to the account of the pragmatics of explanation advanced in section 2. Those who contend that QM provides explanations of individual events must believe that there are why-questions whose topics describe such events to which QM provides complete explanatory answers. Faced with any purported example, the deductive chauvinist must do one of two things: either show that the complete answer is a deductive derivation or demonstrate that the answer is not complete (in the latter case, it will also be helpful to explain the *illusion* of completeness).

Let us start with a situation that is typical of those in which we might think that the probabilistic machinery is brought to bear on individual events. Imagine that a beam of electrons impinges on a potential barrier. For each electron, the probability that it will be reflected is 0.9, the probability that it will tunnel through is 0.1. Consider two electrons, e_1, and e_2. e_1 is reflected; e_2 tunnels through. Can we explain these events?

The question urgently needs disambiguation. Suppose we begin with standard (van Fraassen) why-questions, and consider two obvious candidates, both of which have as their topic the proposition that e_2 tunneled through. The contrast class of the first question contains both the topic and the proposition that e_2 was reflected; the contrast class of the second contains both the topic and the proposition that e_1 tunneled through. Then, I claim, there is no explanatory answer to either question — that is, there is no proposition that stands in the relation of ideal explanatory relevance to topic and contrast class.

Now QM enables us to write down the Schrödinger equation for the system consisting of an approaching electron and the potential barrier, and, by solving this equation we can demonstrate the probabilities of tunneling through and being reflected. Conjoin as many as you like of the propositions that occur in this derivation or consider the entire derivation. Whatever your choice, you will not have shown why e_2 tunneled through, rather than being reflected, or why e_2, rather than e_1, tunneled through. For what those why-questions ask is for a specification of the differences between electrons that tunnel through and electrons that are reflected, and it is, of course, part of the character of QM that there are no such differences to be found. In a sense, the full derivation from the Schrödinger equation is the best possible answer to the questions—it is the best that nature will allow—but it is not an ideal explanatory account. With respect to these questions, there is no ideal explanatory account. The questions are unanswerable. In response to the questions we can only say "It turned out this way, by chance."

There is a superficial difference when we consider the related questions about e_1. If someone inquires why e_1 was reflected rather than tunneling through, or why e_1 rather than e_2 was reflected, one may be tempted to use the fact that the derivation from the Schrödinger equation assigns a relatively high probability (0.9) to the event of e_1's being reflected to suggest that that derivation supplies at least something of an answer to these questions (or, perhaps, to the first). After all, the derivation does show that it was to be expected that e_1 would be reflected. But I think that the situation is symmetrical. We no more understand why e_1 was reflected than we understand why e_2 tunneled through, and, in each case, our failure of understanding rests on the fact that we cannot isolate any distinctive property that separates those that take one course from those that take the other. The symmetry here underscores my earlier diagnosis of how the Hempelian high-probability requirement might be expected to work. High probabilities are useful because they increase our confidence that we are on the track of a deterministic basis. But, when we know from the start that there is no deterministic basis to be found, the probability values are irrelevant.

If this is correct, then what, if anything, can be made of the idea that QM does advance our understanding of episodes like that of electron tunneling and reflection? The answer, I believe, is that we confuse questions that QM can answer with those that it cannot. There are relatives to (van Fraassen) why-questions in the vicinity, and QM can provide answers to some of them.

Sometimes the form of words "Why *P*?" can mean "How is it possible that *P*?" Typically, when this occurs, the question is posed in a context in which certain propositions are taken for granted. Thus, what the questioner is really asking is "Given that *Q* (which I firmly believe) how can it also be the case that *P*?" Sometimes, under such circumstances, corrective answers are called for: *Q* is false and one corrects the question by pointing that out. By analogy with the treatment of why-questions in section 2.1, we can identify a how-possibly question as an ordered pair of propositions $<P,Q>$ where *P* is the topic and *Q* is the *background presupposition* (in some cases *Q* may be null). A noncorrective answer to the how-possibly question is an argument that shows that *P* and *Q* are consistent.

Now imagine a naïve questioner asking why e_2 tunneled through—say in some concrete laboratory setting where the potential barrier is due to an observable object and the electron's tunneling through is revealed by a scintillation on a screen. It is quite possible that the questioner's inquiry arises from astonishment that a particle, an electron, can make its way through a solid object. In responding to this person's question, we would try to tease out the background presuppositions and to show that those presuppositions that are not false are mutually consistent. In other words, we view the apparent why-question as expressing a how-possibly-question — given the character of solid objects and particles, how is it possible that the electron could penetrate? — and we might respond by using the derivation from the Schrödinger equation to show that there is a non-zero probability of tunneling through.

There are other questions that QM can answer. The conclusion of the derivation specifies the probability that an impinging electron will be reflected (tunnel through) and thereby answers the question "What is the probability that an impinging electron will be reflected?" The derivation itself provides an ideal answer to the why-question "Why is the probability that an impinging electron will be reflected 0.9?" (where the contrast class includes all the propositions attributing values in [0,1] to the probability). Of course, the why-question answered here is a question about an entire class of events, and we have the derivation of a generalization. But, as we have already seen (see above p. 427), there are many instances in which answering a why-question completely involves referring the phenomenon to be explained to a general class and then showing why a regularity holds within the class.

So deductive chauvinists should pursue a strategy of divide-and-conquer with respect to the claim that QM can be used to explain properties of individual events. There are some why-questions that are unanswerable and related questions that have ideal, deductive answers. The apparent difficulties for deductive chauvinism, and the apparent need for the *I-S* model of explanation (or some surrogate) arises from the fact that these questions can be presented in the same form of words, and can thus easily be confused.

5.2 The Idealization of Macro-Phenomena

But now the second worry seems to arise with increased force. Consider any deductive explanation of a macroscopic event. Given QM, we know that if enough fundamental particles entered highly improbable states, some of the claims made in the premises of the derivations would be false. Hence it would appear that *all* explanation of individual events needs to be replaced by deductive explanations of why certain kinds of events occur with high probabilities. For example, imagine that we are explaining why a flood occurred in a house in Minnesota during the occupants' winter vacation. The explanation might derive a description of the phenomenon from principles about the relative density of ice and water and about rates of cooling, together with specification of initial and boundary conditions. However, if we think of the house and its environment as a quantum-mechanical system, we recognize that there are highly improbable micro-states that would correspond to macro-states in which there is no flooding of the house (in some of these there is no cooling in the pipes, in some of them the pipes are rejoined once the ice melts, and in some of them the house itself decomposes). So, if acceptable explanations are to involve only principles that we judge to be true, then we ought to specify the probability that there will be a flood and provide a full derivation of the specification of the probability.

I think that examples like this help us understand what occurs in the explanation of individual events. When we explain the behavior of actual objects, the first step is always to achieve an idealized description of those objects. Thus, in the standard explanation of the flood in the Minnesota house, we think of the house and its environment as a certain kind of thermal system, assuming uniformities in temperature gradients, ideal cylindrical pipes, and so forth. The question "Why did this particular object behave in this particular way?" is transformed into the question "Why do ideal objects of this general type exhibit these properties?"[25] We justify the transformation by pointing out that factors we have neglected or have introduced in our idealizing either make definite, but small, differences to the account or are highly unlikely to make any large changes. In effect, we are playing "let's pretend," and giving a deductive account of how things would go in a simpler, cleaner, world (for a view that is similar in some respects, see the "Simulacrum account" offered in Cartwright (1983)). We regard the derivations we give as explaining the actual phenomena, because we can provide justifying arguments for concluding that the actual world is not likely to be signficantly different from the ideal world (the probability that there will be a large difference between the phenomena of the actual world and the phenomena of the pretend world is small).

One of the simplest and most celebrated examples in the history of scientific explanation will make the point clearly. According to popular history, the fusiliers at the Venetian artillery asked Galileo why their guns attained maximum range on a flat plain when they were set at an angle of elevation of 45°. Contemporary students in elementary physics learn how to give the substance of Galileo's answer. Consider the gun as a projector, the cannonball as a point particle, the ambient atmosphere as a vacuum, and the plain as an ideal Euclidean plane. Suppose that the particle is subject only to the force of gravity. *Now* we can represent the components of the resultant motion, for varying angle of projection, and show that maximal range is attained when the angle of elevation is 45°. There is an elegant derivation of the result for an ideal system.

But what about the actual guns and the actual cannonballs? Has their behavior been explained? Yes, but the explanation involves correction. It probably is not true that the fusiliers' guns attained maximum range at 45°. The results probably varied from gun to gun, day to day, cannonball to cannonball, location to location. However, only someone in the grip of a theory of explanation would complain that the presuppositions of the fusiliers' why-questions were false and that the questions could not therefore be answered. So, we can imagine that Galileo, or his contemporary descendants, append to the ideal derivation some remarks about the way in which the actual world and the ideal world are different: in the actual world there are effects of the internal surface of the barrel, of air resistance, of asymmetries in cannonballs, of local inclinations and depressions in the ground; it is possible that the fusiliers have not detected the ways in which these effects produce deviations from the regularity they claim to have found—perhaps the deviations are even too small to be detected with their measuring instruments. The ideal derivation conjoined with an assessment of the extent to which perturbing actual factors would cause deviations from the idealized conclusions gives complete understanding of the actual events.

Here and elsewhere, the idealization ignores both common perturbations with slight effects (such as air-resistance) and major disruptions with negligible probability. The more advanced contemporary physics student knows that the cannonball is a quantum-mechanical system. In consequence, there are extremely small probabilities that unusual motions of its constituent particles will occur, subverting the claims that are made in the equations of the Galilean account. In principle, we could compute the probability that quantum-mechanical effects could generate a significant deviation from the trajectories described in the idealized description—where the standards for significance are set by the magnitude of the macroscopic perturbations we ignore. What is the probability that some QM effect will be more sizable than the effects of air-resistance (say)? Nobody knows the exact answer, but we can make reasonable estimates of order of magnitude and conclude that the probability is very small indeed. Thus part of the defense of our idealization consists in the specification of small perturbations, and the remainder in showing that the probability of a larger difference is negligible.

So the response to the second charge, the charge that the effects of QM percolate up into macrophenomena and subvert the strategy of giving deductive explanations of individual events is that our macroscopic explanations of individual events involve idealizations of the phenomena, deductive derivations that are exactly true of the ideal systems, and assessments of the probable differences between the ideal and the actual systems. The assessments justify our idealizations, and, simultaneously, show us how the topics of our why-questions might be corrected.

5.3 Further Sources of Indeterminism?

It should now be obvious how to respond to the third complaint, the worry that there are sources of indeterminism beyond those of QM.[26] We can consider three possibilities. First, there are no such sources, and the account given in response to the first two objections suffices. Second, there are such sources, but, like those of QM, they only give small probabilities of significant deviations in macro-phenomena. In this case, the sources would have to be handled in just the same way as the QM effects on macro-phenomena, and we would explicitly note a further idealization of actual macro-events. Third and last, there are sources of significant deviations in macro-phenomena, so that, for some class of macro-phenomena C there are no deductive accounts of the behavior of systems that can be defended as idealizations of the phenomena in C. If this last possibility were to occur, then, I suggest, our attitude toward the macro-phenomena in C should be just that I have recommended we take toward the basic phenomena of QM (such as electron tunneling). In other words, should the basis of heart disease (for example) turn out to be irreducibly probabilistic, then we should have to admit that we can no more explain why one person rather than another contracts heart disease than we can explain why one electron rather than another tunnels through a barrier. Hence, in all three cases, the strategies developed in response to the first two objections will suffice to defend deductive chauvinism.

5.4 Two Popular Examples

Some of the points I have been making can be underscored by considering two examples that have figured in the recent literature and that have sometimes been considered to devastate deductive chauvinism. Let us begin with a case that is used by Salmon (1984, 86, 88, 109). A breeding experiment on pea plants produces a filial population in which 0.75 of the plants have red blossoms, 0.25 white. Let b_1 be a plant with red blossoms, b_2 a plant with white blossoms. Salmon argues that we can explain *both* why b_1 has red blossoms and why b_2 has white blossoms by pointing out that the flowers came from the filial population. The case is intended to illustrate a symmetry principle: we understand the improbable outcome just as much – or as little – as we understand the probable outcome. Thus, Salmon contends, there can be probabilistic explanations of individual events, even in cases where, relative to the information given in the explanatory answer, the probability of the event explained is low.

I believe that these conclusions are mistaken and that the mistakes are revealing. First, we should note that the answers to the questions "Why does b_1 have red blossoms?" "Why does b_2 have white blossoms?" are "Because b_1 has genotype RR or Rr (where R is dominant with respect to r and codes for a molecule ultimately producing red pigment)" and "Because b_2 has genotype rr." No appeals to probability enter here, *unless we suppose that there are irreducibly probabilistic factors that enter into the connection between genotype and phenotype.* Thus, if it is impossible to provide a deductive derivation of the explanandum it is because complications owing to QM (or some other source of indeterminism) allow for organisms that bear the *R* allele to fail to produce red blossoms (in the environment, assumed standard, in which the plants are grown). If probabilities enter into the explanation of these individual occurrences it is for just the kinds of reasons I considered above, and, as I contended, those reasons do not compel us to admit probabilistic explanations of individual events (or states).

But the question "Why does b_1 have red blossoms?" may be aimed deeper. The questioner may want to know not only the genetic basis of the trait in b_1 but also why b_1 has the genotype it does. Here we might think that a probabilistic account was appropriate: b_1 is obtained from a cross between *Rr* heterozygotes, so that the probability that it will be *RR* or *Rr* is 0.75. A moment's reflection makes it plain that this does not explain why b_1 rather than b_2 produces red blossoms. So should we conclude, as Salmon suggests (1984, 109) that explanation is not implicitly contrastive? No. For the explanatory value of the probabilistic account just given – and of the corresponding probabilistic account for b_2 – is that they introduce properties of the organisms involved in the process that will figure in a complete explanation. b_1 has genotype *RR* (assuming that that is its genotype) because the fertilization process that gave rise to its zygote involved the fusion of two *R*-bearing gametes; b_2's zygote was formed from two *r*-bearing gametes. There is apparently a fully deductive derivation of the specifications of the genotypes of b_1 and b_2 from statements that describe the events that culminate in the respective fertilizations. Once again, if there are sources of indeterminism in these events, then they would have to be reckoned with along the lines canvassed earlier in this section.

How, then, does probabilistic explanation work in genetics? Answer: along the lines discussed in section 4.6.1. The derivations that are provided by genetics show why certain distributions of genes and traits are expected. Operating against this background of expectations, those who are concerned with particular populations – or with particular individuals – can (in principle) trace the histories of the passage of genes, and so see how the actual cases diverge from (or coincide) with the expectations. By instantiating schemata like **Mendel** we can explain why certain events have specified probabilities. When we want an explanation of the actual events, we have to trace the details of history – but note that the tracing is done along the lines laid down in the theoretical picture.

The second case I want to consider is that of the hapless mayor. Why did the mayor develop paresis? He alone among the townspeople had previously contracted syphilis. But the chance that an individual syphilitic develops paresis is low. Have we, therefore, as Salmon suggests (1984, 31–32, 51–53) given a probabilistic explanation of an event, and one which, furthermore, assigns a low probability to the event explained?

Distinguish possibilities. One reason that we may think that the recognition of the mayor's syphilis has explanatory value is that we think of it as *part* of a complete deductive explanation of the paresis. Syphilitics who have some (possibly complex) property *X* always contract paresis. At the present stage of our knowledge, we can distinguish the mayor from his nonsyphilitic neighbors and answer the question of why he, rather than any of them, contracted paresis. We do not yet know how to distinguish the mayor from his fellow syphilitics who do not contract paresis. To answer the question why it was the mayor rather than any of *them* who got paresis, we would have to know the additional factor *X*. So the information about the mayor is helpful because it enables us to answer one why-question ("Why was it the mayor rather than his [nonsyphilitic] neighbors?"), but more information will be required to answer another ("Why was it the mayor rather than certain other syphilitics?").

But perhaps there is no additional factor to be found. If so then the situation is just like that of electron tunneling. We cannot explain why the mayor, rather than other syphilitics, contracted paresis, any more than we can explain why this electron tunneled through. However, the statement that the mayor had syphilis may answer a different why-question. Suppose that the why-question has an explicit presupposition: "Given that one of the townspeople contracted paresis, why was it the mayor?" Only syphilitics get paresis, and he was the only syphilitic in town. Notice that, in this case, we can deduce the *explanandum* from the presupposition of the question and the information given in the answer. Under these circumstances, we can vindicate the idea that the statement that the mayor had syphilis is part of an explanation of *something like* his getting paresis – but the explanation is deductive.

This last possibility is interesting because it corresponds to a common strategy of scientific explanation (see Sober 1983). Sometimes we show that a system is in state *X* by presupposing that it is in one of the states $\{X, Y_1, \ldots, Y_n\}$ and demonstrating that it cannot be in any of the Y_i. In contemporary evolutionary theory, there are numerous examples of a special case: one shows that if the system starts in any of the Y_i then it ends in *X*. The classic example is that of the sex-ratio. In an evolving population of sexual organisms with variation in the propensity to produce sons and daughters, if the population begins away from a 1:1 ratio (but not so skewed that all the organisms are of one sex!) then selection will bring the population to a 1:1 ratio (Fisher 1931, see Sober 1983 for use of the example). The strategy instantiated here is to answer a question of form "Given *Q*, why *P*?" by producing a derivation of $Q \supset P$.

5.5 Explanation and Responsibility

There is one further point to be addressed before the defense of deductive chauvinism is complete. Salmon has argued that our use of probabilistic explanations of individual events is involved in assignments of responsibility. Suppose that Herman contracts cancer and dies. His widow sues the federal government on the grounds that Herman spent a significant part of his military service in a region used for atomic testing. The government points out that the base rate for Herman's type of cancer in the population is 0.0001 and that the probability of a person contracting cancer, given the period of exposure that Herman had, is 0.02. Most of Herman's army buddies were more fortunate. Arnold, for example, did not contract cancer. Nonetheless, even though there is no factor to which we can point that will distinguish Herman from Arnold, though Herman's getting cancer is, *all things considered*, simply a matter of (bad) luck, nonetheless, we have a strong intuitive conviction that the government has some responsibility for Herman's death, and that Herman's widow has a legitimate case.

Salmon believes that we cannot underwrite this intuition unless we are prepared to endorse probabilistic explanations of events which, *all things considered*, have low probability. I shall assume that the intuition is correct and try to show how it is possible to support it from within the perspective of deductive chauvinism. In exact parallel to examples that we have considered before, I suggest that we cannot answer the question "Why was it Herman, rather than Arnold, who contracted cancer?" However, as usual, there are why-questions in the vicinity that we can answer. Consider the question "Why did Herman have a significantly greater probability of contracting the type of cancer from which he died than do members of the general population?" An incomplete answer to this question is "Because he spent a large part of his military service in a region used for atomic testing." We would complete that answer by introducing details of the conditions, applying principles of atomic physics and employing facts of human physiology. In principle, there is a deductive demonstration that reveals the probability of cancer to be 0.02 for people who share Herman's experience and 0.0001 for those who do not. We can deepen that derivation by replacing the premise stating that Herman spent such-and-such a period in a region of such-and-such a type with a deductive argument that leads from premises about the government's actions to the conclusion that Herman had the kind of exposure that he did. So, ultimately, there are premises that make reference to government actions that figure in an explanation of why Herman's chances of contracting cancer were significantly higher than those of the general population.

I have no good theory of how the existence of an explanation making reference to the actions of X shows that X is responsible for some facet of the *explanandum*. Neither does Salmon. However, part of his case against deductive chauvinism involves the quite plausible idea that our intuition that the government is responsible for Herman's death (at least in part) rests on our seeing that we invoke claims about the government in a probabilistic explanation of Herman's death. Deductive chauvinists should agree with Salmon's (tacit) views about the link between explanation and responsibility. But they should modify the treatment of the example. In one obvious sense there is no explanation of Herman's death. There is, however, an explanation of why Herman was at greater risk than the general population. This explanation involves the actions of the government. Because the actions of the government are explanatorily relevant to the increased probability of Herman's contracting cancer *and* because Herman actually contracted cancer, the government is at least partly responsible.

So there is a chauvinist analog of Salmon's claims about responsibility. Where Salmon sees the responsibility as stemming from the fact that governmental actions are described in premises of a probabilistic explanation of an individual event, I regard it as issuing from the fact that those actions are described in a deductive explanation of why Herman had a greater chance of contracting the cancer from which he died. To put the difference starkly: Salmon holds that the government is responsible because it (partially) caused Herman's death; I hold the government responsible because it caused Herman to be at greater risk for a harm which (by his bad luck) befell him.

Stephen Stich has reminded me that accounts of responsibility will have to come to terms with cases in which the action of one party is pre-empted by the action of another. If the government puts Herman at greater risk for cancer, but Herman succumbs to a heart attack, then the government is not, of course, responsible for his death. Quite evidently, there will have to be clauses that connect explanation and responsibility in more complex cases—both on Salmon's account and on mine—but I see no reason to suspect that the parallelism that I have outlined in the simplest instances cannot be preserved.

I conclude that deductive chauvinism can defeat the obvious challenges. Since acceptance of deductive chauvinism facilitates the development of the idea that explanation is unification, I shall henceforth assume that all derivations in the explanatory store are deductive.

6. Epistemological Difficulties for the Causal Approach

Let us recapitulate. We have seen how the Hempelian models of explanation face apparently insuperable difficulties. Two alternative approaches to explanation have been considered. One of these, the view that scientific explanation identifies the causes of phenomena, is relatively easy to understand, and its virtues, especially in tackling the asymmetries of explanation, are apparent. The other approach, construing explanation in terms of unification, is initially more difficult to formulate. I have tried to show how it might be developed and how some apparent problems with it may be overcome. The rest of this essay will be concerned with what I see as the principal troubles of each program. Besides the smaller problems canvassed in sections 3.2 and 3.3, the causal approach, on the one hand, faces the large question of how the epistemology of causation is to be developed. The unification approach, on the other hand, must show how the constraints on the explanatory store enable us to debar those problematic derivations that engender the problems of asymmetry and irrelevance. The present section will delve into the difficulties of providing an adequate account of our knowledge of causes, once the traditional empiricist idea of making the concept of causation dependent on such notions as law or explanation has been renounced. Subsequent sections will complete my apology for the unification church, by trying to show that *its* central problem is not so recalcitrant as one might fear.

6.1 Hume's Ghost

Sometimes in reading the philosophy of science literature of the 1940s, 1950s, and 1960s, one has the distinct impression that the authors have been spooked by Hume. Causal concepts are viewed with suspicion, if not dread, apparently because they have been placed on the list of empiricistically forbidden notions. The sense that a whiff of sulphur would accompany certain appeals—appeals that would make life so much easier if they only could be made—makes heresy enticing. So, in the wake of logical empiricism, many philosophers of science have made free use of causal concepts, perhaps seeing themselves as shaking off ghostly chains that had seemed to bind their predecessors. Unfortunately, while the mere invocation of Hume's name is not enough to show that such uses are sinful, there are deeper reasons for worrying about causal concepts than a desire to keep one's empiricist conscience pure. We can discover those reasons by returning to Hume's problem of causation in its most general form.

The desire to analyze causation stems from the apparent difficulty of justifying causal judgments. Some of the causal claims that we make are justified. But how does the justification work? Once we have been educated in the causal lore of an ongoing field of science, then it is easy to claim that we simply observe causal relations. But there is very little plausibility to the idea of observing causation, when the observer is a neophyte, say a child. Thus there arises the conviction that we come to make justified causal judgments by observing that certain conditions obtain, and (initially at least) inferring causal claims from the premises that record our observations. Hence we arrive at the project of giving necessary and sufficient conditions for the obtaining of causal relations, formulating those conditions in ways that will dissolve the epistemological mysteries surrounding causation by deploying only concepts whose satisfaction is observationally ascertainable.

With the demise of statistical accounts of causal relations in the late 1970s (see, for example, Cartwright 1979), few of those who want to deploy the concept of causation in analyzing explanation have taken a public stand on the cogency of the line of reasoning rehearsed in the last paragraph. If causal knowledge is observational knowledge, then the apparently implausible implications of that position should be addressed. But, if causal knowledge is inferential knowledge, then we are owed an account of the observational conditions on which causal justifications depend.[27] Yet explicit accounts of causal knowledge are hard to find. Salmon's recent attempt (in his 1984) is a notable exception, and I shall try to show how hard the epistemological problem is by reference to his careful and sensitive discussion.

6.2 Causal Processes and Causal Interactions

Salmon is forthright in his acknowledgment that he needs to "furnish an analysis of the concept of causality or its subsidiary notions," and he views it as an issue of "intellectual integrity" to "face Hume's incisive critique of causal relations and come to terms with the profound problems he raised" (1984, 135). However, he thinks that traditional empiricist approaches to causation have handicapped themselves by starting with the wrong causal concepts. Instead, he proposes that we can formulate causal claims so as to make clear how we know them if we focus on causal processes and causal interactions rather than on the causal connections among events. Ultimately, the project of understanding the causal structure of the world can be regarded as an attempt to specify the relation that must obtain between two events (conceived as spacetime points) just in case the earlier was a causal factor in the occurrence of the later. The simplest type of case is that in which the two points are linked by a causal process that connects an interaction at the earlier point (c) with an interaction at the later point (e). A paradigm will be the breaking of a window by a ball that was struck by a baseball bat. Here the event e is the breaking of the window, the event c the striking of the ball, and the linking process is the flight of the ball. In such cases there are no intermediate interactions (or, at any rate, none that are relevant to the occurrence of e) so that the causal connection seems decomposable into a pair of interactions and a causal process that (spatio-temporally) links them. Inspired by the paradigm, we might claim that at least the central cases of causal connection can be understood, if we can only analyze the notions of causal interaction and causal process.

As a first formulation, let us say that two events are causally related just in case they are causally connectible, where causal connectibility obtains just in case there is a continuous path through spacetime that is a causal process and that terminates in causal interactions at both ends. Now it seems to follow that we have immediately eliminated some examples that some would count as instances of causal connection: eighteenth-century claims about action-at-a-distance and twentieth-century proposals about the possibility of time-travel (see, for example, Lewis 1976) both appear problematic. For the moment, I shall accept the elimination. We can then pose Salmon's problem as follows. We want to state conditions on causal connection that will pick out the causal paths and the causal interactions, and the notions to which we appeal in our specifications must be acceptable to an empiricist. The last constraint does not involve swearing to disavow some index of banned concepts drawn up by the empiricist establishment. The demand is that we do not introduce conditions when we are unable to explain how we can know whether or not they obtain.

Begin with the idea of a connecting causal process. We have a manifold of spacetime points, and, for any two distinct points in the manifold, a large number of paths connecting them. The task is to filter out the paths that are not causal processes. Evidently there is a fairly large collection that are prime candidates for elimination (waiving worries about the possibility of certain kinds of time-travel): suppose that e is not in the future light-cone of c and conversely; then, we might suggest, there is no causal process connecting c to e. Similarly, any path that contains a pair of points such that neither belongs to the future light-cone of the other can be excluded. But this still leaves us with a lot.

Intuitively, there are three kinds of continuous paths through spacetime. There are causal processes, along which information can be transmitted (see Salmon 1984, 141); there are pseudoprocesses (Salmon's paradigms are shadows and moving spots of light on walls), which are incapable of transmitting information, but are relatively respectable; and there are gerrymandered paths, spatio-temporal junk, which do not have enough integrity to lead us to dignify them as processes—one simple, and relatively mild, example would be a segment of the worldline of a mosquito just prior to its biting me and a segment of my worldline immediately following the biting. The last category comprises paths that are continuous but that lack any stability of properties along them. To say just what this amounts to is tricky, but we could succeed in Salmon's project without distinguishing the pseudoprocesses from the spatio-temporal junk. It will be enough if we can separate the genuine causal processes from the rest.

Salmon attacks the problem of distinguishing causal processes by developing ideas of Reichenbach's. "A causal process is capable of transmitting a mark; a pseudo-process is not" (1984, 142). Combining three of Salmon's principles, (MT 148; ST 154; PCI 155), we can define a causal process in terms of epistemologically prior notions as follows:

(CP) P is a causal process iff there are spacetime points c, e, such that P links c and e and it is possible that there should be a modification of P (modifying a characteristic that would otherwise have remained uniform) produced at c by means of a single local interaction and that the modified characteristic should occur at all subsequent points from c to e without any subsequent interaction.

(CP) embodies the idea that causal processes are processes that transmit information because they are *markable*. The modality is obviously needed because not all causal processes are actually marked (think of a universe in which there is a single light signal propagated *in vacuo*). The clause banning subsequent interaction is required because any continuous spatio-temporal path could "transmit a mark" if we were allowed continuous interactions that would impose the marking property upon it at every point.

6.2.1 Some Problems about Processes

Plainly (CP) characterizes the notion of causal process by presupposing the concept of causal interaction. Let us assume that we have the concept of causal interaction in hand, and consider whether the definition of causal process does what it is supposed to do. There seem to be at least four different kinds of problems.

(a) *The problem of pseudomarks.* It is possible for pseudoprocesses to transmit pseudomarks in just the way that causal processes transmit genuine marks. Imagine that a car grazes a stone wall and becomes scratched. The car transmits the scratch: that is, its worldline after the scratch consists of temporal segments that have the property of bearing a scratch. But the shadow transmits the property of being the shadow of a scratched car: its worldline, after the crucial moment, consists of temporal segments that have the property of being a shadow(-stage) of a scratched car.

(b) *The problem of derivative marks.* Suppose that a child traveling in the car puts an arm out the window and holds up a flag. The child's action produces a modification in the shape of the shadow. The modification persists without any further interaction. Provided that the arm is not retracted, the shape of the shadow will continue to be different from what it would otherwise have been.

(c) *The problem of no further interactions.* As I noted above, the clause debarring further interactions is needed to exclude the possibility that pseudoprocesses could "transmit marks" in situations where constant intervention produced the same modification along the spacetime path. But virtually all (all?) *actual* processes are always interacting with other processes. Indeed, there are some causal processes—organisms, for example—for which there could not be any transmission of marks unless there were further interactions. In such cases, the clause that proscribes further interactions will have to be reformulated very carefully. Consider an organism that transmits a mark (a superficial smudge, say). Its actual worldline intersects many other causal processes. We want to say, of course, that those intersecting processes are irrelevant to the persistence of the mark. Even if those particular interactions had been absent the modification would have persisted provided that some interactions of the same general kind had been present. But how do we specify what counts as "the same general kind" here? Some interactions (forceful dowsings, for example) would have removed the smudge. So it appears that if we are to apply (CP) to actual macroscopic cases, we shall need a principled distinction between "relevant" and "irrelevant" interactions—where the concept of relevance is itself a causal notion.

(d) *The problem of fortuitous maintenance.* We can arrange for pseudoprocesses to transmit marks by exploiting the idea that some processes do not require further interventions to keep them going. Imagine that a vehicle equipped with skis is sliding on an ice rink and casting a shadow. A projectile is thrown in such a way that it lands at the edge of the shadow with a horizontal component of velocity equal to that of the shadow of the vehicle. Because the projectile lies across the edge there is an immediate distortion of the shadow shape. Moreover, the distortion persists because the projectile retains its position relative to the vehicle (and to its shadow).

As we shall discover, not all of these difficulties are equally fundamental. However, separately and in combination, they seem to raise serious troubles for Salmon's account of causation—and some of them seem to me to point to deep problems in the epistemology of causation. Let us begin with (a) and (b) where there appears, at first glance, to be an easy solution.

6.2.2 Troubles with Interactions

The obvious suggestion about (a) and (b) is that, in these examples, there is no interaction with the process that is to be "marked." When the shadow takes on the property of being the shadow of a marked car it (the shadow) does not interact with anything—at least it doesn't interact with anything *relevant*. Similarly, when the child thrusts an arm out the window, there is no interaction—at least no *direct* interaction—between the child's arm and the shadow. So we might try to rescue (CP) from examples (a) and (b) by declaring that the alleged markings are pathological and that the pathology consists in the fact that the supposed marking process does not interact with the process which is supposed to be marked.

Tempting as it may seem, this straightforward response is problematic. Recall the point that underlies (c): there are numerous intersections of spacetime paths, and, if the response of the last paragraph is to succeed, then we need to know why none of these intersections can count as a marking process. Specifically, in case (a) there is some intersection between the worldline of the shadow and the worldline of something else (a piece of ground or wall, for example) that occurs at the moment when the car is scratched. In (b) there is an interaction between the child's arm and a beam of sunlight and this interaction transmits properties to the intervening air and ultimately to those molecules that interact with the shadow. Here there is little strain in talking about an *indirect* interaction between the child and the shadow. Thus, if the proposed remedy is to succeed, it will be necessary to explain why the *intersection* of spacetime paths that occurs in (a) does not count as a proper marking and why the apparent *indirect interaction* of (b) should be considered pathological. Notice that even if these explanations are given, there will still be difficulty in accounting for (d), for in that case there is a process (the projectile) that interacts with the shadow.

If (CP) is to be salvaged, then it will be a consequence of the fact that the constraints on causal interactions debar the kinds of pseudomarkings and derivative markings that figure in (a) and (b). So let us investigate Salmon's explicit characterization of causal interactions. This is presented in a principle (CI) (1984, 171):

(CI) Let P_1 and P_2 be two processes that intersect with one another at the spacetime point S, which belongs to the histories of both. Let Q be a characteristic that process P_1 would exhibit throughout an interval (which includes subintervals on both sides of S in the history of P_1) if the intersection with P_2 did not occur; let R be a characteristic that process P_2 would exhibit throughout an interval (which includes subintervals on both sides of S in the history of P_2) if the intersection with P_1 did not occur. Then the intersection of P_1 and P_2 at S constitutes a causal interaction if:

(1) P_1 exhibits the characteristic Q before S, but it exhibits a modified characteristic Q' throughout an interval immediately following S; and

(2) P_2 exhibits the characteristic R before S, but it exhibits a modified characteristic R' throughout an interval immediately following S.

(CI) must be read as claiming only that there are some characteristics Q and R for which these conditions are met, since it would obviously be hopeless to propose that a causal interaction modify all the characteristics that a process would otherwise have retained. We should also note that it is irrelevant for present purposes to demand that the processes P_1 and P_2 be genuine causal processes. Presumably, if they are pseudoprocesses (or worse), they should not be able to satisfy these conditions in any case. However, since the distinction between causal processes and other spatio-temporal lines is what we are trying to draw, we could not appeal to the notion of causal process in (CI), even if we wanted to.

As we saw in the discussion of example (a) above, although when the car scrapes the wall there is no obvious process that can be singled out as intersecting the shadow, there are innumerable contrived spacetime paths that intersect the shadow at the same time. I sketched a strategy for resurrecting the counterexample: select one of these that changes some contrived characteristic just as the shadow takes on the property of being the shadow of a marked car. I shall now develop the strategy in a concrete way and show that (CI) fails to block it.

Assume, for the sake of simplicity, that the shadow of the car is being cast on the wall against which it scrapes, and that there is a stone in the wall, P_2, that is just touched by the shadow of the car at the moment when the car scrapes the wall. If the car had not been so close as to scrape the wall, then the shadow would not have touched that stone. Given that it scraped, then the stone had to be touched by the shadow. Let P_1 be the shadow. Then:

P_1 and P_2 intersect at the moment at which the car scrapes the wall.

Prior to the moment of intersection, P_1 has the characteristic of being the shadow of an unscratched car; after that moment, P_1 has the characteristic of being the shadow of a scratched car.

Prior to the moment of intersection, P_2 has the characteristic of being untouched by the shadow; after the intersection, it has the characteristic of having been touched by the shadow.

Had the intersection not occurred, then, by hypothesis, the car would not have scraped the wall, so that the shadow would not have taken on the property of being the shadow of a scratched car. Had the intersection not occurred, then the stone would not have taken on the characteristic of having been touched by the shadow.

Conclusion: the intersection between P_1 and P_2 meets all the conditions laid down in (CI) for a causal interaction. Hence, we cannot use (CI), Salmon's official explication of the concept of causal interaction, to dismiss the property of being the shadow of a scratched car as a pseudomark.

The discussion so far teaches us how to generalize the problem of which (a) is the prototype. If (CI) exhausts the conditions on causal interactions, then almost any intersection of continuous spacetime paths will constitute a causal interaction. For let P_1 and P_2 be any two continuous spacetime paths, both meeting the condition that for any pair of its points one belongs to the future light-cone of the other, which intersect just once at S. Apparently the intersection at S will count as a causal interaction. For, let Q be the characteristic of not having intersected P_2, R be the characteristic of not having intersected P_1, Q' be the characteristic of having intersected P_2, and R' be the characteristic of having intersected P_1. Then it appears that all the conditions of (CI) are satisfied and that the intersection is a *bona fide* causal interaction.

The most promising rejoinder to the line of argument that I have been developing seems to be to lean heavily on the appeal to counterfactuals. Can we make sense of the idea that, if the intersection with P_2 had not occurred, P_1 would have continued to have Q? That depends on how the processes are picked out. If the identity of P_1 consists in its being that very set of spacetime points (and similarly for P_2) then it is difficult to make sense of the antecedent of the counterfactual. We must impose one further condition to maintain the counterexample: the processes must be identifiable apart from the spacetime points that constitute them. In the case of pseudoprocesses, it is not hard to meet this extra requirement. Imagine, for example, that P_1 and P_2 are intersecting shadows.

Another way to try to save Salmon's account is to place some limitations on the kinds of properties that can serve as marks. Since Goodman's profound investigation of problems about counterfactuals and inductive projection, it has been evident that there are serious difficulties in circumscribing the *genuine* properties (or the genuine natural kinds). The literature on events and "Cambridge changes" (see Geach 1969, Kim 1974) exposes further troubles that must be addressed if cases like (a) and (b) are to be excluded by distinguishing gerrymandered properties and ruling them out as potential marks. Recognizing the kinship between examples like (a) and Goodman's deep puzzles may prove helpful in suggesting an alternative approach to the epistemology of causation, one that would attempt to emulate Goodman's own preferred line of solution to his "new riddle." I shall explore this possibility below.

At best, the maneuvers I have been considering will avail with (a) and its generalization. As already noted, (b) seems to involve some kind of interaction that produces a modification of the shadow, and (d) is an especially pure case in that the modification is direct. Here there seems to be no crucial dependence on contrived "causal interactions," on gerrymandered "marking properties," or on other kinds of Goodmanesque trickery. The shadow takes on a genuine new property: its shape is distorted. Moreover, the distortion would persist without any further interaction—in whatever sense we can give to the idea of macroscopic objects retaining their properties in the absence of further interactions, see (c)—because the projectile retains its position relative to the shadow without any application of force. The example exploits Salmon's strategy of linking causation to the world-picture provided by physics. We don't need further interactions when we have inertial motion, so we can keep a pseudoprocess marked without further interactions by arranging for the mark to undergo inertial motion with the pseudoprocess.

Is it really correct to insist that the mark (the distortion of shadow shape) is preserved without further interactions? After all, the continuation of the shadow requires a surface on which the shadow can be cast, and if this surface were cunningly contoured, there might be compensation for the distortion that the projectile would have produced in the shape of the shadow. But this is simply a special case of a point noted in connection with (c). Ordinary marks on ordinary causal processes—smudges on the cheeks of small children, pieces of chewing gum on balls—persist only if the interactions necessary for the persistence of the marked objects occur. Given sufficiently odd background conditions—driving rain on the child or lubrication of the surface of the ball—the mark would no longer persist. There seems to me to be no obvious basis for distinguishing the perfectly ordinary conditions that help to maintain the distortion of shadow shape from the perfectly ordinary conditions that help to maintain the chewing gum on the baseball or the smudge on the cheek.

Nor is it possible to protest that the interaction between projectile and shadow is a peculiar one in that the worldlines fuse after the intersection. Not only does Salmon explicitly allow for lambda-type interactions (cases in which processes fuse) but we could easily amend the example to ensure that part of the entering process should have a subsequent part of its worldline independent of the process that it intersects. We simply have to suppose that there is a part of the projectile that lands on the interior of the shadow and that is broken off and deflected by the collision. Of course, it would be unreasonable to demand that the *entire* entering process should emerge from the interaction, for that would be at odds with the paradigm cases of marking that Salmon hopes to vindicate: when the chalk marks the object only part of the chalk persists after the interaction.

Cases like (d) provide another clue to the problem of the epistemology of causation. One very obvious point about the situation that I have envisaged is that all kinds of circumstances have to be just right. Consider the general class of cases in which projectiles land on the shadows of moving objects and distort them. In the vast majority of such episodes, the distortion is temporary. The projectile does not move with the shadow in such a way as to maintain the modification of shape. However, even if the velocities of car and projectile are not so exactly coordinated that the projectile maintains a constant relative position to the shadow, there may still be an interval of time throughout which the projectile slides across the shadow in such a way as to distort it. The fact that, at the end of the interval, the shadow reverts to its former shape should not be disconcerting. Many of our paradigms of marks are removed after a finite time in the normal course of events. But isn't it true that an uncoordinated projectile will produce *different* distortions at different points of the time interval during which it distorts the shadow? Yes, but equally, in the ordinary course of events, the shape and thickness of a chalk mark will alter (albeit in ways that are typically imperceptible to us) through the interval in which it marks a ball.

Yet our obvious point suggests something I take to be important. Suppose that we deployed a strategy of attempting to explain the characteristics of shadows by appealing to the properties of processes with which they intersect. That strategy would be doomed to failure because, in the typical case, the processes are not sufficiently well coordinated with the behavior of the shadow for us to be able to account for the subsequent features of the shadow in these terms. The best unification of our beliefs is obtained by incorporating within the explanatory store a pattern of argument in which we derive conclusions about shadow shape from premises about the motions of the objects with respect to sources of light that they partially block and about the characteristics of the surfaces on which the shadows are cast.[28] In general, the motions of objects that cross the shadow are relevant to the shape of the shadow only insofar as they affect the contours of the surface. We could not carry through a strategy of tracing changes in shadow shape to interactions with physical objects that cross the shadow – or, more precisely, we could not do this in a general way. Thus, while *local* violations of Salmon's principle are possible, the recognition that explanation is a matter of instantiating patterns that are applied *globally* enables us to diagnose the source of the trouble. I shall expand on this idea in subsequent sections.

So far I have proceeded by taking some common-sense ideas about macroscopic processes for granted. Strictly speaking, however, macroscopic processes do not conform to Salmon's conditions. The reason is that, as (c) suggests, the characteristics of a macroscopic object are (almost always) sustained by further interactions. Furthermore, the characteristics of such processes are often modified by subsequent interactions, so that, when a process is "marked" (in the ordinary sense) it is likely that the mark will be altered by a later interaction to produce a later mark, which is, in turn, altered to produce a later mark, and so on. Embryological marking furnishes a host of dramatic cases. The embryologist often injects into one of the early cells of an embryo some substance that reacts with the molecules of the cell to produce a new state in which batteries of genes may be switched on or off at quite different times. Consequently, concentrations of proteins may be significantly different from the way they would otherwise have been, so that there may be a new cellular state, an altered tissue geometry, modified intercellular interactions, altered states of neighboring cells, and so on in a cascade of anomalous effects until the biologist finally sees a deviant morphology.

We have here a sequence of causal processes and interactions, and, if we say that the final organism is marked by the initial interaction, that is because we envisage a sequence of marks such that each is transmitted by a causal process that interacts with another process to produce a successor mark. Our attribution is based on our acceptance of a chain of counterfactuals: if P_n had not transmitted M_n then P_{n+1} would not have acquired M_{n+1}. In building up a complex causal process out of elementary causal processes—that is, processes that do not interact with other processes—we need to make heavy use of counterfactuals. The sequence $P_1, \ldots, P_n$ constitutes a complex causal process only if each P_r interacts with P_{r+1} so that if P_r had not been modified to bear Q_r then P_{r+1} would not have been modified to bear Q_{r+1}.

Let us take stock. I have elaborated three kinds of difficulties for Salmon's project, some of which point to quite general troubles in providing an empiricist account of the justification of causal claims. First, there are problems in distinguishing the genuine causal processes from other continuous spatio-temporal paths, and problems in distinguishing causal interactions from mere intersections, all of which are based on the existence of unwanted (spurious?) characteristics—pseudomarks, pseudomodifications, and the like. Second, there are independent problems with both distinctions resulting from the possibility of exploiting the inertia of some processes that can be fortuitously coordinated with pseudoprocesses to "mark" them. Third, Salmon's conditions seem applicable only to ideal (elementary) processes. This requires that we provide an account of the way in which the causal structure of the macroscopic world results from the stringing together of elementary processes. Even if we already had such an account, the emerging picture of our causal knowledge is one in which the justification of *recherché* theoretical claims about idealized processes seems to be fundamental and our ordinary causal knowledge derivative. This is, of course, grist for a mill that I ground above (4.5), when I insisted that our everyday causal knowledge is based on our early absorption of the theoretical picture of the world bequeathed to us by our scientific tradition. But, in any case, as I have argued by reference to the embryological example, our concept of an ordinary macroscopic process involves us in commitment to a large number of counterfactuals.

6.3 Causation and Counterfactuals

It is now time to take up what I view as the most serious trouble of Salmon's project and which I take to threaten any program that tries to use causal concepts to ground the notion of explanation while remaining faithful to an empiricist theory of knowledge. Salmon is very clear in acknowledging that he needs to appeal to counterfactuals in stating his principles (see, for example, 148–49, 171ff.) and in recognizing that he must show how the relevant counterfactuals can be justified. I shall try to show first that the counterfactual commitments of Salmon's theory are far more extensive than he has noted (indeed, I shall claim that the theory naturally evolves into a counterfactual theory of causation), and second that the epistemology for counterfactuals that Salmon outlines encounters grave difficulties.

Return to the question with which we began. We aim eventually to specify the conditions under which two events are causally related, and we have been assuming so far that this can be done by claiming that there is a causal process that links the two events with causal interactions occurring at either end. Now if this assumption were true, then causal explanation would be extraordinarily easy. Take any object that participates in the event for which we want to give an explanation. Its worldline will be a causal process. The considerations that we have drawn from (c) lead us to believe that there will be innumerable causal interactions along this worldline as we trace backward into the past. Any of these earlier points is a terminus of a causal process that is linked to the event that we hope to explain, and at both it and the *explanandum* event we have causal interactions.

The problem of finding causal structure is surely more difficult than this, and the difficulty stems from the fact that we must delineate the *right* causal processes and the *right* causal interactions. Our paradigm, the baseball that hits the window, turns out to be far more intricate than we might have thought. Suppose P is the process that constitutes the window, P_t the point at which the window shatters and $P_t{}'$ some earlier point at which its temperature is changed through the impact of a sudden gust of wind. There is a causal process (the segment of P bounded by $P_t{}'$ and P_t) that links $P_t{}'$ and P_t. There are interactions at both ends: the interaction with the gust of wind at $P_t{}'$ and the interaction with the ball at P_t. (There may even be other interactions at P_t—a second gust of wind, for example.) But the existence of this particular structure of interactions and process is not crucial to the causal history of the event of shattering. To specify the conditions under which c and e are causally related, we need to build into the account the idea that the initial interaction produces the modification that is responsible for the characteristics of the terminal interaction.[29] Intuitively, what is lacking is the kind of articulated structure that I envisaged in the building up of complex processes out of simple ones.

Surely what we want to say about the baseball is something like this. There is an event, the spacetime point c, at which two processes (the bat and the ball) intersect (or, more exactly, overlap). As a result of the intersection, the characteristics of both processes are modified from what they would otherwise have been. Focus on the ball (P_1). We can explicate the dangerously causal-sounding phrase "as a result of" by offering the counterfactual

(A) If the bat had not intersected P_1 then the momentum of P_1 would have been different.

It is in virtue of (A) and related counterfactuals for both bat and ball that we count the intersection of bat and ball as a causal interaction.

Next consider the motion of the ball to the window. During its flight the ball engages in numerous interactions that modify its momentum. What occurs is very like the production of derivative marks in the case of the embryo. We believe many counterfactuals about all these interactions, but, for the sake of simplicity, they may be condensed into

(B) If the momentum of P_1 after its intersection with the bat had been different then the momentum of P_1 just prior to its intersection with P_2 (the window) would have been different.

Now it is because the ball has the momentum it does when it intersects the window that the window breaks in the way that it does. In other words

(C) If the momentum of P_1 just prior to its intersection with P_2 had been different then the properties of P_2 just after the intersection would have been different (specifically, the window would not have broken!).

I take it as evident that these particular counterfactuals can be replaced with weaker counterfactuals about the ball's having non-zero (or appreciable) momentum at different stages of its career. The crucial point is that our claim of a causal relation between c and e depends not simply on the existence of the interactions and the processes but on our acceptance of the counterfactuals (A)–(C) (or of related weaker versions). *We have to invoke counterfactual notions not only in characterizing the concepts of causal process and causal interaction but also in singling out the causal processes and causal interactions that are relevant to particular events.*

But if so much counterfactual machinery is needed to articulate the conditions under which events are causally related, do we really need to talk about processes and interactions at all? Suppose that we simply believed counterfactuals like (A)–(C) – or just that if the bat had not struck the ball then the window would not have been broken. Would this be enough to countenance a causal relation between the striking of the ball and the breaking of the window? Would it be sufficient even if we doubted that there were any intervening process?

Entertaining this possibility takes us in the direction either of past science or of science fiction. The past science comes in with the idea of instantaneous action-at-a-distance. Some philosophers may hold that we would properly maintain that there is a causal relation between two widely separated events if we believed that (i) A acquired property P at t, (ii) B acquired property Q at t, and (iii) if A had not acquired P then B would not have acquired Q. But it is not easy to provide compelling reasons in support of this way of talking, as the eighteenth-century Newtonians discovered.

Science fiction is more promising. Time-travel can be defended as logically possible (see Lewis 1976, Horwich 1987), and certainly time-travel into the future seems less problematic than time-travel into the past. Imagine a time-traveler whose worldline terminates at t, just after he has eaten a peanut butter and jelly sandwich. The time-traveler reappears at a much later time t' with the peanut butter and jelly sandwich in his stomach. He has the sandwich in his stomach at t' because he was ingesting it just before t. There is no continuous process that links the two parts of his worldline, but the pertinent counterfactual holds: if he had not eaten the sandwich then it would not be in his stomach.

In this example there is no continuous process but there are two continuous fragments that surround the big gap in the time-traveler's life. We can make the situation even more dramatic by focusing on complex similarities that hold at the point of departure and at the point of return. As he departs, the time-traveler is quoting verse: "I warmed my hands before the fire of life/It sinks and I . . . "; as he reappears he is still quoting " . . . I am ready to depart." His lips are spaced as they are at the moment of return because that is the way they were when he left. If he had not been in the process of saying "I" when he left, he would not have been in the process of saying "I" when he returned.

I suggest that we can have causation without linking causal processes, and hence causal relations among events at which very peculiar interactions occur. What is critical to the causal claims seems to be the truth of the counterfactuals, not the existence of the processes and the interactions. If this is correct then it is not just that Salmon's account of the causal structure of the world needs supplementing through the introduction of more counterfactuals. The counterfactuals are the heart of the theory, while the claims about the existence of processes and interactions are, in principle, dispensable. Perhaps these notions may prove useful in protecting a basically counterfactual theory of causation against certain familiar forms of difficulty (problems of pre-emption, overdetermination, epiphenomena, and so forth).[30] But, instead of viewing Salmon's account as based on his explications of process and interaction, it might be more revealing to see him as developing a particular kind of counterfactual theory of causation, one that has some extra machinery for avoiding the usual difficulties that beset such proposals.

6.4 Justifying Counterfactuals

But of course many empiricists, including Salmon, worry about the use of counterfactuals in providing accounts of causation. Their anxieties need not simply be the result of an over-reverential attitude toward Hume, Carnap, and their successors. As Goodman (1956) made clear, the straightforward ways of providing a semantics for counterfactuals that will provide them with truth conditions whose presence or absence can be ascertained in ways that empiricists take to be unproblematic encounter apparently insuperable obstacles. Of course, there are illuminating treatments of the semantics of counterfactuals (for example, Lewis 1974, Stalnaker 1968) that enable us to appreciate many points about the logic of counterfactuals that had eluded earlier workers and to formulate controversial issues that had previously been missed. But one thing that semantical accounts in terms of possible worlds and similarity relations fail to do for us is to provide truth conditions that are epistemologically unproblematic. Indeed, the situation with respect to counterfactuals (and modality generally) is distressingly similar to a predicament in the philosophy of mathematics that Benacerraf (1973) presents in compelling fashion: our best semantic accounts and our best epistemological views do not cohere.[31] For the best semantic accounts make reference to possible worlds, our best epistemological views make knowledge (and justification) dependent on the presence of natural processes that reliably regulate belief, and it is (to say the least) unobvious how any natural process could reliably regulate our beliefs about possible worlds.

Salmon does not try to give a semantics for counterfactuals or to resolve the dilemma that I have just indicated. Nor does he endeavor to explain how the counterfactual conditional is reducible to some privileged (empiricistically acceptable) notions. Instead, he pursues the more modest task—which is sufficient for his purposes—of offering an account of how we might justify beliefs in counterfactuals. His account has the virtue of providing a convincing description of the ways in which people do in fact try to justify counterfactuals: no esoteric processes or faculties are invoked. Salmon suggests that we base our counterfactual knowledge on the method of control experiments. To test "If A had not been, then B would not have been" we are to take a sample of test entities, divide it into two subsets under similar conditions, subject one subset to the presence of A and the other to the absence of A. But, of course, the trouble with counterfactuals is that you cannot hold fixed *all* the circumstances of the antecedent while making the antecedent (which is actually false) come true. You have to be selective about what is held constant, and the selection may make a difference to the outcome of the test. Even if the control group and the test group are "similar," the fact that they differ with respect to the presence or absence of A will mean that they differ with respect to many other characteristics. How do we design the *valid* tests, those that do not introduce some crucial difference, unrelated to the presence or absence of *A*, that gives a misleading result?

We can appreciate the general problem by considering a concrete case, arising from an example that Salmon considers (the example is originally due to Patrick Maher and Richard Otte). Imagine that two billiard balls roll across a table with a transparent surface, and that they collide. The table is illuminated from above, so that there is also an intersection of their shadows. Under these circumstances, we will want to claim that the impact of Ball *A* on Ball *B* causes a change in the direction of motion of Ball *B*, but that the crossing of the shadow of Ball *B* by the shadow of Ball *A* does not cause the change in the direction of motion of the shadow of Ball *B*. The problem is to explain how the first causal claim is justified and the second is not.

Call the shadow of Ball *A* P_A and the shadow of Ball *B* P_B. Then the problem arises because, at first sight, both of the following counterfactuals are true

(i) If the worldline of Ball *A* had not intersected that of Ball *B*, then the direction of motion of Ball *B* would not have altered.

(ii) If the worldline of P_A had not intersected that of P_B, then the direction of motion of P_B would not have altered.

Following the method of testing that Salmon suggests, we might confirm (i) by dividing a sample of 100 ball rollings into two types—cases in which there is an intersection of balls on the table (the control cases) and cases in which there is no such intersection (the experimental cases). Call this Experiment I. Allegedly Experiment I confirms (i) because in the cases in which the balls do not meet their directions are unaltered. Now when the balls do not meet neither do their shadows. Hence we might suppose that Experiment I also suffices to confirm (ii)—for does it not show that when the shadows do not meet the directions of motion of the shadows are unaltered?

Salmon claims that Experiment I is the wrong experiment for testing (ii). He urges that our experimental group should comprise cases in which the balls collide, but part of the surface of the floor is illuminated from below the table so that P_A is absent. Call this Experiment II. In Experiment II P_B changes its direction of motion, even in the experimental cases, so we have grounds for saying that II refutes the crucial counterfactual.

The important question is how we justify testing (ii) by performing Experiment II rather than Experiment I. Salmon is surely right to claim that II is the intuitively right experiment to perform, the experiment that accords with our natural ideas about the causal character of the situation. But we are looking for the basis of those ideas, and, in consequence, we need a theory that tells us which controlled experiments are the correct ones.

To see the force of this question, consider an analogous experiment that is proposed to test (i). In the experimental cases, we remove Ball *A* from the table but we "simulate its presence" by subjecting Ball *B* to an appropriately strong impulse at the moment when it would have collided with the absent ball. (Note that the impulse restores certain similarities with the control cases that would otherwise be absent: for example in the motions of air molecules above the table.) Call this Experiment III. Now it is natural to protest that Experiment III involves gratuitous intervention. But isn't it possible to make a similar protest in the case of Experiment II where we introduce a new object (the light source) into the situation? What justifies us in thinking that the interventions involved in II do not confound the probative force of the test while those involved in III do?

The obvious answer (very obvious to anyone who thinks about counterfactuals along Lewis-Stalnaker lines) is that we try to keep the situation of the control group as close to that of the experimental group as possible. But in which respects should we prize similarity? In Experiment I the test group and the control group involve two balls and two shadows; the difference is just that, in the cases in the control group, the balls collide while in those of the test group they do not. In Experiment II the features of the balls, including their collision, are common to both groups; but the groups differ in that the test group shows only one shadow and involves an extra object, the new light source. Finally, in the crazy Experiment III we also introduce a new entity, the source of the impulse, but we may restore certain kinds of similarity between the control and the experimental cases (for example in the air currents above the table). There is a tradeoff here of similarities and differences, and we do not have a theory of the justification of counterfactuals until we have an account of how the trades are made.

In practice, of course, scientists design control experiments by drawing on their background causal knowledge. They endeavor to ensure that the control group and the experimental group are similar in those respects *that they take to be potentially causally relevant*. Once we have some causal knowledge (perhaps a significant amount) then that causal knowledge can be used in the design of control experiments that will test counterfactuals in just the way that Salmon proposes.[32] But if we are looking for a theory of how we justify counterfactuals from scratch, then the appeal to the method of controlled experiments is of no avail.

6.5 Changing the Epistemological Framework

At this point it is reasonable to scrutinize the general epistemological assumptions which have been taken for granted in the framing of our problem, to see if the task that has been set is overly ambitious. For Hume and his logical empiricist successors, providing an account of our causal knowledge required identifying judgments that can be justified on the basis of observation and inferences that would lead from those judgments to justify the causal claims that we accept. However, there are reasons, forcibly presented by Quine, Sellars, Kuhn, and others, for rejecting the tacit idea that each person's knowledge can be reconstructed to reveal a chain of justifying inferences whose ultimate premises are statements that are completely justified by the person's own experiences. Instead, each of us inherits a body of lore from the previous generation, and some, the creative and the talented, modify that corpus and thus affect what is transmitted to the successor generation. There is no question of justifying our knowledge from privileged premises that record our own observations, for each of us is thoroughly indebted to our contemporaries and to the historical tradition.

If we apply this general point to the issue of how we are able to have justifications for causal and counterfactual claims, then we should want to question the idea that somehow each of us begins without any causal knowledge (or any knowledge of counterfactuals) and manages to build up such knowledge without assistance from others. Instead, we are never in the predicament that Hume and the logical empiricists depict. From the start of our conscious experience we absorb causal judgments as part of the lore of our ancestors. Hence it is entirely reasonable to attack the problem of causal knowledge by describing a method for extending such knowledge, *provided that we can show how that method could have been used to build up the knowledge we now take ourselves to possess from the basis with which our prehistoric ancestors began.*[33]

The view I am recommending is exactly parallel to the account of mathematical knowledge that I have defended elsewhere (Kitcher 1983, chaps. 6–10). If we are to understand how the causal and counterfactual claims that we currently make are warranted then we should show that there is a sequence of states of science, beginning with some state for which we can trace a direct justification and ending with our own corpus, such that each state in the sequence is obtained from its predecessor by a justification-conferring transition. Now it is possible that defenders of the causal approach to explanation can complete the project I have outlined, showing how causal knowledge is systematically built up without appealing to methodological principles that govern the acceptance and rejection of putative explanations. The aim of this section has been to identify the problems that they will have to overcome, not to close the books on the causal approach. But it does seem to me that the general epistemological approach that I have been sketching is far more congenial to the view that explanation consists in the systematization of our beliefs.

Consider the following story, highly fanciful and oversimplified, but, I think, one that bears an important moral about the growth of our causal knowledge. Imagine our remote ancestors with a primitive stock of unconnected beliefs about the world. In attempting to systematize their beliefs they arrive at the first picture of the order of nature, coming to see some phenomena as dependent on others. At this stage, their state can be represented by the language they use, the claims that they accept, and the explanatory patterns that they endorse. Successive states of science are generated as the language is modified, the body of claims revised, and the store of explanatory patterns altered. From the very beginning the construction of the explanatory store is guided by the directive to unify belief, in the sense given in section 4 (see sections 7 and 8 for further elaboration of it). At each stage, the explanatory store supplies an ordering for the phenomena and serves as the basis for the introduction of causal concepts. These are absorbed in childhood, sometimes with the giving of scraps of causal information, sometimes simply by learning parts of the language. But the crucial point is that the "because" of causation is always derivative from the "because" of explanation. In learning to talk about causes or counterfactuals we are absorbing earlier generations' views of the structure of nature, where those views arise from their attempts to achieve a unified account of the phenomena.

The final sections will attempt to defend one central idea of this story. I shall try to show that the principle of unification can be formulated so as to give genuine methodological guidance, enabling us to justify modifying the explanatory store in certain ways and not in others. *En passant*, I shall consider the problems of asymmetry and irrelevance.

7. Comparative Unification

There are two distinct contexts in which a methodological principle directing us to unify our beliefs can be expected to operate. In one of these – the simpler of the two – we consider a fixed body of beliefs and use the principle to select that set of derivations (among some class of alternative sets of derivations) that best unifies the belief corpus. The second context is more difficult because it allows for changes in the corpus, and possibly even in the language in which the beliefs are framed. However, we need to consider this context because changes in the state of a science, including changes in belief and in language, are often justified by appeal to the idea that the changes will yield an increase in explanatory power. I shall start with the simpler case, and introduce complications later.

7.1 Comparative Unification without Change of Belief

Let K be the set of statements accepted at some stage in the development of science, and let L be the language used to formulate those statements. Suppose that S and S' are sets of derivations such that all members of each set are acceptable relative to K. The principle that we want to adopt is

(U) S should be chosen over S' as the explanatory store over K, $E(K)$, just in case S has greater unifying power with respect to K than S'.

Note that (U) is formulated as a comparative principle, directing us to make a choice between proposed alternatives. This is because the project is to understand how justified choices are made in the growth of scientific knowledge, and we can safely assume that each choice situation involves a set of proposed alternatives.

The next task is to say how we make judgments of unifying power. Recall from section 4.3 that unifying power depends on paucity of patterns used, size of conclusion set, and stringency of patterns. As I admitted, when the criteria pull in different directions, it will be hard to see how tradeoffs are made. However, this possibility *may* not prove troublesome. Perhaps whenever there is competition between two sets of patterns with different virtues, we can find an acceptable way to combine the virtues. Let us formulate an explicit principle to express optimism about this.

(O) Let U, U' be sets of patterns. Then there is a set of patterns U^* such that

(a) there is a one-one mapping from U^* to U, f, and a one-one mapping from U^* to U, f', such that for each pattern p in U^*, p is at least as stringent as $f(p)$ and at least as stringent as $f'(p)$; (one or both of f, f' may be injections rather than surjections)

(b) let S, S', S^* be the sets of derivations that are the complete instantiations of U, U', and U^* with respect to K; then the consequence sets $C(S)$, $C(S')$, $C(S^*)$ are such that $C(S)$ and $C(S')$ are both subsets (not necessarily proper) of $C(S^*)$.

(A bijection is a one-one mapping. A bijection from A to a subset of B is an injection into B. A bijection that is onto B is a surjection.)

Here clause (a) tells us that U^* does at least as well as its rivals by the criteria of stringency and paucity of patterns and (b) tells us that it does at least as well as generating consequences over K. If (O) is correct—or if it is correct for the bodies of belief that we are interested in considering, then the problem of tradeoffs is unworrying, because in a situation in which rival systematizations have different virtues we can always reject both of them in favor of a systematization that combines their merits. However, as my dubbing of the principle hints, I do not know whether (O)—or some useful restriction of it—is true.

We can formulate a condition on the comparative unifying power of sets of patterns in the obvious way.

(C) Let U, U' be sets of patterns and S, S' their complete instantiations with respect to K. Then U has greater unifying power than U' if one (or both) of the following conditions is met.

(C1) $C(S')$ is a subset of $C(S)$, possibly though not necessarily proper, and there is a one-one mapping f from S to S' such that for each pattern p in S, p is at least as stringent as $f(p)$, and such that either f is an injection or f is a surjection and there is at least one pattern p in S such that p is more stringent than $f(p)$.

(C2) $C(S')$ is a proper subset of $C(S)$ and there is a one-one map f from S to S' (either an injection or a surjection) such that for each p in S, p is at least as stringent as $f(p)$.

(C1) applies if S uses fewer or more stringent patterns to generate the same conclusions as S'. (C2) holds if S does equally well as S' by criteria of stringency and paucity of patterns and is able to generate a broader class of consequences. It is not hard to show that the comparative relation introduced by (C) has the right features to order sets of patterns with respect to unifying power. It is both asymmetric and transitive.

So far I have taken the notion of stringency for granted. As with my approach to comparative unifying power, I shall ignore the problem of tradeoffs and offer conditions for comparing argument patterns with the same structure. Consider first patterns that have a common classification. In this case, one is more stringent than another if corresponding schemata in the first are subject to demands on instantiation that are more rigorous than those in the second. The idea can be made more precise as follows.

(T) Let $<s,i>$ be a pair whose first member is a schematic sentence and whose second member is a complete filling instruction for that sentence, and let $<s',i'>$ be another such pair. Suppose that s and s' have a common logical form. Let g be the mapping that takes each nonlogical expression (or schematic letter) in s to the nonlogical expression (or schematic letter) in the corresponding place in s'. For any schematic letter t occurring in s, $<s,i>$ is tighter than $<s',i'>$ with respect to t just in case the set of substitution instances that i allows for t is a proper subset of the set of substitution instances that i' allows for $g(t)$; $<s,i>$ is at least as tight as $<s',i'>$ with respect to t just in case the set of substitution instances that i allows for t is a subset of the set of substitution instances that i' allows for $g(t)$. $<s,i>$ is tighter than $<s',i'>$ just in case, (i) for every schematic letter occurring in s, $<s,i>$ is at least as tight as $<s',i'>$ with respect to that schematic letter, (ii) there is at least one schematic letter occurring in s with respect to which $<s,i>$ is tighter than $<s',i'>$ or there is a nonlogical expression e occurring in s such that $g(e)$ is a schematic letter, and (iii) for every schematic letter t occurring in s, $g(t)$ is a schematic letter. If only conditions (i) and (iii) are satisfied, then $<s,i>$ is at least as tight as $<s',i'>$

Let p, p' be general argument patterns sharing the same classification. Let $<p_1, \ldots p_n>$ and $<p'_1, \ldots p'_n>$ be the sequence of schematic sentences and filling instructions belonging to p and p' respectively. Then p is more stringent than p' if for each j ($1 \leq j \leq n$) p_j is at least as tight as p'_j and there is a k such that p_k is tighter than p'_k.

There is another way in which one argument pattern might be more stringent than another. One pattern might have a classification that indicated that an inferential transition is to be made by appealing to certain kinds of principles while another might articulate the intervening structure by specifying schematic premises that are to be linked in definite ways.[34] If the latter precludes certain possible instantiations that the former leaves open, then it is appropriate to count it as more stringent. Once again, the idea can be made precise as follows.

(R) Let p, p' be general argument patterns such that the sequence of schematic sentences and filling instructions of p is $<p_1, \ldots p_n>$ and the sequence of schematic sentences and filling instructions of p' is $<p_1, \ldots p_r,$ $q_1, \ldots q_s, p_{r+1}, \ldots p_n>$. Suppose that the classifications differ only in that for p one or more of the p_{r+j} is to be obtained from previous members of the sequence by derivations involving some further principles of a general kind G, while for p' that (or those) p_{r+j} are to be obtained from the same earlier members of the sequence and from some of the q_k by specified inferential transitions. Suppose further that in each case of difference the set of subderivations allowed by p' is a subset of the set of subderivations allowed by p, and that in at least one case the relation is that of proper inclusion. Then p' is more stringent than p.

(T) and (R) together provide analyses of the two basic ways in which one argument pattern may be more stringent than another. I do not pretend that they provide a complete account of relative stringency, but I hope it will be possible to see how to combine them. Of course, there is the worrying theoretical possibility that we may be forced to judge between argument patterns, one of which scores well by the kinds of considerations adduced in (T), while the other is recommended by the kinds of considerations adduced in (R). As with the account of comparative unification, we may hope that when this occurs there will be some acceptable argument pattern that combines the merits of both—but perhaps this is overly optimistic. In any case, the conditions I have given enable us to tackle the problems of comparison that we need to address.

At this point we have the resources to understand how the justification of accepting certain sets of derivations, rather than others, might work. Suppose that we are comparing the merits of two systematizations of K, S and S'. If the best set of patterns we can think of that will generate S fares better in unifying K, as judged by (C) in light of such claims about stringency as (T) and (R), than the best set of patterns we can think of that will generate S', then, according to (U), we are entitled to prefer S to S'. Note that we should allow for justifiable mistakes in overlooking some relatively recondite complete generating set, just as we should allow for justifiable error on the part of a scientist who misses some subtle idea in formulating a theoretical view.

7.2 The Possibility of Gerrymandering

Unfortunately, there is a serious worry about whether the principles that I have introduced will enable us to solve the problems of asymmetry and irrelevance by debarring unwanted derivations. Intuitively, the line of solution that we want to adopt consists in showing that those who accept the wrong derivations are committed to accepting more patterns than they need, or to accepting less stringent patterns than they should, or to generating a more restricted set of consequences. But, as matters now stand, there is room to wonder whether, given any derivation that one wants to incorporate within the explanatory store, one can always discover a set of derivations including it and contrive a set of patterns that will serve as the basis of that set of derivations, in such a way that the contrived set of patterns will score just as well as any rival set of patterns that will generate the derivations that the orthodox accept. To put it bluntly, are the principles that I have assembled toothless?

The difficulty arises because the account I have given lacks resources for convicting those who employ deviant derivations of using deviant patterns. We can formulate a general strategy for exploiting the deficiency as follows. Let us suppose that our accepted ideal explanation of some *explanandum E* is a derivation d that instantiates a pattern p. Someone now proposes that a different derivation d' furnishes the ideal explanation of E. As we shall see shortly, the general strategy of challenging deviance is to show that acceptance of d' would commit us to employing a pattern p', confronting us with the options of either generating the more limited set of consequences yielded by p' or else employing both p and p' and thereby flouting the maxim of minimizing the number of patterns used. But this strategy depends crucially on a view about how patterns are to be individuated. Imagine that the set of derivations proposed as a rival to orthodoxy is just the orthodox set except that d' has been substituted for d. How do we show that the rival set has to use one more pattern than the orthodox set in generating the same consequences?

The obvious and natural approach is this. Call the orthodox set S, the rival S'. S contains many instantiations of p other than d, and S' will contain these derivations too. Hence, if p belongs to the basis of S then p must belong to the basis of S'. But d' doesn't instantiate p, so it must be generated from some other pattern p' in the basis of S'. Now p' cannot belong to the basis of S, for, if it did, then d' would belong to S (recall that an acceptable systematization must contain all instantiations of the patterns in its basis). Therefore p' must be an extra pattern, so that the basis of S' contains all the patterns in the basis of S and one more besides.

But this cannot be quite right. For if S' has *precisely* the pattern p in its basis, then, provided we heed the requirement of completeness, all instantiations of p, and hence d itself, would have to belong to S'. So the basis of S' must contain some doctored version of p, p^d, that yields all the instances of p except for d. Presumably this is accomplished by adding some filling instruction to debar the substitutions that would generate d. Now, of course, it begins to look as though the basis of S' will lose on the paucity of patterns but win on stringency. But this is not the end of the matter. For while we are doctoring, we can surely contrive some gerrymandered pattern $p^{d'}$ that will generate all the instantiations of p except for d together with d'. In effect, we would play with the classification and the schematic sentences until the derivations looked as alike as possible and then contrive filling instructions to let us substitute in just the right way to give the instantiations we want.

We need some requirements on pattern individuation that will enable us to block the gerrymandering of patterns by disjoining, conjoining, tacking on vacuous premises and so forth. The strategy sketched in the last paragraph attempts to disguise two patterns as one, and it does so by making distinctions that we take to be artificial and by ignoring similarities we take to be real. Thus the obvious way to meet the challenge is to demand that the predicates occurring in the schematic sentences, those employed in formulating the filling instructions, and those that figure in the classification all be projectable predicates of the language in which K is formulated. If a pattern fails to satisfy these conditions, then we must decompose it into several elementary patterns that do, and use the number of elementary patterns in our accounting.

7.3 Asymmetry and Irrelevance

The time has now come to put all this abstract machinery to work. Problems of asymmetry and irrelevance take the following general form. There are derivations employing premises which are (at least plausible candidates for) laws of nature and that fail to explain their conclusions. The task is to show that the unwanted derivations do not belong to the explanatory store over our current beliefs. To complete the task we need to argue that any systematization of our beliefs containing these derivations would have a basis that fares worse (according to the principles (U), (C), (T), and (R) stated above) than the basis of the systematization that we actually accept. In practice, this task will be accomplished by considering a small subset of the explanatory store, the derivations that explain conclusions akin to that of the unwanted derivation, and considering how we might replace this subset and include the unwanted derivation. I want to note explicitly that there is a risk that we shall overlook more radical modifications of the explanatory store which would incorporate the unwanted derivation. If there are such radical modifications that do as well by the criteria of unifying power as the systematization we actually accept, then my account is committed to claiming that we were wrong to treat the unwanted derivation as nonexplanatory.

7.3.1 The "Hexed" Salt

Let us start with the classic example of explanatory irrelevance. A magician waves his hands over some table salt, thereby "hexing" it. The salt is then thrown into water, where it promptly dissolves. We believe that it is not an acceptable explanation of the dissolving of the salt to point out that the salt was hexed and that all hexed salt dissolves in water. What is the basis of this belief?

Suppose that $E(K)$ is the explanatory store over our current beliefs, K, and that S is some set of derivations, acceptable with respect to K, that has the unwanted derivation of the last paragraph as a member. One of the patterns used to generate $E(K)$ derives claims about the dissolving of salt in water from premises about the molecular composition of salt and water and about the forming and breaking of bonds. This pattern can be used to generate derivations whose conclusions describe the dissolving of hexed salt and the dissolving of unhexed salt. How does S provide similar derivations? Either the basis of S does not contain the standard pattern or it contains both the standard pattern and a nonstandard pattern that yields the unwanted derivation. In the former case, S fares less well than $E(K)$ because it has a more restricted consequence set, and, in the latter case, it has inferior unifying power because its basis employs all the patterns of the basis of $E(K)$ and one more besides.

It is obviously crucial to this argument that we exclude the gerrymandering of patterns. For otherwise the claim that the basis of S must contain either the nonstandard pattern alone or the nonstandard pattern plus the standard pattern would be suspect. The reason is that we could gerrymander a "pattern" by introducing some such Goodmanian predicate as "x is either hexed, or is unhexed and has molecular structure $NaCl$." Now we could recover derivations by starting from the claim that all table salt satisfies this predicate, by using the principle that all hexed table salt dissolves in water to generate the conclusion from one disjunct and by using the standard chemical derivation to generate the conclusion from the other disjunct. This maneuver is debarred by the requirement that the predicates used in patterns must be projectable from the perspective of K.

Consider next a refinement of the original example. Not all table salt is hexed, but presumably all of it is hexable. (For present purposes, we may assume that hexing requires only that an incantation be muttered with the magician's thoughts directed at the hexed object; this will obviate any concerns that some samples of table salt might be too large or too inaccessible to have the magician wave a hand over them.) Suppose now that it is proposed to explain why a given sample of table salt dissolves in water by offering the following derivation:

a is a hexable sample of table salt.
a was placed in water.
Whenever a hexable sample of table salt is placed in water, it dissolves.
a dissolved.

I take it that this derivation strikes us as nonexplanatory (although it is useful to point out that it is not as badly nonexplanatory as the derivation in the original example). Suppose that S is a systematization of K that contains the derivation. Can we show that S has less unifying power than $E(K)$?

Imagine that *S* had the same unifying power as *E(K)*. Now in *E(K)* the mini-derivation that is most akin to the one we want to exclude derives the conclusion that *a* dissolved from the premise that *a* is a sample of table salt, the premise that *a* was placed in water, and the generalization that samples of table salt that are placed in water dissolve. Of course, this mini-derivation is embedded within a much more exciting chemical derivation whose conclusion is the generalization that samples of table salt dissolve when placed in water. That derivation instantiates a general pattern that generates claims about the dissolving (or failure to dissolve) of a wide variety of substances from premises about molecular structure. In its turn, that general pattern is a specification of an even more general pattern that derives conclusions about chemical reactions and state changes for all kinds of substances from premises about molecular structures and energy distributions. If *S* is to rival *E(K)* then it must integrate the unwanted mini-derivation in analogous fashion.

That can be done. One way to proceed would be to use the standard chemical derivation to yield the conclusion that all samples of table salt dissolve when placed in water and then deduce that all hexable samples of table salt dissolve when placed in water. But now we can appeal to a principle of simplifying derivations to eliminate redundant premises or unnecessary steps. When embedded within the standard chemical derivation, the unwanted mini-derivation is inferior to its standard analog because the latter is obtainable more directly from the same premises. An alternative way of trying to save the unifying power of *S* would be to amend the standard chemical patterns to suppose that they apply only to hexable substances. But since it is supposed that *all* substances are hexable, and since this fact is used throughout *S* to generate derivations to rival those produced in *E(K)*, this option effectively generates a set of derivations that systematically contain idle clauses. Since it is believed that everything is hexable, the outcome is as if we added riders about objects being self-identical or being nameable to our explanations, and again a principle of simplification directs that the idle clauses be dropped.[35]

We can now achieve a diagnosis of the examples of explanatory irrelevance. Citation of irrelevant factors will either commit one to patterns of explanation that apply only to a restricted class of cases or the irrelevancies will be idle wheels that are found throughout the explanatory system. The initial hexing example illustrates the first possibility; the refinement shows the second.

7.3.2 Towers and Shadows

Let us now turn to the asymmetry problem, whose paradigm is the case of the tower and the shadow. Once again, let *K* be our current set of beliefs, and let us compare the unifying power of *E(K)* with that of some systematization *S* containing a derivation that runs from the premises about shadow length and sun elevation to a conclusion about the tower's height. As in the case of the irrelevance problem, there is a relatively simple argument for maintaining that *S* has less unifying power than *E(K)*. There are also some refinements of the original, troublesome story that attempt to evade this simple argument.

Within *E(K)* there are derivations that yield conclusions about the heights of towers, the widths of windows, the dimensions of artefacts and natural objects alike, which instantiate a general pattern of tracing the present dimensions to the conditions in which the object originated and the modifications that it has since undergone. Sometimes, as with flagpoles and towers, the derivations can be relatively simple: we start with premises about the intentions of a designer and reason to an intermediate conclusion about the dimensions of the object at the time of its origin; using further premises about the conditions that have prevailed between the origin and the present, we reason that the object has persisted virtually unaltered and thus reach a conclusion about its present dimensions. With respect to some natural objects, such as organisms, stars, and mountain ranges, the derivation is much more complex because the objects have careers in which their sizes are substantially affected. However, in all these cases, there is a very general pattern that can be instantiated to explain current size, and I shall call derivations generated by this pattern *origin-and-development* explanations.

Now if *S* includes *origin-and-development explanations*, then the basis of *S* will include the pattern that gives rise to these derivations. To generate the unwanted derivation in *S*, the basis of *S* must also contain another pattern that derives conclusions about dimensions from premises about the characteristics of shadows (the *shadow* pattern). In consequence, *S* would fare worse than *E(K)* according to our principles ((U) and so forth) because its basis would contain all the patterns in the basis of *E(K)* and one more. Notice that, once again, the "no gerrymandering" requirement comes into play to block the device of fusing some doctored version of the pattern that generates *origin-and-development* explanations with the shadow pattern. So *S* must foreswear *origin-and-development* explanations.

However, it now seems that *S* must have a consequence set that is more restricted than that of *E(K)*. The reason is that the *shadow* pattern cannot be instantiated in all the cases in which we provide *origin-and-development* explanations. Take any unilluminated object. It casts no shadow. Hence we cannot instantiate the *shadow* pattern to explain its dimensions.

This is correct as far as it goes, but the asymmetry problem cuts deeper. Suppose that a tower is actually unilluminated. Nonetheless, it is possible that it should have been illuminated, and if a light source of a specified kind had been present and if there had been a certain type of surface, then the tower would have cast a shadow of certain definite dimensions. So the tower has a complex dispositional property, the disposition to cast a shadow of such-and-such a length on such-and-such a surface if illuminated by a light-source at such-and-such an elevation above the surface. From the attribution of this dispositional property and the laws of propagation of light we can derive a description of the dimensions of the tower. The derivation instantiates a pattern, call it the *dispositional-shadow* pattern, that is far more broadly applicable than the *shadow* pattern.

But can it be instantiated widely enough? To be sure it will provide surrogates for *origin-and-development* explanations in those cases in which we are concerned with ordinary middle-sized objects. But what about perfectly transparent objects (very thin pieces of glass, for example)? Well, they can be handled by amending the pattern slightly, supposing that such objects have a disposition to be coated with opaque material and then to cast a shadow. Objects that naturally emit light can be construed as having a disposition to have their own light blocked and then to cast a shadow. Objects that are so big that it is hard to find a surface on which their shadows could be cast (galaxies, for example) can be taken to have the dispositional property of casting a shadow on some hypothetical surface.

Yet more dispositional properties will be needed if we are to accommodate the full range of instances in which *origin-and-development* explanations are available. An embryologist might explain why the surface area of the primitive gut (archenteron) in an early embryo is of such-and-such a size by deriving a description of the gut from premises about how it is formed and how modified. To instantiate the *dispositional-shadow* pattern in such cases, we shall need to attribute to the gut-lining a dispositional property to be unrolled, illuminated, and thus to cast a shadow. A biochemist might explain the diameter in the double helix of a DNA molecule by identifying the constraints that the bonding pattern imposes on such molecules both as they are formed and as they persist.[36] Taking a clue from the principles of electronmicroscopy, the *dispositional-shadow* pattern can be instantiated by supposing that DNA molecules have a dispositional property to be coated and irradiated in specified ways and to produce absorption patterns on special surfaces. And so it goes.

Perhaps there are some objects that are too small, or too large, too light sensitive, or too energetic for us to attribute to them any disposition to cast anything like a shadow. If so, then even with the struggling and straining of the last paragraph, the *dispositional-shadow* pattern will still fail to generate derivations to rival those present in $E(K)$. But I shall assume that this is not so, and that for any object whose dimensions we can explain using our accepted patterns of derivation, it is possible to find a dispositional property that has something to do with casting a shadow.

However, if we now consider the critical predicate that appears in the *dispositional-shadow* pattern, we find that it is something like the following: "x has the disposition to cast a shadow if illuminated by a light source or x has the disposition to produce an absorption pattern if x is suitably coated and irradiated or x has the disposition to cast a shadow if x is covered with opaque material or x has the disposition to cast a shadow if x is sectioned and unrolled or x has the disposition to cast a shadow after *x* has been treated to block its own light sources or . . . " At this point it is surely plain that we are cutting across the distinctions drawn by the projectable predicates of our language. Any "pattern" that employs a predicate of the sort that I have (partially) specified is guilty of gerrymandering, for, from our view of the properties of things, the dispositions that are lumped together in the predicate are not homogeneous. I conclude that even if it is granted that we can find for each object some dispositional property that will enable us to derive a specification of dimensions from the ascription of the disposition, there is no *common* dispositional property that we can employ for *all* objects. To emulate the scope of $E(K)$, the basis of S would have to contain a multiplicity of patterns, and our requirement against gerrymandering prohibits the fusion of these into a single genuine pattern.

As in the case of the irrelevance problem, there is a natural diagnosis of the trouble that brings out the central features of the foregoing arguments. Explanation proceeds by tracing the less fundamental properties of things to more fundamental features, and the criterion for distinguishing the less from the more fundamental is that appeal to the latter can be made on a broader scale. Thus an attempt to subvert the order of explanation shows up in the provision of an impoverished set of derivations (as in our original example of the tower and the shadow) or in the attempt to disguise an artificial congeries of properties as a single characteristic (as in our more recent reflections).

7.3.3 When Shadows Cross

At this point it is worth applying the account to the difficulties that emerged in the previous section. Consider the question why we do not consider the intersection of two shadows a causal interaction. The answer is surely that explaining the changes in directions of motion and in shape of a shadow by reference to the relations between it and another shadow would commit us to a pattern of explanation that is far less broadly applicable than that which consists in deriving the properties of shadows from the properties of the objects of which they are the shadows, in accordance with principles about light propagation. To see this, one need only note that we are able to instantiate the latter in situations where there is only one shadow around. Thus, a policy of explaining changes in the properties of shadows in terms of "interactions" among shadows would commit us to a pattern that would yield an impoverished explanatory store—unless, of course, we either admitted the standard pattern as well (thus adopting one more pattern than we need) or else gerrymandered a "pattern" by cutting across the divisions made within our language. I hope it will be apparent that many of the troubles that arose for Salmon's treatment of pseudoprocesses, pseudointeractions, and counterfactuals can be dissolved by invoking our principles about explanatory unification.

If the foregoing is correct, then the unification approach can apparently overcome its most significant obstacle, namely the problems posed by explanatory asymmetry and explanatory irrelevance. But we should not celebrate too soon. As I remarked at the beginning of this section, the principle of explanatory unification should be expected to operate in two distinct contexts. We have been operating within the simpler of these, supposing that we are assessing the merits of two different systematizations of the *same* body of beliefs formulated in the *same* language. The significance of this restriction should be apparent from the fact that I have had to appeal on several occasions to the requirement that separate patterns cannot be fused in artificial ways, and that requirement assumes that the standards of artificiality are set by a *shared* language.

7.4 Comparative Unification and Scientific Change

However, it is a commonplace that scientific change may involve changes in belief and changes in language, both of which are justified on the grounds that the new beliefs and the new language have greater explanatory power than do the old. As a result, the principle of explanatory unification ought to be formulated so as to enable us to decide whether it would be reasonable to modify our scientific practice from $<L, K, E(K)>$ to $<L', K', E(K')>$ on grounds of attaining greater unification in our beliefs. I shall assume that such transitions have sometimes legitimately occurred in the history of science, for example, in the Darwinian revolution and in the birth of electromagnetic theory, and that something of the same kind is currently being envisaged by workers in particle physics. The difficulty is to allow for such changes without undermining our solutions to problems of asymmetry and irrelevance.

To see how such problems might re-emerge, reflect on the refined version of the asymmetry problem. I blocked the device of lumping many different dispositions in a single predicate by insisting that acceptable patterns must employ predicates that conform to the divisions made by the projectible predicates of our language. I have now admitted that linguistic change may be motivated by the desire to achieve explanatory gains. Hence, it would appear possible to defend a systematization whose basis included the *dispositional-shadow* pattern, by arguing that the gerrymandered predicate is a projectable predicate of the new language and recommending the linguistic change. As it stands there is no explanatory *gain* here—only the avoidance of explanatory loss—but once one is in the business of gerrymandering predicates, there will surely be ways of adding some further (unconnected) disjunct to the predicate that figures in the *dispositional-shadow* "pattern" and so generating at least one derivation with a conclusion that is not the conclusion of any derivation in $E(K)$.

Two separate issues arise here. First, we need to specify the conditions under which a systematization S of K provides a better unification of K than a systematization S' of K' does of K'. Second, we need to say when the fact that the best systematization of K', $E(K')$, provides a better unification of K' than the best systematization of K, $E(K)$, does of K gives us reason to make the transition from $<L, K, E(K)>$ to $<L', K', E(K')>$. I shall not try to provide anything like a general account here, but will simply endeavor to offer partial conditions that seem to me to underlie important transitions in the history of science. To address the worry that allowing appeal to explanatory unification as a basis for scientific change, including linguistic change, reintroduces the asymmetry and irrelevance problems, I shall offer an obvious and intuitive suggestion. There is no force to the idea that switching to new language and/or new beliefs would enable one to have a more unified set of beliefs if there are principles that prohibit the kind of linguistic change or belief change envisaged. Thus, I propose that the search for unification of belief is conditional on principles that govern the modification of language and that rule on the acceptability of the proposed beliefs. It is easy to understand that there have to be some such principles, for otherwise unification could run riot over the deliverances of experience. My claim is that the principles are sufficiently powerful to preclude the maneuver envisaged in the previous paragraph in my resurrection of the asymmetry problem.

In considering the first issue, I shall restrict my attention to the special case in which the shift involves no explanatory loss. Assume that our principles (U), (C), (T), and (R) determine for two belief corpora K and K' unique best systematizations $E(K)$ and $E(K')$ respectively. Then the condition that the transition from $<L, K, E(K)>$ to $<L', K', E(K')>$ would involve no explanatory loss can be formulated as follows.

For any statement that occurs as a conclusion of a derivation in $E(K)$ there is an extensionally isomorphic statement that occurs as a conclusion of a derivation in $E(K')$. Two statements are extensionally isomorphic just in case they have the same logical form and the nonlogical expressions at corresponding places refer to the same entities (objects in the case of names and sets in the case of predicates).

The notion of extensional isomorphism is introduced to permit the possibility that the shift from L to L' may involve refixings of reference, the kinds of changes that I have elsewhere (1978, 1982, 1983) taken to underlie the phenomenon of Kuhnian incommensurability. Thus Lavoisier would not accept Priestley's description of the phenomena that the phlogiston theory was used to explain, but he would be able to redescribe those phenomena in his own language. (Whether Lavoisier could also produce an explanatory derivation of a description of each of them, is, of course, a highly controversial matter.)[37]

When there is no explanatory loss (in the sense just characterized), it is not hard to formulate conditions that express the fact that there is a gain in explanatory unification in shifting from $<L, K, E(K)>$ to $<L', K', E(K')>$. Intuitively, there is explanatory gain if we would employ the same number of equally stringent patterns to generate more consequences or fewer or more stringent patterns to generate the same consequences. So, presupposing that there has been no explanatory loss, we can formulate the idea that $E(K')$ unifies K' better than $E(K)$ unifies K as follows.

(C') Suppose there is a one-one mapping f from the basis of $E(K')$ to the basis of $E(K)$ such that for each p in the basis of $E(K')$ p is at least as stringent as $f(p)$; and (i) f is an injection, or (ii) there is some p in the basis of $E(K')$ such that p is more stringent than $f(p)$, or (iii) there is some statement in the consequence set of $E(K')$ that is not extensionally isomorphic to any statement in the consequence set of $E(K)$. Then $E(K')$ unifies K' better than $E(K)$ unifies K.

To apply (C') it is necessary to make sense of the notion of relative stringency for patterns that are not formulated in the same language. The way to make these comparisons can be suggested by some of the major examples in which large changes in science have been defended by appealing to the explanatory advantages of introducing the new language and the new theoretical claims. Consider Maxwellian electromagnetic theory. This supplies a variety of patterns for generating explanatory derivations within geometrical optics, the theory of diffraction, electrostatic interactions, and so forth. The old patterns of explanation are isomorphic to subpatterns of the new patterns, and unification is achieved because the same underlying pattern can be partially instantiated in different ways to generate patterns that were previously viewed as belonging to different fields. Similarly, in the Darwinian revolution, previously available patterns for explaining biogeographical distribution (such as they were) by appealing to the migrations of groups of organisms, were embedded within the Darwinian pattern with the initial premise describing an act of creation at some special center giving way to a derivation of a conclusion about the results of descent with modification. I shall therefore propose that (C') can be satisfied by meeting condition (i) if there is one (or more) pattern p of $E(K')$ such that there are at least two patterns of $E(K)$ that are extensionally isomorphic to subpatterns of p and if all other patterns of $E(K')$ — that is patterns that are not partially instantiable by isomorphs of patterns of $E(K)$ — are themselves extensionally isomorphic to patterns of $E(K)$. It seems to me likely that many episodes from the history of science will require comparisons of unifying power that are based on more subtle conceptions of relative stringency, but I shall not try to pursue this difficult issue further here.

Let us now turn to the second half of the problem that I posed above. Conceptual revision plainly occurs in the history of the sciences, leading to revocation of prior judgments about which predicates are projectable, which collections are natural, and which distinctions are artificial. If we allow the principle of explanatory unification full rein and suppose that advances in unifying power can be achieved through the strategy of embedding just described, then it would seem possible to make cheap explanatory improvements. Take any two unconnected patterns in the basis of the current explanatory store. Form a new "pattern" in which both can be embedded. Typically, this new "pattern" will employ predicates that cut across the boundaries marked out by the projectable predicates of current language. Hence, *provided that the language is not itself amended*, the new "pattern" will be debarred by the ban on gerrymandered patterns. However that ban can be circumvented by proposing the adoption of a language in which the new predicates are taken to be projectable. As a result, if we allow that appeal to explanatory unification can serve as the basis of a defense of conceptual change—and I believe that such allowance is needed to cope with examples like that of the development of electromagnetic theory and of the Darwinian revolution—then we shall have to show how similar appeals will not justify spurious fusion of unconnected patterns.

I suggest that appeals to explanatory unification can underwrite transitions from $<L, K, E(K)>$ to $<L', K', E(K')>$ only subject to the proviso that the shifts from K to K' and the shifts from L to L' are defensible. This does not require that K and K', L and L' be identical—that would defeat the point of the current enterprise—but, roughly, that there are no strong arguments from the perspective of $<L, K, E(K)>$ against the shifts envisaged. Consider the simpler case first. Modification of K to K' may involve the addition of statements for which there was previously no positive evidence but which were not precluded by strong arguments from well-established principles of K (or conversely, such modification may involve abandoning statements in such a way that the prior view that there was evidence in favor of such statements is explained as illusory). When such modifications occur, or are proposed, I shall say that K is relatively *neutral* toward the changes. Contrast this with cases in which there are arguments using premises that are common both to K and to K' either against statements that would be added or in favor of statements that would be dropped. In the latter cases, where K is *negative* toward the changes, the proviso for the appeal to the increased unification is not met. However, when K is simply neutral toward the changes, the fact that the new corpus would permit greater unification of belief justifies the transition.

Darwin's argument in the *Origin* shows how the appeal to explanatory unification must be supplemented with a demonstration that the proviso is satisfied. As I have argued (Kitcher 1985a), the argument-strategy of the *Origin* consists in showing that certain kinds of modification of belief would enable a significant increase in unification. But Darwin is sensitive to the need for showing that accepted forms of reasoning *against* the modifications he would introduce—appeals to the uselessness of incipient complex structures, to the gappiness of the fossil record, and to the apparent stability of organisms—can be rebutted. If any of these major lines of objection had been left unanswered, then opponents could have responded that the theory of evolution by natural selection was an attractive fantasy, promising the unification of belief at the cost of riding rough-shod over well-established facts.

The reception of the theory of continental drift in the 1920s and 1930s shows what happens when the proviso is not satisfied. Wegener and his early supporters could parallel *part* of Darwin's argument. They could show that if the assumptions about wandering continents were correct then considerable unification of geological, biogeographical, and paleometeorological beliefs could be obtained. Unfortunately, there was an apparently compelling argument against the possibility of moving continents. Wegener's own attempt to respond to the argument was unsuccessful, and a later effort at rebutting it (due to Arthur Holmes) was either unheeded or rejected by the geological community. In the absence of a demonstration that the proviso was satisfied, proponents of continental drift (such as du Toit) expanded the inventory of the advantages of unification in vain. Because geologists "knew" that it was impossible to move the continents, the unification described by Wegener and du Toit looked to them like an attractive fantasy that came to grief on well-established "facts" about the earth's crust.[38]

There is an analogous proviso about the modification of language. If we alter our language so as to change judgments about projectability, then we must respond to any existing arguments against the projectability of predicates that the new language takes to be projectable (or in favor of the projectability of predicates that the new language regards as unprojectable). Now in some cases, L is effectively neutral toward the projectability of new predicates: before the transition has been proposed, scientists have regarded certain phenomena as separate simply because they have seen nothing in common between them. But, in the transition, some predicate already viewed as projectable from the standpoint of L is taken to cover these phenomena. So, for example, in the birth of electromagnetic theory, both light and electromagnetic radiation are subsumed under the (projectable) predicate "transverse wave propagated with velocity c." Before the proposed subsumption, phenomena in the propagation of light would have been seen as unconnected with (what we see as analogous) phenomena involving electromagnetic effects, simply because light seemed to have nothing in common with either electricity or magnetism. Maxwell offered a common standpoint from which all these kinds of phenomena could be viewed—introducing a recognizably projectible predicate that would cover all of them—and there were no further negative arguments to be met.

However, in the artificial examples that threaten to resurrect the problems of asymmetry and irrelevance, there will be arguments against the projectability of the predicates that are employed in the new "patterns." Consider the trick of gerrymandering a new "pattern" by disjoining predicates. Let A, B, be disconnected predicates of L and suppose that L' proposes to treat "$Ax \lor Bx$" as a projectable predicate. Suppose that C, D are predicates such that "$(x)(Ax \supset Cx)$" and "$(x)(Bx \supset Dx)$" are generalizations accepted both in K and K', and such that *C, D*, like *A* and *B*, are disconnected predicates of *L*. Now I take it that part of the reason for thinking that "*Ax v Bx*" is not a projectable predicate is that one could not confirm the generalization "*(x)((Ax v Bx) ⊃ (Cx v Dx))*" by observing a sample consisting of instances of *A* that are also instances of *C*. To make the proposed transition from *L* to *L'*, it is not sufficient simply to declare that "*Ax v Bx*" is now to be counted as a projectable predicate. One will also have to answer arguments based on past inferential practice. Any such answer will, I believe, involve the modification of views about confirmation in such a way as to yield widespread changes in the corpus of beliefs—*K'* will have to differ from *K* in systematic ways—and some of the proposed changes will fall afoul of the proviso governing the modification of *K* to *K'*.

My tentative solution to the problem of how to allow for conceptual change based on an appeal to explanatory unification, while avoiding a revival of the asymmetry problem, is thus to insist that the appeal to advances in unification is contingent on the satisfaction of certain conditions. To *declare* that a predicate is projectable is not sufficient; one must respond to previously accepted reasons for not projecting the predicate. In doing so, it appears that there will have to be a large-scale revision of views about what confirms what. The change in inferential practice will lead to alterations in the corpus of beliefs that cannot be sustained in light of the arguments from premises that continue to be accepted.

We began this section with the worry that the idea of explanation as unification could not handle the problems of asymmetry and irrelevance. I have tried to show that, within the relatively simple context in which we consider a single language and a single corpus of beliefs, the unification criterion can be articulated to resolve major cases of these problems. With respect to the more complex context in which the possibility of change in belief and of conceptual change is allowed, matters are more complicated. I hope at least to have forestalled the complaint that the unification approach has no prospects of solving the apparently more recalcitrant asymmetry and irrelevance problems that may be posed within this context. Perhaps at this stage it is appropriate to see the burden of proof as resting with those who claim that the asymmetry and irrelevance problems doom any approach to explanation that does not explicitly invoke causal concepts in the analysis of explanation.

The discussion of appeals to unification and their role in the growth of scientific knowledge points to a more important project than resolving artificial problems of asymmetry and irrelevance. As I noted at the end of the last section, the demise of traditional foundationalist ideas about human knowledge leads to a reformulation of the question how our causal claims are justified. Ideally, we should show how our picture of the causal structure of the world has been built up, by tracing a chain of transitions leading from a hypothetical primitive state to the present. According to the approach to explanation as unification, principles like those presented in this section play a part in the dynamics of scientific knowledge. Thorough vindication of the approach would require that those principles be articulated in greater detail, that they be embedded in a full set of principles for rational change in science, and that it be shown how such principles have been systematically operative in the growth of our scientific knowledge and, in particular, of our view of the causal structure of the world. Evidently, I am in no position to complete so massive a task, but the treatment sketched here offers some hope that the task might in principle be completed and that we can avoid the deep problems about causal knowledge that I tried to expose in the preceding section.

8. Metaphysical Issues

There is one final respect in which the two approaches to explanation need to be compared. Proponents of the causal approach to explanation can readily distinguish between acceptable and correct explanations. A true or correct explanation is one that identifies the causal structure that underlies the *explanandum* phenomenon. An acceptable explanation, for a person whose beliefs constitute a set K, identifies what it would be rational, given K, to take as the causal structure underlying the phenomenon. However, on the unification approach, the focus of the discussion has been on acceptable explanations. I have said what conditions must be met for a derivation to be an acceptable explanation of its conclusion relative to a belief corpus K, and (in the previous section) I have briefly discussed conditions governing the modification of belief and the modification of views about acceptable explanations. But I have not said what constitutes a true or correct explanation.

8.1 Correct Explanation

One obvious way of filling this gap is to say that d is a correct explanation of its conclusion just in case d belongs to $E(K)$ where K is the set of all true statements. But this will not do for several reasons. First, we would want K to be something like the set of all true statements in some particular language. Which language? Might it make a difference—particularly if the languages differ on their views about projectability of predicates? Second, if K includes *all* the true statements in a particular language, then it will include all true causal statements, all true statements about what explains what. So we could apparently shortcut our work by saying that d correctly explains its conclusion just in case a statement to that effect belongs to the set of all true English statements.

8.2 "What If the World Isn't Unified?"

I shall approach the problem of characterizing correct explanation more obliquely, starting with a difficulty that David Lewis has forcefully presented against views that take unification to be crucial to explanation. The difficulty can be formulated in a short question: suppose the world isn't unified, what then? The idea is that proponents of the unification approach are committed to viewing factors as explanatorily relevant if they figure in a unified treatment of the phenomena. However, if the world is messy, then the factors that are causally relevant to the phenomena may be a motley collection and the account of explanatory structure may not reveal the causal structure. The unification approach thus runs the risk of providing an account of explanation on which giving explanations is divorced from tracing causes or of imputing to nature a priori a structure that nature may not have.

Let us elaborate the problem a little. The unification approach is apparently committed to claiming that some factor F is explanatorily relevant to a phenomenon P if there is a derivation belonging to the best unifying systematization of the complete set of truths about nature, a derivation in which a description of P is derived from premises that refer to F. To respond to the worries about the complete set of truths about nature that were raised at the beginning of the section, let us say that an *ideal Hume corpus* is a set of beliefs in a language whose primitive predicates are genuinely projectable (pick out genuine natural kinds) that includes all the true statements in that language that do not involve causal, explanatory, or counterfactual concepts. Then we can formulate one crucial premise of the objection as follows:

(1) F is explanatorily relevant to P, according to the criteria of explanatory unification, provided that there is a derivation of P that belongs to the best unifying systematization of an ideal Hume corpus such that there is a premise of the derivation in which reference to F is made.

A second important idea behind the objection is that explanations of at least some phenomena trace causes. Some proponents of the causal approach (Lewis, though not Salmon or Railton) hold that all singular explanation involves the tracing of causes. I have offered reasons for dissenting from this view (see section 3.2), but I have also suggested that it would be foolish for the unification approach to break the link between explanation and causation. Indeed, I have been emphasizing the idea (favored by Mill, Hempel, and many other empiricists) that causal notions are derivative from explanatory notions. Thus I am committed to

(2) If F is causally relevant to P then F is explanatorily relevant to P.

Now (1) and (2) are supposed to conflict with the genuine possibility that our world is messy and that there is a heterogeneous collection of basic causal factors. The idea of the possibility of a nonunified world can be presented as

(3) It is possible (and may, for all we know, be true) that there is a factor F that is causally relevant to some phenomenon P such that derivations of a description of P belonging to the best unifying systematization of an ideal Hume corpus would not contain any premise making reference to F.

(3) conjures up the vision of the obsessive unifier producing derivations of descriptions of *P* within a unified systematization of an ideal Hume corpus when the real causal factors underlying *P* are quite distinct from those that figure in the derivations. Unity is imposed where none is really to be found.

Now provided that the unification approach is conceived as offering (1) and (2) as necessary truths about explanatory and causal relevance, there is a genuine conflict among (1), (2), and (3). For (1) and (2) close off a possibility that (3) asserts to be open. One apparent way out would be to claim only that (1) and (2) formulate contingent conditions and that the unification approach is committed only to the idea that the world is actually unified. But this is not attractive, since the motivation for the unification approach rests on doubts about what explanatory relevance could be if it did not involve unification and about what causal relevance could be if it was not the projection of explanatory relevance. So it looks as though the approach must defend the *prima facie* implausible thesis that the world is necessarily unified—or, more exactly, that (3) is false.

Now the line of argument that I have been rehearsing not only raises difficulties for the attempt to combine the views that explanation consists in the unification of beliefs and that causal structure is explanatory structure; it also tells against efforts to co-opt some of the virtues of the unification approach for the causal approach. Salmon is sensitive to the fact that theoretical explanation often provides unification of the phenomena. In the final section of his (1984), he writes

> The ontic conception looks upon the world, to a large extent at least, as a black box whose workings we want to understand. Explanation involves laying bare the underlying mechanisms that connect the observable inputs to the observable outputs. We explain events by showing how they fit into the causal nexus. Since there seem to be a small number of fundamental causal mechanisms, and some extremely comprehensive laws that govern them, the ontic conception has as much right as the epistemic conception to take the unification of natural phenomena as a basic aspect of our comprehension of the world. *The unity lies in the pervasiveness of the underlying mechanisms* upon which we depend for explanation. (276)

It is clear from Salmon's earlier discussions that theoretical explanations are often defended on the grounds that they offer a unified treatment of phenomena previously regarded as diverse. But, once it is accepted that there is a causal structure of the world independent of our efforts to achieve a unified vision of it, then it is purely a contingent truth (supposing that it is indeed a truth) that there is a limited number of basic mechanisms. We are then forced to ask what evidence we may have for believing this supposed truth, and how we are able to appeal to it in our theoretical postulations.

On the causal approach, any use of a principle of explanatory unification to underwrite the revision of theory raises questions about how such a methodological principle is to be justified. Salmon's contention that the ontic conception of explanation has as much right as the epistemic conception to take unification as a *basic* aspect of our comprehension of the world seems unfounded: for, on the version of the epistemic conception developed here (the unification approach) unification is *constitutive* of explanation, while on Salmon's version of the causal approach, unification is at best a *contingent concomitant* of the tracing of causal structure. The moral of Lewis's objection, elaborated in the tension among (1)–(3), seems to be that we must choose between two visions of what is central to explanation: the unification of phenomena or the identification of causal structure.

However, the discussion of Salmon's attempt to work ideas about unification into the causal approach indicates a different way of developing the metaphysics of explanation and of saving (1) and (2). What motivates our acceptance of (3) is the idea that there is an independent causal order, so that it is a purely contingent matter whether there are a few pervasive basic mechanisms or a motley assortment of fundamental causal factors. The heart of the unification approach is that we cannot make sense of the notion of a basic mechanism apart from the idea of a systematization of the world in which as many consequences as possible are traced to the action of as small a number of basic mechanisms as possible. In short, on the unification approach, the basic mechanisms must be those picked out in the best unifying systematization of our best beliefs, for if they were not so picked out then they would not be basic.

How can we defend the idea that (3) is false? Surely not by taking up Copernican ideas of the harmony of nature or Newtonian principles to the effect that nature "is wont to be simple and consonant to herself." Theological solutions that ultimately trace the necessary unity of nature to divine providence will not do. Instead, I recommend rejecting the idea that there are causal truths that are independent of our search for order in the phenomena. Taking a clue from Kant and Peirce, we adopt a different view of truth and correctness, and so solve the problem with which we began.[39]

8.3 Correct Explanation Again

Conceive of science as a sequence of practices, each of which is distinguished by a language, a body of belief, and a store of explanatory derivations. Imagine this sequence extending indefinitely forward into the future, and suppose that its development is guided by principles of rational transition, including the principles about unification outlined in the previous section. Instead of thinking that a language is ideal provided that its primitive predicates are projectable (or pick out genuine kinds) let us say that a predicate is correctly projectable if it is counted as a projectable predicate of the language of science in the limit as our sequence of practices develops according to the principles of rational transition. Similarly, true statements are those that belong to the belief corpus of scientific practice in the limit of its development under the principles of rational transition. Finally, and most important for present purposes, *correct* explanations are those derivations that appear in the explanatory store in the limit of the rational development of scientific practice.

Is there a worrying circularity in the account that I have offered? Consider: the concept of projectability figured heavily in my discussion of the principles that govern rational modification of practices and the acceptability of argument patterns; yet now I am apparently planning to use those principles of rational modification to characterize the notion of a projectable predicate; surely something has gone wrong here. I reply that the circularity is only apparent. Recall that the enterprise of the sections up to and including 7 was to specify conditions on *acceptable* explanation. Thus, in focusing on the rational modification of scientific practices, I was concerned to understand what kinds of patterns of argument might be acceptable as explanatory against the background of judgments about natural kinds, judgments expressed in commitment to the projectability of various predicates. Hence, I used the concept of a predicate's *being accepted as projectable* to characterize that of an argument pattern's *being accepted as explanatory*. The dynamics of acceptance of predicates and of explanatory patterns was admittedly incomplete, but I believe that the principles governing rational modification of language, of beliefs, and of the explanatory store can be formulated independently of notions of correctness (of a predicate's being *actually projectable*, a belief's being *true*, and so forth). The proposal of the present section is to extend the envisaged account of rational modification of practices to an account of the corresponding concepts of correctness, by viewing correctness as that which is achieved in the ideal long run when the principles of rational modification are followed. There is no circularity here, although, as I would concede, there is plenty of work to be done before the program has been brought to completion.

Now we can respond to Lewis's worry about the unification approach and simultaneously articulate the idea that this version of the epistemic conception makes unification basic to explanation in a way that it is not basic to Salmon's version of the ontic conception. Given our new metaphysical perspective, (1) must be modified as follows

(1′) *F* is explanatorily relevant to *P* just in case, in the limit of the development of scientific practice under principles of rational transition, the explanatory store comes to contain a derivation in which a description of *P* is derived from premises, at least one of which makes reference to *F*. Among the principles of rational transition is a *static* principle of unification that directs that the explanatory store with respect to a body of belief *K* consists of those derivations that best unify *K*, and a *dynamic* principle that directs us to modify practice so as to achieve advances in unification. The dynamic principle is subject to provisos, as indicated in the previous section.

The connection between explanatory relevance and causal relevance can go over unchanged. So we have

(2) If *F* is causally relevant to *P* then *F* is explanatorily relevant to *P*.

Finally, the claim that previously proved troublesome becomes

(3′) It is possible (and may, for all we know, be true) that there is a factor *F* that is causally relevant to some phenomenon *P* such that no derivation occurring in the explanatory store in the limit of scientific practice derives a description of *P* from premises that make reference to *F*.

The solution to Lewis's objection is to reject (3′) on the grounds that there is no sense to the notion of causal relevance independent of that of explanatory relevance and that there is no sense to the notion of explanatory relevance except that of figuring in the systematization of belief in the limit of scientific inquiry, as guided by the search for unification.

The difference between the role of unification on Salmon's approach and on mine can now be appreciated. It consists in the fact that Salmon would accept (3′) while I reject it. However, it is important to note that, although the problem posed by Lewis has been resolved, there is a sense in which the unification approach, in its new guise, can tolerate the possibility that the world might prove to be a messy place. Explanatory relevance emerges in the limit of our attempts to achieve a unified view of the world, but there is no a priori guarantee of how successful we shall be in achieving unification. Thus, while we try to make the phenomena as unified as we can, it is possible that there should be two different worlds, in one of which there were far fewer basic mechanisms than in the other. This would be reflected in the fact that the modification of scientific practice, guided in both worlds by the principles of rational transition, including both the static and dynamic principles of unification, produced in the one case a limit explanatory store whose basis contained only a few patterns and in the other an explanatory store with a far more prodigal basis. Hence, while my version of the unification approach makes it constitutive of explanatory relevance that there be no basic explanatory (or causal) mechanisms that are not captured in the limit of attempts to systematize our beliefs, it does *not* make it a necessary truth that this limit achieves any particular degree of unification. The vision of the obsessive unifier is dissolved.

8.4 Conclusions

As Railton clearly recognizes (see this volume) differences in views about scientific explanation connect to differences in metaphysics. The causal approach is wedded to a strong version of realism in which the world is seen as having a structure independent of our efforts to systematize it. It should be no surprise that the metaphysical extras bring epistemological problems in their train (see section 6). I have been trying to show that we can make sense of scientific explanation and our view of the causal structure of nature without indulging in the metaphysics. The aim has been to develop a simple, and, I think, very powerful idea. The growth of science is driven in part by the desire for explanation, and to explain is to fit the phenomena into a unified picture insofar as we can. What emerges in the limit of this process is nothing less than the causal structure of the world.

Notes

1. This approach to pragmatic issues has been articulated with considerable sophistication by Peter Railton. See his (1981) and his unpublished doctoral dissertation.

2. Achinstein's theory of explanation, as presented in his (1983) is extremely complex. I believe that it ultimately suffers from the same general difficulty that I present below for van Fraassen. However, it is eminently possible that I have overlooked some subtle refinement that makes for a disanalogy between the two versions.

3. It would not be difficult to extend the account of ideal Hempelian questions to include questions that are answered by the provision of I-S explanantia. In this way we could mimic the entire Hempelian approach to explanation in the framework of the theory of why-questions.

4. As should now be evident, the second of the four problem-types that beset the Hempelian account assumes a derivative status. We may be able to manage the theory of explanation without a characterization of laws if we can distinguish the genuine relevance relations without invoking the notion of lawlike dependence. I shall articulate an approach below on which this possible strategy is attempted.

5. I am extremely grateful to Isaac Levi for raising the issue of the goals of a theory of explanation, by inquiring whether we can expect there to be a single set of relevance relations that applies for all sciences at all times.

6. The distinction between originating selection and maintaining selection is crucial, for the pressures need by no means be the same. A classic example is the development of wing feathers in birds, hypothesized to have been originally selected for their advantages in thermoregulation and subsequently maintained for their advantages in flight. The distinction is clearly drawn in Niko Tinbergen's celebrated account of the four "whys" of behavioral biology (see his 1968), and has been succinctly elaborated by Patrick Bateson (1982). Philosophical theories of teleology clearly need to use the contrast between maintaining selection and originating selection—see, for example, the treatments offered by Larry Wright (1976) and John Bigelow and Robert Pargetter (1987).

7. It should not, of course, be assumed that all functional/teleological questions are legitimized by the account that I have given. My concern is with the status of a class of why-questions, and it is compatible with my claims to suppose that the concepts of function and adaptation have been far too liberally applied in recent evolutionary theory. See, for example, Gould and Lewontin 1979, Gould and Vrba 1982, Kitcher 1985b, chapter 7, and Kitcher 1987a.

8. The distinction between global and local methodology is drawn in more detail in chapter 7 of (Kitcher 1983). It is only right to note that some scholars have challenged the idea that there is any very substantive global methodology. See, for example, Laudan 1984.

9. The version of the causal view that I present here is my own reconstruction of ideas put forward in seminar discussions by Railton and Salmon. Although I have some uneasiness about the mammoth causal histories that appear to be envisaged and about their status as "ideal" answers to "ideal" questions, I shall not press this line of objection. My main concerns about the causal approach lie elsewhere.

10. Plainly, this diverges from the Hempelian idea that the essence of an adequate explanation is that it show that the *explanandum* was to be expected. For discussion of this idea and motivation for an opposing viewpoint, see Salmon 1984, Railton 1978, and Railton's dissertation.

11. This apparent casualness can easily be understood by reviewing some of the literature on statistical explanation. In Salmon (1970), the Hempelian requirement of high-probability was replaced with a requirement of statistical relevance. Unfortunately, as it became clear during the 1970s, the notion of statistical relevance cannot do duty for the concept of causation – a *locus classicus* for the point is Cartwright (1979). Thus Salmon and others (e.g., Railton [1978, 1981], Humphreys [1981, 1982]) have explicitly introduced the notion of causation into the account of statistical explanation. In fact, the introduction can lead to more detailed models of statistical explanation than I consider here. See for example, the cited articles by Humphreys and Railton, and Salmon (1984), chapters 2 and 3.

12. Examples of this type have been suggested by Sylvain Bromberger in conversation and in his (1986).

13. Similar examples have been suggested by Peter Railton, who imagines explaining a person's failure to put a left-handed glove on the right hand.

14. This example has been used by Elliott Sober in discussions of causal explanation (1983). Although my treatment of it differs from Sober's in some matters of detail, we concur in viewing it as problematic for the causal approach.

15. We might think of the systematization approach as covering an entire family of proposals among which is that based on the view of systematization as unification. Since it appears to me that the latter view provides the best chances of success, I shall concentrate on it and ignore alternative possible lines of development.

16. See my (1986) for a reconstruction of Kant's views that tries to defend this attribution.

17. " . . . our total picture of nature is simplified via a reduction in the number of independent phenomena that we have to accept as ultimate" (Friedman 1974), 18. There is an interesting recapitulation here of T. H. Huxley's summary of Darwin's achievement. "In ultimate analysis everything is incomprehensible, and the whole object of science is simply to reduce the fundamental incomprehensibilities to the smallest possible number." (Huxley 1896) 165

18. I think it entirely possible that a different system of representation might articulate the idea of explanatory unification by employing the "same way of thinking again and again" in quite a different – and possibly more revealing – way than the notions from logic that I draw on here. Kenneth Schaffner has suggested to me that there is work in AI that can be deployed to provide the type of account I wish to give, and Paul Churchland has urged on me the advantages of connectionist approaches. I want to acknowledge explicitly that the adaptation of ideas about logical derivation may prove to be a ham-fisted way of developing the idea of explanatory unification. But, with a relatively developed account of a number of facets of explanation available, others may see how to streamline the machinery.

19. Carlson (1966) is an extremely valuable source for those who wish to interpolate. Extrapolation into the present is tricky, although Watson (1987) provides a helpful guide.

20. See the final section of Kitcher (1984), and Oster and Alberch (1982) for an account of *some* developmental complexities. Recent work on *Drosophila* only serves to underscore the point that epigenesis is hierarchically organized, and that changes near the top of the hierarchy can issue in a cascade of effects.

21. Since, as we shall see below, the primary Darwinian pattern of the more ambitious type is a pattern that traces the presence of prevalent traits to their selective advantages, the major debates concern the omnipresence and power of selection. Some critics (for example, Ho and Saunders 1984) contend that explanations in terms of natural selection are conceptually confused, even unintelligible. This radical claim, which lies behind much of the rhetoric about the demise of neo-Darwinism, does not seem to me to survive serious scrutiny, and is, I think, best viewed as the hypertrophied form of a sensible line of criticism that emphasizes the difficulties of identifying the workings of selection in particular cases. Those who pursue this more moderate line are quite willing to allow that explanations instantiating the primary Darwinian pattern can be given in particular cases, but they believe that many biologists are carried away by enthusiasm for paradigm cases in which the primary Darwinian pattern has proved successful, so that they overlook alternative forms of explanation in studies

where rival possibilities deserve to be taken seriously. Cautionary notes of this kind can be found in (Gould and Lewontin 1979) and in (Bateson 1982). I have tried to articulate the moderate criticism in my (1985b).

22. Aficionados of sorites problems will recognize that the use of induction in the context of reasoning about "almost all" members of a group is dangerous. The deficiency could be made up by reformulating the reasoning in probabilistic terms, or simply by adding a premise to the effect that lapses from universality are noncumulative. In the interests of preserving something that is fairly close to Darwinian arguments, I shall avoid such niceties here.

23. This picture of the structure of contemporary evolutionary theory is akin to that provided by Sober (1984). Ecology can be seen as providing principles about the sources of the factors that are cited in the equations of population genetics. Chapters 2 and 3 of Kitcher (1985b) develop the picture in the context of evolutionary/functional studies of animal behavior.

24. The version of the deductivist gambit presented here is quite close to that offered in Papineau (1985). Besides Papineau, Levi has contributed important ideas to the defense of deductive chauvinism. See Levi forthcoming.

25. Ironically, this is very close to Hempel's paradigm example of singular explanation in his 1965 (232). The similarity reinforces the idea that explanation is implicitly global, the point from which Friedman (1974) began his seminal discussion of unification, and which I am trying to develop further here.

26. Here I use standard talk of determinism and indeterminism to distinguish QM from other parts of science. John Earman (1986) has shown in lucid and thorough detail how loose some of this talk is, but the inexactness will not affect the present arguments.

27. Behind the dilemma enunciated here there are significant epistemological presuppositions. I shall consider a possible way of altering the epistemological framework at the end of this section (see 6.5).

28. This point will be elaborated briefly in 7.3.3.

29. Intuitively, we have no interest in an initial interaction that produced a mark in a process that transmitted that mark and interacted at the *explanandum* event to produce some completely irrelevant property. We are concerned with the interaction that gave rise to the very property whose presence is to be explained, and whatever other markings and transmittings are going on at the point of the *explanandum* event or along the line that constitutes its causal history can be ignored.

30. See Lewis (1973), both for an elegant statement of a counterfactual theory of causation and for a survey of difficult cases. Loeb (1974) endeavors to cope with the problem of overdetermination.

31. Others have seen that a relative of Benacerraf's dilemma applies to modal discourse. See, for example, Mondadori and Morton (1976). Unfortunately, the significance of the point does not appear to be widely appreciated.

32. Some accounts of causation, for example those of Cartwright (1979) and Sober (1984), provide recursion clauses for enabling some causal judgments to be generated from others. Valuable as this is, it does not explain the basis of our practice of causal justification.

33. This line of attack was forcefully presented in seminar discussions by Paul Humphreys, who maintains that it can be developed to overcome the obstacles to Salmon's program (and programs, like Humphreys's own, which are akin to Salmon's) that I have presented here.

34. This can easily occur in cases of explanation extension. See the examples from section 4.6.

35. Notice that derivations that systematically contain idle clauses are not so clearly nonexplanatory as the kind of irrelevant derivation with which we began. It seems to me that this is because the unwanted mini-derivations are viewed as giving us information about the structure of a full, ideal, derivation, and the natural implication is that the properties picked out in the premises will play a key role. Once we see that these properties are inessential, and that predicates referring to them figure throughout all our derivations, then we may feel that cluttering up the explanatory store does nothing more than add a harmless irrelevancy. The resultant derivations are untidy, but I think that there is reason to argue about whether they should be counted as nonexplanatory.

36. For this example, it is important to recognize that the *origin-and-development* pattern must allow for explanatory derivations in which we appeal to general constraints that keep a system close to an equilibrium state throughout its career. See the discussion of the sex-ratio example in 3.2.

37. Numerous scholars have contended that examples like this involve what has come to be known as "Kuhn-loss." In the present case, it is sometimes suggested that Priestley's phlogiston chemistry could explain something that Lavoisier could not, namely what all the metals have in common. I shall take no stand on this vexed question here.

38. For a brief, comprehensible account of the main ideas and the course of the debate, see Hallam (1973). Comparing the arguments of du Toit with the responses of the leading opponents of continental drift in the 1930s and 1940s makes it quite apparent that the appeal to unifying power had to be subject to provisos. However, in the spirit of Kuhn (1977), it seems that scientists sometimes differ with respect to the advantages of unifying and the severity of the arguments against unifying. In consequence, there are sources of diversity within the scientific community, and this, I would contend, is a good thing from the point of view of the *community's* enterprise.

39. The view that I shall develop from hints of Kant and Peirce is more closely connected with Kantian texts in my (1986). It also has affinities with ideas presented in Rescher (1970) and in the recent writings of Hilary Putnam.

References

Achinstein, Peter. 1983. *The Nature of Explanation*. New York: Oxford University Press.

Bateson, P. P. G. 1982. Behavioural Development and Evolutionary Processes. In *Current Problems in Sociobiology*, eds. King's College Sociobiology Group. Cambridge: The University Press.

Benacerraf, Paul. 1973. Mathematical Truth, *Journal of Philosophy*, 70: 661–79

Bigelow, John, and Pargetter, Robert. 1987. Functions, *Journal of Philosophy*, 84: 181–96.

Bromberger, Sylvain. 1963. A Theory about the Theory of Theory and about the Theory of Theories. In *Philosophy of Science: The Delaware Seminar*, ed. W. L. Reese. New York: John Wiley.

———. 1966. Why-Questions. In *Mind and Cosmos*, ed. R. Colodny. Pittsburgh: University of Pittsburgh Press.

———. 1986. On Pragmatic and Scientific Explanation: Comments on Achinstein's and Salmon's Papers. In *PSA 1984*, Volume II, eds. Peter Asquith and Philip Kitcher. East Lansing: Philosophy of Science Association

Carlson, E. O. 1966. *The Gene: A Critical History*. Philadelphia: Saunders.

Cartwright, Nancy. 1979. Causal Laws and Effective Strategies, *Noûs*, 13: 419–37. (Reprinted in Cartwright 1983.)

———. 1983 *How the Laws of Physics Lie*. Oxford: the University Press

Coffa, J. Alberto. 1974. Hempel's Ambiguity, *Synthèse* 28: 141–64.

Dretske, Fred. 1973. Contrastive Statements, *Philosophical Review*, 82: 411–37.

Earman, John. 1986. *A Primer on Determinism*. Dordrecht, Holland: D. Reidel.

Feigl, Herbert. 1970. The 'Orthodox' View of Theories: Remarks in Defense as well as Critique. In *Minnesota Studies in the Philosophy of Science*, Volume IV, eds. M. Radner and S. Winokur. Minneapolis: University of Minnesota Press.

Fisher, R. A. 1931. *The Genetical Theory of Natural Selection*. Oxford: the University Press. (Second Edition, New York: Dover, 1958)

Friedman, Michael. 1974. Explanation and Scientific Understanding, *Journal of Philosophy*, 71: 5–19.

Garfinkel, Alan. 1981. *Forms of Explanation*. New Haven: Yale University Press.

Geach, Peter T. 1969. *God and the Soul*. London: Routledge and Kegan Paul.

Goodman, Nelson. 1956. *Fact, Fiction, and Forecast*. Indianapolis: Bobbs-Merrill.

Gould, S. J., and Lewontin, R. C. 1979. The Spandrels of San Marco and the Panglossian Paradigm: A Critique of the Adaptationist Programme. Reprinted in *Conceptual Issues in Evolutionary Biology*, ed. Elliott Sober. Cambridge MA.: Bradford Books/MIT Press, 1983.

Gould, S. J., and Vrba, E. 1982. Exaptation: A Missing Term in the Science of Form, *Paleobiology*, 8: 4–15.

Hallam, Anthony. 1973. *A Revolution in the Earth Sciences*. Oxford: the University Press.

Hempel, C. G. 1965. *Aspects of Scientific Explanation*. New York: Free Press.

Hempel, C. G., and Oppenheim, P. 1948. Studies in the Logic of Explanation. Chapter 9 of Hempel (1965).

Ho, Mae-Wan, and Saunders, Peter T. 1984. *Beyond Neo-Darwinism: An Introduction to the New Evolutionary Paradigm*. New York: Academic Press.

Horwich, Paul. 1987. *Asymmetries in Time*. Cambridge MA: MIT Press/Bradford Books.

Humphreys, Paul. 1981. Aleatory Explanations, *Synthèse*, 48: 225–32

——. 1982. Aleatory Explanations Expanded. In *PSA 1982*, eds. Peter Asquith and Thomas Nickles. East Lansing: Philosophy of Science Association.

Huxley, Thomas Henry. 1896. *Darwiniana*. New York: Appleton.

Jeffrey, Richard. 1969. Statistical Explanation vs. Statistical Inference. In *Essays in Honor of Carl G. Hempel*, ed. Nicholas Rescher. Dordrecht, Holland: D. Reidel.

Kim, Jaegwon. 1974. Noncausal Connections, *Noûs*, 8: 41–52

Kitcher, Philip. 1975. Bolzano's Ideal of Algebraic Analysis, *Studies in the History and Philosophy of Science* 6: 229–71.

——. 1976. Explanation, Conjunction, and Unification, *Journal of Philosophy* 73: 207–12.

——. 1978. Theories, Theorists, and Theoretical Change, *Philosophical Review*, 87: 519–47.

——. 1981. Explanatory Unification, *Philosophy of Science*, 48: 507–31.

——. 1982. Genes, *British Journal for the Philosophy of Science*, 33: 337–59

——. 1983. *The Nature of Mathematical Knowledge*. New York: Oxford University Press.

——. 1984. 1953 And All That. A Tale of Two Sciences, *Philosophical Review*, 93: 335–73.

——. 1985a. Darwin's Achievement. In *Reason and Rationality in Science*, ed. N. Rescher. Washington: University Press of America.

——. 1985b. *Vaulting Ambition: Sociobiology and the Quest for Human Nature*. Cambridge MA.: MIT Press.

——. 1985c. Two Approaches to Explanation, *Journal of Philosophy*, 82: 632–39.

——. 1986. Projecting the Order of Nature. In *Kant's Philosophy of Physical Science*, ed. Robert Butts. Dordrecht: D. Reidel.

——. 1987a. Why Not The Best? In *The Latest on the Best: Essays on Optimality and Evolution*, ed. John Dupre. Cambridge MA.: Bradford Books/MIT Press.

——. 1987b. Mathematical Naturalism. In *Essays in the History and Philosophy of Modern Mathematics*, eds. William Aspray and Philip Kitcher. Minneapolis: University of Minnesota Press, 293–325.

——. 1988. Mathematical Progress. To appear in *Revue Internationale de Philosophie*.

——, and Salmon, Wesley. 1987. Van Fraassen on Explanation, *Journal of Philosophy*, 84: 315–30.

Kuhn, Thomas S. 1970. *The Structure of Scientific Revolutions*. Chicago: University of Chicago Press.

——. 1977. Objectivity, Value Judgment, and Theory Choice. In *The Essential Tension*. Chicago: University of Chicago Press.

Kyburg, Henry. 1965. Comment, *Philosophy of Science*, 35: 147–51.

Laudan, Larry. 1984. *Science and Values*. Berkeley: University of California Press.

Levi, Isaac. (forthcoming) Four Types of Statistical Explanation. To appear in *Probabilistic Causality*, eds. W. Harper and B. Skyrms. Dordrecht, Holland: D. Reidel.

Lewis, David. 1973. Causation, *Journal of Philosophy*, 70: 556–67.

——. 1974. *Counterfactuals*. Oxford: Blackwell.

——. 1976. The Paradoxes of Time-Travel, *American Philosophical Quarterly*, 13: 145–52.

Loeb, Louis. 1974. Causal Theories and Causal Overdetermination, *Journal of Philosophy*, 71: 525–44.

Mondadori, Fabrizio, and Morton, Adam. 1976. Modal Realism: The Poisoned Pawn, *Philosophical Review*, 85: 3–20.

Oster, George, and Alberch, Pere. 1982. Evolution and Bifurcation of Developmental Programs, *Evolution*, 36: 444–59.

Papineau, David. 1985. Probabilities and Causes, *Journal of Philosophy*, 82: 57–74.

Pauling, Linus. 1960. *The Nature of the Chemical Bond* (Third Edition), Ithaca: Cornell University Press.

Railton, Peter. 1978. A Deductive-Nomological Model of Probabilistic Explanation, *Philosophy of Science*, 45: 206–26.

——. 1981. Probability, Explanation, and Information, *Synthèse*, 48: 233–56.

Rescher, Nicholas. 1970. Lawfulness as Mind-Dependent. In *Essays in Honor of Carl G. Hempel*, ed. N. Rescher. Dordrecht, Holland: D. Reidel.

Salmon, Wesley. 1970. Statistical Explanation. In *The Nature and Function of Scientific Theories*, ed. R. Colodny. Pittsburgh: University of Pittsburgh Press.

———. 1984. *Scientific Explanation and the Causal Structure of the World*. Princeton: Princeton University Press.

Scriven, Michael. 1959. Definitions, Explanations, and Theories. In *Minnesota Studies in the Philosophy of Science* Volume II, eds. H. Feigl, M. Scriven, and G. Maxwell. Minneapolis: University of Minnesota Press.

———. 1962. Explanations, Predictions, and Laws. In *Minnesota Studies in the Philosophy of Science* Volume III, eds. H. Feigl and G. Maxwell. Minneapolis: University of Minnesota Press.

———. 1963. The Temporal Asymmetry between Explanations and Predictions. In *Philosophy of Science. The Delaware Seminar*, Volume I, ed. B. Baumrin. New York: John Wiley.

Sober, Elliott. 1983. Equilibrium Explanation, *Philosophical Studies*, 43: 201–10.

———. 1984. *The Nature of Selection*. Cambridge, MA: Bradford Books/MIT Press.

Stalnaker, Robert. 1968. A Theory of Conditionals. In *Studies in Logical Theory*, ed. N. Rescher. Oxford: Blackwell.

Steiner, Mark. 1978. Mathematical Explanation, *Philosophical Studies* 34: 135–51.

Tinbergen, Niko. 1968. On War and Peace in Animals and Man. Reprinted in *The Sociobiology Debate*, ed. Arthur Caplan. New York: Harper and Row, 1978.

van Fraassen, Bas. 1980. *The Scientific Image*. Oxford: the University Press.

Watson, J. D. 1987. *Molecular Biology of the Gene* (Fourth Edition). San Francisco: Benjamin.

Wright, Larry. 1976. *Teleological Explanation*. Berkeley: University of California Press.

Contributors

Nancy Cartwright is professor of philosophy at Stanford University. She received her Ph.D. from the University of Illinois at Chicago Circle. She is the author of *How the Laws of Physics Lie* and *Nature's Capacities and Their Measurement*.

Paul W. Humphreys is associate professor of philosophy at the University of Virginia. Author of numerous articles on explanation, causation, and probability, his most recent work is *The Chances of Explanation* (Princeton University Press, 1989). His current research interests include computer simulation methods in the physical sciences.

Philip Kitcher is currently professor of philosophy at the University of California, San Diego. He has also taught at Vassar College, the University of Vermont, and the University of Minnesota, where he served as director of the Minnesota Center for Philosophy of Science from 1984 until mid-1986. He earned his doctorate in the history and philosophy of science at Princeton University in 1974. Kitcher is the author of *Abusing Science: The Case Against Creationism*, *The Nature of Mathematical Knowledge*, and *Vaulting Ambition: Sociobiology and the Quest for Human Nature*; he was co-editor, with William Aspray, of *History and Philosophy of Modern Mathematics* (Minnesota, 1988). He has contributed articles to numerous journals, among them the *Philosophical Review*, *Philosophy of Science*, the *Journal of Philosophy*, and the *British Journal for the Philosophy of Science*.

David Papineau received his Ph.D. from Cambridge University, where he is now on the faculty in history and philosophy of science. He has published articles on epistemology, representation, and causation, and is the author of *For Science in the Social Sciences*, *Theory and Meaning*, and *Reality and Representation*.

Peter Railton is associate professor of philosophy at the University of Michigan, Ann Arbor. His writings in the philosophy of science have focused on explanation, probability, and objectivity. His current research concerns the explanation of human behavior and the fact/value distinction.

Merrilee H. Salmon is professor of history and philosophy of science, University of Pittsburgh. She received her Ph.D. from the University of Michigan. She is the author of *Philosophy and Archaeology* and *An Introduction to Logic and Critical Thinking*, and editor of the forthcoming *The Philosophy of Logical Mechanism*. She will be editor-in-chief of *Philosophy of Science* beginning January 1990.

Wesley C. Salmon is University Professor of Philosophy at the University of Pittsburgh. Among his books are *Scientific Explanation and the Causal Structure of the World*; *Space, Time, and Motion: A Philosophical Introduction*; and *The Foundations of Scientific Inference*. He edited *Hans Reichenbach: Logical Empiricist* and *Zeno's Paradoxes*. He has taught at UCLA, Washington State College, Northwestern University, Brown University, Indiana University, and the University of Arizona, with visiting appointments at Bristol (England), Melbourne (Australia), Bologna (Italy), and the Minnesota Center for Philosophy of Science. He has served as president of the Philosophy of Science Association and of the American Philosophical Association (Pacific Division). He received his Ph.D. in philosophy from the University of California, Los Angeles. He is a fellow of the American Academy of Arts and Sciences and of the American Association for the Advancement of Science.

Matti Sintonen is a research fellow at the Academy of Finland and teaches philosophy at the University of Helsinki and University of Turku, where he now is acting professor of theoretical philosophy. He was a Florey Student in Queen's College, Oxford, in 1977–80, and worked as a research associate and Fulbright fellow at the Boston Center for the Philosophy and History of Science in 1986–87. He is mainly interested in philosophy of science, especially biology, as well as philosophy of law and philosophy of the social sciences. He has contributed to *Synthese*, *PSA Proceedings*, *Philosophy of Science*, *Rechtstheorie*, *Pittsburgh Studies in the Philosophy of Science*, and several anthologies.

James Woodward is associate professor of philosophy at the California Institute of Technology. He has published articles on causation and scientific explanation and is currently working on a book on explanation.

Index

Compiled by Charlotte A. Broome